T0226529

An Elsevier Energy Compendium

EMISSIONS REDUCTION: NO$_x$/SO$_x$ SUPPRESSION

Elsevier Science Internet Homepage - http://www.elsevier.com

Consult the Elsevier homepage for full catalogue information on all books, journals and electronic products and services.

Related Journals/Products

Free specimen copy gladly sent on request. Elsevier Science Ltd, The Boulevard, Langford Lane, Kidlington, Oxford, OX5 1GB, UK

Fuel
Fuel Processing Technology
Combustion and Flame
Progress in Energy and Combustion Science

Fuelfirst (http://www.fuelfirst.com)

To Contact the Publisher

Elsevier Science welcomes enquiries concerning publishing proposals: books, journal special issues, conference proceedings, etc. All formats and media can be considered. Should you have a publishing proposal you wish to discuss, please contact, without obligation, the publisher responsible for Elsevier's energy and power programme:

Victoria Thame
Publishing Editor
Elsevier Science Ltd
The Boulevard, Langford Lane   Phone: +44 1865 843402
Kidlington, Oxford   Fax: +44 1865 843920
OX5 1GB, UK   E.mail: v.thame@elsevier.co.uk

General enquiries, including placing orders, should be directed to Elsevier's Regional Sales Offices – please access the Elsevier homepage for full contact details (homepage details at the top of this page).

An Elsevier Energy Compendium

EMISSIONS REDUCTION: NO$_x$/SO$_x$ SUPPRESSION

A collection of papers from the journals
Fuel, Fuel Processing Technology and
Progress in Energy and Combustion Science
1999–2001

Edited by

AKIRA TOMITA

Tohoku University, Sendai, Japan

ELSEVIER
2001
AMSTERDAM – LONDON – NEW YORK – OXFORD – PARIS – SHANNON – TOKYO

ELSEVIER SCIENCE Ltd
The Boulevard, Langford Lane
Kidlington, Oxford OX5 1GB, UK

First edition 2001

British Library Cataloguing in Publication Data

Emissions reduction : NOx/SOx suppression : a collection of
  papers from the journals Fuel, Fuel processing technology
  and Progress in energy and combustion science, 1999-2001. -
  (An Elsevier energy compendium)
  1.Air - Pollution 2.Combustion gases 3.Nitrogen oxides
  4.Sulphur oxides
  I.Tomita, A.
  628.5'32

  ISBN 0080440894

The paper used in this publication meets the requirements of ANSI/NISO Z39.48-1992 (Permanence of Paper).
Transferred to digital printing 2005
Printed and bound by Antony Rowe Ltd, Eastbourne

PREFACE

Introduction

Over the past decade the topic of emissions reduction and control has remained an important area of research due to the enforcement of various Government policies in an attempt to minimize the impact on the environment. One area in which a great deal of research has been conducted to address this policy is NO_X/SO_X suppression. However, despite the progress that has been made over this time period, further research into the most effective method of reducing is NO_X/SO_X emissions is still urgently required. In developed countries, a more stringent requirement in the level of emissions (such as is NO_X/SO_X component of less than 10ppm) will be enforced in the near future. Developing countries will also need a new technology that is effective and that is suited to each countries needs. Additional research and development efforts are thus necessary to meet such requirements.

This compendium contains a collection of key papers themed around is NO_X/SO_X emissions from combustion of hydrocarbon resources and the attempts to secure an efficient and effective method for reducing these emissions. These key papers are taken from the journals *Fuel, Fuel Processing Technology* and *Progress in Energy and Combustion Science* and have all been published with the last 3 years.

In the following paragraphs of this Preface, the highlights of the selected papers are overviewed. The topics covered here can be divided into four parts: new analytical techniques, fundamental chemistry of NO_X/SO_X-related reactions, modeling of NO_X/N_2O emission, and R&D activities in large-scale facilities.

New Analytical Techniques

The analysis of gaseous species in a combustion system is essential in understanding the reactions that take place in a combustor. It is very desirable if the technique is accurate, as many species as possible are analysed, and *in situ* conditions are applied. Several papers detail approaches along this line.

Colson *et al.* attempted to use the FTIR technique to continuously monitor in-flame concentrations of CO, CO_2, CH_4, C_2H_6, C_2H_4, C_2H_2 and NO in a 15 MW dual fuel burner. The comparison of such *in situ* concentration with that in stack gas is beneficial in the development of more efficient technology to suppress NO_X emissions. Ruão *et al.* used ultraviolet/visible spectroscopy in their laboratory scale furnace with a propane or ethylene swirl burner. From photometric data of spontaneous emission of OH and CH radicals from various regions of the flame, NO_X formation was found to be strongly dominated by the prompt mechanism. An excellent correlation was observed between the NO_X emissions and the OH + CH photometric data. Higgins *et al.* measured OH radical concentration in fuel-lean, premixed, laminar flames by using chemiluminescence. Such information can be used to minimize NO_X and CO emissions as an active-feedback-control parameter. Zhuiykov developed a potentiometric sensor based on zirconia solid electrolyte. This can be used to simultaneously monitor O_2 and SO_2 concentrations under industrial *in situ* conditions.

Fundamental Chemistry of NO_X and SO_X-Related Reactions

In spite of many efforts to understand the chemistry of NO_X and SO_X-related reactions, further clarification is still required. Some of the papers in this compendium contain some new results on the precursors for NO_X and SO_X emissions, reaction kinetics, and secondary reactions in gas phase as well as over solid materials.

Friebel and Köpsel examined in detail the fate of nitrogen during pyrolysis of coals. Several groups have investigated this topic, but this remains an important area for study. Tan and Li presented a series of papers concerning the formation of NO_X and SO_X precursors during the pyrolysis of coal and biomass. The co-processing of these resources is now one of several very popular topics. They clarified the effects of reactor type, coal rank and heating rate on the formation and transformation of intermediates like HCN, NH_3 and H_2S.

Hampartsoumian et al. investigated the effect of SO_2 doping on NO production in pulverised coal flames, and found that the change in the NO emission due to SO_2 addition, is dependent upon the stoichiometry of the flame. They pointed out that pyritic sulfur content is one of most influential factors in predicting NO_X emission. Liu and Gibbs examined the influence of calcined limestone on NO_X and N_2O emissions in fluidized bed combustors. They observed some catalytic effect of limestone on the conversion of volatile-N to NO_X. However, more important is the role of limestone in enhancing the conversion of char-N to NO_X, which is likely to take place by coupling with the homogeneous reaction of HCN. Tomita reviewed the present status on the suppression of nitrogen oxides emission by carbonaceous reductant. In particular, the reaction mechanism between solid carbon and NO was described in detail, and the importance of surface nitrogen species as a reaction intermediate was emphasized.

In order to effectively capture CO_2 from coal combustion plants, one possible solution would be to use high O_2 concentration gas in place of air. Liu et al. examined the direct sulfation behaviour of $CaCO_3$ under high CO_2 concentration. The sintering of $CaCO_3$ is less than that observed for CaO-SO_2 system, and this explains a high sulfation degree in this direct sulfation system. Hu et al. studied the kinetics of CO_2, NO_X and SO_2 emissions from the combustion of coal with high O_2 concentration gases at relatively high temperature. The NO_X emission level was much less in the CO_2-based inlet gas than in N_2-based inlet gas when compared at the same O_2 concentration.

Modeling

Modeling of combustion process is one of the most important subjects in designing more efficient and cleaner processes. A review by Hill and Smoot focused on the NO_X formation and destruction during coal combustion. This includes current NO_X control technologies, NO_X reaction processes, and techniques to calculate chemical kinetics in turbulent flames. Comparisons of measured and predicted values of NO_X concentrations are shown for several full-scale and laboratory-scale systems. NO_X models are also applied to developing technologies, such as the use of over-fire air, swirling combustion air streams, use of reburning fuels, and others. Xu et al. made a numerical simulation of the flow and combustion process in their furnace of a pulverized coal fired utility boiler of 350 MWe. The measured furnace outlet temperature, O_2/NO_X concentrations, and total heat energy transferred to walls/superheaters were generally in good agreement with calculated values. Mathematical modeling of fluidized bed combustion was presented by Chen et al. The single char particle reaction-diffusion model was successfully integrated into a three-phase hydrodynamic description of the fluidized bed reactor.

Research and Development in Commercial Scale Facilities

Much progress has also been made using pilot plant and commercial scale combustion systems, where NO_X/SO_X emission and flue gas clean up are of major interests. A review written by Beér provides a comprehensive overview on the development of combustion technology from the pre-environmental era (target: complete combustion) through the environmental era (reduction of pollutant) to the present and near future (CO_2 emissions mitigation). There is an extensive discussion of the promising clean coal technologies, such as pressurized fluid bed combustion, air heater gas turbine-steam combined cycle, integrated gasification combined cycle and others. Maly *et al.* investigated the effectiveness of advanced reburning using alternative fuel, such as biomass, coal pond fine, refuse-derived fuel and others. Injection of NH_3 together with promoter like sodium compounds into reburn fuel was found to be effective in reducing NO_X emission. Tree and Clark also confirmed the effectiveness of advanced reburning, where NH_3 was added after reburning fuel. Under certain conditions, over 95 % NO reduction was achieved. A technical, environmental and economic analysis of NO_X reduction technologies in supercritical pulverized coal fired power station was made by McCahey *et al.* The additional cost due to selective catalytic reduction, reburning, use of coal micronizer was assessed using the ECLIPSE process simulator. NO_X emission characteristics in a fluidized bed combustion system are also of significant importance. Sänger *et al.* studied the fluidized bed incineration of semi-dried sewage sludge, and the results were compared with available data on wet and dry sludge. NO_X emissions were slightly higher than those from wet sludge but much lower than those from dry sludge.

The above papers are all related to *in situ* NO_X/SO_X reduction. In addition, flue gas treatment is also quite effective in suppressing pollutant emission. In the selective catalytic reduction of NO_X emission from diesel engine, Krijnsen *et al.* proposed a new tool using the inferential feedforward control technique. This is capable of accurately and quickly predicting the fluctuating NO_X emission and also capable of predicting the optimum reductant/NO_X ratio. Olson *et al.* presented details of their industrial process, where activated coke was used to simultaneously reduce NO_X and SO_X from flue gas. They claimed that, in coal-fired fluidized bed boilers, the removal rate reached >99% for SO_X, >80% for NO_X, and 99% for other toxic materials such as Hg, HCl and polychlorinated aromatic compounds. Haase and Koehne reviewed the state-of-the-art of so-called "condensing boilers", the concepts of which are based on the idea of using the condensate for washing the flue gas. The flue gas is brought into contact with already neutralized condensates. This process allows the wet separation of noxious matter produced in rather small boilers.

Summary

Although SO_X emissions are now controlled to a considerable extent, it is necessary to introduce these established technologies or additional new technologies to developing countries. The reduction of NO_X would be a tough challenge to cope with given the future of environmental regulation, which is expected to be more stringent than the present one. Much excellent fundamental and applied research is collected here, together with several review papers. This compendium certainly represents the present state-of-the-art of this area and therefore provides a good basis for future research and development.

Akira Tomita
Institute of Multidisciplinary Research for Advanced Materials,
Tohoku University,
Sendai 980-8577, Japan

CONTENTS

Modelling

Research and Development in Commercial Scale Facilities

New Analytical Techniques

Emissions Reduction: NO_x/SO_x Suppression
A. Tomita (Editor)
© 2001 Elsevier science Ltd. All rights reserved

Experimental in-flame study of a 15 MW dual fuel gas/oil burner

G. Colson[a,*], F. Peeters[b], J. De Ruyck[a]

[a]*Vrije Universiteit Brussel, Department of Mechanical Engineering, Building Zw 106, Pleinlaan 2, B-1050 Brussels, Belgium*
[b]*Flemish Institute for Technological Research, Boeretang 200, 2400 Mol, Belgium*

Received 25 November 1998; accepted 29 March 1999

Abstract

Extensive in-flame measurements have been carried out on a 15 MW dual fuel burner, in order to obtain more information on the influence of different burner head configurations on the profiles of chemical species in the flame. Together with stack measurements of NO_x concentrations, this will improve our knowledge on the formation of NO_x in this installation. In a half plane of the installation at four axial distances radial concentration profiles were recorded. For these measurements an FTIR analyser has been configured to allow the simultaneous quantification of CO, CO_2, CH_4, C_2H_6, C_2H_2, C_2H_4 and NO concentrations on a continuous base. The calibration of the analyser has been carried out by the multivariate PLS1 algorithm, with calculated standard spectra, which avoids elaborate lab calibration procedures. The comparison of concentration profiles of CH_4 and C_2H_6 close to the burner mouth for different burner head configurations, gives a good indication of how the fuel is injected in the installation and together with the C_2H_4, C_2H_2 concentrations gives more information about the combustion itself. © 1999 Elsevier Science Ltd. All rights reserved.

Keywords: Combustion; FTIR; NO_x

1. Introduction

The study and the development of low NO_x burners, requires flexible and accurate analysers for concentration measurements of numerous chemical components in the stack and in the flame of the installation. FTIR analysers are nowadays extensively used to study the composition of flue gases generated by the combustion of fossil fuels in cars [1] and textiles [2] in research laboratories, to control the combustion at incinerator plants [3,4], and to study fluidised bed combustion [5].

The goal of the work described in this paper is to study the influence of different burner head configurations of a 15 MW burner on the flame structure and its effect on the production of NO. For this the profiles of temperature, velocity and concentration of stable species were measured for flames selected, based on preliminary measurement of stack NO emissions. These profiles will be used to give a qualitative explanation about the ongoing combustion process in the installation and on the interaction between burner geometry and aerodynamics and their effect on the concentration profiles of stable components. Furthermore the experimental results will serve as a validation set for future simulations of the combustion process. The chemical species measured are CH_4, C_2H_6, C_2H_4, C_2H_2, CO, CO_2, NO_x and O_2.

2. Configuration of an FTIR for in-flame measurements

When using an FTIR for combustion research, the following problems have to be to be dealt with:

1. The analyser should both be accurate and fast, since it will be used for studying emissions produced during stationary operations and during transition between operating points. For high measuring speeds the FTIR, will use lower resolutions, which will decrease the maximal optical path difference. Consequently different selected resolutions require different sets of standard spectra.
2. The gas samples coming from a flame can contain various IR active products which interfere with each other and thus cannot be quantified with simple peak height measurements.
3. In-flame measurements result in gas samples with a wide range of concentrations. This not only requires many standard spectra for calibration, but also causes difficulties in the selection of a spectral region for the quantification of a specific molecule. One has to make sure that the absorption peaks in the selected region do not saturate for the imposed concentration range.

---

* Corresponding author. Tel.: + 32-2-629-2805; fax: + 32-2-629-2865.
E-mail address: guy.colson@vub.ac.be (G. Colson)

Reprinted from *Fuel* **78 (11)**, 1253-1261 (1999)

4

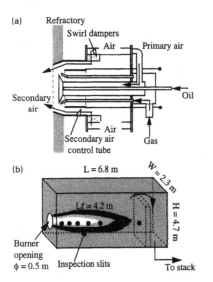

(a) Refractory
     Swirl dampers
        Air ── Primary air
Secondary
air   Oil
        Air
     Secondary air   Gas
     control tube

(b)   L = 6.8 m
        Lf = 4.2 m
Burner
opening
φ = 0.5 m   Inspection slits   To stack

Fig. 1. (a) The Elco-Mat RPD60 burner; (b) the VUB 20 MW test furnace's dimensions.

Taking into account the above mentioned problems we used calculated standard spectra for calibrating the FTIR. By doing this we avoided the elaborate collection of standard spectra in the laboratory for the species of interest. The univariate quantification techniques will not work properly for in-flame concentration measurements, because of interferences between various components encountered in a natural gas flame. Therefore the multivariate PLS1 algorithm has been used to calculate the concentration values of the CO, CO_2, CH_4, C_2H_6, C_2H_2, C_2H_4 and NO water was also included since it was impossible to eliminate all the water from the gas samples.

The multivariate PLS1 method is based on the factorisation of the absorption data matrix, containing the calibration spectra, into a score and loading matrix. An implicit regression step relates the score matrix to the concentration matrix

and calculates the regression vector. This has to be repeated for every component in the calibration set. The prediction of an unknown spectrum then consists of calculating the score vector and multiplying it with the regression vector. An important advantage of the multivariate methods is the possibility of reconstructing the unknown spectrum and consequently evaluating the residual spectrum on spectral artefacts such as interferents not included in the calibration model [6,7].

2.1. Calculation of IR spectra

The spectra were calculated from the absorption line parameters given by the HITRAN92 database, which includes line parameters for 31 individual atmospheric gases [8]. Considering these line parameters together with Doppler and pressure broadening, it is possible to calculate the IR spectrum of a pure component or a mixture of components, without instrumental influence. The influence of the FTIR analyser is included by introduction of the instrument line shape function which itself is the convolution of the apodisation function and a rectangular function, the width of which depends on the divergence of the collimated beam in the in the interferometer [9,10]. For apodisation the Norton–Beer–Strong apodisation function was used [11]. For further details about the configuration of the FTIR analyser we refer to another Ref. [12].

3. Burner, furnace and fuel

The burner is an Elco–Mat RPD60 dual fuel burner, which has a nominal heat output of 15 MW for both gas and oil combustion. It has separate primary and secondary air control mechanisms, twelve alignable gas tubes for natural gas injection (arranged in six pairs), a central oil nozzle and an externally controllable swirl generators in the secondary air duct. The secondary air flow pattern can be altered by setting the position of the secondary air control tube, i.e. a cylindrical tube placed in the secondary air tube. In Fig. 1 the most important parts of the burner are described

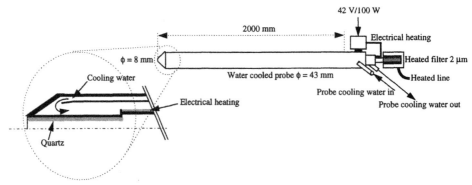

Fig. 2. The IFRF Quartz-Lined Heated NO_x/SO_x water cooled sampling probe.

Table 1
Composition of slochteren gas

Component	Mole%
Methane	83.467
Ethane	3.287
Propane	0.584
Iso-butane	0.094
Normal-butane	0.117
Iso-pentane	0.031
Normal-pentane	0.032
Hexane and higher	0.066
Oxygen	0.000
Nitrogen	11.085
Carbon dioxide	1.183
Helium	0.055

schematically. The diameter of the burner before the convergent section is 0.5 m, the exit diameter is 0.42 m. For more details on the installation we refer to previous work [13].

The burner was installed on a 20 MW test furnace of the university (VUB). The boiler is a Baumgarten water tube type of boiler in which pressurised water (14 bar) enters at 120°C and is heated up to 150°C. The released energy during the tests is used to heat up the university campus. A second 20 MW boiler is placed in parallel to the test boiler and automatically anticipates on the heat demand of the campus, making the test boiler independent of heat demand. Except for the floor, all walls in the fire room consist of 75 mm diameter tubes welded to each other. At the end of the fire room a wall of tubes separates the fire room from the convective part of the boiler in which the cold entering water is preheated by convective heat transfer. The flue gas is then collected by a funnel and lead to the stack. In order to access the furnace with probes, inspection slits were installed at seven axial distances as shown in Fig. 2.

Natural gas—so called "Slochteren gas"—was taken from the Brussels city net. Its lower heating value is 33 MJ/N m^3. The main components of this gas are CH_4 (83.47 mol%), C_2H_6 (3.287 mol%), C_3H_8 (0.584 mol%), N_2 (11.085 mol%) and CO_2 (1.18 mol%). A more detailed composition is given in Table 1. For the conversion of the measurements into appropriate units, a standard wet air composition was used for the calculations. The gas flow measurements were carried out with a diaphragm (venturi flange), the accuracy of which was estimated to be better than 2%. A second independent, but less accurate gas flow measurement with the flow counter of the gas company and a chronometer corresponded within 5% with the venturi measurements.

4. Experimental set-up

Gas species concentration measurements of CO, CO_2,

CH_4, C_2H_6, C_2H_2, C_2H_4 and NO were achieved by an analysis chain consisting of a sampling unit, a gas conditioning unit, and an FTIR analyser to record the IR spectra in the spectral range between 4000 and 400 cm^{-1}. To measure O_2 a paramagnetic analyser was used (Oxynos 100 Rosemount).

Gas samples from the flame were taken by a Quartz-Lined Heated NO_x/SO_x probe, manufactured by the IFRF[1], this is a water cooled convective quench type of probe (Fig. 2). The quenching tube is quartz coated to prevent reactions of NO with the tube. The inner and outer diameter of the probe are 8 and 43 mm, respectively, the operating length is 2 m. The quenching section has a length of about 100 mm and rapidly chills the gas to about 500 K. After this quenching section the sample transport tube is electrically heated to avoid condensation of water vapour. The probe exit is connected to a heated filter and a heated umbilical line, which leads the gas to the conditioning unit. After extraction, the gas sample is conditioned and transported to the analysers whilst taking care not to affect the representativeness of the sample. At this point water vapour and particles are removed from the sample, since they can contaminate the analysers and thus strongly bias the measurements. The conditioning system used for this work comprises a membrane pump to extract 6 Nl/min of gas through the probe, leading it to a jet stream cooler to remove water vapour and obtain a dry gas (dew point = 5°C). The dry gas is then passed over a filter and a water detection system to secure the analysers from accidental condensation or fouling by particles. Due to the jet stream condenser, rapid chilling of the sample is achieved, resulting in a minimum contact between the condensate and the sample gas. Solution of some gaseous compounds in the condensate is therefore avoided or minimised, removing a major source of uncertainty in the conditioning phase [14]. The condensate itself is removed through two peristaltic pumps. The uncertainty introduced by the conditioning system was found to be smaller than 1% for the measurement of NO and/or NO_x.

4.1. FTIR analyser

The FTIR-analyser used for our measurements is the PE system 2000 FTIR analyser. Its resolution can be set from 0.2 to 16 cm^{-1}. The detector is a DTGS-detector. The source is a voltage-stabilised air cooled wire coil, which operates at a temperature of 1350 K. The beam splitter is a Ge/KBr beam splitter, and the gas cell (Infrared Analysis Inc.) is a multiple-pass gas cell, which can be heated with a temperature-controlled heating jacket up to 200°C. The total path length is 4.2 m and the cell volume is 0.325 l. The gas cell is sealed with infrared transparent KBr-windows, and the cell mirrors are coated with gold. The body of the cell is constructed in glass. The spectra were collected at a

---

[1] IFRF: International Flame Research Foundation.

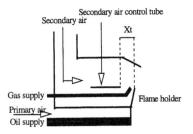

Fig. 3. Definition of the position Xt of the secondary air control tube.

resolution of 1 cm^{-1}, with a velocity of 0.2 cm/s, each spectrum is the result of 16 accumulated spectra.

4.2. Set-up for velocity measurements

Three dimensional velocity measurements in the flame were performed with a water cooled 5-hole Pitot tube, constructed and calibrated by the IFRF. The probe consists of a 1950 mm long, 43 mm diameter water cooled stem and a 60 mm long, 12 mm diameter perpendicular mounted hemispherical head with five 2 mm diameter pressure taps, one on the probe axis and the four others at 90° intervals around the axis and placed at an angle of 33°.

4.3. Set-up for temperature measurements

Flame temperature measurements were obtained by means of a bare 0.5 mm diameter platinum/6% rhodium–platinum/30% rhodium thermocouple (type B). The wires are contained in a 1600 mm long ceramic tube with at the end a 300 mm stainless steel stem, resulting in a total operating length of 1.9 m. A similar (but smaller) reference thermocouple (at 0°C) was connected to one end of the main thermocouple to compensate for the cold junction error. The temperature measurements, obtained after processing the voltage data by interpolation of the NIST

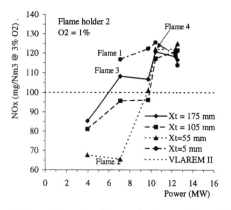

Fig. 4. Influence of the position of the secondary air control tube on the NO$_x$ emissions.

tables, were corrected for radiation and convection heat transfer in the furnace.

5. Results and discussion

The experimental results of this study will be discussed in two parts. First the in stack emission measurements are dealt with. These will consequently lead to the selection of flames corresponding with a certain burner configuration, for which in-flame measurements of temperature, velocity and concentration profiles will be carried out in order to gain a better insight in the trend of the observed emission measurements.

5.1. Emission measurements

In order to establish base line NO$_x$ emissions for the commercially available RPD60 burner, a series of emission measurements have been carried out. For these measurements the influence of power, oxygen excess and secondary air control tube position has been tested. The position of the secondary air control tube, which affects the flow pattern of the secondary air in both the burner and furnace, is determined by the distance between the flame holder and the end of the tube (Fig. 3).

Preliminary measurements [13] showed that NO$_x$ increases with both power and oxygen excess levels. The oxygen excess in these tests was controlled by the secondary air, since the use of primary air was shown to lead to an increasing CO emission and soot formation for gas combustion. Fig. 4 shows that up to 10 MW the position of the secondary air control tube, and therefore the flow pattern of the secondary air significantly affects the NO$_x$ formation. The O$_2$ excess for the cases in Fig. 4 is 1%. Four flames came up as interesting for further investigation: a high NO$_x$ flame at 7 MW (115 mg/Nm^3, Flame 1), a low NO$_x$ flame at 7 MW (65 mg/Nm^3, Flame 2), a manufacturer's settings flame at 7 MW (108 mg/Nm^3, Flame 3) and a flame at 11 MW. For these flame configurations concentrations of O$_2$, CO, CO$_2$, CH$_4$, C$_2$H$_6$, C$_2$H$_2$, C$_2$H$_4$ and NO together with temperature and velocity profiles were measured in the flame. With the information gathered this way we will try to give an explanation for the observed differences in the emission measurements and to provide more information about the influence of the burner head configuration on

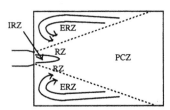

Fig. 5. Location of the different flow zones in the flames.

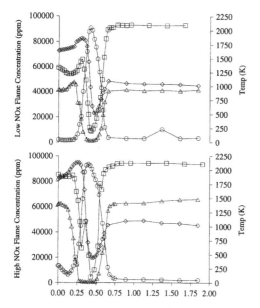

Fig. 6. Profiles of CO_2 (□), $O_2/2$ (○), $NO_x \times 10^3$ (△) and T(◇) at $x/D = 0.26$ as a function of radial distance (r/D).

the combustion process in the installation. Furthermore the data collected this way will serve as a validation set for future simulations. In the present paper only the high NO_x and the low NO_x flame at 7 MW will be compared with each other, based on temperature, concentration and velocity measurements (Figs. 6–12) carried out at the two axial positions closest to the burner ($x/D = 0.26$ and $x/D = 1.15$, with $D = 0.42$ m and x the axial distance from the burner exit).

5.2. In-flame measurements

In general four flame zones can be distinguished for the type of burner we use: the internal recirculation zone (IRZ) just beyond the flame holder, the external recirculation zone (ERZ) between the boiler wall and the outer boundary of the flame jet, the reaction zone (RZ) close to the burner nozzle and the post combustion zone (PCZ) in the fully developed part of the flame jet (Fig. 5). The localisation of these zones in our installation is based on the velocity measurements. In the IRZ it is impossible to measure the velocities accurately with the 5-hole Pitot probe. This is because of the difficulties to find the main direction of the flow, since the velocities are low and strongly fluctuate. Nevertheless this allows us to determine whether we are in the central IRZ or not. In the ERZ a similar remark can be made, here the velocity is small (<5 m/s) and reverse furthermore from the concentration measurements we see that the gas composition in the ERZ is similar to that of the flue gas in the stack. The temperatures, in contrast, are respectively high (>1700 K) and moderate to low (<1000 K) in the IRZ and ERZ. The

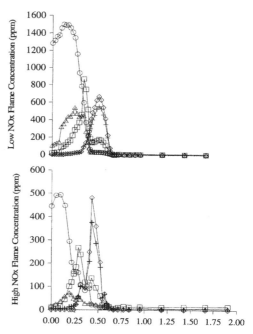

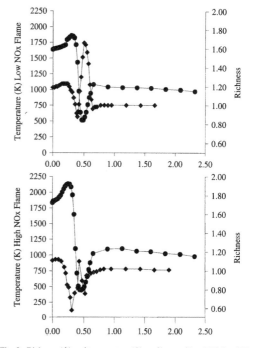

Fig. 7. Profiles of CO/50 (○), C_2H_2 (△), C_2H_4 (□), $CH_4/200$ (◇), $C_2H_6/10$(+) at $x/D = 0.26$ as a function of radial distance (r/D).

Fig. 8. Richness (◆) and temperature (●) profiles at $x/D = 0.26$: low NO_x case (top), high NO_x case (bottom) as a function of radial distance (r/D).

8

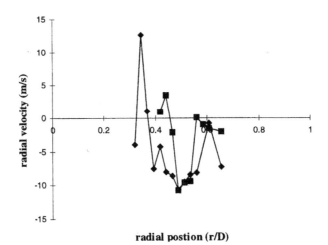

radial postion (r/D)

Fig. 9. Radial velocity profiles of high NO$_x$ (◆) and low NO$_x$ (■) flame at axial distance x/D of 0.26.

main combustion reactions take place in the RZ of the flame between the outer and inner boundary of the flame jet, close to the burner. Velocity in the RZ ranges from 30 to 60 m/s and the temperature is lower than 1500 K. In the PCZ the flame jet is fully developed to one single jet with velocities from 20 to 30 m/s and high temperatures. In the PCZ mainly further oxidation of CO to CO$_2$ takes place. From the

velocity measurements we can conclude that the internal recirculation zone extends over axial positions beyond the second probing position for both flames, but not beyond probing position 3.

In Fig. 6 temperature NO$_x$, O$_2$ and CO$_2$ concentration profiles at an axial distance of 0.26 burner diameters for the low and high NO$_x$ flames are compared with each

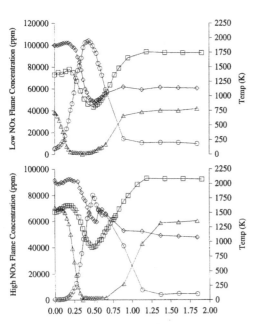

Fig. 10. Profiles of CO$_2$ (□), O$_2$ (○), NO$_x$ × 10^3 (△) and T(◇) at x/D = 1.15 as a function of radial distance (r/D).

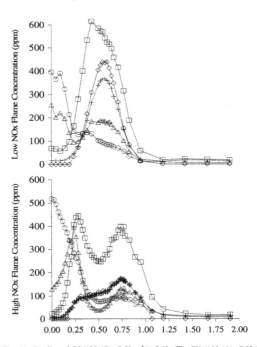

Fig. 11. Profiles of CO/100 (○), C$_2$H$_2$ (△), C$_2$H$_4$ (□), CH$_4$/100 (◇), C$_2$H$_6$/5(+) at x/D = 1.15 as a function of radial distance (r/D).

9

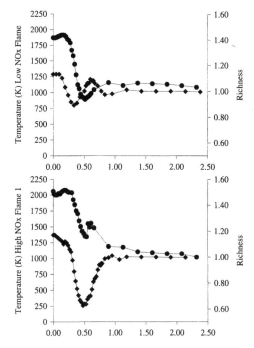

Fig. 12. Richness (◆) and temperature (●) profiles at $x/D = 1.15$: low NO_x case (top), high NO_x case (bottom) as a function of radial distance (r/D).

other. This was the axial position closest to the burner we could measure at. The temperature in the IRZ ($r/D < 0.3$) is about 200 K lower for the low NO_x flame, furthermore the O_2 concentration in the IRZ of the low NO_x flame is zero while for the high NO_x flame this is not the case (1–4%). Consequently in this zone conditions exist which are favourable for production of thermal NO_x. From $r/D = 0.75$ to the outside wall of the combustion chamber no differences are observed between the concentration and temperature profiles of both flames, this is no surprise since this zone corresponds with the ERZ in which the combustion has finished.

The concentration profiles of unburned hydrocarbon species and CO at $x/D = 0.26$, demonstrated in Fig. 7 show that the low NO_x flame has concentrations of CO which are about three times higher than for the high NO_x flame (60 000 ppm against 25 000 ppm) for radial positions close to the burner mouth (IRZ). For ethylene and acetylene which are formed in the flame we observe respectively 500 and 900 ppm in the IRZ of the low NO_x flame against 80 and 300 ppm for the high NO_x flame. When looking deeper into the profiles of C_2H_4, C_2H_2 and the richness profiles (Fig. 8) at $x/D = 0.26$ for both flames we see that the maximum concentration coincides with the maximum temperatures and that there is a smaller peak which coincides with the peaks of CH_4 and C_2H_6 which correspond with the fuel inlet.

The richness is defined as the ratio of mixture fractions f/f_{st}, using the massfraction of carbon as extensive property, f_{st} is the mixture fraction at stoichiometry. The maximum C_2H_4 concentration for the low NO_x flame is about three times higher than for the high NO_x flame. For C_2H_2 in the low NO_x flame a broad peak (500 ppm) is observed between 0.05 diameters and the position of the maximum C_2H_4 concentration ($r/D = 0.3$), in the high NO_x flame the same is observed but again the concentration is lower (80 ppm) compared to the low NO_x flame. These differences in CO, C_2H_4 and C_2H_2 levels can be explained by the higher richness of the low NO_x flame in this zone (Fig. 8).

In the combustion theory in literature [15] it is stated that for rich methane combustion the C2-pathway becomes more important. This is caused by the recombination of CH_3 radicals, which in another step give rise to C_2H_4 and further reacts to C_2H_2, which is an important species in the formation of prompt NO_x. The profiles of C_2H_2 and C_2H_4 can thus identify zones where the prompt NO_x pathway becomes important. In Fig. 6 we see that the low NO_x flame has a NO_x peak that practically coincides with the maximum C_2H_2 (Fig. 7) values in the IRZ, which can be an indication of the formation of prompt NO_x.

The profiles of CH_4 and C_2H_6 which are the main contributors of the fuel show peak values of 150 000 ppm for the low NO_x flame while for the high NO_x flame this is around 100 000 ppm (Fig. 7). Which indicates that mixing of fuel and air is delayed in the low NO_x flame, which is also demonstrated in the richness plots (Fig. 8).

For the velocity profiles it was impossible to perform accurate measurements in the IRZ and ERZ as stated above. This explains why the velocity profiles were only measured in limited regions for axial distances close to the burner mouth. Nevertheless an important difference is observed between the radial velocity profiles of both flames at an axial distance of 0.26 diameters. For the high NO_x flame the radial component between $r/D = 0.35$ and $r/D = 0.6$ is negative, which means that there is a flux directed to the centre. For the low NO_x flame the radial velocity is negative over a narrower range in the vicinity of $r/D = 0.5$ (Fig. 9). This difference is caused by the interaction between the gas and air jet with the secondary air control tube, which for the high NO_x flame is in the maximum forward position. This results in entrainment of the gas by the surrounding air jet, leading the mixture towards the centre of the flame and enhancing mixing.

When comparing the temperature and concentration profiles further downstream ($x/D = 1.15$) in Fig. 10 we see again that for the high NO_x flame the temperatures close to the axis of the installation are high (2000 K) and coincide with oxygen concentrations varying between 0 and 3%. For the low NO_x flame the same situation is observed but the maximum temperatures still are about 200 K lower than for the high NO_x flame. This reduces the thermal NO formation, which again explains the difference (25 ppm) in observed NO_x levels of both flames.

10

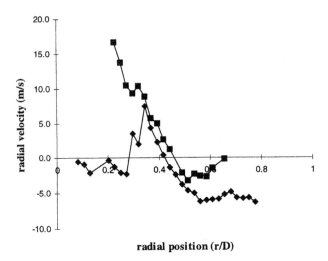

radial velocity (m/s)

radial position (r/D)

Fig. 13. Radial velocity profiles of high NO$_x$ (◆) and low NO$_x$ (■) flame at axial distance x/D of 1.15.

The profiles of unburned hydrocarbon species further downstream are demonstrated in Fig. 11 and now show maximum CO concentrations of 50 000 ppm for the high NO$_x$ flame against 40 000 ppm for the low NO$_x$ flame. This corresponds with the richness plot in Fig. 12, which shows higher richness values for the high NO$_x$ case in the central zone with maximum CO concentrations. The concentrations of CH$_4$ and C$_2$H$_6$ are spread much wider over a radial range between 0.1 and 0.9 in the high NO$_x$ flame, while for the low NO$_x$ flame the profiles of these species are situated between radial distances of 0.3 and 0.8. This is another indication for the occurrence of enhanced mixing in the high NO$_x$ flame. Due to this enhanced mixing NO$_x$ formation is increased. This is similar to the situation at $x/D = 0.26$, and is a result of the interaction between the gas and air jet with the secondary air control tube. The radial velocity profile for the high NO$_x$ flame similar as at $x/D = 0.26$ is negative over a wider range than for the low NO$_x$ flame (Fig. 13).

The profiles of C$_2$H$_2$ and C$_2$H$_4$ show two peaks (Fig. 11) for the high NO$_x$ flame, one coincides with the maximum of the CH$_4$, C$_2$H$_6$ profiles at $r/D = 0.8$ and shows maximum C$_2$H$_4$ and C$_2$H$_2$ values of 400 and 100 ppm, respectively. The other peak is situated at a radial position of $r/D = 0.3$, on the edge of the rich central zone and shows maximum values of C$_2$H$_4$ and C$_2$H$_2$ of 450 and 400 ppm, respectively. The latter coincides with the maximum temperature.

In the low NO$_x$ flame the C$_2$H$_2$ concentration profile has a plateau of maximum values of 200 ppm at the centre and also has a peak, which coincides with the maximum of CH$_4$ and C$_2$H$_6$. Finally for C$_2$H$_4$ the concentration profile is somewhat broader than those of CH$_4$ and C$_2$H$_6$ with a maximum of 600 ppm at $r/D = 0.3$.

In general it can be said that mixing is delayed in the low NO$_x$ flame, resulting in lower temperatures at the flame axis

close to the burner and less overlap between hot and oxygen rich regions. The delayed mixing explains the existence of higher fuel peaks (CH$_4$, C$_2$H$_6$) for the low NO$_x$ flame. The difference in NO$_x$ levels between both flames is mainly caused by reducing the production of thermal NO$_x$ by decreasing the rate of the heat release.

6. Conclusions

From the point of view of combustion the in-flame measurements in the selected gas flames reveal the reasons for different NO$_x$ emissions observed by flue gas measurements. Conclusions can be drawn and suggestions can be made to modify burner geometry in order to reduce NO$_x$ emissions from this specific burner.

The aerodynamic structure of the flame close to the burner determines to a large extent the structure of the flame as a whole and the formation of NO$_x$. Therefore, controlling the aerodynamics close to the burner, means controlling the NO$_x$-formation. From the low NO$_x$ flame it was observed visually and experimentally from concentration measurements of fuel species (CH$_4$, C$_2$H$_6$) and the calculation of the richness that the fuel jets did not mix very rapidly, inducing a kind of fuel staging. For this specific type of gas spud burner, fuel staging can easily be realised by creating a "finger flame", that is the fuel jets should maintain their jet character and not mix too fast with the air. Reducing the number of gas spuds and increasing the momentum of the gas contribute to this effect.

Considering the profiles of C$_2$H$_2$ it is possible to identify zones where the production of prompt NO$_x$ can become important.

From the point of view of concentration measurements it can be stated that the use of an FTIR for analysing species coming from in-flame sampling has shown to have a lot of perspective in this field. Using this technique together with an oxygen analyser made it possible to generate a set of experimental data that will be used for the validation of future CFD simulations of the installation. Up until now CO_2, CO, NO, CH_4, C_2H_6, C_2H_4, C_2H_2 were taken into account for detection, in the future N_2O and CH_2O will be added to this list. Finally the use of calculated standard spectra has proven to be a good alternative for experimentally recorded standards.

Acknowledgements

First of all our gratitude goes to Elcomat N.V. who supported this work and granted their permission to publish the results and to show some details of the burner. Further we thank the staff of Cofreth for their technical support.

References

[1] Hanst PL, Stephens ER. Spectroscopy International 1989;2:44.

[2] Pottel H. Fire and Materials 1995;19:221.

[3] Zhuoxiong M, Demirgian JC. Waste Management 1995;15:233.

[4] Stapf D, Leuckel W. Chemie Ingenieurs Technik 1992;64:860.

[5] Amand LE, Tullin C. In: Proceedings of the Nordic Seminar on Gas Analysis in Combustion, Tampere Finland, 4–5 October, 1994.

[6] Haaland DM, Thomas EV. Analytical Chemistry 1988;60:1193.

[7] Martens H, Naes T. Multivariate Calibration, New York: Wiley, 1989.

[8] Rothman LS, et al. Journal of Quant Spectrosc Radiat Transfer 1992;48:469.

[9] Griffiths PR, De Haseth JA. Fourier transform infrared spectrometry, New York: Wiley, 1986.

[10] Griffith DWT. Applied Spectroscopy 1996;50:59.

[11] Norton RH, Beer R. Journal of Optical Society 1976;66:259.

[12] Colson G, De Ruyck J. FTIR spectroscopy for the in-flame study of a 15 MW dual fuel gas/oil burner, Journal of Applied Spectroscopy 1999, in press.

[13] Peeters F, Konnov AA, Colson G, De Ruyck J. Proceedings of the Fourth European Conference on Industrial Furnaces and Boilers. 1–5 April, Porto, vol. 2, 1997.

[14] Jahnke JA. Continuous emission monitoring, New York: Van Nostrand Reinhold, 1993.

[15] Warnatz J. Brennstof Wärme Kraft 1985;37:11.

Emissions Reduction: NO_x/SO_x Suppression
A. Tomita (Editor)
© 2001 Elsevier science Ltd. All rights reserved

A NO_x diagnostic system based on a spectral ultraviolet/visible imaging device

M. Ruão, M. Costa*, M.G. Carvalho

Mechanical Engineering Department, Instituto Superior Técnico/Technical University of Lisbon, Av. Rovisco Pais, 1049-001 Lisbon, Portugal

Received 9 November 1998; accepted 17 February 1999

Abstract

The present investigation is aimed to develop and test a NO_x diagnostic system based on a spectral ultraviolet/visible imaging device. The study was performed in a small-scale laboratory furnace fired by a propane or ethylene swirl burner. The data reported include flue-gas concentrations of O_2, CO_2, CO, unburnt hydrocarbons and NO_x, obtained with conventional techniques, and photometric data of spontaneous emission of OH and CH radicals, obtained with the imaging device, from various regions of the flame. All these measurements have been obtained for twenty three furnace operating conditions which quantify the effects of the gaseous fuel (propane and ethylene), flue-gas recirculation, heat losses through the furnace walls to the surroundings and excess air. Against this background, it was possible to obtain a wide range of NO_x emissions. In addition, the analysis of detailed near burner in-flame data of local mean major gas-phase species and gas temperatures collected for four furnace operating conditions has indicated that NO_x formation was strongly dominated by the prompt (or Fenimore) mechanism. The results reveal that there is an excellent correlation between the NO_x emissions from propane or ethylene flames and the OH + CH photometric data, which can be mathematically expressed as a logarithmic function, provided that the radicals images are collected from a flame zone close to the burner exit. © 1999 Elsevier Science Ltd. All rights reserved.

Keywords: Propane; Ethylene; Nitrogen oxides; Diagnostic system; Spectral ultraviolet/visible imaging device

1. Introduction

Spectral studies in the ultraviolet (UV) and visible ranges yield detailed information of combustion processes and have been the subject of recent investigations, not only in internal combustion engines [1–3] but also in industrial combustion systems [4,5]. When using the flue-gas composition as the input value for a closed-loop control, the time scale for the control system becomes much longer than the combustion time scales [4]. The present investigation aims to develop and test a NO_x diagnostic system based on a spectral ultraviolet/visible imaging device which will allow a faster response.

The tests were performed in a small-scale laboratory furnace. Conventional techniques were used to gather flue-gas and in-flame data for several furnace operating conditions, which combined the use of two gaseous fuels (propane and ethylene), flue-gas recirculation (FGR), furnace walls with and without insulation and two excess air levels. In this way, it was possible to obtain for the present combustion system a wide range of NO_x emissions. Simultaneously, a spectral ultraviolet/visible imaging device was used to detect OH and CH spontaneous emissions from various flame zones for all the conditions studied with conventional techniques. The NO_x emission data were then confronted against the photometric data to establish a correlation between the NO_x flue-gas concentrations and both the OH and CH spontaneous emission.

2. Indicator radicals for NO_x formation in gas flames

In the combustion of fuels not containing nitrogen compounds, NO_x is formed through the thermal and prompt (or Fenimore) mechanisms. The formation of thermal-NO_x takes place through the extended Zel'dovich mechanism [6]:

$$O + N_2 \leftrightarrows NO + N \tag{1}$$

$$N + O_2 \leftrightarrows NO + O \tag{2}$$

$$N + OH \leftrightarrows NO + H \tag{3}$$

with reaction (1) controlling the NO_x formation. In addition,

* Corresponding author. Tel.: + 351-1-8417186; fax: + 351-1-8475545.
E-mail address: mcosta@navier.ist.utl.pt (M. Costa)

Reprinted from *Fuel* **78 (11)**, 1283-1292 (1999)

14

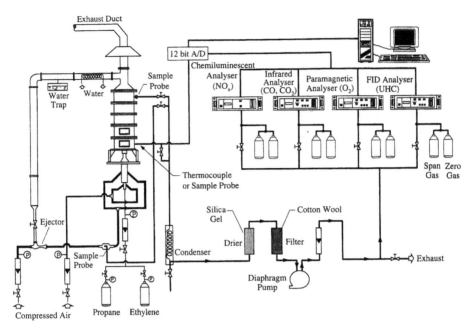

Fig. 1. Schematic of the furnace and measurement equipment.

because reaction (1) is strongly temperature dependent due to its high activation energy, thermal-NO_x formation is significant only at high temperatures.

Prompt-NO_x in hydrocarbon flames is formed primarily by a reaction sequence described by Miller and Bowman [7] as follows:

$$CH + N_2 \leftrightarrows HCN + N \qquad (4)$$

$$HCN + O \leftrightarrows NCO + H \qquad (5)$$

$$NCO + H \leftrightarrows NH + CO \qquad (6)$$

$$NH + H \leftrightarrows N + H_2 \qquad (7)$$

$$N + OH \leftrightarrows NO + H \qquad (3a)$$

Note that the last reaction in both NO_x mechanisms (reactions (3) and (3a)) is the same.

Owing to the impossibility to detect spontaneous NO emission in flames of fuels not containing nitrogen compounds, there is a need to find indicators for its formation in such flames. In the case of the Zel'dovich mechanism the radical OH is the only possible NO indicator based on spontaneous emission. The prompt mechanism is initiated through reaction (4) which involves the radical CH. It is obvious to take its spontaneous emission as a possible prompt-NO_x indicator. Other possible indicator of the prompt mechanism is the radical OH, reaction (3), in common with the thermal mechanism. Let us then assume

that the spontaneous emission of the radical OH and of both the radicals OH and CH are linked, in same way, to the NO_x formation via the thermal and prompt mechanisms, respectively.

3. The experimental facility

A schematic of the experimental facility is shown in Fig. 1. It comprises a small-scale laboratory furnace up-fired by a swirl burner equipped with facilities for flue-gas recycling. The combustion chamber is cylindrical in shape and consists of five interchangeable steel segments each 0.2 m in height and 0.3 m in internal diameter. Each segment can be insulated using a ceramic fibre blanket. Each one of the two lower segments has four rectangular ports for probing and/or optical access which are closed with quartz or steel inserts. The burner geometry, shown in Fig. 2, consists of a burner gun (i.d. 8 mm, o.d. 12 mm) and an air supply in a conventional concentric configuration. The latter has two separated air streams, both having very thin guide vanes for inducing swirl. In the present work, the inner and outer streams were fitted with guide vanes of constant cord and angle of 60 and 30°, respectively.

Before entering the burner, the air flows through an ejector in order to suck flue-gas directly from the exhaust duct of the furnace as illustrated in Fig. 1. The remainder of the flue-gas is exhausted from the test furnace. The flue-gas withdrawn for recirculation purposes is controlled by a valve and

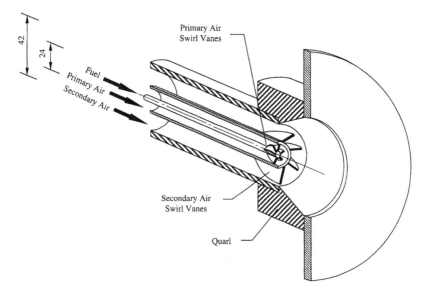

Fig. 2. Schematic of the burner arrangement.

cooled by a water coil placed in the recycling duct after which the condensate is removed. The oxidant mixture (fresh air + flue-gas) is then directed to the burner. A probe is permanently placed just before the burner to measure the oxidant composition so that recirculation rates can be calculated.

4. Experimental method

4.1. Flue-gas measurements

The sampling of flue-gas for the measurement of O_2, CO_2, CO, unburnt hydrocarbons (UHC) and NO_x concentrations was achieved using an aerodynamically quenched quartz probe. The probe design and dimensions were similar to those used by Drake et al. [8]. The probe was mounted on a traverse mechanism which enabled radial movement across the entire furnace at the exit sampling location ($x/D = 23.8$—x is the axial distance from the quarl exit plane and D is the inner diameter of the secondary air tube).

A schematic of the gas analysis system is also shown in Fig. 1. The gas sample was drawn through the probe and the sampling system by an oil-free diaphragm pump. A condenser removed the main particulate burden and condensate. A drier and a filter removed any residual moisture and particles so that a constant supply of clean dry combustion gases was delivered to each instrument through a manifold to give species concentrations on a dry basis. The analytical instrumentation included a magnetic pressure analyser for O_2 measurements, non-dispersive infrared gas analysers for CO_2 and CO, a flame ionisation detector for UHC measure-

ments and a chemiluminescent analyser for NO_x measurements. Zero and span calibrations with standard mixtures were performed before and after each measurement session. The maximum drift in the calibration was within $\pm 2\%$ of the full scale.

At the furnace exit, radial profiles showed that the concentration of the species were uniform so that probe effects were likely to be negligible and errors arose mainly from quenching of chemical reactions, sample handling and analysis. Adequate quenching of the chemical reactions was achieved and NO_2 removal by acid formation within the probe and sampling system was negligible. Repeatability of the data was, on average, within 5%.

4.2. In-flame measurements

Concentration measurements of O_2, CO_2, CO, UHC and NO_x in the near burner region were made on a dry basis as described earlier. However, a water-cooled probe was used instead of the quartz probe. It comprised a centrally located 2 mm i.d. tube through which quenched samples were evacuated, surrounded by two concentric tubes for probe cooling. In the near burner region the major sources of uncertainties were associated with the quenching of chemical reactions and aerodynamic disturbance of the flow. Quenching of the chemical reactions was rapidly achieved upon the samples being drawn into the central tube of the probe because of the high water cooling rate in its surrounding annulus—our best estimate indicated quenching rates of about 10^6 K/sec. As in flue-gas measurements, uncertainty because of NO_2 removal by acid formation within the probe and sampling system was negligible. No attempt was made

16

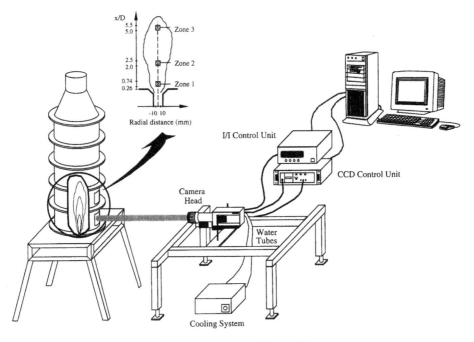

Fig. 3. Schematic of the imaging device.

to quantify the probe flow disturbances. The repeatability of the data was, on an average, within 10%.

Gas temperature measurements were obtained using uncoated 25 μm diameter Pt/Pt:13% Rh thermocouples. The hot junction was installed and supported on 300 μm wires of the same material as that of the junction. The

300 μm wires were located in a twin-bore alumina sheath with an external diameter of 4 mm and placed inside a stainless steel tube. As flame stabilisation on the probe was not observed, interference effects were unlikely to have been important and, hence, no effort was made to quantify them. Radiation losses represent the major source of

IMAGE INTENSIFIER   **CCD-CHIP**

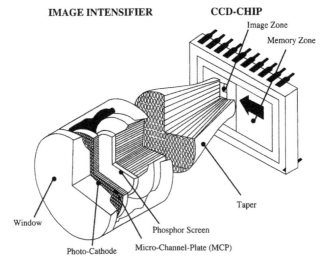

Fig. 4. Overview of the intensifier/taper/CCD-sensor configuration.

Table 1
Furnace operating conditions and flue-gas data

Group	Flame	Fuel	Temperature of the furnace walls for R = 0%	Oxidant (dry volume) O_2 (%)	CO_2 (%)	λ	R (%)	Flue-gas composition (dry volume) O_2 (%)	CO_2 (%)	CO (ppm)	UHC (ppm)	NO_x (ppm, 3% exc. O_2)
I	P1	Propane (C_3H_8)	180°C	20.9	0.0		0.0	1.4	12.6	236	21.5	32.0
	P2		210°C	20.3	0.3		2.9	1.4	12.5	285	11.1	29.4
	P3		240°C	18.4	1.5	1.07	14.3	1.3	12.7	97	16.1	20.2
	P4		280°C	17.7	1.9		18.9	1.3	12.7	123	15.8	17.8
	P5		210°C	17.2	2.3		24.1	1.4	12.6	430	29.0	14.0
II	P6	Propane (C_3H_8)	670°C	20.9	0.0		0.0	1.4	12.9	71	1.0	49.2
	P7		800°C	19.5	0.7		7.4	1.3	13.1	38	0.7	42.4
	P8		840°C	17.6	2.1	1.07	20.2	1.5	12.8	54	0.5	33.3
	P9		900°C	16.8	2.5		26.4	1.5	12.9	65	0	27.1
	P10		870°C	16.2	3.0		31.4	1.5	12.6	43	0	18.6
	P11			15.6	3.1		36.8	1.5	12.7	16	0	16.3
	P12			14.7	3.8		44.5	1.3	13.1	0	0	8.8
III	E1	Ethylene (C_2H_4)	190°C	20.9	0.0		0.0	2.9	12.7	0	1.5	61.8
	E2		220°C	18.6	1.5		14.9	3.5	12.4	0	0.5	58.0
	E3		270°C	17.2	2.9	1.15	26.3	3.2	12.6	0	0	41.9
	E4		310°C	16.4	3.1		33.8	3.2	12.6	0	0	32.1
	E5		240°C	15.6	3.7		42.6	3.2	12.6	0	6.5	24.0
IV	P13	Propane (C_3H_8)	185°C	20.9	0.0		0.0	3.0	12.1	0	0	44.2
	P14		215°C	20.2	0.5		4.1	3.0	12.0	0	0	37.3
	P15		255°C	19.2	1.1	1.15	10.3	2.9	12.1	0	0	29.8
	P16		290°C	18.5	1.4		15.3	3.1	11.9	0	0	29.3
	P17		220°C	17.5	2.1		22.9	2.9	12.1	0	0	21.6
	P18			16.8	2.8		29.0	2.8	11.9	0	5.2	17.1

18

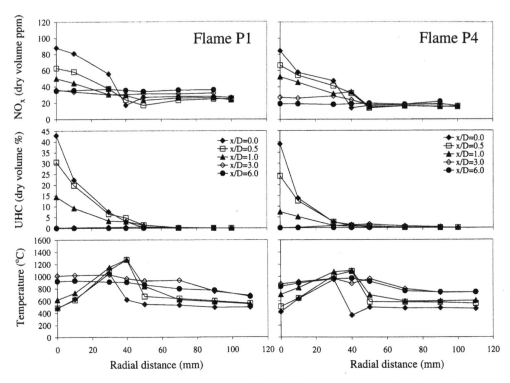

Fig. 5. Radial profiles of local mean concentrations of NO$_x$ and UHC and gas temperatures for flames P1 and P4.

uncertainty in the mean temperature measurements. An attempt made to quantify them on the basis of a theoretical expression developed by De [9] led to uncertainties of about 5% in the regions of highest temperatures and lower elsewhere.

Both probes were mounted on a 3-D computer controlled traverse mechanism which allowed for axial and radial movements throughout the furnace. The analogue output of the analysers and of the thermocouple were transmitted via A/D boards to a computer where the signals were processed and the mean values computed. No thermal distortion of the probes was observed and the positioning of the probes in the furnace was accurate to within ±0.5 mm.

4.3. Images acquisition

The images were acquired using an intensified CCD camera system [10], equipped with an UV objective (UV Nikkon, 105 mm, f/4.5). Fig. 3 shows a schematic of the imaging device arrangement. The camera consists of an image intensifier that is optically coupled via a fibre optic imaging bundle (taper) to a CCD chip whose configuration is shown in Fig. 4. The incident light on the photo cathode, which has sufficient spectral sensitivity in a wide

wavelength range from 200 to 900 nm, causes emission of electrons proportional to the number of incident photons. These photoelectrons are accelerated by a strong electric field from the cathode to the micro-channel-plate (MCP), where the electron current is intensified by a secondary electrons emission in the channels of the plate, maintaining the spatial resolution. The gain depends on this voltage across the MCP and both the gain and gating time were kept constant throughout the present study for the purpose of contrasting the data. The electron stream is accelerated again between the MCP and the phosphor screen before hitting the phosphor screen, resulting in the emission of visible light. The coupling of the image intensifier and CCD chip is made via a taper, which has a much higher transmission when compared to a lens coupling. The CCD chip (TH7863, Thomson) has 576 × 384 (horizontal × vertical) pixels and is divided into two identical arrays. The upper part of the chip is the image zone, which receives the optical information, and the lower half is the memory zone, into which the image field is transferred for subsequent read out by an 12-bit A/D-converter and stored under software control within the RAM of the computer. The CCD chip is cooled by a Peltier element at 10°C to minimise thermal dark current.

To eliminate the dark signal in the present investigation,

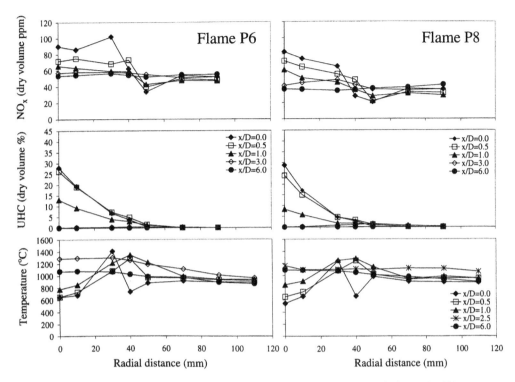

Fig. 6. Radial profiles of local mean concentrations of NO$_x$ and UHC and gas temperatures for flames P6 and P8.

background images were taken with capped camera, under the same integration time and gain of the measured images, for subsequent subtraction from the original measured images. The OH radical was detected using a UG11 glass filter and the CH radical was filtered using a band pass filter centred at 431 nm with a bandwidth of 10.3 nm. All the photometric images of OH and CH radicals were averaged over 30 times to provide statistical mean. For the calibration

of the spatial resolution, a paper map was placed in the centre of the furnace and images of the map were taken at three different zones, which are shown in Fig. 3. Each image obtained, which maintains the spatial resolution, showed an area of 20 mm × 20 mm. The attenuation of the intensity due to soot deposition on the optical access of the furnace was also taken into account. To this end, an image of a tungsten lamp placed in the centre of the furnace was

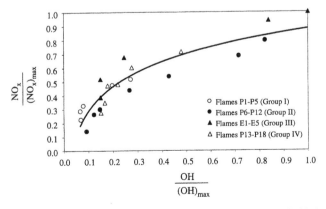

Fig. 7. Variation of the NO$_x$ emission, normalised by its maximum, with the OH intensity, normalised by its maximum value.

20

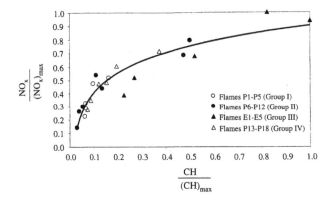

Fig. 8. Variation of the NO$_x$ emission, normalised by its maximum, with the CH intensity, normalised by its maximum value.

taken before and after every measurement to quantify the change in the transmissivity of the quartz window. The maximum correction associated with the intensity attenuation occurred in zone 1 (see Fig. 3): 2.5% in the case of the CH for the ethylene flames.

To take into account the soot continuous radiation, the flame temperature was estimated by the two-colour pyrometry method [11,12]. The correct temperature can only be measured when band spectra do not exist in the measured wavelengths. For this purpose, we have used two filters centred at 530 and 660 nm with a bandwidth of approximately 8 nm and assumed that the spectral emissivity of the flame is the same at the two wavelengths. The entire system was calibrated by means of a black-body furnace which could operate in the range 300–1500°C. For each flame, images with the 660 nm filter were collected and knowing that OH-radical has its main head at 306 nm and CH at 431 nm [13], we used the calibration curves I_{306}/I_{660} and I_{431}/I_{660}, where I_λ is the luminous intensity obtained using a filter centred at λ, to subtract the soot radiation to the OH and CH band emission. The maximum soot background correction occurred in zone 1

(see Fig. 3): 1.5% in the case of the CH for the ethylene flames.

Our best estimated indicated uncertainties of less than 5%, considering the uncertainties of the tungsten lamp, the transmissivity change of the optical access and the soot continuous radiation correction. Repeatability of the photometric data was, on average, within 5%.

5. Test conditions

The furnace operating conditions are summarised in Table 1 and encompass twenty three experimental flames. It should be noted that the present investigation has involved four different groups of flames, corresponding to two gaseous fuels (propane and ethylene), furnace combustion chamber walls with and without insulation and two excess air coefficients ($\lambda = 1.07$ and 1.15). In all the tests performed, the primary air flow was about 20% of the total air flow. The thermal input (≈15 kW) was kept constant for all flames studied and for each group of them the fresh air flow rate was held constant being the flue-gas

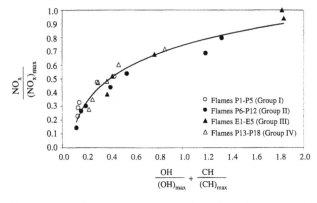

Fig. 9. Variation of the NO$_x$ emission, normalised by its maximum, with the OH + CH intensity, each normalised by its maximum value.

flow rate gradually increased. In this work the FGR rate is defined as [14]:

$$R = \frac{\dot{m}_{rec}}{\dot{m}_a + \dot{m}_f} \times 100, \qquad (8)$$

where $\dot{m}_{rec}$ is the mass flow of recycled flue-gas products per unit time.

6. Results and discussion

6.1. The relative importance of the prompt and thermal mechanisms in the present flames

Table 1 lists the furnace operating conditions used in the present work along with flue-gas data revealing a wide range of NO_x emissions, as initially intended. Note that the values of the NO_x emissions given in the table have been reduced to 3% of O_2 in the combustion products. The effects of the gaseous fuel and FGR on pollutants emissions are fully discussed elsewhere [15,16]. In addition to the wide range of NO_x emissions required, it is important to examine, at least qualitatively, the sources of NO_x formation (prompt and/or thermal) from the present flames in order to investigate the relative importance of reactions (3), (3a) and (4). To this end we have conducted detailed near burner in-flame measurements for four furnace operating conditions: flames P1 and P4 of group I and flames P6 and P8 of group II (see Table 1).

Fig. 5 shows radial profiles of local mean concentrations of NO_x, UHC and gas temperatures for flames P1 and P4 at five axial locations. For both flames it can be observed that the NO_x concentration maxima occur in their central region close to the burner quarl exit (axial locations of $x/D = 0$ and 0.5). In this region, the high UHC concentrations and the relatively low gas temperatures measured clearly indicate that the prompt mechanism is the dominant source of NO_x in both flames. Further, because thermal-NO_x formation is strongly temperature dependent, as mentioned earlier, thermal-NO_x is significant only at temperatures greater than 1500°C [17]. The maximum temperatures measured for flames P1 and P4 barely exceed 1200 and 1000°C, respectively. Owing to turbulent fluctuations the instantaneous temperature may be sufficiently high to promote thermal-NO_x formation, but the value of the maximum mean temperature reveals that the probability of an instantaneous temperature above 1500°C is small. Hence, it is concluded that thermal-NO_x plays a smaller role in the formation of NO_x in these flames, as compared to the prompt-NO_x.

Fig. 6 shows radial profiles of local mean concentrations of NO_x, UHC and gas temperatures for flames P6 and P8 at five axial locations. As a consequence of the combustion chamber being insulated, flames P6 and P8 present higher temperature levels: the maximum mean temperatures measured for flames P6 and P8 are about 1400 and 1300°C, respectively. These maximum temperatures are

still too low to promote extensive NO_x-thermal formation even considering the turbulent fluctuations. Moreover, as in flames P1 and P4, near the flame axis, the high NO_x and UHC concentrations and the relatively low gas temperatures measured indicate that the prompt mechanism is still an important, if not dominant, source of NO_x in flames P6 and P8.

We have not conducted detailed in-flame data for any of the flames of groups III and IV. As a consequence of the combustion chamber not being insulated (see Table 1) we firmly believe that, once again, NO_x formation was dominated by the prompt mechanism in all of these flames. In conclusion, the flames that have been established present a wide range of NO_x emissions to which the prompt mechanism contributed in a dominant way. Possible exceptions can be flame P6–P12—we shall return to this point in the following subsection.

6.2. Diagnostic system

Photometric images of OH and CH radicals were obtained for all furnace operating conditions included in Table 1 from different flame zones as described in a previous section. The flame zones analysed with the imaging device have been selected according to the in-flame data. Zone 1 (see Fig. 3) has proved to be the most adequate for the purposes of this study not only because it presented the highest intensities of spontaneous emission of the OH and CH radicals but also because exhibited their greatest variations accordingly with the different flames. This suggests that zone 1 controls the NO_x formation/emission in agreement with the in-flame data. In the present article only the photometric data obtained from zone 1 is included.

Figs. 7–9 represent the variation of the NO_x emission, normalised by its maximum, with the OH, CH and OH + CH intensities, each normalised by its maximum value, respectively. The need to normalise the intensity of each radical by its maximum value results from the cathode efficiency wavelength dependence and from the different transmissivity and bandwidth of the filters. In each figure, the solid line represents the best curve-fit—a logarithmic function— obtained using the least-squares approximation. The similar trends of the data plotted in Figs. 7–9 clearly support the choice of the two radicals as indicators of the NO_x formation in the present flames.

In Fig. 7 the correlation of the normalised NO_x emissions with the normalised OH spontaneous emission is relatively poor (correlation factor of 0.83). A systematic deviation from the solid line occurs for flames P6–P12. For these flames the normalised values of OH are all located below the solid line. This appears to confirm the doubts expressed in the end of the previous subsection, i.e. flames P6–P12 present the greatest thermal/prompt formation ratio of all flames that have been studied.

In Fig. 8 the correlation of the normalised NO_x emissions with the normalised CH spontaneous emission does not present systematic deviations from the solid line for any group of flames being the correlation factor 0.89. This suggests that the NO_x formation in the present flames is dominated by the prompt mechanism as the spontaneous emission of the radical CH is solely associated with this mechanism—reaction (4).

Figs. 7 and 8 suggest that an improved correlation can be obtained by plotting the normalised NO_x emissions with the normalised OH + CH spontaneous emission which is showed in Fig. 9. Indeed an improved correlation is found which can be mathematically expressed through the following function:

$$\frac{NO_x}{(NO_x)_{max}} = 0.26 \ln\left[\frac{OH}{(OH)_{max}} + \frac{CH}{(CH)_{max}}\right] + 0.74, \quad (9)$$

the correlation factor being 0.94. In spite of this factor being good, surely it could be improved if the relation between the local rates of NO_x formation and the local emission intensities of the two radicals were firmly determined. This is simply impossible to accomplish with the present imaging device owing to the nature of its measurements (line-of-sight measurements). In view of this along with the impossibility to quantify exactly the prompt and thermal contributions for each flame, the degree of success is noteworthy.

The non-intrusive NO_x diagnostic system developed and tested in the present investigation appears to be a powerful tool to control in real time NO_x emissions showing promising features to be applied to a wide range of gas-fired combustion equipment provided that the prompt mechanism is an important route for the NO_x formation. The system is currently being applied to an 0.5 MW pilot furnace firing gas where NO_x is formed mainly via the thermal mechanism.

7. Conclusions

An experimental study has been performed in a small-scale laboratory furnace fired by a propane or ethylene swirl burner to develop and test a NO_x diagnostic system based on a spectral ultraviolet/visible imaging device. The data reported include flue-gas concentrations of O_2, CO_2, CO, UHC and NO_x, and photometric data of spontaneous emission of OH and CH radicals from various regions of the flame. These data have been obtained for twenty three furnace operating conditions which quantify the effects of the gaseous fuel (propane or ethylene), flue-gas recirculation, heat losses through the furnace walls to the surroundings and excess air. The main conclusions drawn from the present results are the following:

1. As a result of the broad furnace operating conditions studied it was possible to obtain a wide range of NO_x emissions. Further, detailed in-flame data of local mean major gas-phase species and gas temperatures collected for four furnace operating conditions showed that the NO_x formation is dominated by the prompt mechanism.
2. The results reveal that there is an excellent correlation between the NO_x emissions from propane or ethylene flames and the OH + CH photometric data, which can be mathematically expressed as a logarithmic function, provided that the radical images are collected from a flame zone close to the burner exit.
3. The present non-intrusive NO_x diagnostic system appears to be a powerful tool to control in real time NO_x emissions and shows promising features to be applied to a wide range of gas-fired combustion equipment provided that the prompt mechanism is the dominant route for the NO_x formation.

Acknowledgements

Financial support for this work was partially provided by the European Commission-DGXII (JOULE Programme)—and is acknowledged with gratitude. The authors wish also to thank technicians Manuel Pratas and Vasco Fred for their valuable contributions to the experiments and are grateful for the cooperation of Jorge Coelho during the preparation of the figures.

References

[1] Nagase K, Funatsu K. SAE Techn. Paper Ser., No. 881226, 1989.
[2] Nagase K, Funatsu K. SAE Techn. Paper Ser., No. 901615, 1990.
[3] Block B, Möser P, Hentschel W. Submitted to Optical Engineering.
[4] Leipertz A, Obertacke R, Wintrich F. Twenty-sixth symposium (International) on combustion, The Combustion Institute, Pittsburgh, PA, 1996:2869–2875.
[5] Luczak A, Eisenberg S, Knapp M, Schluter H, Beushausen V, Andresen P. Twenty-sixth symposium (International) on combustion, The Combustion Institute, Pittsburgh, PA, 1996:2827–2834.
[6] Zel'dovich Ya. B, Sadovnikov P Ya, Frank-Kamenetskii D A. Oxidation of nitrogen in combustion, translated by Shelef M, Academy of Sciences of USSR, 1947.
[7] Miller JA, Bowman CT. Prog Energy Combust Sci 1989;15:287–338.
[8] Drake MC, Correa SM, Pitz RW, Shyy W, Fenimore CP. Combust Flame 1987;69:347–365.
[9] De DS. J Inst Energy 1981;54:113–116.
[10] LaVision 2D-Messtechnik GmbH. FlameStar Instruction Manual. Göttingen, 1993.
[11] Tourin RH. Spectroscopic gas temperature measurement. Amsterdam: Elsevier, 1966.
[12] Bach JH, Street PJ, Twamley CS. J Physics E: Scientific Instruments 1970;3:281–286.
[13] Gaydon AG. The spectroscopy of flames. London: Chapman and Hall, 1974.
[14] Godridge AM. J Inst Energy 1988;61:38–54.
[15] Costa M, Ruão M, Carvalho MG. Archivum Combustionis 1996;16:77–86.
[16] Baltasar J, Carvalho MG, Coelho P, Costa M. Fuel 1997;76:919–929.
[17] Nimmo W, Hampartsoumian E, Clarke AG, Williams A. Proceedings of the Second European Conference of Industrial Furnaces and Boilers, Algarve (Portugal), 1991.

Emissions Reduction: NO$_x$/SO$_x$ Suppression
A. Tomita (Editor)
© 2001 Elsevier science Ltd. All rights reserved

Systematic measurements of OH chemiluminescence for fuel-lean, high-pressure, premixed, laminar flames

B. Higgins, M.Q. McQuay*, F. Lacas, J.C. Rolon, N. Darabiha, S. Candel

Laboratoire EM2C, CNRS, Ecole Centrale Paris, 92295 Chatenay-Malabry, France

Received 30 October 1999; revised 31 March 2000; accepted 7 April 2000

Abstract

Systematic measurements are reported of OH chemiluminescence from a premixed laminar flame at pressures and equivalence ratios ranging from 0.5 to 2.5 MPa and 0.66 to 0.86, respectively. The objective was to obtain non-existent experimental data and to determine the viability of using OH chemiluminescence as an active-control parameter for high-pressure, premixed flames. The signal from the first electronically excited state of OH to ground (at 305.4 nm) was detected through a band-pass filter with a photo-multiplier tube. For constant mass flow rate, OH emission decreased significantly with increasing pressure. Emission also monotonically increased with the equivalence ratio. A linear relationship was observed between increasing mass flow and increasing chemiluminescence. These trends support the conclusion that suitable resolution and dynamic range exist for a high-pressure flame to be adequately controlled to minimize both NO$_x$ and CO emissions. Results are presented for a simple, active-control system using OH chemiluminescence measurements to demonstrate flame stabilization for equivalence ratio disturbances at fixed, elevated pressures. Finally, only qualitative agreement was observed between the measurements and numerical predictions of OH chemiluminescence using an adiabatic, freely propagating premixed flame. Although more work is required on the chemical-kinetic mechanism of OH chemiluminescence, the modeling effort supports the use of OH chemiluminescence for active-feedback-control applications. © 2000 Elsevier Science Ltd. All rights reserved.

Keywords: Lean flames; Industrial application; Premixed flames; High pressure; Active control; OH chemiluminescence

1. Introduction

Increasingly stringent, NO$_x$ emission regulations remain one of the major driving forces for gas-turbine combustion research [1,2]. This technological challenge is accentuated by the fact that increased combustion temperatures and/or overall pressure ratio improve engine thermal efficiency. Any effort to reduce combustion temperatures to minimize NO$_x$ emission should be carried out under the constraint of maintaining the present level of engine reparability, operating costs, reliability, performance, and safety. Although the best single-annular technology is capable of meeting current legislation, further reductions in permissible NO$_x$ levels has required the introduction of either fuel-staged combustors [3,4] or lean-premixed technology [5,6] for ultra-low NO$_x$. In the case of lean-premixed combustion, low NO$_x$ and CO emission levels are obtained in a narrow range of equivalence ratios and it is then necessary to continuously monitor

the combustion process in order to assure the reliable operation in a desired engine operating envelope. Equivalence ratios in excess of this operational window lead to increased NO$_x$ emissions and equivalence ratios at the other extreme can lead to unstable combustor operation, indicated by elevated CO emissions, possibly resulting in engine failure.

Present research efforts suggest that chemiluminescence of flame species such as CH, C$_2$, OH, and CO$_2$ can be used as qualitative and/or quantitative combustion diagnostic tools [7–19]. This is further substantiated by the recent advances in optical fiber technology, such as increased transmission properties in the ultraviolet (UV) wavelengths, and the development of more sensitive UV detectors [8]. The objective of this experimental study has been to parametrically study OH chemiluminescence as a function of operating pressure, overall fuel/air flow rate, and equivalence ratios of a fuel-lean, laminar, premixed flame. The purpose is to obtain unavailable experimental data and to investigate both experimentally and numerically the possibility of developing an active-control system for the flame. Active control can be realized if the following three criteria are met: (1) there is a direct correlation between OH chemiluminescence and equivalence ratio; (2) NO$_x$ and CO emissions are

* Corresponding author. Present address: Brigham Young University, Mechanical Engineering Department, 435 CTB, Provo, UT 84602, USA. Tel.: +1-801-378-4980; fax: +1-801-378-5037.

E-mail address: mq01@byu.edu (M.Q. McQuay).

Reprinted from *Fuel* **80 (1)**, 67-74 (2001)

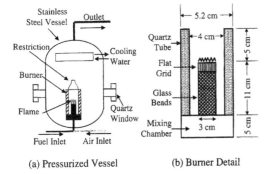

(a) Pressurized Vessel   (b) Burner Detail

Fig. 1. Schematic of experimental apparatus illustrating: (a) high-pressure vessel, coarse-grid burner, and optics; and (b) detail of the coarse-grid burner.

functionally related to equivalence ratio; and (3) a relationship between these variables can be developed that includes mass flow rate and pressure. The main contribution of the work presented here is: (1) to report non-existent data obtained from experiments in a realistic range of operating pressures under suitably controlled operating conditions; and (2) to illustrate the feasibility of flame active-control using those measurements. This effort is part of a larger project (BRITE/EURAM-SENSICOME [20]) aimed at: (1) researching, manufacturing, and validating the operation of non-intrusive sensors to measure process temperatures within natural-gas-fired, lean-burn combustors, and other sensors to measure exhaust emissions; and (2) demonstrating, under realistic operating conditions, that the output from such sensors can be used within an electronic management system to regulate the combustion process and thus minimize the emissions. Existing commercial combustors using lean-premixed technology can require expensive means to assure that operation is within the desirable levels of NO_x and CO emissions (e.g. in-line FTIR spectroscopy systems) [21].

Several studies exist in the open literature dealing with the combustion control systems based on chemiluminescence measurements either as primary or secondary variables in premixed or diffusion combustion. Zabielski et al. [22] developed a fuel/air control system for an industrial fiber-matrix burner using a correlation between the intensity of the optical emission from the fiber surface and the carbon monoxide concentration in the exhaust products at given fuel/air ratios. Brouwer et al. [23] investigated the use of a closed-loop scheme to control the performance of a gas-turbine combustor by monitoring the emission of carbon monoxide and carbon dioxide, the radiative heat flux to the combustor liner associated with soot, and combustor stability. In their study, nozzle atomizing airflow rate was used as an input to the control system. John and Samuelsen [24] studied the use of optimal, active-control techniques to minimize NO_x emissions and maximize combustion

efficiency in an industrial, swirl-stabilized, natural-gas-fired burner.

An adaptive combustion control strategy was developed by Padmanabhan et al. [25] to optimize the performance of a laboratory-scale combustor with respect to volumetric heat release and pressure fluctuations of a premixed flame using CH and C_2 emission detected by photodiodes. Allen and coworkers [26] used a neural network approach to control a spray-fired laboratory flame with a CCD camera monitoring CH and OH emissions. By the use of infrared thermography, Shuler et al. [27] have developed a control system for waste incinerators with good time response in order to detect furnace temperature fluctuations on a time scale of a few minutes. This information was then used to fine-tune the total under-fire airflow. Farias and Ngendakumana [28] have used visual and infrared spectrometer measurements (mainly CH and soot emissions) to control the efficiency, emissions of CO and NO_x, and flame stability in an oil-fired boiler with a power output of 370 kW. The measured spectra were processed in order to estimate the relative intensity of CH emission taking into account the broadband emission due to soot and CO_2.

Although much work has been done to study flame emission, only now are the numerical models able to accurately predict these phenomena. Abid et al. [29] have shown that through modeling of OH chemiluminescence, pertinent details of the combustion event can be correlated to other measured parameters (i.e. flame-ball radii). This article seeks to methodologically address our experimental observations in light of scaling arguments and presently existing, numerical-modeling capabilities.

2. Experimental measurements

The high-pressure combustion facility, shown schematically in Fig. 1, is described in detail elsewhere [30]. This system is comprised of a 40 l-capacity, stainless-steel chamber capable of pressures up to 3.0 MPa. All flows were metered externally by mass flow controllers and were carried into the pressure vessel by stainless-steel flexible tubes. The pressure in the vessel was controlled by a 2.5 cm-diameter, computer-controlled valve located 3.5 m downstream of the combustion chamber. The pressure vessel had four quartz windows (6 cm diameter; 4 cm thick), which allowed horizontal optical access perpendicular to the burner axis at four angles (from above: 0, 90, 180, and 225°). These windows were individually purged with a jet of nitrogen to minimize water condensation. Within the pressure vessel a platform was constructed to allow OH chemiluminescence measurements from a coarse-grid, lean, methane/air, laminar flame, Fig. 1.

The coarse grid consisted of a series of parallel, notched, stainless-steel plates (0.5 mm thick) inserted into an opposed series rotated 90°, thereby forming a grid with square holes approximately 1.3 mm on a side with 0.5 mm

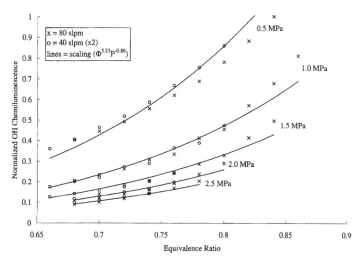

Fig. 2. Measured OH chemiluminescence intensity (305.4 nm) as a function of equivalence ratio. Symbols represent OH emission normalized by the initial mass flow rate and isobaric lines are obtained by use of the correlation $h\nu \propto \dot{m}\Phi^{5.23}P^{-0.86}$.

spacing between the holes. This arrangement was cut into a cylinder (30 mm diameter; 11 mm thick) and inserted into the end of a stainless-steel tube filled with ceramic beads. At elevated pressures (over 0.30 MPa), the methane/air flame amounted to a mesh of nearly 200 Bunsen-type flames, each stabilized at the end of a grid point. We found this to be more stable at high pressures than a single Bunsen flame and the range of operating conditions was larger than that of a sintered-steel flat-flame configuration. At high pressures, the flame was isolated from pressure oscillation within the chamber by a 64 mm diameter quartz tube containing a stainless-steel constriction downstream of the flame. This was required to reduce resident pressure coupling between the pressure chamber and the flame and to eliminate buoyant interactions with the quiescent surroundings. In this arrangement, the flame holder was only steady at pressures above 0.3 MPa and the experimental results at atmospheric pressure were not available.

For the two flow conditions in this study, we calculate Reynolds numbers of 3400 and 1700 for flows of 80 and 40 slpm through the coarse-grid flame holder. These numbers are calculated by assuming that the flow has the properties of air at atmospheric temperature and pressure. Increasing pressure for constant mass flow will not change significantly the Reynolds number ($Re = 4\dot{m}/\pi d\mu$) because viscosity (μ) only weakly depends on pressure. The flow through each of the small channels is approximately fourteen times smaller; resulting in Reynolds numbers of 240 and 120 for flows of 80 and 40 slpm, respectively. One may then consider that all flows in this study are laminar.

The OH chemiluminescence signal was measured with a band-pass interference filter (centered at 307.1 nm; 11.2 nm full-width half-height; 16.5% maximum transmission) in conjunction with a photo-multiplier tube (PMT).

Background radiation from soot incandescence near 300 nm was assumed to be relatively small compared with the OH emission because soot incandescence is small at UV wavelengths and the emission bandwidth of OH chemiluminescence is nearly the same as the transmission width of the filter. This assumption has been made by other researchers [17,31] for similar experiments and has been reconfirmed in this experiment by measuring the OH emission spectrum with a spectrometer.

The time-resolved signal from the PMT was passed through a programmable low-pass filter set at one-half the sampling frequency ($\sim$2000 Hz) of our computer data acquisition system. The data acquisition system consisted of a board for analog-to-digital conversion and data collection. Care was taken to shield all the cables to the experimental apparatus and to the sampling hardware to reduce the amount of electrical noise. Considering the noise level in the overall data acquisition system and the PSD of several preliminary measurements, a sampling frequency of 4000 Hz was used in all tests, thus avoiding aliasing by meeting the Nyquist criterion. Root-mean-square noise in the OH signal was low, indicating little fluctuation in the time-resolved signal in this laminar flame. Only average values are reported in this manuscript. Experimental errors primarily come from signal variation due to flow fluctuations under high-pressure conditions. We found that the repeatability (as an indicator of the experimental variation) to be within $\pm5\%$. Systematic errors, resulting in a shift of the measured amount of OH chemiluminescence, were small since the data were normalized.

In an attempt to investigate other inexpensive instruments to perform the measurements obtained here by use of PMT tubes, particularly considering the cost and special handling needs (e.g. high voltage) of these instruments, we pursued

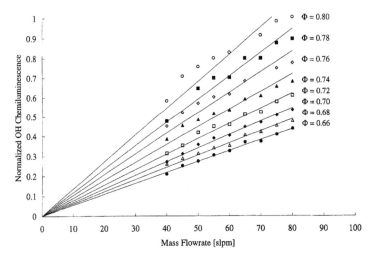

Fig. 3. Measured OH chemiluminescence intensity (305.4 nm) at 1.0 MPa as a function of mass flow rate for various equivalence ratios. Symbols represent OH emission normalized by the initial mass flow rate and straight lines are linear regressions through the origin for each equivalence ratio.

duplication of the results later presented here with a UV-sensitive photodiode. Nonetheless, the scale of this flame was sufficiently small, making photodiode measurements subject to prohibitive levels of electronic noise interference. A cost-effective method, which combines large-band-pass, optical filters with photodiodes could be developed, requiring a larger scale flame study and a more extensive development sequence.

3. Experimental results

Fig. 2 shows the experimental OH-chemiluminescence data as a function of equivalence radio for five pressures (0.5, 1.0, 1.5, 2.0, and 2.5 MPa) and two flow rates (40 and 80 slpm). The correlation lines also shown in this figure are described later in the paper in association with Fig. 4. The results are first divided by the flow rate through the burner and then normalized by the highest measured value, which was obtained at an equivalence ratio of 0.84 and a pressure of 0.5 MPa. There are several notable trends: (1) a change by a factor of two in flow rate gives a proportional increase in OH chemiluminescence; (2) for any fixed equivalence ratio, OH chemiluminescence decreases with increasing pressure; and (3) for any fixed pressure, OH

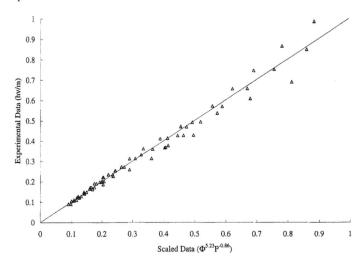

Fig. 4. Individual experimental data points from Fig. 2 plotted against their corresponding values predicted by the correlation $h\nu \propto \dot{m}\Phi^{5.23}P^{-0.86}$. The line with unity slope represents ideal agreement between the correlation and experimental data.

chemiluminescence increases with increasing equivalence ratio.

Further investigating the first trend, that OH chemiluminescence is directly proportional to flow rate, the mass flow for eight equivalence ratios at one elevated pressure (1.0 MPa) was varied. The results are plotted in Fig. 3. The symbols are measured values and the lines are linear regression, fit through the origin for each equivalence ratio. In general, all points fit well although larger deviations are observed for those experimental conditions corresponding to the highest equivalence ratios and flow rates. Nonetheless such behavior would not prevent the implementation of a control algorithm based on these observed variations. In general, OH chemiluminescence (hν) for these conditions is proportional to the mass flow rate ($\dot{m}$), i.e. h$\nu \propto \dot{m}$.

The second trend, that OH chemiluminescence decreases with increasing pressure (0.5–2.5 MPa), can be attributed primarily to increased collisional quenching at higher pressures. Plotting the data from Fig. 2 on a log–log plot versus pressure gives linear results for each equivalence ratio. That is, there is a power law dependence of pressure on OH chemiluminescence.

The third trend, that OH chemiluminescence increases with increasing equivalence ratio (from 0.65 to 0.85), can be attributed to a temperature effect. An increase in the equivalence ratio towards stoichiometry increases the adiabatic flame temperature. Again there is a power law dependence that can be seen on a log–log plot.

The power dependence of equivalence ratio and pressure on OH chemiluminescence can be found with a correlation of the form:

$$\log(h\nu) = A_0 + A_1 \log(\Phi) + A_2 \log(P) + A_3 \log(\dot{m}) \qquad (1)$$

where A_0 is arbitrary and A_3 has been shown to be unity. Fitting the data from Fig. 2 in the above form gives $A_1 = 5.23$ and $A_2 = -0.86$, or:

$$h\nu \propto \dot{m}\Phi^{5.23}P^{-0.86}. \qquad (2)$$

The quality of this correlation can be visualized in Fig. 4 by plotting the scaled OH chemiluminescence versus the measured data. The error of the correlation can be indicated by the R-squared value ($R^2 = 0.991$) or as a standard estimate of error (2.7%); both of which are relatively good. In Fig. 2, the correlation in Eq. (2) is used to plot the isobaric lines shown. The fit is adequate over the factor of five increase in pressure (0.5–2.5 MPa) and lean equivalence ratios from 0.66 to 0.86 for this combustor configuration.

In summary, the experimental results indicate that for these conditions OH chemiluminescence monotonically increases for increasing equivalence ratios and decreasing pressures. OH chemiluminescence appears to exhibit significant dynamic range, particularly with equivalence ratio. That is, small changes in equivalence ratio have an exponential effect on light emission.

4. Application

To illustrate the potential of OH chemiluminescence as an active-control feedback parameter, a simple experiment was devised where a desired equivalence ratio could be fixed for an arbitrarily adjusted airflow. That is, the computer would compensate for airflow changes with fuel flow changes, keeping the equivalence ratio constant. In order to accomplish this task OH chemiluminescence for equivalence ratios from 0.65 to 0.85 at one pressure and airflow rate were recorded. The PMT output, the airflow meter, and the pressure transducer signals were then controlled by the computer using a simple control algorithm that would take these inputs and the recorded OH chemiluminescence and proportionally control the fuel flow to give a fixed equivalence ratio. Externally the airflow was varied with a valve and the computer was able to keep the equivalence ratio constant for a variable mass flow rate.

Although this simple example only used proportional feedback control, a more complicated proportional-integral-derivative (PID) control system could have been implemented to dampen oscillations and to provide a quicker response time. Additionally, pressure variation could have been added including the pressure variation developed in the previous section. The utility of this example is to illustrate that at high pressure, active-feedback control is possible, using only OH chemiluminescence. Including the chemiluminescence from other species (e.g. CO_2, C_2, and CH) can add layers of confidence and dynamic range, further supporting the use of chemiluminescence as an active-control parameter.

5. Numerical analysis

In this section, the experimental data are modeled as an adiabatic, one-dimensional, laminar flame with multispecies transport and detailed chemical kinetics. As previously mentioned, given the required limiting assumptions, the main purpose of these simulations is to obtain a qualitative support for the measured trends. Specifically, the premix code from the Chemkin III software package (Reaction Design, Inc.) was utilized. The one-dimensional assumption allows detailed chemical-kinetic calculations, essential to model adequately the radical-pool chemistry needed to predict chemiluminescence. The validity of the one-dimensional model to our experimental conditions is strengthened by the observation that variations in the mass flow of our laminar flame only produced linear deviations in the measured chemiluminescence.

We used a combination of two chemical-kinetic submechanisms in our analysis: (1) the OH^* chemiluminescence sub-mechanism of Dandy and Vosen [17]; and (2) the methane-oxidation sub-mechanism from GRIMECH 2.11 [32] excluding all nitrogen chemistry. None of the reaction rates have been adjusted to fit the experimental

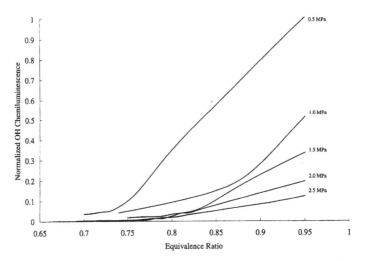

Fig. 5. Modeling results on an arbitrary scale for five pressures (isobaric lines) as a function of equivalence ratio.

data. The OH-chemiluminescence sub-mechanism contains 22 reactions and was developed by Dandy and Vosen [17] by assuming that the production of excited-state OH radicals (OH^*) occurs through the reaction:

$$CH + O_2 \leftrightarrow CO + OH^*. \tag{3}$$

This reaction produces OH^* radicals, which then react through several channels: (1) spontaneous emission (chemiluminescence) via the reaction:

$$OH^* \rightarrow OH + h\nu, \tag{4}$$

where $h\nu$ represents one photon at 305.4 nm, not accounting for broadening due to changes in rotational and/or vibrational energy levels; (2) collisional relaxation of OH^* to OH via:

$$OH^* + M \leftrightarrow OH + M, \tag{5}$$

where the reaction rate depends on the reacting partner M. The reaction rates for these reactions with typical combustion products (CO_2, O_2, H_2O, CO, CH_4, H_2, and N_2) were derived by Crosley et al. [33] for laser-induced fluorescence studies; and (3) inclusion of duplicate OH reactions where OH has simply been replaced by OH^* because OH^* radicals can react similarly to OH. It should be noted that Dandy and Vosen found, as we do, that this route is small and practically negligible. The thermodynamic data for OH^* was

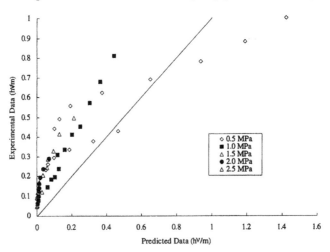

Fig. 6. Individual experimental data points from Fig. 2 plotted against their corresponding values predicted by the numerical model. The line with unity slope represents ideal agreement between the numerical predictions and experimental data.

derived with the assistance of Nick Marinov of Lawrence Livermore National Laboratories.

The methodology for directly comparing emitted light from the experiments to the one-dimensional, numerical-modeling results is now described. From the experimental results, the PMT and interference-filter combination is assumed to give a voltage linearly related to the net photon flux from the flame [photons s^{-1}]. This flux is then scaled by the total mass flow rate [kg s^{-1}], giving results proportional to the number of photons per mass of reactants consumed [photons kg^{-1}]. Since the collection efficiency, which is a linear function of the filter, the optics, and the PMT, is not known, the results are normalized based on the maximum value measured. Ideally, experiments similar to those of Dandy and Vosen, where the photon flux at the PMT was estimated, were to be performed; however, the high-pressure environment makes those types of experiments exceedingly difficult.

To compare the numerical and experimental results, the integrated molar flux through the spontaneous-emission reaction, $OH^* \rightarrow OH + h\nu$, scaled by the initial mass flow rate ($m = \rho S_L A$), is assumed proportional to the experimentally measured chemiluminescence, and is estimated by:

$$\frac{h\nu}{\dot{m}} = \frac{1}{\rho S_L} \sum [(Q_i + Q_{i-1})(x_i - x_{i-1})] \qquad (6)$$

where Q_i is the molar flux through the reaction ($OH^* \rightarrow OH + h\nu$) at node i [mol cm^{-3} s^{-1}]; ρ, the density [kg cm^{-3}]; S_L, is the laminar flame speed [cm s^{-1}]; x_i is the position of node i [cm]; and the integration is approximated by a trapezoidal summation across the entire numerical grid.

Fig. 5 shows the results of the OH chemiluminescence predictions normalized by the method just described. One finds a poor agreement between the experimental results and the numerical results. In particular, the OH^* chemiluminescence is predicted poorly for low equivalence ratios and high pressures. This is illustrated best in Fig. 6 where the experimental data is plotted versus the predicted values. Ideally, the data points would fall on the 45° line. The obvious disagreement is most likely due to several factors. The one-dimensional assumption of the numerical model may not be valid for this physical arrangement, and, more importantly, the chemical-kinetic mechanism is probably not tuned to work under these conditions (i.e. burner stabilized flames at high pressure).

However, in Fig. 5, the trend of increasing OH chemiluminescence for increasing equivalence ratio at constant pressure is captured by the predictions. Also predicted is the trend of decreasing OH chemiluminescence for increasing pressure at constant equivalence ratio. Qualitatively, the monotonic relationships among OH chemiluminescence, equivalence ratio, and pressure are predicted and it is these relationships that provide the basis for the use of OH chemiluminescence as an active-feedback-control parameter. Although a better chemical-kinetic mechanism and a

numerical model are desirable, the objective was to provide a basis for using OH chemiluminescence for feedback control; a premise that is supported by these predictions. Application of these results to other geometries and flow conditions will necessitate more work on the chemical-kinetic mechanism describing high-pressure OH^* reactions.

6. Conclusion

Measuring the OH-chemiluminescence from CH_4/air combustion as a function of equivalence ratio (0.66–0.86) and pressure (0.5–2.5 MPa), we find that the functional relationship is monotonically decreasing for: (1) decreasing equivalence ratio; and (2) increasing pressure. Further, OH chemiluminescence is linearly proportional to the total mass consumption. From these results we have shown that, for a given system, one can deduce the equivalence ratio if the airflow rate and pressure are known.

A well-founded modeling effort of the emission phenomenon should allow application of these results to other geometries and flames, particularly strained flames. Using the chemical-kinetic, OH-chemiluminescence sub-mechanism of Dandy and Vosen [17], combined with the high-pressure, methane-chemistry of GRIMECH2.11, only the general trends of OH chemiluminescence were predicted. More work on the model is certainly needed, particularly for high pressures. However, the trends of: (1) increasing OH chemiluminescence for increasing equivalence ratio at constant pressure; and (2) decreasing OH chemiluminescence for increasing pressure at constant equivalence ratio are predicted. Qualitatively, the monotonic relationships among the OH chemiluminescence, equivalence ratio, and pressure are predicted, providing the basis for the use of OH chemiluminescence as an active-feedback-control parameter.

While increasing equivalence ratio produces elevated NO_x emissions, decreased equivalence ratio leads to unstable combustion and increases CO emissions. The results of the present study illustrate that, with a good in situ measurement of equivalence ratio, an active-control algorithm based on OH chemiluminescence can reduce both NO_x and CO emissions, resulting in a low-emitting, stable operating mode for a lean, premixed, laminar flame.

Acknowledgements

This work was sponsored by the European Community under Project no. BRPR960217 (BRITE/EURAM 3). B. Higgins and M.Q. McQuay are also grateful for the financial support received from other sources, including Elf Aquitaine, Ecole Centrale Paris, the Chateaubriand Fellowship, and the France-Berkeley Fund.

References

[1] Lefebvre AH. Gas turbine combustion. New York: McGraw-Hill, 1983.

[2] Bahr DW. Aircraft turbine engine combustors—development status/challenges. In: Duro DFG, et al., editors. Combustion flow diagnostics, NATO/ASI series. Dordrecht: Kluwer, 1989. p. 357–74.

[3] Segalman I, McKinney RG, Sturgess GJ, Huang L-M. Reduction of NO_x by fuel-staging in gas turbine engines—a commitment to the future. AGARD Conference Proceedings 536, 1993. p. 29/1–17.

[4] Sturgess GJ, McKinney RG, Morford SA. Modification of combustor stoichiometry distribution for reduced NO_x emission from aircraft engines. Trans ASME 1983;115:570–80.

[5] Corbett NC, Lines NP. Control requirements for the RB 211 low emission combustion system, ASME International Gas Turbine and Aeroengine Congress and Exposition, Cincinnati, OH, paper 93-GT-12, 1993.

[6] Bahr DW. Aircraft turbine engine NO_x emissions abatement. In: Culick F, et al., editors. Unsteady combustion, NATO/ASI series. Dordrecht: Kluwer, 1995. p. 243–64.

[7] Locke RJ, Ocknuzzi KA. OH imaging in a lean burning high-pressure combustor. AIAA J Tech Note 1996;34(3):622–3.

[8] Roby RJ, Hamer AJ, Tilstra SA, Burt TJ. Improved method for flame detection in combustion turbines. ASME J Engng Gas Turbine Power 1995;117:332–40.

[9] Obertacke R, Wintrich H, Wintrich F, Leipertz A. A new sensor system for industrial combustion monitoring and control using UV emission spectroscopy and tomography. Combust Sci Tech 1996;121:133–51.

[10] Chou T, Patterson DJ. In-cylinder measurement of mixture maldistribution in a L-head engine. Combust Flame 1995;101:45–57.

[11] Edwards T, Weaver DP. Investigation of high pressure solid propellant combustion chemistry using emission spectroscopy. J Propulsion 1986;2(3):228–34.

[12] Hertz HM, Faris GW. Emission tomography of flame radicals. Opt Lett 1988;13(5):351–3.

[13] Kuwahara K, Ando H. Analysis of barrel-stratified lean-burn flame structure by two-dimensional chemiluminescence measurement. JSME Int J 1991;37(3):260–4.

[14] Yoshida A, Narisawa M, Tsuji H, Hirose T. Chemiluminescence emission of C_2, CH and OH radicals from opposed jet burner flames. JSME Int J 1995;38(2):222–9.

[15] Kauranen P, Andersson-Engels S, Svanberg S. Spatial mapping of flame radical emission using a spectroscope multicolor imaging system. Appl Phys 1991;53:260–4.

[16] Haile E, Delabroy O, Lacas F, Veynante D, Candel S. Structure of Acoustically Forced Turbulent Spray Flames. Twenty-sixth International Symposium on Combustion. Pittsburgh: The Combustion Institute, 1996. p. 1663–70.

[17] Dandy DS, Vosen SR. Numerical and experimental studies of hydroxl radical chemiluminescence in methane–air flames. Combust Sci Tech 1992;82:131–50.

[18] Samaniego J-M, Egolfopoulos FN, Bowman CT. CO_2^* chemiluminescence in premixed flames. Combust Tech 1995;109:183.

[19] Smooke MD, Xu Y, Zurn RM, Lin P, Frank JH, Long MB. Computational and Experimental Study of OH and CH Radicals in Axisymetric Laminar Diffusion Flames. Twenty-fourth International Symposium on Combustion. Pittsburgh: The Combustion Institute, 1992. p. 813–21.

[20] Lines NP, et al. Integrated Sensor Techniques for Industrial Combustion Monitoring and Control, BRITE/EURAM 3 Project Reference no. BRPR960217 (November, 1996–March, 2000).

[21] Roe PJ, BRITE/EURAM 3 Principal Investigator, Rolls Royce Industrial and Marine Gas Turbines Ltd, Personal Communication, 1997.

[22] Zabielski MF, Freihaut JD, Egolf CJ. Fuel/air control of industrial fiber matrix burners using optical emission. ASME Fossil Fuel Combust 1991;PD-vol. 33:41–8.

[23] Brouwer J, Ault BA, Bobrow JE, Samuelsen GS. Active Control for Gas Turbine Combustor. Twenty-Third International Symposium on Combustion. Pittsburgh: The Combustion Institute, 1990. p. 1087–92.

[24] St John D, Samuelsen GS. Active, Optimal Control of a Model Industrial, Natural Gas-Fired Burner. Twenty-fifth International Symposium on Combustion. Pittsburgh: The Combustion Institute, 1994. p. 307–16.

[25] Padmanabhan KT, Bowman CT, Powell JD. An adaptive optimal combustion control strategy. Combust Flame 1995;100:101–10.

[26] Allen MG, Butler CT, Johnson SA, Lo EY, Russo F. An imaging neural network combustion control system for utility boiler application. Combust Flame 1993;94:205–14.

[27] Schuler F, Rampp F, Martin J, Wolfrum J. TACCOS-A thermography-assisted combustion control system for waste incinerators. Combust Flame 1994;99:431–9.

[28] Farias O, Ngendakumana P. Flame spectroscopy and the NO_x formation mechanisms in fuel oil boilers. Bull Soc Chim Belges 1996;105(9):545–54.

[29] Abid M, Wu MS, Liu JB, Ronney PD, Ueki M, Maruta K, Kobayashi H, Niioka T, Vanzandt DM. Experimental and numerical study of flame ball IR and UV emissions. Combust Flame 1998;116:348–59.

[30] Versaevel P. Laminar two-phase combustion: experimental and numerical Study, PhD thesis, 1996-20. Ecole Centrale Paris, Chatenay-Malabry, 1996.

[31] Peeters J. Key reactions in the oxidation of acetylene by atomic oxygen. Bull Soc Chim Belges 1997;106:337.

[32] Bowman,CT, Hanson RK, Davidson DF, Gardiner WC Jr, Lissianski V, Smith GP, Golden DM, Frenklach M, Goldenberg M. http://www.me.berkeley.edu/grimech/.

[33] Garland NL, Crosley DR. On the Collisional Quenching of Electronically Excited OH, NH and CH in Flames. Twenty-First International Symposium on Combustion. Pittsburgh: The Combustion Institute. 1986. p. 1693–702.

Development of dual sulfur oxides and oxygen solid state sensor for "in situ" measurements

S. Zhuiykov*

Analyt Instruments Pty. Ltd., 15/594 Inkerman Rd., Caulfield North, Vic. 3161, Australia

Received 11 December 1998; received in revised form 20 October 1999; accepted 10 November 1999

Abstract

Dual SO_x/O_2 "in situ" potentiometric sensor based on zirconia solid electrolyte and a composition of metal sulfates was investigated for simultaneous measurement of both oxygen and sulfur oxide emissions in combustion gas. The $BaSO_4-K_2SO_4-SiO_2$-based electrochemical cell of the sensor exhibited excellent sensing characteristics for SO_x measurement within a reasonably wide working temperature range of 650–1000°C and measuring SO_x concentrations (18–10,000 ppm). Carbon dioxide, oxygen and nitrogen oxides had no measurable effect on the SO_x sensing properties of the sensor. Typical response times at 700°C were in the range of 45–80 s. The sensor also showed good correlation between the measuring SO_x concentration and the output EMF in accordance with the Nernst equation. The installation of the probe based on dual SO_x/O_2 sensor in control loops can provide a better and a more effective way towards fuel saving and efficiency. © 2000 Elsevier Science Ltd. All rights reserved.

Keywords: Sulfur oxides sensor; Solid electrolytes; Combustion efficiency

1. Introduction

Large coal-fired and oil-fired utilities are under heavy pressure to reduce both sulfur oxide (SO_x) and nitrogen oxide (NO_x) emissions [1,2]. Sulfur is a principal organic impurity in fossil fuels [3]. The interest in sulfur recovery and in monitoring of SO_x emissions continues to remain high because it is market-driven by stringent government regulations [4]. While the industry is still deciding on the best method to reduce emissions and to monitor SO_x concentrations, a recent development in SO_x/O_2 solid state sensors promises hope for a better way to simultaneously monitor both SO_x and O_2 concentrations in the stack.

In various combustion processes SO_x volume measurements must be referenced to some oxygen concentration in the flue gas to be meaningful. Because the oxygen level indicates how much the flue gas sample has been diluted with air, it is meaningless to report the measured SO_x concentration unless the oxygen content is given. The higher the excess air in the combustion gas the greater the energy or fuel loss. However, the excess of oxygen by itself is an index of excess air ratio, and not an index of combustion quantity or completeness of combustion.

Therefore, a combination of both O_2 and SO_x measurements would provide the best technical approach to the control of efficiency of the coal-fired and oil-fired utilities.

The zirconia (ZrO_2) solid electrolyte sensor is one of the most common and accepted sensors for excess oxygen measurement [5,6]. The sensor operates at a temperature range from 650°C up to 1600°C [6]. Major advances in this sensor are required in terms of low cost, selectivity, high sensitivity, durability and reliability. The possibility of determining other gas constituents besides oxygen using sensors with oxide-ion-conducting electrolytes arises from the thermodynamic equilibria that are set up with oxygen in the gas phase or between particles in the gas phase and the solid electrolyte. However, for oxygenic gases such as SO_x, NO_x and carbon dioxide (CO_2) with more complex molecular structures, this type of practical ZrO_2-based sensors is hardly available. Using an auxiliary phase, which is sensitive to one of the measuring oxygenic gases, is very attractive way for the practical development of the electrochemical solid state sensors. Furthermore, the combination of zirconia electrolyte with on auxiliary phase through the heterojunction formed in between can mitigate many difficulties encountered with use of oxyanion salts alone as an electrochemical cell. Recent breakthrough in the development of this kind of sensors [7] provides an opportunity to measure very low concentrations with high

* Tel.: +61-3-9527-4014.
  E-mail address: sz98uk@hotmail.com (S. Zhuiykov).

Reprinted from *Fuel* **79 (10)**, 1255-1265 (2000)

Table 1
Performance of newly developed zirconia-based gas sensors

Gas	Solid electrolyte	Auxiliary phase	Operating temperature (°C)	Measured concentration (ppm)	Response time (s)	References
SO_x	Zirconia	Li_2SO_4–$CaSO_4$	700	2–200	< 50	[26]
		Li_2SO_4–$CaSO_4$–SiO_2	650–800	2–200	< 50	[27]
		$BaSO_4$–K_2SO_4–SiO_2	650–1000	18–10000	< 45	[29]
NO_x	Zirconia	$Ba(NO_3)_2$–$CaCO_3$	450	4–1000	< 30	
		$CdCr_2O_4$	550	20–200	< 30	[7]
		$CdMn_2O_4$	500–600	20–600	< 30	
CO_2	Zirconia	$LiCO_3$	550–600	100–2000	< 10	[7]
		WO_3, CdO–SnO_2	550–600	20–4000	< 10	[6]

sensitivity and selectivity. Moreover, these ZrO_2-based sensors do not require specific reference gases with fixed concentrations of measuring gases, which are expensive and impractical. Most of them use air as a reference gas. Table 1 illustrates the performance of these newly developed gas sensors.

Potentiometric ZrO_2-based gas sensors are based on the determination of electromotive force (EMF) of solid state galvanic cell. A simple relationship can be derived between the EMF and the gas phase composition that remains in contact with one or both cell electrodes.

The simplest electrochemical gas sensor is an oxygen conductive cell that allows the determination of oxygen partial pressure. The basic components of this cell include yttria-stabilized zirconia, as a solid electrolyte, and Pt electrodes (Fig. 1):

$$p(O_2)^1, \ Pt/ZrO_2\text{–}Y_2O_3/Pt, \ p(O_2)^2 \tag{1}$$

Its EMF is determined by oxygen partial pressure on both electrodes and, according to the Nernst law, could be expressed by the equation:

$$E = E_0 \frac{RT}{nF} \ln \frac{p(O_2)^1}{p(O_2)^2}, \tag{2}$$

where E is an EMF, E_0 an EMF in standard conditions, R the gas constant, T the absolute temperature, n the number of electrons accompanying the cell reaction, F the Faraday constant and both $p(O_2)^1$ and $p(O_2)^2$ are the oxygen partial pressures on the measuring and reference electrodes, consequently. Assuming $p(O_2)^1$ as a known reference $p(O_2)$, which is usually 20.9 vol% O_2 in ambient air, the determination of $p(O_2)^2$ may be done from Eq. (2). Due to the constant value of the reference $p(O_2)$, the accuracy of the potentiometric method of measurement for industrial oxygen probe is in the range of $\leq 2\%$ of reading or $\pm 0.1\% \ O_2$, whichever is greater [8].

Recently, the majority of metal sulfates, e.g. K_2SO_4 [9], Na_2SO_4 [10,11], Li_2SO_4 [12,13], Ag_2SO_4 [14–16], Li_2SO_4–$CaSO_4$ [17,18], $Na(Ag)$–β''-alumina [19–23] and NASICON ($Na_2Zr_2Si_3PO_{12}$) [24,25] were investigated as solid electrolytes in SO_x sensors. These sensors repeatedly

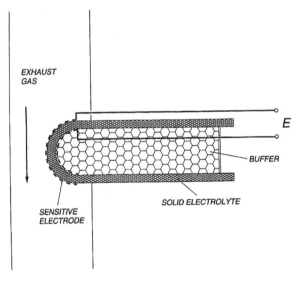

Fig. 1. Solid electrolyte electrochemical oxygen cell.

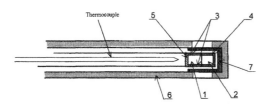

Fig. 2. Cross-sectional view of SO_x/O_2 probe assembly based on dual sulfur oxides and oxygen solid electrolyte sensor: (1) oxygen electrochemical cell; (2) SO_x electrochemical cell; (3), (4) measuring electrodes; (5) reference electrode; (6) stainless steel protective sheath; (7) ceramic diffusion element.

responded well to SO_x, but they appeared to suffer from sluggish responses and from the chemical instability of the solid electrolytes in atmospheres containing SO_x. Furthermore, all sensors reported to date require an independent determination of the oxygen potential in order to calculate the concentration of SO_3.

More recently it has been found that the stabilized ZrO_2 tube coated with a Li_2SO_4–$CaSO_4$–SiO_2 can be utilized as an SO_x sensor with a reasonably good response and a stable output [26,27]. However, both the chemical and mechanical stability of these sensors proved faulty during protracted tests. Furthermore, these sensors have a restricted working temperature range from 600 to 800°C and are still far away from industry's demand for a reliable sensor.

From a practical point of view, no work has been done on a solid-state sensor or probe to measure SO_x concentration under industrial conditions; instead, most investigations have focused on electrochemical sensors which measure equilibrium SO_x concentration in air, from which the SO_2 concentration is then calculated from the SO_2–SO_3 equilibrium constant [28].

However, a dual SO_x/O_2 solid state sensor with auxiliary phase sensitive to SO_x [29] can avoid many problems associated with SO_x sensors based on pure metal sulfates. The sensor is a composite of barium sulfate, potassium sulfate and silica in conjunction with a ZrO_2-based solid electrolyte. This sensor has two electrochemical cells. The first of these cells provides a signal proportional to the SO_x concentration in the stack. The second cell measures the excess of oxygen in combustion flue gas. The sensor has a wide range of industrial applications. The dual SO_x/O_2 solid electrolyte sensor has no moving parts to break down and it is not subject to the interfering absorptions so characteristic of the harsh and corrosive stack environment. The main advantage of this device is that it provides EMF signals which are linearly proportional to both partial pressures of O_2 and SO_x. Even for well-instrumentated industrial power plant facilities with multiple gas detectors, it would be advantageous to have a single instrument to monitor SO_x and oxygen. This sensor would also be useful for monitoring the above-mentioned gases at incineration facilities which have a complex mix of household and industrial wastes. Therefore, this paper focuses on the performance of a dual SO_x/O_2 sensor.

2. Experimental

2.1. Combined oxygen and sulfur oxides sensor

The starting high purity (99.9 wt%) materials were $BaSO_4$, K_2SO_4, SiO_2 and ZrO_2. The spectroscopically pure Y_2O_3, with an average grain sizes <1 μm, was selected as a sintering aid. Both sulfates and zirconia were weighed and well mixed in a ball mill and separately sintered into two electrochemical cells. The combustion dual SO_x/O_2 sensor, consisting of two electrochemical cells 1 and 2 (Fig. 2), was assembled into the probe. Both cells were positioned at the end of an elongated alumina or $MgAl_2O_4$ spinel tube which was exposed to the furnace atmosphere. Electrochemical cell 1 measures the excess of O_2 in a combustion gas stream and electrochemical cell 2 measures the SO_x concentration in a combustion gas. The outer sensor dimensions are: i.d. 5 mm, o.d. 8 mm. The length of the sensor can be varied from 100 up to 1300 mm. The four-bore tube inside the sensor typically has a diameter of about 4.5 mm with each of the four bores being typically 1.1 mm in diameter. Three of the bores are provided with wiring (the inner current conductor and the thermocouple wires) that extends through the bores. The other bore provides a passageway for reference O_2 atmosphere which is usually air. A built-in K-type thermocouple was used for tests. Usually, the corrosion of platinum electrodes of the sensor occurs when the SO_x is presenting in combustion gas [30,31]. To overcome these corrosion problems, the inner and external surfaces of the electrochemical cell 1 and external surface of electrochemical cell 2 were coated with Au.

The melting point of two-phase metal sulfates is about 1150°C. Use of Au in this sensor limits the maximum temperature to 1050°C. One of the advantages of the above-mentioned sensor is that it can be used either for SO_x or for simultaneous SO_x and O_2 measurements. The sensor needs only two current conductors for SO_x measurements and it could be employed for some practical application even without a protective metal sheath. In the case of SO_x/O_2 measurements, the dual sensor needs three current conductors. Therefore, it is essential to assemble this sensor into the probe using a metal sheath as one of the current conductors. In a dual SO_x/O_2 sensor the combustion gas contacts the measuring electrode 3 of the electrochemical cell 1 and the measuring electrode 4 of the electrochemical cell 2. The reference electrode 5 is a common electrode for both cells 1 and 2 and is contacted by a reference gas of known O_2 concentration. Due to the differences in O_2 and SO_x partial pressures of the gas contacting each electrode, EMF are established between the electrodes 3 and 5 and between the electrodes 4 and 5. Measurement of the EMF between electrodes 3 and 5 enable the O_2 concentration of the combustion gas to be calculated. Similarly, measurement of the EMF between electrodes 4 and 5 enable the SO_x concentration of the combustion gas to be determined. Thus, the simultaneous measurement both O_2 and

SO_x levels enable the efficiency of combustion to be optimized.

The probe based on the dual SO_x/O_2 sensor includes MA 253 (Sandvik Australia Pty. Ltd.) stainless steel protective sheath 6 and ceramic diffusion element 7. The ceramic diffusion element 7 is used for filtration of the combustion gas from ash, soot and other particles which are common in coal- or oil-fired furnaces and kilns. The reduction in corrosion rate at industrial applications apparently can be achieved because the low amount of excess air did not permit oxidation of vanadium and sulfur to their highest states of oxidation and thus prevented formation of the most corrosive oil-ash compounds. Low-temperature corrosion can also be virtually eliminated by maintaining excess air at 0.1–2% [32].

Standard digital multimeters 179 TRMS with an accuracy of ± 0.1 mV and with an input impedance $R > 10^{10}$ Ω were used for EMF measurements of both the electrochemical cells and the thermocouple.

The SO_x level is dependent upon the burner conditions, the fuel and the type of desulfurization process. For most fuels, burners and loads, optimum combustion is obtained when fuel gas concentrations of SO_x are in the range of 10–200 ppm [33]. These current firing conditions include the type of fuel or fuels being burned, degree of burner fouling, imbalance of fuel or imbalance of air between burners, etc. If the control of combustion efficiency by the dual SO_x probe is able to measure the SO_x concentration from approximately 10 up to 10,000 ppm and to maintain approximately 100–200 ppm of SO_x for industrial applications, the deviations from expected air/fuel ratio will indicate any needed operator attention, such as cleaning burners, adjusting fuel temperature, adjusting air distribution registers, etc.

Thus all following tests were targeted on the above-mentioned SO_x level and were made for measurement of SO_x concentration from 18 up to 10,000 ppm and for the measurement of excess O_2 from 0.1 up to 4 vol%.

Three sensors were prepared for the trial. SO_x-sensing properties were measured in a laboratory furnace with conventional flow systems in the temperature range of 600–1000°C. Sample gas mixtures containing SO_2 at varying levels under a constant nitrogen concentration of 99.9% were prepared by diluting a parent gas (10,000 ppm SO_2 in N_2) with dry nitrogen. Oxygen in ambient air (20.9 vol%) was used as a reference gas. To investigate the dependence of EMF on O_2 concentration, sample gases containing different O_2 concentrations under a fixed concentration of SO_2 (18 ppm) were prepared by mixing the parent SO_2 gas (diluted in N_2) with O_2 and N_2. The sample gases were also mixed with CO_2, NO or NO_2 for cross-sensitivity tests. Each sample gas was allowed to flow over the sensor at a rate of 100 cm^3/min. The flow rate of the reference gas was also 100 cm^3/min.

To determine the EMF–temperature relationship at various SO_2 partial pressures, the probe was initially heated to a temperature between 650 and 950°C at different SO_2

concentrations. Data were recorded during several heating and cooling cycles at approximately 25°C temperature intervals. To establish a relationship between the cross-sensitivity of the sensor to different combustible gases, the probe was heated to 720°C. The furnace atmosphere was changed by switching the solenoid valve to the sample gases contained NO_x or CO_2. These measurements were made at one or more temperatures between 18 and 1000 ppm of SO_x.

Some response rate measurements were also made in order to determine the response and recovery time of the sensor to the changes of the SO_x concentrations. A simple step change in sulfur dioxide concentration was used for this purpose. The change in the EMF signal as a function of time was recorded on a chart recorder. To establish a relationship between the aging process and the characteristics of electrochemical cells, the sensor was taken to the highest measurement temperature, left for 30 days and cooled to the desired temperature. The sensor was also used in extended-life tests.

3. Results and discussion

3.1. SO_2-sensing properties of yttria stabilized zirconia-based sensors

The O_2-sensing properties of yttria stabilized zirconia electrolyte are well known and were not a subject of this investigation. The SO_2-sensing properties of the potentiometric sensor using yttria stabilized zirconia electrolyte and a composition of $BaSO_4$–K_2SO_4–SiO_2 were examined. The sensor detected SO_2 concentration within a range of 18 to 10,000 ppm. Fig. 3 shows that the output EMF of ZrO_2-based dual sensor as a function of SO_2 concentration (Fig. 3a) and working temperature (Fig. 3b). The full line represents ideal (Nernstian) behavior. The lack of accurately prepared standard-gas mixtures has limited the inlet SO_2 concentration down to 18 ppm. Therefore, the minimum SO_2 concentration limits for reliable operation of the sensor have not been established yet. However, previous research [34] appears to indicate that the lower SO_2 concentration limit for reliable measurements is set by the thermodynamic decomposition of $BaSO_4$–K_2SO_4–SiO_2 composite, which is around 10^{-3} ppm SO_2 in air at 650°C. The upper limit is more difficult to estimate, because the lifetime of this composite depends upon kinetic factors which vary with cell design. However, it appears that the sensor lifetime in atmospheres having high SO_x concentrations is related to the ease or difficulty of transporting SO_2 and/or SO_3 from the gas mixture to the metal particles embedded in the two-phase electrolyte [35]. Analysis of the sensing properties of both the above-mentioned sensor and similar SO_x sensors, which employed the magnesia-stabilized zirconia as a solid electrolyte and either Li_2SO_4–$CaSO_4$–SiO_2 [26] or Li_2SO_4–MgO [27] auxiliary phases for SO_x measurement, show that they have a different sensitivity. Yan [27] concluded that the yttria stabilized zirconia-based electrolytes with auxiliary composition of the metal sulfates have a poor sensing performance and are not suitable for the development of SO_x sensors. However, the results of this investigation do not support Yan's results. The sensor based on yttria stabilized zirconia electrolyte and a composition of $BaSO_4$–K_2SO_4–SiO_2, provides higher stable EMF output and has a good sensing performance. Fig. 3 shows that the dual sensor has a higher EMF output and can accurately work within wider temperature range than the sensors previously described [26,27]. To achieve these results in the present investigation, the SO_x electrochemical cell was specifically designed to provide the maximum surface area of contact between the ZrO_2 and the metal sulfates [29].

Typical response and recovery times of the above-mentioned dual SO_x/O_2 sensor with $BaSO_4$–K_2SO_4–SiO_2 auxiliary phase at different temperatures are shown in Fig.

4. The response and recovery times of the two-phase sulfate electrochemical cell used to measure SO_x concentrations depend upon the equilibration time for the SO_2–O_2–SO_3 gas mixture [34]. For the majority of SO_x sensors based on metal sulfate electrolytes the response and recovery times were found to be less than that required for industrial applications [35]. Thus the dual SO_x/O_2 sensor was designed to provide gas-flow conditions and catalyst-surface area which minimize the equilibration reaction. Typical response time was fast and varied from 45 to 80 s at 700°C. The recovery time was still slow, i.e. the 90% recovery time to 98 ppm of SO_2 at 745°C was more than 4 min. The reversibility and recovery characteristics were faster at higher temperatures. Nevertheless, the complete recovery at 850°C was more than 9 min. However, it was faster than the recovery time for magnesia stabilized

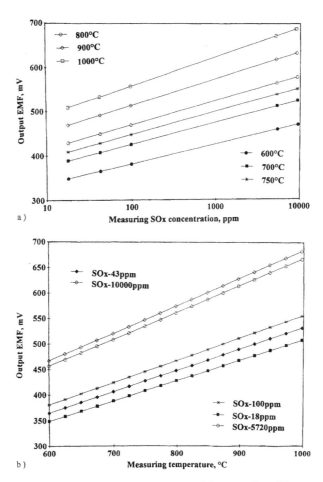

Fig. 3. (a) EMF of yttria stabilized zirconia-based solid electrolyte sensors as a function of SO_2 concentration at different temperatures. The full line represents ideal (Nernstian) behavior. (b) EMF of yttria stabilized zirconia-based solid electrolyte sensors as a function of working temperature at different SO_x concentrations.

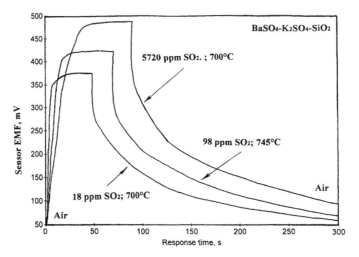

Fig. 4. Response and recovery times of the sensor based on yttria stabilized zirconia with $BaSO_4$–K_2SO_4–SiO_2 phase at different temperatures.

zirconia solid electrolyte and Li_2SO_4–$CaSO_4$–SiO_2 auxiliary phase [27].

From an industrial point of view, the cross-sensitivity to different combustible gases is another important problem. Fig. 5 shows cross-sensitivity effects of SO_x sensor, operating at 720°C, produced by the presence of CO_2 ($\approx$13 vol%), and NO_2 ($\approx$1000 ppm). The working temperature and concentrations of the gases were similar to industrial conditions. Fortunately, many of the hydrocarbons or related reducing species that are present in the combustion gas in high concentrations (CO, CH_4, CO_2, etc.) do not increase the Au work function very much in their interaction with the surface [36]. As clearly shown in this figure, the EMF of the sensor is unaffected during the tests by the co-existence of CO_2 (10 ppm, 9%). The influence of the carbon dioxide on the output EMF began measurable only when the CO_2

concentrations were more than 9.5% (Fig. 5a). However, the change in the output signal of the sensor was not significant and remained within 3% of the EMF values. The results of these tests also show that in the presence of NO_2, the EMF of the sensor did not change during the whole test period at the different SO_2 concentrations (Fig. 5b). Therefore, it appears that the dual sensor is unaffected by the presence of CO_2 and NO_x in the gas mixture and can be used for SO_x detection in the exhaust gases from burners or industrial furnaces.

The increase in electrical resistivity of solid electrolytes (aging) as a result of long time annealing at a temperature of $\approx$ 1000°C has an influence on the major characteristics of the electrochemical sensor. The accuracy and stability of the sensor may be effected during the high temperature measurements as a result of aging. The results of

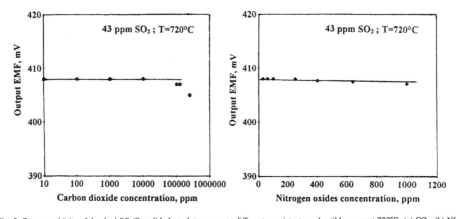

Fig. 5. Cross-sensitivity of the dual SO_x/O_2 solid electrolyte sensor to different coexistent combustible gases at 720°C: (a) CO_2; (b) NO_x.

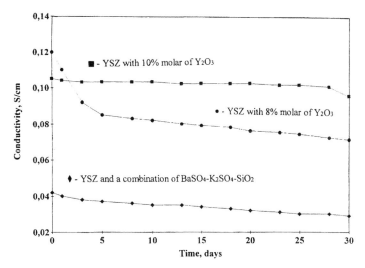

Fig. 6. Conductivities of yttria stabilized zirconia and two-phase sulfate electrolytes of the dual SO_x/O_2 sensor at 1000°C in air versus time of annealing at 1000°C: (●) yttria stabilized zirconia with 8% molar of Y_2O_3; (■) yttria stabilized zirconia with 10% molar of Y_2O_3; (♦) yttria stabilized zirconia and a combination of $BaSO_4$–K_2SO_4–SiO_2.

experimental investigations of ZrO_2-based solid electrolytes [37] indicate that the aging kinetics has a different nature for single- and two-phase polycrystalline electrolytes. Moreover, it is dependent upon which stabilizing oxide (Y_2O_3, MgO, Yb_2O_3, Sc_2O_3, etc.) was selected for the preparation of ZrO_2 electrolyte. It is also dependent upon the molar percentage rate between solid electrolyte and stabilizing oxide. For example, ZrO_2–Sc_2O_3 electrolytes show higher rate of ageing than ZrO_2–Y_2O_3 electrolytes [37].

The increase in electrical resistivity of yttria stabilized zirconia is attributed to segregation of an yttria-rich layer near the grain boundaries and triple points *electrode–electrolyte-gas*, formation of tetragonal phase of ZrO_2 at temperature above 950°C and retention of this metastable phase along with the stabilized fluorite phase on cooling to lower temperature as well as ordered domains forming within the disordered fluorite phase. Since ZrO_2 electrolyte is a part of the SO_x measuring electrochemical cell in the dual SO_x/O_2 sensor, it was important to investigate the aging processes for both the SO_x and O_2 electrochemical cells. Fig. 6 shows a typical time dependent change of the conductivity for yttria stabilized zirconia with 8 and 10 mol% of Y_2O_3 and two-phase sulfate electrolyte of the dual SO_x/O_2 sensor at 1000°C. In industrial applications the lower the resistivity of the probe the greater the electrode contact area in the assembly. For example, a value below 15 kΩ at temperatures above 820°C is acceptable for a ZrO_2 probe [38]. A higher figure usually indicates a problem. Therefore, an impedance measurement has been one of the most common methods of probe testing in industry for years. In our investigation ZrO_2-based solid electrolytes with 8 and 10 mol% of Y_2O_3 were annealed for 30 days at 1000°C. As shown in

Fig. 6, the major drop in conductivity occurred during the first 5 days of experiment for the yttria stabilized zirconia with 8 mol% of yttria. 10 mol% of yttria was found to be less affected by the annealing at high temperature. Fig. 7 shows the results of the conductivity measurements for the electrolytes with different content of yttria before and after annealing. The results of present investigation independently supported the fact that the aging process does not occur before 1000°C in these electrolytes with a concentration of the stabilizing additive Y_2O_3 of 10 mol% [39]. Furthermore, the present investigations of the aging process have shown that after annealing of ZrO_2 electrolytes, the level of conductivity for yttria stabilized zirconia with 10 mol% of yttria is higher than the level of conductivity for that with 8 mol% of yttria. With a lower than 10 mol% concentration of Y_2O_3 the cubic solid solution gradually breaks up into two-phases, thus leading to a drop in its electrical conductivity [40]. Consequently ZrO_2 electrolytes containing 10 mol% of Y_2O_3 have been used in the SO_x measuring electrochemical cell of the newly designed dual SO_x/O_2 sensor [29].

The investigation of the aging of the two-phase sulfate electrolyte was also studied. Changes in the conductivity of this cell are also shown in Fig. 6. It is noted that the intensity of the aging processes for the SO_x measuring electrochemical cell is less than for a ZrO_2-based O_2 measuring electrochemical cell. This suggests that a two-phase sulfate glass is partially formed in the fabricating process. However, it is considered that the role of the sulfates as K^+ and Ba^{2+} conductors is very important. The main cause of aging of solid electrolytes in the two-phase region appears to be the precipitation and growth of the second low-conductive

38

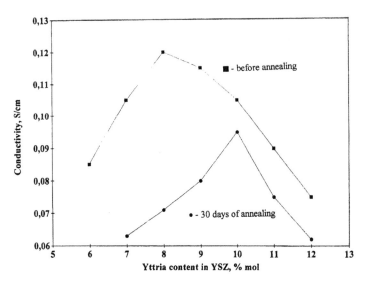

Fig. 7. The conductivities of yttria stabilized zirconia solid electrolytes at 1000°C versus Yttria content. (■) before annealing; (●) 30 days of annealing.

phase. The kinetic of the aging process is determined by the kinetic of two processes running independently of each other: nucleation of the number of second phase centers and the growth of their bulk. In the case of polycrystalline solid electrolytes the process of the second phase growth is connected with the diffusion of solid solution components. The cation diffusion could be considered as the limiting stage of the ageing process in composite solid electrolyte because the vacancy concentration in the cation sublattice is small in comparison with the anion one. Therefore, the SO_x measuring electrochemical cell is less affected by the aging process than the O_2 measuring electrochemical cell.

The life of the probe is very important for both practical and economic reasons. A further consideration for electrochemical devices based on ionic conductors is that conductivity of the solid electrolyte should be unaffected by prolonged heat treatments. Therefore, the dual SO_x/O_2 sensor was used in extended-life tests [34,36]. The results for one of these long-term tests (120 days), in which the SO_2 concentration in the gas at temperature of 749°C was changed after every 20–30 days, are shown in Fig. 8. The

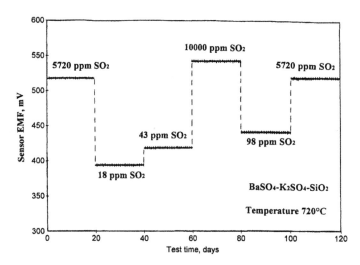

Fig. 8. Long-term test of the EMF response of the SO_x measuring electrochemical cell of the dual SO_x/O_2 sensor.

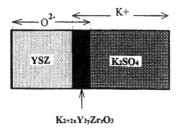

Fig. 9. Model of the heterojunction of electrochemical chain for yttria stabilized zirconia-based electrolyte attached with K_2SO_4–$BaSO_4$–SiO_2.

measured EMF were within ± 5 mV of the calculated values, which are shown in parentheses. The measured values appear to be stable even at the end of a 120 day test. No phase transition was observed in the sulfate composite electrochemical cell as the SO_2 concentration was changed from 18 to 10,000 ppm during the test. After such long-term tests, the other probe components showed no observable chemical or mechanical degradation which would limit the probe's life. It would appear that the dual solid electrolyte SO_x/O_2 sensor has a high level of reliability, good SO_2-sensing properties and selectivity over a long period of time and long-term chemical stability.

3.2. Electrochemical constitution of dual sensor

The new designed dual SO_x/O_2 sensor has utilized the following two cell structures:

air, Au//ZrO_2–Y_2O_3//Au, measuring O_2 in gas

(the O_2-measuring electrochemical cell) (3)

air, Au//ZrO_2–Y_2O_3/$BaSO_4$–SiO_2//Au, measuring SO_x in gas

(the SO_x-measuring electrochemical cell) (4)

The sensing mechanism for ZrO_2 electrolyte with auxiliary composition of metal sulfates was described previously [27] for Li_2SO_4–$CaSO_4$–SiO_2 auxiliary phase. The sensing mechanism for dual SO_x/O_2 sensor is similar. The half-cell reaction at the electrode-gas interface is the equilibration of the SO_4^{2-} ion in the two-phase sulfate electrolyte when the SO_3 is present in the gas. In this sensor, the Au measuring electrode 4 of the electrochemical cell 2 (Fig. 2) also works as a catalyst and sulfur dioxide is catalytically oxidized to SO_3 in combustible gas:

$$SO_2 + \frac{1}{2}O_2 \leftrightarrow SO_3 \tag{5}$$

and the equilibrium constant, K, of reaction (5) is expressed as

$$K = P_{SO_3}/(P_{SO_2} \cdot P_{O_2}^{1/2}). \tag{6}$$

With a catalyst, the cell potential is established by the equilibrium pressure of SO_3, which can be related to the inlet SO_2 pressures using Eq. (6), where K is the equilibrium constant for reaction (5)

$$P_{SO_2}(in) = P_{SO_2} + P_{SO_3} = P_{SO_2}(1 + 1/KP_{O_2}^{-1/2}). \tag{7}$$

If the electrical current flows through a sulfate-electrolyte cell, the half-cell reaction at the gas electrode–electrolyte interface is the equilibration of the SO_3 in the gas with the SO_4^{2-} ion in the electrolyte:

$$SO_3 + \frac{1}{2}O_2 + 2e^- \leftrightarrow SO_4^{2-}. \tag{8}$$

The measuring electrode 4 and K_2SO_4-based two-phase sulfate electrolyte form an SO_x-sensitive half-cell (SO_3 electrode) according to the following reaction:

$$2K^+ + SO_3 + \frac{1}{2}O_2 + 2e^- \leftrightarrow K_2SO_4. \tag{9}$$

The reference electrode and the stabilized ZrO_2 confirm a well-known O_2-sensitive half-cell reaction for both electrochemical cells (3) and (4):

$$\frac{1}{2}O_2 + 2e^- \leftrightarrow O^{2-} \tag{10}$$

The cell reaction for electrochemical cell (4) is obtained by combining the half-cells through a heterojunction between the O^{2-} conductor and the K^+ conductors (Fig. 9). To achieve an electrochemical chain for the heterojunction, one must assume the formation of an interfacial compound, which contains both conducting ions of K^+ and O^{2-}. Presumably, a mixed oxide derived from K_2O, BaO, ZrO_2 and Y_2O_3, expressed as $K_{2+2x}Y_{3y}Zr_zO_3$, provide an ionic bridge between the two half-cells by means of the following reaction:

$$2K^+ + O^{2-} + K_{2x}Y_{3y}Zr_zO_2 \leftrightarrow K_{2+2x}Y_{3y}Zr_zO_3 \tag{11}$$

and thus completes an electrochemical chain for the SO_x-measuring electrochemical cell (4) as described below.

Electrochemical equilibrium for reactions (9)–(11) are expressed as follows:

$$2\mu'_{K^+}m + \mu^m_{SO_3} + \frac{1}{2}\mu^m_{O_2} + 2\mu'_{e^-}m = \mu^m_{K_2SO_4}, \tag{12}$$

$$\frac{1}{2}\mu^r_{O_2} + 2\mu'_{e^-}r = \mu'_{O^{2-}}r \tag{13}$$

$$2\mu'_{K^+}i + \mu'_{O^{2-}}r + \mu^i_{K_{2x}Y_{3y}Zr_zO_2} = \mu^i_{K_{2+2x}Y_{3y}Zr_zO_3}, \tag{14}$$

where μ' and μ represent an electrochemical and chemical potential, respectively. The superscripts m, r and i represent the measuring electrode, the reference electrode and the interfacial compound. The electrochemical potential of the conducting ions should be constant throughout the conducting regions at equilibrium [27], i.e.

$$\mu'_{K^+}m = \mu'_{K^+}i \qquad \mu'_{O^{2-}}i = \mu'_{O^{2-}}r. \tag{15}$$

Thus the difference between the electrochemical potential of electrons for both measuring and reference electrodes can be expressed as

$$E = \frac{1}{2F}(\mu^i_{K_{2+2x}Y_{3y}Zr_zO_3} - \mu^i_{K_{2x}Y_{3y}Zr_zO_2} - \mu^m_{K_2SO_4} + \mu^0_{SO_3})$$
$$+ \frac{RT}{4F} \ln \frac{P^m_{O_2}}{P^r_{O_2}} + \frac{RT}{2F} \ln P_{SO_3}. \qquad (16)$$

In Eq. (16) P_{SO_3} and P_{O_2} are partial pressures of SO_3 and O_2, $\mu^0_{SO_3}$ is the standard chemical potential of SO_3, R is the gas constant, F is Faraday's constant and T is temperature (K). The term in parenthesis of Eq. (16) is the standard Gibbs free-energy change (ΔG^0) of the reaction

$$K_2SO_4 + K_{2x}Y_{3y}Zr_zO_2 = K_{2+2x}Y_{3y}Zr_zO_3 + SO_3. \qquad (17)$$

The first term of Eq. (16) is thus constant (E_0) allowing to express EMF of the cell (4) as:

$$E = E_0 + \frac{RT}{4F} \ln \frac{P^m_{O_2}}{P^r_{O_2}} + \frac{RT}{2F} \ln P_{SO_3}. \qquad (18)$$

This equation shows that the EMF is contributed to, not only by the two gas electrodes, but also by the thermodynamic properties of the interfacial compound. In case when $P^r_{O_2}$ and $P^m_{O_2}$ are constant, Eq. (18) is simplified to

$$E = E'_0 + \frac{RT}{2F} \ln P_{SO_3}. \qquad (19)$$

Thus EMF is linear to the logarithm of inlet SO_2 concentration with the gradient corresponding to $n = 2$. The theoretical results given above were in excellent agreement with the experimental results.

A number of very important lessons were learnt from this series of trials. Firstly, the two electrochemical cells can be connected electrochemically in one solid electrolyte sensor through an interfacing compound formed in between. The interfacing compound is estimated to be a mixed oxide containing K_2O, ZrO_2 and Y_2O_3. The EMF dependence of the SO_3 electrode on P_{SO_2} and P_{O_2} is essentially the same as that reported for a metal sulfate membrane-based SO_2 sensor [27].

Secondly, it was essential to protect the measuring electrodes of both cells from dust or soot using a porous ceramic filter and it was also essential to develop a probe design that allowed ash or soot to be burnt off the filter during continuous operation. One of the major advantages of this SO_x/O_2 sensor is that the sensor's design allows the reference ambient air to be common for both SO_x and O_2 measuring electrochemical cells. This feature provides alternative calibration of the setpoints for both electrochemical cells to be made without an addition analysis.

Thirdly, it was demonstrated that the SO_x/O_2 probe provides an accurate and reliable measurement of O_2 and SO_x in a range from 0.01 to 100 vol% for O_2 and from 18 to 10,000 ppm for SO_x. The probe can also work in a wider temperature range (600–1050°C) than any other "in situ" solid electrolyte SO_x probe available on the market today.

Finally, a two-phase sulfate electrolyte combined with silica has improved mechanical strength and is not as susceptible to microcrack formation as other single-phase sulfate electrolytes. The superiority of the investigated composition of a two-phase sulfate electrolyte enabled accurate measurements to be obtained.

4. Conclusions

To achieve an accurate and simultaneous measurement of the SO_x emissions and oxygen level in the combustion gas stream, the dual SO_x/O_2 "in situ" solid electrolyte sensor, based on stabilized ZrO_2 and a composition of two-phase metal sulfates with silica, has been designed and developed.

From practical experience of using solid electrolyte sensors for O_2 trim control and maintenance, it can be seen that the control of air/fuel ratio by measuring oxygen and the control of effectiveness of combustion by measuring SO_x can be a very effective way of reducing fuel costs while improving furnace performance. Improvements can be made in general quality of the process due to decreased fluctuations in the O_2 partial pressure and temperature and steadier waste gas composition. Attention to probe care can maintain the repeatability of measurements and hopefully extend probe life. The results of this investigation suggest that by utilizing a single probe the benefits of both the oxygen measuring electrochemical cell and the SO_x measuring electrochemical cell can be combined. This would provide a better control of combustion atmospheres, especially those where assumptions regarding sulfur oxides levels do not always hold true.

Although a large number of oxygen and sulfur dioxide probes have been separately installed for monitoring O_2 and SO_x in combustion gas, the installation of a dual SO_x/O_2 probe in control loops can provide a better and a more effective way towards fuel saving and efficiency.

Acknowledgements

The author wishes to thank Mr V. Primissky (Ukranalyt Ltd., Ukraine) for his support and help during cross-sensitivity and extended-life tests of the sensors and for useful discussions of the results. The author also wishes to thank Mr J. Walton for his valuable support during the preparation of this paper.

References

[1] Gulyurtlu I, Lopes H, Cabrita I. Fuel 1996;75(8):940.
[2] Ayers GP, Granek H. Clean Air 1997;31(1):38.
[3] Kakaras E, Vourliotis P. J Inst Energy 1995;68:22.
[4] Lunt R, Little AD. Chem Engng Prog 1996;2:11.
[5] Gauthier M, Chamberland A. J Electrochem Soc 1977;124:1579.

[6] Moseley PT. Meas Sci Technol 1997;8:223.

[7] Yamazoe N, Miura N. Solid State Ionics 1996;86–88:987.

[8] Zirconia Oxygen Analyser Systems. ABB Kent-Taylor Catalogie 1998:5.

[9] Gauther M, Bellemare R, Belander A. J Electrochem Soc 1981;128:371.

[10] Jacob KT, Rao DB. J Electrochem Soc 1979;126:1842.

[11] Saito Y, Maruyama T. Solid State Ionics 1984;14:273.

[12] Imanaka N, Yamaguchi Y, Adachi G, Shiokawa J. Solid State Ionics 1986;20:153.

[13] Fedorov PP. Solid State Ionics 1996;86–88:113.

[14] Mari CM, Bechi M, Pizzini S. Sensors and Actuators B 1980;2:51.

[15] Liu Q, Sun X, Wu W. Solid State Ionics 1990;40/41:456.

[16] Kirchnerova J, Bale CW, Skeaff JM. Solid State Ionics 1996;91:257.

[17] Akila R, Jakob KT. Sensors and Actuators 1989;16:311.

[18] Maruyama T, Sato Y, Matsumoto Y, Tano Y. Solid State Ionics 1985;17:281.

[19] Rog G, Kolowska-Rog A, Zakula K. J Appl Electrochem 1991;21:308.

[20] Adachi G, Imanaka N. Handbook on the chemistry of rare earths, vol 21, 1995:179.

[21] Yan Y, Shimizu N, Miura N. Chem Lett 1992;83:635.

[22] Yang PH, Yang JH, Chen CS, Peng DK, Meng GY. Solid State Ionics 1996;86–88:1095.

[23] Nafe H. Solid State Ionics 1997;93:117.

[24] Yan Y, Shimizu Y, Miura N, Yamazoe N. Sensors and Actuators B 1993;12:77.

[25] Slater DJ, Kumar RV, Fray DJ. Solid State Ionics 1996;86–88:1063.

[26] Yan Y, Shimizu Y, Miura N, Yamazoe N. Sensors and Actuators B 1994;20:81.

[27] Yan Y, Miura N, Yamazoe N. J Electrochem Soc 1996;143(1):609.

[28] Skeaff JM, Dubreuil AA. Sensors and Actuators B 1993;10:161.

[29] Zhuiykov S. Australian Patent No.706,931 (1999).

[30] Copcutt RC, Maskell WC. Solid State Ionics 1994;70/71:578.

[31] Whelan PT, Borbisge WE. J Appl Electrochem 1988;18:188.

[32] Rowe JG. 4. Metals handbook, 4. ASM International, 1991. p. 569.

[33] Rapp RA, Zhang TS. JOM 1994;12:47.

[34] Worrell WL. In: Sieyma T, editor. Chemical senor technology, 1. New York: Elsevier, 1988. p. 97.

[35] Azad AM, Akbar SA, Mhaisalkar SG, Birkefeld LD, Goto KS. J Electrochem Soc 1992;139(12):3690.

[36] Anderson DC, Galwey AK. Fuel 1995;74(7):1031.

[37] Vlasov AN, Perfiliev MV. Solid State Ionics 1987;25:245.

[38] Robinson K. Heat Treatment of Metals 1996;2:43.

[39] Blumental RN. In: Proceedings of the Second International Conference on Carburizing and Nitriding with Atmospheres, Cleveland, 6–8 December 1995:17.

[40] Moghadam FK, Yamashita T, Stevenson DA. In: Heuer AH, Hobbs LW, editors. Science and technology of zirconia, 3. Columbus: The American Ceramic Society, 1981. p. 364.

Fundamental Chemistry of NO$_X$ and SO$_X$ Related Reactions

Emissions Reduction: NO$_x$/SO$_x$ Suppression
A. Tomita (Editor)
© 2001 Elsevier science Ltd. All rights reserved

The fate of nitrogen during pyrolysis of German low rank coals — a parameter study

J. Friebel, R.F.W. Köpsel[*]

Institute of Energy Process Engineering and Chemical Engineering, Freiberg University of Mining and Technology, Reiche Zeche, 09596 Freiberg, Germany

Received 25 August 1998; received in revised form 4 January 1999; accepted 7 January 1999

Abstract

Proceeding from the knowledge of the functional forms of nitrogen in coals it should be possible to draw conclusions on the mechanisms of its release during devolatilization. Pyrolysis experiments were carried out with a series of lignite from the main German mining districts in a temperature range from 673 to 1173 K. Nitrogen functionalities of the coals and the obtained chars were determined by XPS. The nitrogen content of the residual pyrolysis chars was found to be dependent on a wide variety of parameters: coal type, temperature, residence time, mineral matter. An increasing N/C ratio was detected for chars obtained under conditions of oxopyrolysis. The conversion of fuel-bound nitrogen to NO$_x$-precursors (HCN, NH$_3$) is also strongly dependent on coal type. The mineral matter of the parent coals was found to affect both the primary devolatilization process and the reaction paths to form ammonia. The balance of the nitrogen distribution during pyrolysis shows a large amount of N$_2$ to which the coal-N is converted. © 1999 Elsevier Science Ltd. All rights reserved.

Keywords: Nitrogen; Low rank coal; Pyrolysis

1. Introduction

The mechanisms of nitrogen release during pyrolysis are of great importance for the prediction of NO$_x$ formation in coal combustion. It is well known that emissions of these hazardous gases have decreased as a result of the application of modern facilities and processes (low-NO$_x$ burners, air and fuel staging). What remains is the nitrogen input into the combustor as part of the coal. This fuel bound nitrogen is still the main source of nitrogen oxides. Much of the published results in the recent years have been related to high rank coals because of their economic importance world-wide. Compared with that the utilization of brown coal is an essential part of power and heat generation in Germany. The results of the studies carried out with a series of these low rank coals are presented in this paper.

From the chemical point of view it should be possible to conclude from the nitrogen bonding in the macromolecule of the coal on its conversion ratio to intermediates (HCN, NH$_3$) and finally to NO$_x$. The functional forms of the organic nitrogen are well investigated for a wide range of coals from peat and lignite to anthracites [1–3]. Most of the obtained results originate from XPS and XANES spectroscopy. The

pyrrole-type was correspondingly found to predominate [2, 4,5]. Some changes within the distribution of the functionalities occur during pyrolysis and gasification. The five-membered ring of the pyrrole is suggested to convert partially to the thermally more stable pyridine under severe conditions [3,6,7]. However, some researchers found N-5 in high temperature chars of model compounds and coals also indicating a high stability [8, 9]. Beside this the occurrence of another two six-membered ring functionalities is of great interest — the quaternary nitrogen (N-Q) and pyridone (N-6(O)). N-Q is closely related to the pyridine-type both in coals, where the N-atom interacts with functional groups in the environment and in chars as part of the graphene layer [6]. Pyridone is more difficult to detect with XPS because of the binding energy of the N 1s electron which is similar to that of pyrrole. Only the application of XANES in the recent years threw some more light on structural analysis. However, it is still hard to receive reliable information on the actual ratio of five- to six-membered ring nitrogen.

The reaction paths of the nitrogen released with the volatiles and of those remaining in the char are considerably understood. Davidson and others [2,3,10,11] gave a comprehensive survey of fundamental mechanisms of nitrogen distribution on solid and gaseous phase during pyrolysis. A wide variety of parameters (coal type, temperature, heating rate, residence time, pressure, oxygen level) has been

* Corresponding author. Tel.: + 49-3731-394526; fax: + 49-3731-394555.

Reprinted from *Fuel* **78 (8)**, 923–932 (1999)

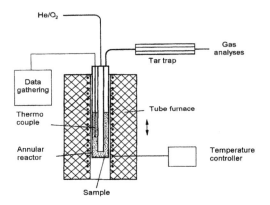

Fig. 1. Schematic diagram of the pyrolyser.

investigated. In spite of these studies there remain some uncertainties about the role which the mineral matter of the coals plays in the early stage of devolatilization, whereas the influence of the ash is well known in combustion [12–14]. A few of the inherent components are able to promote the formation as well as the reduction of NO_x. Carrying out pyrolysis experiments, Ohtsuka, Hayashi and others [15–17] compared the release of nitrogen from parent, demineralized and with catalysts doped coals. They found a selective removal of the char-N in presence of iron. Another important fact is the release of nitrogen compounds as NO_x precursors (HCN and NH_3). It is noted that from the distribution of the coal nitrogen on the char and the volatiles can be concluded on the final emission. The reaction paths of both sources differ significantly and there are a few variables being of importance, first of all heating rate and temperature. Hydrogen cyanide has proved to be the primary gaseous compound evolved from tar and char nitrogen [18,19]. Only in some low rank coals ammonia was found to be released in an earlier stage than HCN [12,20]. Evaluating the significance of these gases not only their potential for the formation of NO_x but also the ability for its reduction have to be considered.

2. Experimental

2.1. Apparatus

Pyrolysis experiments were carried out in a temperature range from 673 to 1173 K. A fixed bed annular reactor which can be shifted vertically was used to realise two different heating rates during pyrolysis (0.25 and 5 K/s). The high heating rate was achieved by preheating the tube furnace before putting in the reactor. A schematic of the pyrolyser is shown in Fig. 1. The stainless steel covered reactor allows a maximum pressure of 1 MPa at 1273 K. The samples were heated under high purity helium (99.999 vol.-% He) atmosphere to the final heat treatment temperature (HTT) which was held for 5 and 30 min, respectively. Under conditions of oxopyrolysis an oxygen/helium ratio of 1 : 20 was adjusted. This corresponds to a total excess air coefficient of $\lambda \approx 0.1$. The flushing gas is led from the inlet at the top of the apparatus through the inner tube of the reactor for preheating. Afterwards it flows in inverse direction through the sample placed within the annulus. The high boiling volatiles were captured in a water cooled tar trap.

2.2. Sample preparation

A series of three lignites from the largest mining areas in Germany were used for the pyrolysis experiments. They represent different types of soft brown coals. The analytical data of the samples are given in Table 1. The coals were dried and sieved to a grain size of 0.63/1.25 mm. Part of them was demineralized by boiling in hydrochloric acid. This procedure has been described elsewhere [13]. It is supposed to have no effects on the organic structure. As can be seen from the ultimate analyses, about 90% of the coal ash could be removed by acid washing. The residue mainly consisted of silicon oxide which is suggested to be considerably ineffective in accelerating chemical reactions catalytically. The achieved level of ash removal only using HCl allowed to avoid a further washing in hydrofluoric acid or nitric acid to eliminate even more inorganic matter.

Table 1
Characterisation of the coal samples (wt.-%, daf unless otherwise stated)

	Lusatia(L)		Central Germany (CG)		Rhineland (R)	
	Parent coal	Demineralized coal	Parent coal	Demineralized coal	Parent coal	Demineralized coal
Ash (db)	6.7	0.9	10.5	1.2	5.0	1.2
Volatiles	56.0	55.5	61.5	58.4	52.4	51.4
C	67.0	65.6	70.2	70.0	68.1	66.8
H	5.0	5.0	5.9	4.9	4.8	4.5
N	0.7	0.7	0.7	0.6	0.8	0.8
S	0.9	0.8	4.0	3.9	0.4	0.4
O[a]	26.2	27.9	19.2	20.6	25.8	27.5

[a] By difference.

Table 2
Distribution of nitrogen functionalities of coal and char samples (determined by XPS)

	Sample	N_{char}/N_{coal}	N-5	N-6	N-Q
L	Parent coal		0.68	0.11	0.21
	Char (673 K)	0.71	0.58	0.32	0.10
	Char (873 K)	0.57	0.55	0.39	0.06
	Char (1173 K)	0.26	0.22	0.40	0.40
	Coal methylated		0.55	0.15	0.30
CG	Parent coal		0.53	0.27	0.20
	Char (673 K)	0.80	0.55	0.34	0.11
	Char (873 K)	0.65	0.46	0.34	0.20
	Char (1173 K)	0.61	0.35	0.00	0.65
R	Parent coal		0.44	0.13	0.43
	Char (673 K)	0.81	0.60	0.31	0.09
	Char (873 K)	0.74	0.53	0.43	0.04
	Char (1173 K)	0.38	0.32	0.35	0.33

A part of the Lusatian lignite was separated by extraction of the humic acids with dimethyl sulfoxide. The coal sample was demineralized and debituminized to achieve an enrichment in the nitrogen content. Both humic acids and the solid residue were methylated using methyl iodide to quaternarize a fraction of the nitrogen. This procedure was also applied by Wallace et al. [21] for investigating coal derived liquids. As a result of the low nitrogen content of the coals a better assignment of functionalities was expected from the analysis of the obtained derivatives.

2.3. Coal and char analyses

CHN analyses were carried out for all char and tar samples obtained after pyrolysis. Special attention was given to the solid residues whereas the elemental analyses of the tar were only required for balancing the nitrogen distribution.

The XPS measurements were performed on a VG ESCA-LAB 220i XL spectrometer with MgKα radiation (12 kV and 20 mA). The energy scale was calibrated to reproduce the binding energies of Cu $2p_{3/2}$ (932.65 eV) and Au $4f_{7/2}$ (84.00 eV). Sample charging was corrected by the C 1s peak: E_B (C 1s) = 284.8 eV. The core-level spectra were fitted with the software UNIFIT2.1 for Windows using a

least-squares procedure [22]. An analytical model of the peak shape (Gaussian multiplied by Lorentzian) was used. The FWHMs of N 1s were fixed at 1.8 eV and the Gauss/Lorentz mixing ratios were fixed at 0.5 and 0.7. The distribution of the nitrogen functionalities of the coal and char samples determined by XPS are summarized in Table 2.

The analysis of the mineral matter was required to achieve conclusions on the catalytic activity of several ash components. Therefore the samples of the parent and demineralized coals were incinerated and the received residues were analysed by X-ray fluorescence spectroscopy. Table 3 contains the results of these analyses for the most important components.

There are some significant differences between the several coals on the one hand and between original and demineralized samples on the other. Especially the contents of calcium and iron differ strongly. From that some changes in the pyrolysis behaviour and hence in the nitrogen release could be expected. Beside this it is suggested that the effect of the HCl treatment for ash removal depends on some coal characteristics, e.g. origin of the minerals, incorporation into the organic matter. It is noted that the elimination of iron was most successfully with the Rhenish lignite whereas nothing could be removed with the CG lignite.

2.4. Gas analysis

The pyrolysis gas was passed through a filter system and washing bottles with H_2SO_4 and NaOH to absorb NH_3 and HCN, respectively. The determination of the amounts released was carried out by alcalimetric and argentometric titration. NO_x and N_2O evolved during pyrolysis were determined by NDIR gas analysers. N_2 was measured by gas chromatography with thermal conductivity detector.

A total balance of nitrogen release was determined for the whole temperature range with a heating rate of 0.25 K/s and a soak time of 30 min. The balance error was about 5%–7% at temperatures lower than 973 K and about 2% at higher values.

3. Results and discussion

3.1. Nitrogen functionalities in coals and pyrolysis chars

The pyrrole-type was found to represent the most frequent functionality in low rank coals as it has been

Table 3
Composition of mineral matter (wt.-%, X-ray fluorescence analysis)

Ash component	L		CG		R	
	Original	Demineralised	Original	Demineralised	Original	Demineralised
SiO_2	6.7	50.1	6.1	23.7	10.5	41.5
Fe_2O_3	29.3	22.3	0.5	5.9	17.1	4.7
Al_2O_3	5.0	5.8	9.4	0.6	2.3	0.7
CaO	29.6	10.8	39.4	25.4	34.1	13.1
SO_3	21.1	8.8	38.4	35.0	14.6	19.0

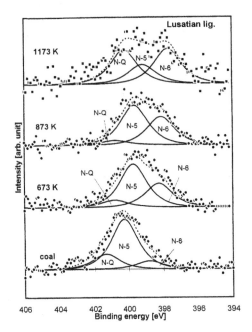

Fig. 2. XPS spectra (N 1s) of Lusatian lignite and its pyrolysis chars (0.25 K/min, 30 min).

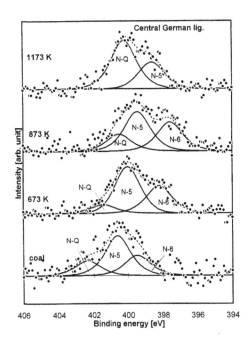

Fig. 3. XPS spectra (N 1s) of Central German lignite and its pyrolysis chars.

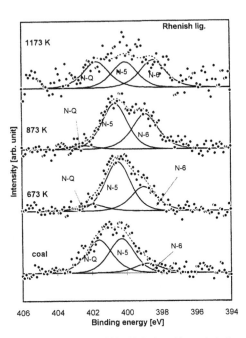

Fig. 4. XPS spectra (N 1s) of Rhenish lignite and its pyrolysis chars.

already shown by other authors [1–4]. In all coal samples three components – pyrrolic (N-5), pyridinic (N-6) and quaternary nitrogen (N-Q) – at energy positions well documented in the literature were required to achieve acceptable fits in the XPS spectra. The N 1s binding energies of the coals and the chars obtained at 673, 873 and 1173 K (heating rate: 0.25 K/s; soak time: 30 min) are shown in the Figs. 2–4. Although the ultimate and elemental analyses of the Lusatian and Rhenish lignite gave rise to a few similarities, the results indicated remarkable differences. There is a strong peak at 401.5 eV in the N 1s spectrum of the RL sample which is assigned to quaternary nitrogen. The occurrence of this component in coals has already been described by Wojtowicz and others [1,8]. Pels et al. [6] suggested that protonated pyridine-N is the origin of N-Q in low rank coals. This peak occurs in the Rhenish coal with the same intensity as the N-6 peak. The observed phenomenon is suggested to be caused by a wide variety of functional (hydroxyl, carboxyl, phenol) groups adjacent to the nitrogen. However, comparing the hydrogen and oxygen content of the coals such a correlation cannot be confirmed.

During pyrolysis at medium temperatures the peak at 401.5 eV disappears partially owing to decomposition of these functional groups. The decrease corresponds to an increase in pyridine-type functionalities. The chars obtained under mild pyrolysis conditions ($T < 873$ K) are likely to represent the fractions of five- and six-membered rings better than the spectra of the parent coal. It is also concluded from the minor changes in the N 1s energy distribution as a

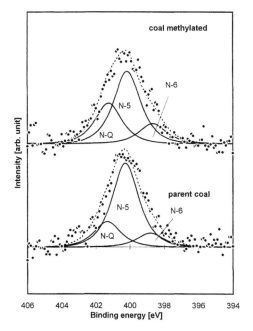

Fig. 5. XPS spectra (N 1s) of parent and methylated Lusatian lignite.

Comparing all XPS data of the coals and the low temperature chars a small increase in the fraction of pyrrole-type nitrogen is also obvious. This can be attributed to the destruction of some functionalities (amino-type) which are supposed to occur in low rank coals [1,12,20]. They could not be determined by XPS because of their low concentration and the expected peak position at ∼ 399.4 eV between the two large N-5 and N-6 peaks. At low temperatures these functionalities are preferably released leading to an apparent increase in the fraction of the pyrrole type.

The analyses of the high temperature pyrolysis chars indicated a high stability of N-5. It has been detected in all 1173 K chars with a fraction of about 30% of char-N. As a result of the pyrolysis conditions it is quite unlikely that the component at a binding energy around 400.5 eV represents a considerable part of pyridone-type nitrogen. Pyridone is suggested to be decomposed because of high temperature treatment under inert atmosphere. Another essential finding is that the uniformity of the spectra of the low-temperature chars disappeared after further pyrolysis. Whereas the fraction of pyrrolic nitrogen in all char samples is in the same order of magnitude the ratio of N-6 to N-Q varies from 1 (LL and RL) to zero (CGL). It is concluded that the incorporation of the six-membered ring nitrogen into the pre-graphitic structure of the char is influenced by coal-specific properties.

For a better assignment of the nitrogen functional forms humic acids and the residual coal of Lusatian lignite were prepared by quaternarization with methyl iodide (CH$_3$I). The parent substances and the derivatives were also analysed using XPS. As an example the spectra of the demineralized and the methylated coal are presented in Fig. 5. Surprisingly the N-6 peak of the coal survived after methylation, instead the pyrrole-type disappeared partly causing an increase of the fraction of N-Q. A comparison of the distribution of the functionalities both for the coal and the humic acids (HA) is shown in Fig. 6. A part of the component at the binding energy of ∼ 400.3 eV has been quaternarized. As pyrrolic nitrogen cannot be methylated under these conditions to form N-Q another functionality, which can be quaternarized easily, must be contained in the peak in

result of pyrolysis from 673 to 873 K. Proceeding from these data and by adding the quaternary nitrogen to pyridine type according to Pels et al. [6] it is observed that the N5/N-6 ratio is approximately 1.4. This calculated value is lower than those reported in the literature particularly for lignite [2,4]. Moreover, if another nitrogen functionality — pyridone — is taken into consideration this ratio becomes even smaller. Pyridone has been an important object to be examined in the recent years [2,23]. It cannot be identified only using XPS because of its binding energy being similar to that of pyrrole. The application of XANES improved the knowledge, but a quantitative determination is still difficult. Nevertheless it is generally accepted that part of the XPS pyrrole peak in low rank coals is caused by pyridone.

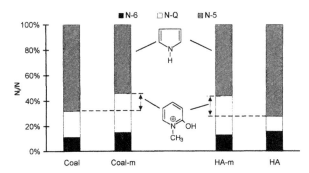

Fig. 6. Changes in the occurrence of nitrogen functionalities in L lignite caused by methylation (m) of coal and humic acids (HA).

50

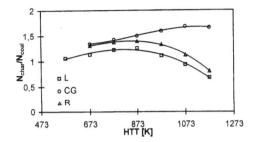

Fig. 7. Normalised nitrogen content of the chars (0.25 K/s).

question. It is suggested that this component is pyridone (N-6(O)) which is also known to be the origin of peak in the XPS spectrum at about 400.3 eV. The reaction of the nitrogen atom in pyridone with CH_3I is favoured because of its higher basicity. In spite of the fact that not all of the N-6(O) but maybe some of the N-6 molecules have been methylated, the real ratio of five- to six-membered ring nitrogen must be corrected. It decreases from the above-mentioned 1.4 to 1.2 so that it reaches values which are known from bituminous coals. From these results it can be concluded that the N-5/N-6 ratio of coals should not deviate with rank as much as previously reported.

3.2. Nitrogen retention in chars

3.2.1. Coal type and heating rate

The normalised nitrogen content of the chars of Lusatian and Rhenish lignite follows the typical shape running through a maximum at about 900 K (Fig. 7). It exceeds the nitrogen content of the parent coal in the first stage up to this temperature by a factor of 1.5. At higher temperatures the nitrogen loss increases as a result of the less thermal stability of nitrogen heterocyclic compounds resulting in a strong decrease of the normalised N/C ratio of the chars. In contrast the N-content of the CG chars remains nearly constant for all samples produced at temperatures higher than 800 K. It has been noted that the nitrogen release occurs equivalent to total mass loss. Thus from these data

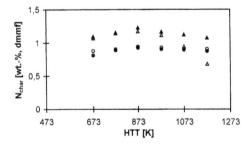

Fig. 8. Changes in char nitrogen as a result of demineralization. (symbols see Fig. 7, filled – demineralized coals).

a dependence of nitrogen retention in the char on coal type can be concluded.

Comparing the results from XPS and elemental analyses no direct correlation was found between functionality and nitrogen release which could explain the differences in the pyrolysis behaviour of the coals. Proceeding from the analytical data only the following mechanism can be suggested. The CG lignite possesses the lowest nitrogen content of all samples. This can be attributed to its older age in comparison to the other coals. Some readily volatile nitrogen has possibly been removed during the later stage of carbonisation. The remainder, therefore, consists of a higher fraction of more stable structures incorporated in larger aromatic clusters. This assumption could be confirmed by total nitrogen contents of the high temperature chars. There is only a small difference between the Central German and the Lusatian lignite (CG: 4.0, R: 3.2 mg_{char-N}/g_{coal}). Moreover the absence of pyridinic nitrogen in the CG char implies that this functionality is the main source of high temperature N release, unless it has been converted to N-Q. However, there remain some doubts concerning this conclusion.

The range where the heating rate had been varied was found to be to small to obtain significant dependencies. The only observation significant to note was the lower the heating rate the more nitrogen is retained in the char.

3.2.2. Mineral matter

From comparative studies of the parent and demineralized coals conclusions can be drawn, whether the mineral matter already effects devolatilization, i.e. if it promotes or suppresses primary release of nitrogen by catalytic reactions. Besides the ultimate and elemental analyses of the samples the char yield has been considered to be conclusive that parent and demineralized coals are comparable. The char yields of both CG samples (dmmf basis) correspond very well within the whole temperature range. As it is shown in Fig. 8 for Rhenish lignite there is a tendency to retain a larger amount of the fuel nitrogen (FN) in the chars with the demineralized samples, especially at higher temperatures. A similar course of the nitrogen content was observed for the Lusatian lignite. In contrast the values of the original and acid-treated CG lignite are complete identically (see also Fig. 8). This indicates that some of the ash components give rise to a preferential release of nitrogen. Looking for the elements causing these significant changes a few constituents have to be considered. As can be seen in Table 3 the most important ones are silicon, iron, calcium, magnesium, aluminium and sulphur, too. Silicon compounds occurring as quartz are proved to behave ineffectively to a great extent. Sulphur — both part of the organic and the inorganic matter of coals — is not considered, either. About 80% of this element has been removed during acid washing without any consequences for nitrogen retention in char. The same applies for aluminium and calcium. However, the content of the latter in the demineralized CG ash remains 2–3 times higher than in the other

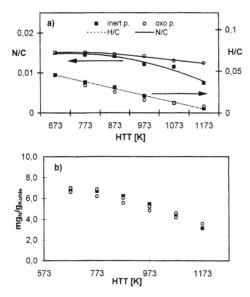

Fig. 9. Atomic ratios (a) and nitrogen content (b) of inert and oxopyrolysis chars (R lig., 0.25 K/s).

magnesium. The most distinct changes in magnesium content have been observed with Rhenish and Lusatian lignite. However, the Mg/N atomic ratios of all coals are nearly similar both for original (0.5) and demineralized samples (0.03). From the earlier discussion it can be concluded that the main component in ash which promotes nitrogen release has to be iron. This metal was removed by HCl washing to a larger extent in the Rhenish than in the Lusatian samples and not in the CG ones. This fact corresponds to the decreasing differences in the nitrogen content of the original and demineralized chars in the same order. Comparing the Fe/N atomic ratios the remaining differences in the nitrogen release of the demineralized coals can be explained. The values are still the highest in the Lusatian char (0.09), 3–4 times more than in the other samples.

3.2.3. Oxopyrolysis

Partial gasification (oxygen : helium: 0.05 and 0.2) of the coals was carried out to study the effect of a low oxygen partial pressure simulating the conditions at the particle surface during combustion of the volatiles ($\lambda \approx 0.1$–0.2). The results shown in Fig. 9(a) indicate a selective oxidation of hydrocarbons, whereas nitrogen is enriched in the residual char. The comparison of the H/C atomic ratios shows a similar trend both for inert- and oxopyrolysis. In contrast, there is an increasing difference in the N/C ratios between both experimental series. This phenomenon was found with all coals at temperatures higher than 973 K. If the char yield is taken into consideration it has been found that the total nitrogen content of the residue is similar for all samples (Fig. 9(b)). From this it can be concluded that aliphatic structures, side chains and functional groups of ring molecules containing scarcely any N-atoms are preferably oxidised. Aromatic hydrocarbons and nitrogen-containing aromatic structures are not involved in these reactions. The results also imply that aliphatic nitrogen compounds if they occur in the parent coal are already released during initial stage of pyrolysis. Another explanation for the enrichment of nitrogen except a selective oxidation can be given as follows. Oxygen occupies active sites at the surface which are responsible for prior nitrogen release. That means it inhibits some catalytic reactions being of importance in coal nitrogen chemistry.

coals. Regarding the N/Ca atomic ratios the difference becomes even more clear. The influence of calcium compounds on primary nitrogen devolatilization cannot be excluded because of its high concentration level. But it should be responsible for the observed pyrolysis behaviour only to a minor extent. A smaller effect could also be attributed to

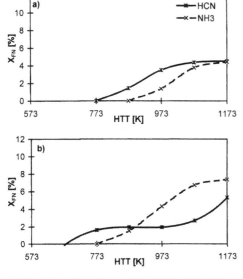

Fig. 10. Conversion of fuel nitrogen (FN) to HCN and NH₃ (L lig.), (a) 0.25 K/s, 30 min, (b) 5 K/s, 5 min.

3.2.4. Pressure

A series of experiments under elevated pressure (1 MPa) were performed at both heating rates with CG lignite. Only a minor influence of this parameter was found concerning the nitrogen retention in the char. At a low heating rate (0.25 K/s) no differences were obtained between chars produced at 0.1 and 1 MPa, respectively. This could be assigned to chemical reactions proceeding almost completely because of long residence time of primary products within the porous structure of the solid even at low pressure. If the heating rate has been increased by a factor of 20 secondary reactions of volatiles have led to a more incomplete reincorporation of nitrogen into the char. In this case the

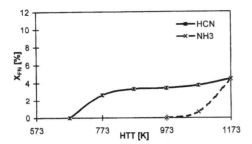

Fig. 11. FN-conversion of demineralized coal to HCN and NH_3 (L lig., 5 K/s).

rise in pressure results in a slight intensified nitrogen retention in char.

3.3. Release of gaseous NO_x-precursors (HCN and NH_3)

3.3.1. Coal type and heating rate

Hydrogen cyanide is the primary product of nitrogen devolatilization. It has been detected in all experimental series at lower temperatures than ammonia. The formation of HCN starts at 673 K and its profile can be divided into two parts: the release at an early stage as a result of the decomposition of the tar nitrogen and the elimination of CN^- groups, and in a second — high-temperature — stage as a result of ring cleavage of heteroaromatic compounds. Ammonia formation has been detected at temperatures higher than 773 K. The total amounts of NH_3 released up to 1173 K exceed those of HCN for all coals and all experimental series. In the figures discussed in the following the fuel nitrogen conversion (X_{FN}) to NO_x-precursors is described as summation function of the whole temperature range up to the HTT. The variation of the heating rate in a range from 0.25 to 5 K/s was found to have only little (coals L and R) or no effect (coal CG) on the conversion of FN. The results are shown for Lusatian lignite in Fig. 10. The graph describing the evolution of HCN shifts towards lower temperatures. From this it follows a corresponding shift in the formation of NH_3. The minor differences in the graph shapes are ascribed to the low heating rate

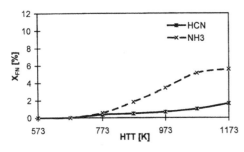

Fig. 12. FN-conversion to HCN and NH_3 as a result of oxopyrolysis (L lig., 5 vol.-% O_2, 5 K/s).

compared with other investigations in fluidized bed and drop tube pyrolysers [12,24]. The residence time of the primary gaseous products is obviously sufficient to enable that middle-temperature HCN is almost complete converted to ammonia via radical reactions. It was found that the direct hydrogenation of HCN is responsible for the formation of NH_3 only at temperatures below 973 K. Corresponding with this results the conversion of hydrogen cyanide to ammonia discontinues at slightly higher temperatures. Thus the main paths are suggested to involve radical reactions. The released HCN from the char is converted to ammonia only to a minor extent at temperatures higher than 1123 K because of the absence of such H/OH radicals.

Fuel nitrogen was found to react to HCN at somewhat lower temperatures with Central German lignite. This phenomenon can be attributed to the high tar yield of this sample. Hydrogen cyanide originates from early devolatilization products which are thermally cracked during heat treatment at low temperatures. It is noted that the conversion of FN to both NO_x-precursors has been the highest for the coal with the lowest nitrogen content.

3.3.2. Mineral matter

As to be seen from the first part of this paper the mineral matter of the coal plays an important role in the mechanisms of nitrogen release. Therefore and from the knowledge of the catalytic efficiency of several ash components in the formation and reduction of nitrogen oxides a corresponding influence on the reactions of primary devolatilized compounds could be expected. The results obtained from pyrolysis of demineralized coals show according to theoretical thoughts a predominating effect on gas composition, indeed. The formation of ammonia was found to decrease strongly and the detection limit shifted to higher temperatures (Fig. 11). This behaviour was observed to be independent of heating rate but dependent on coal type. The most significant changes occurred as a result of pyrolysis of CG lignite. With this sample no ammonia has been detected, whereas for both other demineralized coals its evolution has started at much higher HTT. It is interesting to note that the decrease in NH_3 is not accompanied by a corresponding rise in HCN formation. The following explanations are possible: the results indicate that the conversion of primarily formed hydrogen cyanide is strongly promoted by the mineral matter of the coal. From the ash analyses it can be concluded that not one single element but a wide variety of compounds are responsible for catalysing this reaction path. Comparing the experimental data of the original with the demineralized coal samples a change in the preferred conversion reactions is obvious. Without the presence of certain minerals the formation of NH_3 is only possible at high temperatures and NCO as an intermediate compound is supposed to become more important. However, the reactions both via the amino-pool and the NCO-pool result in the formation of N_2. Especially at high

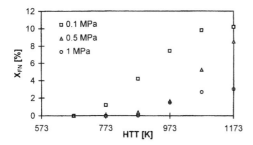

Fig. 13. NH₃ evolution at elevated pressure (CG lig., 0.25 K/min).

residence times other products (nitriles) as suggested by several authors [24, 25] should be of minor significance.

3.3.3. Oxopyrolysis

As can be seen from the total nitrogen content of the chars (Fig. 9(b)) the amounts of nitrogen retaining in the solid residue are independent of inert- or oxopyrolysis processes. It has been suggested that most of the reactions in the conversion process of the NO_x-precursor HCN and NH_3 are radical ones. From this point of view a decrease of both primary products could be expected. The results obtained from partial gasification at a heating rate of 5 K/s confirm this assumption. A decrease in the conversion ratio both to HCN and NH_3 has been noticed as indicated in Fig. 12 for Lusatian lignite. Even at relatively low temperatures

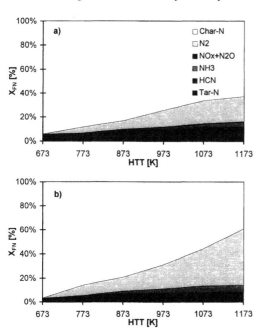

Fig. 14. Nitrogen distribution on pyrolysis products (R lig., 0.25 K/s) (a) parent coal, (b) demineralized coal.

an influence of the oxygen is obvious. In contrast measurements of nitrogen oxides in the pyrolysis gas have shown that only a minor part of the released nitrogen has been oxidised. Although the conversion ratio of the FN to the precursors decreased by 4% only 1% has been oxidised additionally. It was found that only nitrous oxide had been formed as a result of the increase in oxygen supply. Although NO should be the primary product of oxidation N_2O is a result of reduction reactions because of the low excess air coefficient. The secondary reactions of HCN and NH_3 are promoted by formation of oxygen-containing radicals resulting in a higher conversion rate to N_2.

3.3.4. Pressure

In contrast to the results obtained with nitrogen retention in the char some differences have been observed concerning the evolution of hydrogen cyanide and ammonia. HCN formation has been found to be nearly unaffected by elevated pressure whereas the NH_3 release has strongly decreased. The greatest changes have occurred at low heating rates because of increased residence time of the primary devolatilization products (Fig. 13). According to the over-all reaction

$$2HCN + 6H_2 \leftrightharpoons 2NH_3 + 2CH_4$$

ammonia formation should be promoted under high pressure. However, it has been found that this reaction is only of importance at temperatures below 873 K. From this aspect the conversion of ammonia mainly to N_2 must dominate with increasing pressure. Apart from the catalytic decomposition of ammonia, reactions with hydrogen radicals have to be considered to understand the chemical mechanisms underlying these reactions.

3.3.5. Balancing the nitrogen distribution during pyrolysis

The nitrogen balance for low-rate pyrolysis of German lignite results in a conversion of the FN as follows: 30%–50% retaining in the char, 10%–20% released as HCN and NH_3, 10%–15% as tarN, 1%–2% as NO/N_2O and 30%–50% as N_2. Such a remarkable conversion ratio of nitrogen has also been described by Ohtsuka and Wu et al. [15,17]. The formation of oxides during inert pyrolysis has been found to be independent of coal type. It has been assigned to the high oxygen content of the coals, which can be partly associated with the organic nitrogen. In Fig. 14 (a) and (b) the total nitrogen balances are clearly shown for original and demineralized Central German lignite. They illustrate char-N and N_2 being the main contributors. The comparison of the figures also indicates the temperature range where catalyst-promoted reactions become more important. The strong rise in the N_2 formation with the L sample at temperatures greater than 1023 K is assigned to such reactions. The molecular nitrogen originates mainly from char-N. In this connection must be added that the mechanism of the formation of N_2 from the solid bound nitrogen is scarcely investigated. Above all the high reactivity of the new arisen char surface should be responsible for the observed high conversion level.

4. Conclusions

Comprehensive studies were carried out to investigate the fate of coal nitrogen of German lignite during non-isothermal pyrolysis at low heating rates. The XPS data of the coals and low temperature chars give rise to only small differences in the nitrogen functionalities. The occurrence of the binding forms in the spectra depends on the chemical environment especially on the presence of functional groups adjacent to the N atoms. A connection between functionalities and nitrogen retention in char as well as the release of gaseous intermediates as suggested by other authors could not be proved. The fixed bed conditions are characterised by an almost complete conversion of the volatile nitrogen to N_2 because of long residence time and sufficient supply of reactants. HCN and NH_3 are of minor importance, at most 20% of the FN is converted to these NO_x-precursors. From the comparison of original and demineralized coals, it can be suggested that not only the primary release of nitrogen but also the following conversion to intermediates and final products are affected by heterogeneous catalytic reactions. The evolution of char-N was found to be promoted mainly by iron compounds. The nitrogen additionally evolved from the original compared with the demineralized coal appears as N_2. The strong decrease in the formation of ammonia during the pyrolysis of demineralized coals has been attributed to the entirety of the ash.

Elevated pressure up to 1 MPa has no effects on nitrogen retention in char. The caused increase of the residence time of the primary volatiles entirely leads to more intense secondary reactions converting the intermediates (NH_3). Under conditions of partial gasification with only a small burn-out of the chars the oxidation of hydrocarbons occurs selective. Hence the nitrogen is enriched in char as a result of oxidative pyrolysis. The conversion rate of volatile-N increases in presence of oxygen. Nevertheless the dominating species remains N_2.

Acknowledgements

This work was supported by the German Federal Ministry of Education, Research, Science and Technology (BMBF) – Contract-No. 0326785B. The authors grateful acknowledge the help of Dr Peter Streubel and Mr Ronald Hesse from Leipzig University for carrying out the XPS analyses.

References

[1] Wojtowicz MA, Pels JR, Moulijn JA. Fuel 1995;74:507.
[2] Davidson RM. Nitrogen in coal, Perspectives, IEA Coal Research, 1994.
[3] Thomas KM. Fuel 1997;76:457.
[4] Burchill P. Int. Conf. on Coal Science [Proc.]. Maastricht, 1987. pp. 5–8.
[5] Kelly MD, Buckley AN, Nelson PF. 6th Int. Conf. on Coal Science [Proc.]. Newcastle-upon-Tyne. 1991. pp.356–359.
[6] Pels J R, Kapteijn JA, Moulijn JA, Zhu Q, Thomas KM. Carbon 1995;33:1641.
[7] Spracklin CJ, Thomas KM, Marsh H, Edwards IAS. 6th Int. Conf. on Coal Science [Proc.]. Newcastle-upon-Tyne. 1991. pp. 343–346.
[8] Boavida D, Gulyurtlu I, Lobo LS. 8th Int. Conf. on Coal Science [Proc.]. Oviedo. 1995. pp. 751–754.
[9] Watt M, Allen W, Fletcher T. 8th Int. Conf. on Coal Science [Proc.]. Oviedo. 1995. pp. 1685–1688.
[10] Leppälahti L, Koljonen T. Fuel Processing Technology 1995;43:1.
[11] Wojtowicz MA, et al. 8th Int. Conf. on Coal Science [Proc.]. Oviedo. 1995. pp. 771–774.
[12] Johnsson JE. Fuel 1994;73:1398.
[13] Köpsel RFW, Halang S. Fuel 1997;76:345.
[14] Wojtowicz MA, Pels JR, Moulijn JA. Fuel 1994;73:1416.
[15] Ohtsuka Y, Mori H, Nonaka K, Watanabe T, Asami K. Energy & Fuels 1993;7:1095.
[16] Hayashi J, Kusakabe K, Morooka S, Furimsky E. 8th Int. Conf. on Coal Science [Proc.]. Oviedo. 1995. pp. 1697–1700.
[17] Wu Z, Ohtsuka Y, 9th Int. Conf. on Coal Science [Proc.]. Essen (Germany). 1997. pp. 1143–1146.
[18] Baumann H, Kadelka P, Romey I. 6th Int. Conf. on Coal Science [Proc.]. Newcastle-upon-Tyne. 1991. pp. 339–342.
[19] Norrman J, Pourkashanian M, Williams A. Fuel 1997;76:1201.
[20] Aho MJ, Hämäläinen JP, Tummavuori JL. Fuel 1993;72:837.
[21] Wallace S, Bartle KD, Perry DL. Fuel 1989;68:1450.
[22] Hesse R. Software UNIFIT2.1 for Windows, Leipzig, 1997.
[23] Zhu Q, Money SL, Russell AE, Thomas KM. Langmuir 1997;13:2149.
[24] Kambara S, Takarada T, Toyoshima M, Kato K. Fuel 1995;74:1247.
[25] Nelson PF, Buckley AN, Kelly MD. 24th Symp. (Int.) Comb. [Proc.]. The Combustion Institute. 1992. p. 1259.

Formation of NO$_x$ and SO$_x$ precursors during the pyrolysis of coal and biomass. Part I. Effects of reactor configuration on the determined yields of HCN and NH$_3$ during pyrolysis

Li Lian Tan, Chun-Zhu Li*

Department of Chemical Engineering, P.O. Box 36, Monash University, Vic. 3800, Australia

Abstract

The formation of HCN and NH$_3$ during the pyrolysis of a biomass (bagasse) and a set of rank-ordered coal samples has been studied in a novel reactor. The reactor has some features of both a drop-tube reactor and a fixed-bed reactor: the reactor allows the coal/biomass particles to be heated up rapidly as well as to be held for a pre-specified period of time at peak temperature. The experimental results obtained suggest that a considerable amount of the nitrogen in the nascent char could be converted into HCN and NH$_3$ if the char is held at high temperatures for long time. The formation of NH$_3$ from the thermal cracking of char was seen to last for more than an hour even at temperatures as high as 700–900°C. The formation of HCN went to completion much more rapidly than that of NH$_3$. Compared with the results in the literature from the pyrolysis of coals in a fluidised-bed reactor, the reactor configuration used in this study allows the effects of fuel rank to be studied on an unbiased basis towards the type of fuel. The yields of HCN and NH$_3$ from the present study decrease with increasing rank. The experimental results suggest that the differences in reactor configurations used by various researchers would account at least partially for some of the discrepancies in the literature regarding the formation of HCN and NH$_3$ during the pyrolysis of coals. © 2000 Elsevier Science Ltd. All rights reserved.

Keywords: HCN; NH$_3$; NO$_x$ precursors; SO$_x$ precursors; Coal; Biomass; Pyrolysis

1. Introduction

NO$_x$/N$_2$O and SO$_x$ contribute to the formation of photochemical smog and acid rain, to the enhancement of greenhouse effects and to the enhanced depletion of stratospheric ozone. Emissions of oxides of nitrogen (NO, NO$_2$ and N$_2$O) and oxides of sulphur (SO$_x$) from power generation using coal are and will be an important environmental problem. During pyrolysis, the nitrogen in coal, as part of coal organic matter, is converted into NO$_x$ precursors such as NH$_3$, HCN and HNCO as well as the nitrogen in tar and the nitrogen in char [1–11]. These NO$_x$ precursors (particularly the nitrogen in char) may then be converted into either NO$_x$/N$_2$O or N$_2$ during subsequent gasification/combustion [12–17]. The conversion efficiency of these NO$_x$ precursors into NO$_x$/N$_2$O depends strongly upon the type of NO$_x$ precursor. Similarly, the organic sulphur in coal may be converted into gaseous SO$_x$ precursors (e.g. H$_2$S) or retained in char during pyrolysis [18–20]. The interactions between the gaseous SO$_x$ precursors (e.g. H$_2$S) and absorbent (e.g. lime) in a

fluidised-bed reactor would largely depend on the timing of the release of SO$_x$ precursors during pyrolysis and subsequent combustion/gasification. Clearly, the reductions in the emissions of NO$_x$/N$_2$O and SO$_x$ would ultimately rely on our understanding of the reaction pathways leading to the formation and destruction of NO$_x$/N$_2$O, SO$_x$ and their precursors during the whole pyrolysis, gasification and combustion. As the first paper in this series on the formation of NO$_x$ precursors and SO$_x$ precursors, this paper discusses the effects of reactor configuration on the yields of NO$_x$ precursors during the pyrolysis of a set of rank-ordered coals and a biomass (bagasse).

The mechanism by which the nitrogen in coal is converted into the NO$_x$ precursors during pyrolysis is imperfectly understood at present. There have been a lot of disagreements reported in the literature, especially for the formation of NH$_3$ [4,6,9,11]. Many different types of reactors have been used in the study of NO$_x$ precursor formation during pyrolysis. The results seem to depend strongly on the configurations of the reactors used. Even for the same set of coals, entirely different trends were observed for the formation of HCN and NH$_3$ when the coals were pyrolysed in different types of reactors. For example, when a suite of

---

* Corresponding author. Tel.: +61-3-9905 9623; fax: +61-3-9905-5686.
  E-mail address: chun-zhu.li@eng.monash.edu.au (C.-Z. Li).

Reprinted from *Fuel* **79 (15)**, 1883-1889 (2000)

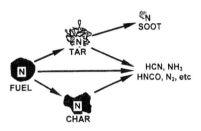

Fig. 1. A schematic diagram showing the major routes for the formation of NO$_x$ precursors during the pyrolysis of coal.

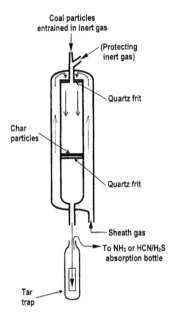

Fig. 2. A schematic diagram of the reactor used.

rank-ordered American coals were pyrolysed in an entrained-flow reactor, very little NH$_3$ was detected in the product gas [8]. However, when a similar suite of rank-ordered coals were pyrolysed in a thermogravimetric analyser (TGA) by the same group, NH$_3$ was found to be one of the most important nitrogen-containing products from the nitrogen in coals [9]. Even for the pyrolysis of simple model compounds such as pyrrole and pyridine, no agreement has been reached in the literature. For example, while no NH$_3$ was detected when pyrrole and pyridine were pyrolysed in a shock-tube reactor [21,22], NH$_3$ was found to be an important product when the same compounds were pyrolysed in an entrained flow reactor [23] or in a tubular reactor [24].

Our recent studies [25–29] have indicated that NH$_3$, HCN and HNCO may be formed during the primary pyrolysis of coal and/or during the secondary thermal cracking of volatiles and char, as is schematically shown in Fig. 1. Considering the very different particle time–temperature histories (e.g. heating rate and holding time) in various reactors used in the past by various researchers, part of the disagreement in the literature may be due to the use of different reactor configurations. For example, while the residence time of solid particles may be extremely short in an entrained-flow reactor [8], their residence time would be very long in the TGA [9]. Even the average residence time of char particles in a fluidised-bed reactor [25] differs very considerably from caking coals to non-caking coals (also see Section 3 in this paper). The variation in time–temperature history may change the relative importance of individual routes (Fig. 1), leading to the formation and destruction of NO$_x$ precursors. Moreover, our recent study [30] has suggested a possible artefact that NH$_3$ can be lost (to N$_2$) through interactions with many kinds of

materials often used in the construction of reactors such as stainless steel, quartz and zircon sands. This finding further complicates the evaluation of the existing literature data.

The purpose of this study was to gain experimental evidence on the importance of reactor configuration for the observed yields of HCN and NH$_3$ during pyrolysis, particularly during the thermal cracking of nascent char in a novel reactor. The reactor has some features of a drop-tube reactor and of a fixed-bed reactor, allowing for the study of the formation of HCN and NH$_3$ during the in situ thermal cracking of nascent char formed from the rapid pyrolysis of coal and biomass.

2. Experimental

2.1. Coal and biomass samples

A small set of rank-ordered coal samples was used in the

Table 1
Properties of the coal and biomass samples studied

Fuel	Rank	Particle size range, μm	Ash	Volatile matter	C	H	N	S (total)	O (by diff.)
			(Wt%, dry)	Wt%, daf basis					
Bagasse	Biomass	90–150	9.7	82.4	47.6	5.5	0.31	0.28	46.3
Loy Yang	Brown	106–150	1.0	51.5	68.5	4.8	0.55	0.32	25.8
Drayton	High-volatile bituminous	90–130	5.4	n.a.[a]	82.1	6.0	1.83	0.95	9.1
Pocahontas No. 3	Low-volatile bituminous	75–106	5.0	n.a.[a]	91.0	4.7	1.27	0.90	2.1

[a] n.a., not available.

present study representing the main parts of the coal rank spectrum. The properties of these coal samples are given in Table 1. The brown coal (Loy Yang) was from the Latrobe Valley in Victoria, Australia. The bituminous coal (Drayton) was a typical Australian high-volatile bituminous coal. The American coal (Pocahontas No. 3) was chosen for its low volatile yield, representing a high-rank bituminous coal/semi-anthracite. A biomass sample, sugarcane bagasse from Queensland in Australia, was also used in this study. The bagasse is considered to represent the solid fuel of "lowest rank" in this study.

2.2. Pyrolysis

A schematic diagram of the quartz reactor used is shown in Fig. 2. The reactor was made of two concentric quartz tubes. The inner diameters of the inner and outer quartz tubes were about 45 and 70 mm, respectively. During an experiment, the reactor was heated externally with an electrical furnace. A stream of ultra-high-purity argon (sheath gas) was fed into the reactor from the bottom. The gas was heated up while it moved up through the jacket formed by the inner and outer quartz tubes. The gas stream then passed through a quartz frit (acting as a flow straightener) and then moved down along the inner reactor tube. The purpose of this stream of sheath gas is to prevent the possible reactions [30] of char and volatiles with the quartz reactor wall. Under current experimental conditions, there were very little deposits of carbonaceous materials on the inner wall of the inner quartz tube. Coal particles were entrained in a feeder that was a modified version of the feeder used by Hayashi and co-workers [31]. Further details about the feeder may be found elsewhere [32]. The entrained particles from the feeder were fed into the reactor from the top (see Fig. 2) after the reactor system had been heated to the required temperature. The coal particles were heated up rapidly in a similar manner to that in a drop-tube pyrolysis reactor. A stream of gas flowing outside the feeding tube was also used in order to protect coal particles from being heated up before entering the reactor during some experiments. However, the heating rates of the coal particles in the current reactor might be somewhat slower than those in a drop-tube reactor, because the flow rates (about $0.95 \, \mathrm{l \, min^{-1}}$) of the cold gas coming together with the coal particles were relatively large compared with the flow rate (about $0.05 \, \mathrm{l \, min^{-1}}$) of the "sheath gas".

Another quartz frit was also installed inside the inner reactor tube (see Fig. 2). The frit was located within the isothermal zone inside the furnace. While the volatiles passed through this frit, the nascent char particles formed during pyrolysis were retained by the frit. Therefore, as soon as coal particles were heated up rapidly, the status of the char particles in the reactor simulated that in a fixed-bed reactor. However, the amount of coal or biomass (~0.5 g, weighed accurately) fed into the reactor during an experiment was relatively small, the thickness of the coal/char bed is very thin, much less than 1 mm. Clearly, the char particles would interact with volatiles, particularly with the volatiles generated from the coal particles subsequently fed into the reactor (see Part 2 for more detailed discussion). Therefore, the reactor has some features of both a drop-tube reactor (in terms of high heating rates) and a fix-bed reactor (somewhat in terms of the interactions between volatiles and char). A thermocouple was inserted into the reactor to measure the temperature of the frit located in the centre of the reactor before the feeding of coal was started. This measured temperature of the frit was taken as the nominal reaction temperature because all the gas flow rates remained almost unchanged before, during and after the coal was fed into the reactor (this is an unique feature of the feeder). The thermocouple was not in the reactor during the pyrolysis experiment in order to avoid possible reactions [30] of volatiles with the thermocouple.

The volatiles exiting the bottom of the reactor passed through a glass trap fitted with a thimble (similar to that described by Tyler [33]) to filter out the heavier tar components. Following the procedures outlined elsewhere [11,25], the tar trap was heated with hot water in a vacuum flask throughout an experiment in order to minimise the condensation of HCN and NH_3 inside the trap. After the removal of heavy tar components, the product gas passed through a bubbler containing an absorption solution for the collection of NH_3 or HCN separately. NH_3 was collected in a bubbler containing $0.02 \, \mathrm{M} \, CH_3SO_3H$ solution. HCN was collected in a separate experiment in a bubbler containing an aqueous solution of $0.1 \, \mathrm{M} \, NaOH$, $0.7 \, \mathrm{M} \, CH_3COONa$ and 0.5% $H_2NCH_2CH_2NH_2$. Prior experiments with more bubblers in series containing the solutions confirmed that the NH_3 or HCN collected in extra bubblers were negligible. In our earlier work [34], the acid solution and the alkaline solution were put in series. This has caused serious underestimate of the HCN yield due to the high solubility of HCN in the acid solution preceding the alkaline solution.

2.3. Quantification of NH_3 and HCN

NH_3 and HCN absorbed in the solutions were quantified with a Dionex 500 ion chromatograph equipped with a ED 40 electrochemical detector. The quantification of NH_3 was carried out with electrical conductivity detection following separation in a Dionex CS12 column with $0.02 \, \mathrm{M} \, CH_3SO_3H$ as eluent. It should be noted that the NH_3 yields quantified in this study may also include some contribution from the hydrolysis (in the absorption solution) of HNCO which has been observed as a nitrogen-containing species from the pyrolysis of coals [11]. However, such contribution to NH_3 from the hydrolysis of HNCO *in the bubbler* may be very small because the yields of HNCO *in the product gas* were always much smaller than those of HCN and NH_3 [11]. HCN was quantified with amperometric detection using a silver electrode following separation in a Dionex AS7 column with an aqueous solution of $0.1 \, \mathrm{M} \, NaOH$, $0.7 \, \mathrm{M}$

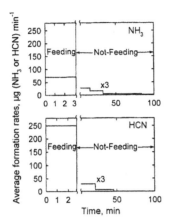

Fig. 3. Average formation rates of NH_3 and HCN as a function of time during the pyrolysis of Loy Yang brown coal at 800°C. Nominal coal feed rate in the feeding period: 185 mg min^{-1}, total gas flow rate: 1.1 l min^{-1} (under ambient conditions). Note the changes in scales.

CH_3COONa and 0.5% $H_2NCH_2CH_2NH_2$ as eluent. The calibration of CN$^-$ was done with standard solutions purchased from Alltech.

3. Results and discussion

3.1. Formation of NH_3 and HCN from the thermal cracking of nascent char

Fig. 3 shows a typical example of the average rates of formation of NH_3 and HCN during the pyrolysis of the Loy Yang brown coal at 800°C. A total gas flow rate of 1.1 l min^{-1} (measured at ambient conditions) was used. The "feeding" period in Fig. 3 refers to the time when the coal was continuously fed into the reactor at a (nominal) coal feed rate of 185 mg min^{-1}. The "not-feeding" periods in Fig. 3 refer to the time after the feeding of coal had stopped. In other words, NH_3 and HCN formed during pyrolysis were collected as "reaction-time-resolved" fractions. It should be noted that the time scales in the figure are different for the "feeding" and "not-feeding" periods. Each

step in Fig. 3 represents an average formation rate calculated from the quantification of NH_3 (or HCN) in a bubbler. The most important feature of the data shown in Fig. 3 is that a considerable amount of NH_3, about 60% of the total NH_3 formed during the whole experiment shown in Fig. 3, was formed after the feeding of coal had stopped. Initially, it was suspected that the NH_3 detected in the "not-feeding" periods might have been due to the release of NH_3 that was adsorbed/absorbed inside the trap, for example on the thimble, in the "feeding" periods. Experiments were then carried out where the product gas from the reactor in the "not-feeding" periods, bypassing the tar trap, was sent directly to the absorption bubbler. The results showed that the NH_3 detected in the "not-feeding" periods was not due to the release of adsorbed/absorbed NH_3 in the tar trap during the "feeding" periods but directly came from the reactions taking place inside the reactor [35].

It has been found that the percentage of NH_3 detected in the "not-feeding" periods strongly depends on the feed rate of coal in the "feeding" period, as is shown in Table 2. In fact, the data in Fig. 3 suggest that the NH_3 formation rate decreased rapidly with time. In addition to the formation of NH_3 concurrent with the release of tar (i.e. during primary pyrolysis, Fig. 1), NH_3 must have continuously formed from the thermal cracking of the nascent char. Therefore, a significant amount of NH_3 detected in the "feeding" period in fact must have also been due to the thermal cracking of the nascent char. In other words, the char formed during the earlier stages in the "feeding" period was continuously cracked while coal particles were still being fed into the reactor at the later stages in the "feeding" period. The NH_3 observed in the "not-feeding" periods represented only a "tail" of the NH_3 formation from the thermal cracking of char.

Increasing temperature decreased the relative importance of the NH_3 formation "observed" in the "not-feeding" periods, as is shown in Table 3. The data in Table 3 are in agreement with increased reaction rates of NH_3 formation at high temperature.

Compared with the formation of NH_3, the rate of HCN production decayed very rapidly with time (Fig. 3). In the case of data shown in Fig. 3, the amount of HCN produced in the "not-feeding" periods constituted only small fractions (25% in Fig. 3) of the total HCN yield during the whole

Table 2
NH_3 formation observed during the pyrolysis of Loy Yang brown coal in the "feeding" and "not-feeding" periods at 800°C. [Total gas flow rate: 1.1 l min^{-1} (measured under ambient conditions)]

Feeding rate (mg min^{-1})	Time required to feed 0.5 g of sample	NH_3 observed in the "feeding" period, % of the total NH_3 in the whole experiment	NH_3 observed in all the "not-feeding" period, % of the total NH_3 in the whole experiment
170	2.9	40	60
95	5.3	65	35
55	9.1	65	35
45	11.1	75	25
25	20.0	75	25

Table 3
NH$_3$ formation observed during the pyrolysis of Loy Yang brown coal in the "feeding" and "not-feeding" periods as a function of temperature. [Nominal coal feed rate: 90–110 mg min^{-1}. Total gas flow rate: 1.0 l min^{-1} (measured under ambient conditions)]

Temperature (°C)	NH$_3$ observed in the "feeding" period, % of the total NH$_3$ in the whole experiment	NH$_3$ observed in all the "not-feeding" period, % of the total NH$_3$ in the whole experiment
600	15	85
700	20	80
800	65	35
900	50	50
1000	75	25

experiment. As will be discussed elsewhere [36], the yield of HCN observed in the "feeding" period is a strong function of coal feeding rate and gas flow rate. It is nevertheless important to note here that a significant amount of HCN was indeed observed during the "not-feeding" periods. With the same reasoning given above for the formation of NH$_3$, what was observed in the "not-feeding" periods for the formation of HCN represented only a "tail" of the formation of HCN from the thermal cracking of the nascent char. The thermal cracking of char is also an important source of HCN during the pyrolysis of the Loy Yang brown coal (Fig. 3).

Similar trends to the pyrolysis of Loy Yang brown coal shown in Fig. 3 were seen with the pyrolysis of the bagasse, the Drayton high-volatile bituminous coal and the Pocahontas No. 3 low-volatile bituminous coal, as is shown in Figs. 4–6 for the formation of NH$_3$. In all cases, significant amounts of NH$_3$ (and HCN, not shown) were formed in the "not-feeding" periods during pyrolysis, suggesting that thermal cracking of char is an important source of NH$_3$ and HCN during the pyrolysis of these coals and the bagasse. The observations described here confirmed our earlier conclusions [25–29] summarised in Fig. 1. As is shown in Table 4 for the case of pyrolysis at 800°C, the relative importance of NH$_3$ observed in the "not-feeding" periods to the total NH$_3$ production increased with increasing rank

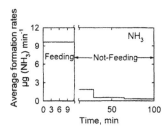

Fig. 4. Average formation rates of NH$_3$ as a function of time during the pyrolysis of bagasse at 800°C. Nominal bagasse feed rate in the feeding period: 36 mg min^{-1}, total gas flow rate: 1.1 l min^{-1} (under ambient conditions). Note the changes in scales.

from the brown coal to the semi-anthracite. This is thought to mean that the formation of NH$_3$ becomes increasingly slow with rank, making the formation of NH$_3$ in the "not-feeding" periods increasingly important (see Part III [37] for more details).

For all the samples studied (Figs. 3–6), while the thermal cracking of the nascent chars is an important source of NH$_3$ and HCN, the formation rates of NH$_3$ and HCN decreased dramatically with time. This means that some nitrogen has converted into more stable forms in the char. Indeed, pyrolysing a lignite and model compounds, Pels and co-workers [1] have found that, under severe pyrolysis conditions of 1173 K for 1 h in a fluidised-bed pyrolyser, all nitrogen exists in char in six-membered (pyridinic or quaternary) rings as part of graphene layers. Further release of nitrogen from the char becomes very difficult.

3.2. Effect of coal rank on the yields of HCN and NH$_3$ during pyrolysis

From the data presented above, there appear to be a rapid and a slow process of HCN and NH$_3$ formation during pyrolysis. A slow process of HCN and NH$_3$ formation follows the rapid formation of HCN and NH$_3$ during the primary pyrolysis and/or during the initial stage of char/tar cracking. The formation of HCN may go to completion much more rapidly than that of NH$_3$. This finding would seem to present at least partial explanation for the discrepancies in the yields of NO$_x$ precursors (particularly NH$_3$) reported in the literature. If the experimental procedures and the reactor configuration allow for holding the nascent char particles at high temperature for long time as in the case of a TGA apparatus, significant amounts of NH$_3$ may be detected. Conversely, if the residence time of the solid particles in the reactor is very short as in the case of an entrained-flow reactor, primary pyrolysis and/or initial thermal cracking of char (and volatiles) would be the main sources of HCN and NH$_3$. Based on this consideration, the effect of coal rank on the formation of NO$_x$ precursors should only be evaluated where coals of varying rank are subjected to the same experimental conditions. Not only the residence time of the volatiles but also the residence time of the nascent char particles would be important aspects of experimental conditions to be considered.

Take the pyrolysis of a set of rank-ordered coals in a fluidised-bed reactor as an example [25,38]. In a fluidised-bed reactor, a large proportion of the char particles from the non-caking (e.g. brown) coals would be elutriated out of the reactor directly. On the other hand, the majority of the char particles from the caking (e.g. bituminous) coals would tend to agglomerate with sand particles and remain in the reactor until the conclusion of the experiment. Due to the difference in the residence time of char particles in the reactor in a fluidised-bed reactor, the NH$_3$ and HCN yields of the non-caking coals may be underestimated compared with those of the caking coals. Fig. 7 compares the yields of HCN and

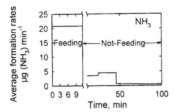

Fig. 5. Average formation rates of NH₃ as a function of time during the pyrolysis of Drayton high-volatile bituminous coal at 793°C. Nominal coal feed rate: 30 mg min^{-1}, total gas flow rate: 1.1 l min^{-1} (under ambient conditions). Note the changes in scales.

Table 4
NH₃ formation observed during the pyrolysis of solid fuels in the "feeding" and "not-feeding" periods as a function of rank. [Temperature: 793–800°C. Nominal coal feed rate: 25–35 mg min^{-1}. Total gas flow rate: 1.1 l min^{-1} (measured under ambient conditions)]

Coal substrate	NH₃ observed in the "feeding" period, % of the total NH₃ in the whole experiment	NH₃ observed in all the "not-feeding" period, % of the total NH₃ in the whole experiment
Bagasse	60	40
Loy Yang	75	25
Drayton	50	50
Pocahontas #3	40	60

NH₃ from the pyrolysis of a typical set of rank-ordered coals in the fluidised-bed reactor [25,38] with those in the reactor used in this study. Due to the difficulties involved in quantifying the nitrogen content of coal [39] as well as the HCN and NH₃ produced, care should be exercised not to compare the absolute yields in the literature [25,38] with those in this study. Instead, comparison should only be focused on the relative effects of coal rank seen with the two reactors. The data from the pyrolysis in the fluidised-bed reactor seem to suggest that the yield of HCN goes through a maximum with increasing rank (carbon content). On the contrary, the data from the present study suggest that the yields of both HCN and NH₃ decrease monotonically with increasing rank. In order to extend the observation in the current reactor, the data from the pyrolysis of bagasse, having even "lower rank" than the brown coal, are also included in the comparison in Fig. 7. The discrepancies between the effects of coal rank observed with the fluidised-bed reactor and with the current reactor are mainly due to the differences in the residence time of the char particles in the two reactors for the non-caking coals and biomass. In the current reactor, the volatiles from the pyrolysis of all the coals and biomass had the same residence time in the heated zone. Moreover, the char particles from the pyrolysis of all the coals and biomass had the same residence time on the heated frit. In the fluidised-bed reactor, the yields of HCN and NH₃ from the pyrolysis of non-caking fuels (bagasse and brown coal) were seriously underestimated.

It is also noted that the pyrolysis of the bagasse sample gave NH₃ yields that are lower than "expected" from its "rank". The exact reasons remain unknown. However, its "low" NH₃ yields are at least partially related to its extremely high (82.4 %, see Table 1) volatile yield. The majority of the bagasse organic matter became volatiles upon heating to 800°C. As will be discussed in detail elsewhere [37], the formation of NH₃ during pyrolysis mainly takes place in the solid phase. There may be little nitrogen left in the char from the pyrolysis of the bagasse available for the formation of NH₃. It should also be pointed out that the possible high contents of amino acids in the bagasse does not necessarily lead to high yields of NH₃. The pyrolysis of amino acid model compounds such as proline and glutamic acid [40] did give high HCN yields via N-heterocycle formation and low NH₃ yields. The amino acids possibly present in the bagasse might have produced HCN via a similar route of N-heterocycle formation during pyrolysis.

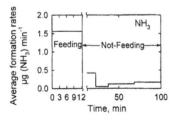

Fig. 6. Average formation rates of NH₃ as a function of time during the pyrolysis of Pocahontas No. 3 low-volatile bituminous coal at 793°C. Nominal coal feed rate: 30 mg min^{-1}, total gas flow rate: 1.1 l min^{-1} (under ambient conditions). Note the changes in scales.

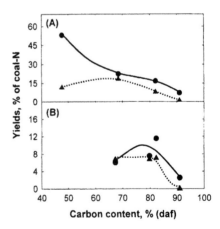

Fig. 7. Effects of "rank" on the yields of HCN and NH₃ from the pyrolysis of coal and biomass (A) in the reactor used in the present study at 800°C with feeding rates of about 25–35 mg min^{-1} (NH₃) or 45–75 mg min^{-1} (HCN) and (B) in a fluidised-bed reactor [25,38]. ● and —, HCN yields; ▲ and - - -, NH₃ yields.

4. Conclusions

The experimental results from this study show that, in addition to the primary pyrolysis process, the thermal cracking of (nascent) char is an important source of NH_3 and HCN during pyrolysis. In particular, the formation of NH_3 from the thermal cracking of char was seen to last for more than an hour even at temperatures as high as $700-900°C$. The formation of HCN went to completion much more rapidly than that of NH_3. The experimental results indicate that the differences in reactor configurations, particularly in terms of char residence time, would account at least partially for some of the discrepancies in the literature about the formation of NO_x precursors during pyrolysis. The reactor configuration used in this study allows for the comparison among coals of varying rank on an unbiased basis. Based on the data in this study, it is concluded that both HCN and NH_3 yields from the pyrolysis of coals decreased with increasing rank.

Acknowledgements

We gratefully acknowledge the financial support of this study by the Australian Research Council (ARC) and the New Energy and Industrial Technology Development Organisation (NEDO) in Japan. We thank Dr P.F. Nelson of CSIRO Division of Energy Technology for providing the Pocahontas No. 3 coal sample. We also thank Dr T.F. Dixon and Mr J. Joyce of the Sugar Research Institute for providing the bagasse sample. L.L. Tan gratefully acknowledges the scholarships (MGS and OPRS) from Monash University.

References

[1] Pels JR, Kapteijn F, Moulijn JA, Zhu Q, Thomas KM. Carbon 1995;33:1641.
[2] Nelson PF, Buckley AN, Kelly MD. 24th Symp (Intl) Combust, The Combustion Institute, Pittsburgh, PA, 1992. p. 1259–67.
[3] Varey JE, Hindmarsh CJ, Thomas KM. Fuel 1996;75:164.
[4] Baumann H, Möller P. Erdol und Kohle–Erdgas–Petrochemie vereinigt mit Brennstoff-Chemie 1991;44:29.
[5] Cai H-Y, Güell AJ, Dugwell DR, Kandiyoti R. Fuel 1993;72:321.
[6] Kambara S, Takarada T, Yamamoto Y, Kato K. Energy and Fuels 1993;7:1013.
[7] Leppälahti J. Fuel 1995;74:1363.
[8] Solomon PR, Hamblen DG, Carangelo RM, Krause JL. 19th Symp (Int) Combust, The Combustion Institute, Pittsburgh, PA, 1982. p. 1139–49.
[9] Bassilakis R, Zhao Y, Solomon PR, Serio MA. Energy and Fuels 1993;7:710.
[10] Leppälahti J, Koljonen T. Fuel Process Technol 1995;43:1.
[11] Nelson PF, Li C-Z, Ledesma E. Energy and Fuels 1996;10:264.
[12] Wang W, Brown SD, Thomas KM, Crelling JC. Fuel 1994;73:341.
[13] Deandres AIG, Thomas KM. Fuel 1994;73:635.
[14] Hindmarsh CJ, Wang WX, Thomas KM, Crelling JC. Fuel 1994;73:1229.
[15] Wang W, Brown SD, Hindmarsh CJ, Thomas KM. Fuel 1994;73:1381.
[16] Thomas KM. Fuel 1997;76:457.
[17] Middleton SP, Patrick JW, Walker A. Fuel 1997;76:1195.
[18] Calkins WH. Energy and Fuels 1987;1:59.
[19] Calkins WH. Fuel 1994;73:475.
[20] Davidson RM. Fuel 1994;73:988.
[21] Mackie JC, Colket III MB, Nelson PF. J Phys Chem 1990;94:4099.
[22] Mackie JC, Colket III MB, Nelson PF, Esler M. Int J Chem Kinetics 1991;23:733.
[23] Hämäläinen JP, Aho MJ, Tummavuori JL. Fuel 1994;73:1894.
[24] Sugiyama S, Arai N, Hasatani M, Kawamura S, Kudou I, Matsuhiro N. Environmental Sci Technol 1978;12:175.
[25] Li C-Z, Nelson PF, Ledesma EB, Mackie JC. 26th Symp (Int) Combust, The Combustion Institute, Pittsburgh, PA, 1996. p. 3205–11.
[26] Li C-Z, Buckley AN, Nelson PF. Proc 14th Annual Int Pittsburgh Coal Conf, 1997, in the form of a CD-ROM.
[27] Ledesma EB, Li C-Z, Nelson PF, Mackie JC. Energy and Fuels 1998;12:536.
[28] Li C-Z, Ledesma EB, Buckley AN, Nelson PF. Proc 7th Australian Coal Sci Conf, Australian Institute of Energy, 1996. p. 171–8.
[29] Li C-Z, Buckley AN, Nelson PF. Fuel 1998;77:157.
[30] Li C-Z, Nelson PF. Fuel 1996;75:525.
[31] Hayashi J-i. Personal communications, 1998.
[32] Mody D. PhD thesis, Monash University, in preparation.
[33] Tyler RJ. Fuel 1979;58:680.
[34] Tan LL, Mathews JF, Li C-Z. Proceedings of the 10th International Conference on Coal Science (Prospects for Coal Science in the 21st Century), vol. II. Taiyuan, China: Shanxi Science and Technology Press, 1999. p. 1509–12.
[35] Li C-Z, Pang Y, Li X-G. Proc 15th Annual Int Pittsburgh Coal Conf, 1998, in the form of a CD-ROM.
[36] Tan LL, Li C-Z. (Part II of this series) Fuel 2000;79:1891.
[37] Li C-Z, Tan LL. (Part III of this series) Fuel 2000;79:1899.
[38] Li C-Z, Kelly MD, Nelson PF. Proceedings of the Australian Symposium on Combustion and the Fourth Australian Flame Days, Gawler, South Australia. 9–10 November 1995. Paper C2-2.
[39] Davidson RM. Nitrogen in coal. IEAPER/08. IEA Coal Research, 1994, London, UK.
[40] Haidar NF, Patterson JM, Moors M, Smith Jr WT. J Agric Food Chem 1981;29:163.

Emissions Reduction: NO$_x$/SO$_x$ Suppression
A. Tomita (Editor)

Formation of NO$_x$ and SO$_x$ precursors during the pyrolysis of coal and biomass. Part II. Effects of experimental conditions on the yields of NO$_x$ and SO$_x$ precursors from the pyrolysis of a Victorian brown coal

Li Lian Tan, Chun-Zhu Li*

Department of Chemical Engineering, P.O. Box 36, Monash University, Vic. 3800, Australia

Abstract

A Victorian brown coal was pyrolysed in a quartz reactor. The reactor has some features of a drop-tube reactor and of a fixed-bed reactor, capable of operating at fast and slow heating rates. The yield of HCN was found to change with gas flow rate and coal feeding rate, indicating that HCN and/or its precursors could interact significantly with the nascent char to be incorporated into char as soot or to form N$_2$. Experimental results indicated that HCN does not significantly convert to NH$_3$, either on the char surface or in the gas phase, at least during the pyrolysis of the brown coal in this study. The yields of HCN and NH$_3$ were both sensitive to changes in heating rate. The reduction in the yields of HCN and NH$_3$ with decreasing heating rate is mainly due to the lack of radicals at the slow heating rate, which are required to initiate the opening of the N-containing rings. The carbonisation/condensation reactions also make the N-containing heteroaromatic ring systems increasingly stable during the extended holding at high temperatures at the slow heating rate. Experimental results appear to suggest that there are two types of organic sulphur-containing structures in the brown coal with very different thermal stability. The first type could be converted into H$_2$S at low temperatures (<600°C). The other type was stable at temperatures up to 1000°C. The changes in heating rate or coal feeding rate did not affect significantly the formation of H$_2$S. © 2000 Elsevier Science Ltd. All rights reserved.

Keywords: HCN; NH$_3$; H$_2$S; NO$_x$ precursors; SO$_x$ precursors; Pyrolysis; Brown coal

1. Introduction

The nitrogen and sulphur in coal can be released during pyrolysis to form NO$_x$ precursors (e.g. HCN and NH$_3$) [1–18] and SO$_x$ precursors (e.g. H$_2$S) [19–21] in a series of overlapping processes. A simplified reaction scheme for the release of nitrogen in coal is shown schematically in Fig. 1 in Part I of this series [22]. Firstly, HCN, NH$_3$ and H$_2$S can be formed during the initial "primary" pyrolysis together with the release of the majority of the volatiles. Secondly, the thermal cracking of volatiles and char provide two important sources of HCN, NH$_3$ and H$_2$S [22]. Clearly, the precursors to HCN, NH$_3$ and H$_2$S can interact strongly during pyrolysis. The complexity of this reaction system and the high reactivity of the N- and S-containing intermediates would necessarily mean that the final yields of HCN, NH$_3$ and H$_2$S would be affected greatly by the experimental conditions. The development of optimum NO$_x$/SO$_x$ abatement strategies for retrofitting the current pf combustors or for developing new gasification/reforming-based power generation technologies would benefit from a better understanding of the reactions leading to the formation/destruction of important NO$_x$ and SO$_x$ precursors, viz HCN, NH$_3$ and H$_2$S.

Considerable studies have been reported on the effects of experimental conditions on the formation of these NO$_x$ and SO$_x$ precursors. It is generally agreed that the yield of HCN increases with increasing temperature [8–15]. However, the trend in the yield of NH$_3$ is still not very clear. While Kambara and co-workers [14] showed that the yield of NH$_3$ increased monotonically with increasing temperature in a pyroprobe, the data from a fluidised-bed reactor often showed maxima in the NH$_3$ yield at around 800°C [10,12]. It should, however, be pointed out that the yields of NH$_3$ are often complicated by a possible artefact that NH$_3$ can be lost to N$_2$ through interaction with such materials as stainless steel and quartz [23] that are commonly used to construct reactors. Comparing the data from an entrained-flow reactor [16] with those from a thermogravimetric analyser (TGA) [17], Bassilakis and co-workers [17] concluded that the yields of HCN and NH$_3$ can be greatly affected by heating rate. Leppälahti [18] found that the increasing heating rate resulted in an increase in secondary reactions such as tar

* Corresponding author. Tel.: +61-3-9905-9623; fax: +61-3-9905-5686.
E-mail address: chun-zhu.li@eng.monash.edu.au (C.-Z. Li).

Reprinted from *Fuel* **79 (15)**, 1891-1897 (2000)

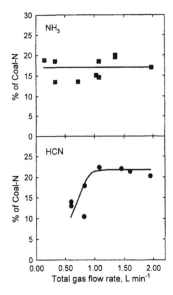

Fig. 1. Yields of NH₃ and HCN as a function of gas flow rate from the pyrolysis of Loy Yang brown coal at 800°C at the fast heating rate. Coal feeding rates: 90–120 mg min⁻¹.

cracking and produced more HCN and NH₃ in the case of high volatile peat. Many researchers have considered the possibilities of inter-conversion/interaction between various NO_x precursors (e.g. see Refs. [8,12,17]). However, there has been no direct evidence about the nature of the inter-conversion and/or interaction.

The purpose of this study is to investigate the effects of various experimental conditions on the yields of NO_x and SO_x precursors during pyrolysis in a novel reactor. The experimental results provide some better insights into the inter-conversion/interaction between S- and N-containing intermediates leading to the formation and destruction of NO_x and SO_x precursors during pyrolysis.

2. Experimental

2.1. Coal

A brown coal (Loy Yang, −106 + 150 μm; C, 68.5; H, 4.8; N, 0.6; S, 0.3 wt% on daf basis) from the Latrobe Valley in Victoria, Australia, was used in this study. The coal contains very little mineral matter and gives only a 1.0 wt% ash yield on dry basis. The brown coal melts to a very limited degree during pyrolysis both at slow (1 K s⁻¹) and high (2000 K s⁻¹) heating rates [24]. The particles remain porous throughout the pyrolysis. The use of this coal thus allows for a better chance to study the interaction of the N- and S-containing species with the inner surface of

the char particles during pyrolysis without the complications caused by the formation of a metaplast.

2.2. Pyrolysis

A detailed description of the quartz reactor system used has been given previously [22]. Two types of pyrolysis experiments were carried out in this study. In addition to experiments at the high heating rate described previously [22], the brown coal was also pyrolysed at a slow heating rate in this study. In the latter case, the coal particles were firstly fed into the reactor at room temperature. The coal particles stayed on the quartz frit located in the centre of the reactor [22]. The reactor was then heated up with the external electric furnace at 6.7 K min⁻¹ from room temperature to the required final temperature with pre-set holding time. Throughout a slow heating rate experiment, a thermocouple was used to measure the temperature of the frit on which coal/char particles stayed. The interactions between the volatiles and the thermocouple were minimal as the volatiles formed only within the thin coal/char bed and only a few particles had contact with the thermocouple.

2.3. Quantification of NH₃, HCN, H₂S and H₂

The quantification of NH₃ and HCN has been described elsewhere [22]. H₂S was also quantified in the same experiment as HCN. The product gas stream exiting the bubblers was collected in gasbags and analysed for H₂ content using a HP 5890 gas chromatograph equipped with a thermal conductivity detector. For experiments with long holding time, product gas could only be sampled into gasbags at certain time intervals for the quantification of H₂. The total H₂ yield from an experiment was then *estimated* by extrapolating the data and then carrying out integration over the whole period of experimental time.

3. Results and discussion

3.1. Effects of gas flow rate and coal feeding rate on the yields of NH₃, HCN, H₂S and H₂ at the fast heating rate

Fig. 1 shows the variation in the yields of NH₃ and HCN with the total gas flow rate from the fast heating rate experiments at 800°C. Each datum point represents an integral yield from both the "feeding" and "not-feeding" periods [22]. Coal feeding rates between about 90 and 120 mg min⁻¹ were used. The yield of NH₃, with a relatively big scatter, seemed to remain relatively constant over the range of flow rate studied. However, the yield of HCN increased with increasing gas flow rate. This is surprising, as the increases in the total gas flow rate had been expected, if any thing, to decrease the HCN yield as a result of reduced extents of secondary cracking of volatiles due to the reduced residence time. Therefore, some other

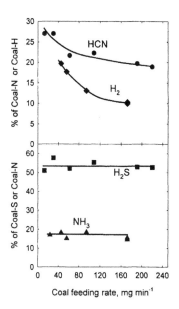

Fig. 2. Yields of NH_3, HCN, H_2S and H_2 as a function of coal feeding rate from the pyrolysis of Loy Yang coal at 800°C at the fast heating rate. Gas flow rate: 1.1 L min^{-1}.

mechanisms must have operated, leading to the increases in the HCN yield with increasing gas flow rate.

The set up of the reactor system in the present study does not allow for the direct determination of N_2. However, it is

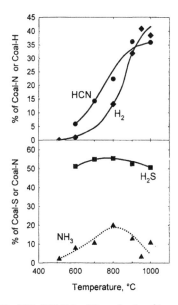

Fig. 3. Yields of NH_3, HCN, H_2S and H_2 as a function of temperature at the fast heating rate. Total gas flow rate: 1.1 L min^{-1}; coal feeding rates: 90–135 mg min^{-1}.

believed that considerable amount of N_2 must have formed during the pyrolysis based on the information in the literature [14,25,26]. Hayashi and co-workers [25] showed that the formation of N_2 from the pyrolysis of Yallourn brown coal (from the same valley as the Loy Yang brown coal used in the present study) started at temperatures higher than 600°C. With the current experimental rig set up [22], the pyrolysing char particles came into contact with the volatiles both in the suspension before the particles met the frit and on the frit when the volatiles passed through the thin coal/char bed on the frit. Decreasing the total gas flow rate in the present study would result in increasing contact time between HCN (and/or its precursors) and char. During the extended contact between the volatiles and char both in the suspension and on the frit, HCN (and/or its precursors) might have been incorporated into the char through soot formation or converted into N_2 (or other species) at char surface. The formation of N_2 must have been a result of preferential reaction between N-containing sites on the char surface and N-containing volatile species such as HCN or its precursors. Considering the low content of nitrogen in coal, without a catalyst [25,26] or an agent to mobilise the nitrogen atoms, it is very unlikely for two nitrogen atoms in the solid coal/char to combine and form a N_2 molecule. It should be noted within this context that the decreases in the yield of HCN did not seem to be reflected by any increases in the yield of NH_3.

Fig. 2 shows the effects of coal feeding rate on the yields of HCN, NH_3, H_2S and H_2 from the pyrolysis of the brown coal at 800°C at the fast heating rate. A total gas flow rate of 1.1 L min^{-1} was used for the experiments shown in Fig. 2. The yield of HCN was seen to decrease with increasing coal feed rate. This observation may again be explained by considering the relative contact between the pyrolysing coal/char particles and the volatiles. As was stated above, the pyrolysing coal/char particles were in contact with the volatiles in the suspension before the volatiles and the particles met the frit in the centre of the reactor. Increasing coal feeding rate at constant gas flow rates would thus increase the concentrations of volatiles and coal/char particles in the suspension, leading to the increased contact between the volatiles and the particles. Furthermore, the devolatilised char particles are expected to distribute evenly on the frit in the centre of the reactor. The char would rapidly age while it was being held on the frit at high temperatures. Increasing coal feeding rate would thus also mean that HCN and other volatile species and char were generated increasingly rapidly. This made it more likely for the HCN and its precursors, in passing through the thin coal/char particle bed and the quartz frit, to come into contact with increasingly fresh (reactive) char before the char aged. Therefore, in agreement with the effect of flow rate, the decreases in the yield of HCN were a result of the interaction of HCN (or its precursors) with char both in the suspension and on top of the frit in the centre of the reactor. Again, it should be noted that the decreases in the

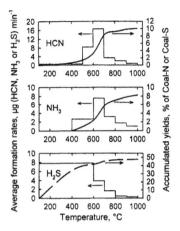

Fig. 4. Formation of NH_3, HCN and H_2S as a function of temperature at slow heating rate. Total gas flow rate: 1.1 L min^{-1}.

yield of HCN are not reflected by any increases in the yield of NH_3.

The conversion of coal-H to H_2 follows a similar trend to that of HCN (Fig. 2). H_2 is produced from the thermal cracking (including condensation) of (hetero)aromatic ring systems as well as aliphatic chains in char or volatiles. Similar measurement to those on HCN and NH_3 described previously [22] showed that char thermal cracking is also a

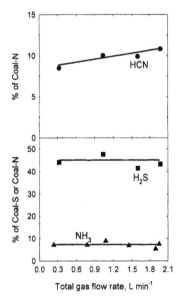

Fig. 5. Effects of the total flow rate of the gas passing through the coal/char bed on the yields of HCN, NH_3, H_2S and H_2 from the pyrolysis of the Loy Yang brown coal by heating the coal at the slow heating rate of 6.7 K min^{-1} from room temperature to 1000°C.

significant source of H_2. It is thought that the decreases in the yield of H_2, similar to those of HCN, at high feeding rates were also due to the incorporation of its precursors into char, possibly through soot formation.

The data in Figs. 1 and 2 showed that the changes in the yields of H_2 and HCN were not reflected by any discernible changes in the yield of NH_3. Taken together, it may be fair to conclude that, at least in the case of this brown coal, the data in Figs. 1 and 2 do not support the assumptions that HCN is converted into NH_3 either on the inner char surface [17] or in the gas phase [8].

Experiments similar to those on NH_3 and HCN [22] also showed the thermal cracking of char to be an important source of H_2S. However, the data in Fig. 2 seem to suggest that, like NH_3, H_2S and/or its precursors did not react with the fresh char to any significant degree.

3.2. Effects of temperature and heating rate on the yields of NH_3, HCN, H_2S and H_2

Fig. 3 shows the yields of NH_3, HCN, H_2S and H_2 as a function of temperature at fast heating rate. A total gas flow rate of 1.1 L min^{-1} was used in all these experiments. The yield of HCN increased with increasing temperature, in agreement with the literature [8–15]. The yield of NH_3 increased with increasing temperature up to 800°C and then decreased with further increase in temperature. Whilst the exact reasons for the apparent decreases in the NH_3 yield at high temperatures still remain unclear, it is very likely that the decreases were caused by the interaction of NH_3 with the quartz frit [23] in the centre of the reactor.

Significant amounts of H_2 could only form at temperatures much higher than 600°C, in agreement with the fact that the cleavage of the bonds (mainly C–H) to form H_2 requires high temperatures. About half of the sulphur in the brown coal were converted into H_2S even at temperatures as low as 600°C.

Fig. 4 shows the formation of NH_3, HCN and H_2S as a function of temperature during experiments when the brown coal was heated at 6.7 K min^{-1} from room temperature to 1000°C. Similar trends were also observed with different total gas flow rates ranging from 0.24 to about 2.0 L min^{-1} (also see later). During such an experiment, the products were also collected into "reaction-time-temperature-resolved" fractions in separate absorption bottles. The accumulated yields of HCN, NH_3 and H_2S are shown in Fig. 4 by the spline curves fitting the experimental data. In agreement with the data shown in Fig. 3 for experiments carried out at the fast heating rate, only negligible amounts (but not zero!) of HCN and NH_3 were formed at temperatures lower than 500°C. However, more than 30% of the sulphur in the brown coal was converted into H_2S when the coal was heated to 500°C, again in agreement with the observation at the fast heating rate.

In sharp contrast to the case of the fast heating rate experiments, increasing the total flow rate of the gas passing

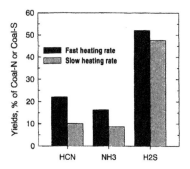

Fig. 6. Yields of HCN, NH₃ and H₂S as a function of heating rate. Peak temperature: 800°C; total gas flow rate: 1.1 L min⁻¹; nominal coal feeding rate during the fast heating rate experiments: 60 mg min⁻¹.

through the coal/char bed had very little effects on the yields of HCN, NH₃ and H₂S at the slow heating rate, as is shown in Fig. 5. Furthermore, comparison of the data in Figs. 1 and 5 indicates that the yields of NH₃ and HCN decreased with decreasing heating rate. It was initially thought that the decreases in the HCN and NH₃ yields with decreasing heating rate were due to the reduced holding time at high temperatures in the slow heating rate experiments. Experiments were then carried out by heating the coal particles at 6.7 K min⁻¹ to the required temperature (e.g. 800°C) and then held at the temperature for a long period until the release of NH₃ and HCN was negligible. The yields of HCN and NH₃ from such experiments were still much lower than those obtained at the fast heating rate, as is shown in Fig. 6 for the final temperature of 800°C. It is perhaps worth to note that all other variables (reactor configuration, total gas flow rate, holding time at the final temperature, etc.) were kept unchanged except from the heating rate itself in making the comparison in Fig. 6. The yield of H₂S (and H₂, not shown) did not show significant sensitivity to changes in heating rate.

With the current experimental set up, the extents of secondary tar cracking reactions decrease with decreasing heating rate. During experiments at the slow heating rate, the majority of the volatiles would have been released from the coal/char particles before the temperature reached 600°C [24]. The volatiles were soon carried out of the reactor and thus were hardly thermally cracked due to the short residence time at low temperatures. Therefore, the contributions by the thermal cracking of volatiles (tar) to the observed HCN yield were minimal at the slow heating rate. However, the reductions of the yields of HCN and NH₃ can not be entirely attributed to the reduced extents of volatile cracking reactions. In fact, closer examination of the data in Figs. 3 and 4 revealed a noteworthy contrast between the results from the fast and slow heating rate experiments. While the HCN yield increased sharply from 800 to 900°C in the case of the fast heating rate experiments (Fig. 3), the HCN yield increased only very slightly when the char was further heated up from 800 to 1000°C in the case of the slow heating

rate experiments (Fig. 4). This observation indicates that the char produced by heating the coal at the slow heating rate to high temperatures, unlike the corresponding char produced at the fast heating rate, was not reactive in terms of the release of HCN. Similar findings were also reported by Hayashi and co-workers [25] who pyrolysed a similar brown coal (Yallourn) in a fixed-bed reactor at a heating rate of 10 K min⁻¹. The evolution of HCN at temperatures higher than 700 or 800°C was negligible.

Recent studies [12,13,15,27,28] have suggested that the thermal cracking of volatiles gave mainly HCN. Only negligible amounts of NH₃ could be produced from the thermal cracking of volatiles [12,13,15,27,28]. Therefore, the reduction in the NH₃ yield with decreasing heating rate may not be at all due to the reduced tar cracking reactions.

Ring opening initiated by free radicals is an important step in the formation of HCN and NH₃. The reduced NH₃ and HCN yields with decreasing heating rate are believed to be mainly due to the mismatch between the production of radicals and the opening of heteroaromatic ring systems. In particular, the formation of NH₃ requires the in situ generation of free radicals that can migrate to the N-containing site to result in the full hydrogenation of the N-site to NH₃ [29–31]. At slow heating rate, the majority of volatiles are released at temperatures lower than 600°C [24]. With the further increase in temperature, two competing reactions take place. On the one hand, char thermal cracking reactions continuously generate free radicals, resulting in the rupture of N-containing heteroaromatic ring systems for the formation of HCN and NH₃. On the other hand, the heteroaromatic ring systems (with and/or through the formation of free radicals) recombine/condense into larger and more stable ring systems. This is a carbonisation process. During the prolonged holding at elevated temperatures in a slow heating rate experiment, the carbonisation process would have converted some N-containing heteroaromatic ring system into "inert" ones by the time the temperature was raised to about 800 or 900°C. Indeed, XPS and XANES studies [1,2,32–34] of chars from the pyrolysis/carbonisation of coal and model compounds did indicate that, under severe pyrolysis conditions, various nitrogen functionalities originally in coal or model compounds tended to transform to more stable forms. After severe pyrolysis, all nitrogen is eventually present in chars in six-membered rings located at the edges of the graphene layers as pyridinic-N or in the interior as quaternary-N [1]. In the meantime, when the coal/char was heated at the slow heating rate of 6.7 K min⁻¹, the generation of free radicals would have also been exhausted by the time the temperature reached 800 or 900°C. The stabilised heteroaromatic ring systems and the lack of reactive radicals at slow heating rate combine to make the formation of HCN and NH₃ much more difficult at slow heating rate than at fast heating rate. The data in Fig. 4 do suggest that the release of NH₃ and HCN at temperatures higher than 800°C at slow heating rate was negligible. Hayashi and co-workers [25] also showed the release of N_2 to be very small at temperatures higher than 900°C from the pyrolysis of a similar Victorian brown coal at 10 K min^{-1}.

It should also be pointed out that extra free radicals might have also been generated from the thermal cracking of volatiles while the volatiles passed through the (fresh) char bed during an experiment at the fast heating rate. The similar process of free radical generation would be almost absent during an experiment at the slow heating rate. Clearly, these extra free radicals generated during an experiment at the fast heating rate, as a "self-gasification" process of char by the volatiles, would also help to produce HCN and NH_3.

The formation of H_2S required much lower temperatures than HCN, NH_3 or H_2. Even at 600°C at the fast heating rate (Fig. 3), more than 50% of sulphur in the brown coal has been converted into H_2S. The data in Fig. 4 further indicate that more than 30% of the sulphur in coal could be converted into H_2S at temperatures up to 500°C at the slow heating rate. Heating the brown coal to temperatures higher than 700°C, both at the fast (Fig. 3) and at the slow (Fig. 4) heating rates, did not seem to produce significantly more H_2S. In sharp contrast to the yields of NH_3 and HCN, the yield of H_2S was little affected by increasing heating rate or increasing gas flow rate. Analysis of the Loy Yang coal sample used in this study showed that the coal contained very little pyritic sulphur (<0.02 wt%). Sulphur mainly exists in the coal as organic sulphur (~0.28 wt% on dry basis). Whilst the detailed information about the sulphur functionalities in this brown coal is not known, it is likely from the literature [19–21,35–37] that the organic sulphur is present in such forms as aliphatic mercaptans, sulphides and disulphides as well as aromatic sulphides and thiophenes. Taken together, these data appear to suggest that there are two types of organic sulphur in the brown coal. One type of sulphur-containing structures seems to be very thermally unstable. This type of sulphur may be of aliphatic structures or in small S-containing heteroaromatic ring systems. This type of sulphur can be released almost completely at temperatures of about 600°C (Figs. 3–5). This is in agreement with the literature about the presence of thermally labile S-containing structures in coal [20,35]. The other type of sulphur-containing structures (possibly thiophenes) appears to be very thermally stable: the data in Fig. 4 indicate that heating from 700 to 1000°C only resulted about another 3% of the sulphur in the coal as H_2S.

4. Conclusions

The observations in this study on the pyrolysis of a Victorian brown coal can be summarised as follows:

1. The yield of HCN changes with the total gas flow rate and the coal feeding rate, indicating that HCN and/or its precursors can interact with the nascent char significantly to form soot or N_2.

2. The data from the present study indicate that HCN is not hydrogenated inside the pores in the char or in the gas phase to any significant degree to be an important source of NH_3 during pyrolysis.

3. The yields of HCN and NH_3 are both sensitive to changes in heating rate. The reduced HCN and NH_3 yields at slow heating rate are partly due to the lack of enough free radicals required to initiate the opening of the N-containing rings. The carbonisation/condensation reactions also make the N-containing heteroaromatic ring systems increasingly stable during extended holding at high temperatures in a slow heating rate experiment.

4. The yield of H_2S is mainly affected by temperature and is little affected by heating rate, total gas flow rate or coal feeding rate.

5. There appear to be two types of organic sulphur-containing structures in the brown coal in terms of the thermal stability during pyrolysis. The first type is not thermally stable and can be converted into H_2S at temperatures around 600°C. The other type is very thermally stable up to 1000°C.

Acknowledgements

We gratefully acknowledge the financial support of this study by the Australian Research Council (ARC) and the New Energy and Industrial Technology Development Organisation (NEDO) in Japan. L.L.T. gratefully acknowledges the scholarships (MGS and OPRS) from Monash University.

References

[1] Pels JR, Kapteijn F, Moulijn JA, Zhu Q, Thomas KM. Carbon 1995;33:1641.
[2] Thomas KM. Fuel 1997;76:457.
[3] Varey JE, Hindmarsh CJ, Thomas KM. Fuel 1996;75:164.
[4] Davidson RM. Nitrogen in coal. IEAPER/08, IEA Coal Research, London, UK, 1994.
[5] Cai H-Y, Güell AJ, Dugwell DR, Kandiyoti R. Fuel 1993;72:321.
[6] Leppälahti J, Koljonen T. Fuel Processing Technology 1995;43:1.
[7] Nelson PF, Li C-Z, Ledesma E. Energy and Fuels 1996;10:264.
[8] Baumann H, Möller P, Erdol und Kohle—Erdgas—Petrochemie vereinigt mit Brennstoff-Chemie 1991;44:29.
[9] Nelson PF, Kelly MD, Wornat MJ. Fuel 1991;70:403.
[10] Nelson PF, Buckley AN, Kelly MD. 24th Symp. (Int) Combust., The Combustion Institute, Pittsburgh, PA, 1992, p. 1259–67.
[11] Chen JC, Niksa S. Energy and Fuel 1992;6:254 (see also p. 265).
[12] Li C-Z, Nelson PF, Ledesma EB, Mackie JC. 26th Symp. (Int) Combust. The Combustion Institute, Pittsburgh, PA, 1996. p. 3205–11.
[13] Li C-Z, Buckley AN, Nelson PF. Fuel 1998;77:157.
[14] Kambara S, Takarada T, Yamamoto Y, Kato K. Energy and Fuels 1993;7:1013.
[15] Ledesma EB, Li C-Z, Nelson PF, Mackie JC. Energy and Fuels 1998;12:536.
[16] Solomon PR, Hamblen DG, Carangelo RM, Krause JL. 19th Symp.

(Intl) Combust. The Combustion Institute, Pittsburgh, PA, 1982. p. 1139–49.

[17] Bassilakis R, Zhao Y, Solomon PR, Serio MA. Energy and Fuels 1993;7:710.

[18] Leppälahti J. Fuel 1995;74:1363.

[19] Calkins WH. Energy and Fuels 1987;1:59.

[20] Calkins WH. Fuel 1994;73:475.

[21] Davidson RM. Fuel 1994;73:988.

[22] Tan LL, Li C-Z. (Part I of this series) Fuel 2000;79:1883.

[23] Li C-Z, Nelson PF. Fuel 1996;75:525.

[24] Sathe C, Pang Y, Li C-Z. Energy and Fuels 1999;13:748.

[25] Hayashi J-i, Kusakabe K, Morooka S, Nielsen M, Furimsky E. Energy and Fuels 1995;9:1028.

[26] Wu Z, Ohtsuka Y. Energy and Fuels 1997;11:902.

[27] Li C-Z, Buckley AN, Nelson PF. Proceedings of the 14th Annual Pittsburgh Coal Conference, 1997, in the form of a CD-ROM.

[28] Li C-Z, Ledesma EB, Buckley AN, Nelson PF. Proceedings of the Seventh Australian Coal Science Conference. Australian Institute of Energy, 1996. p. 171–8.

[29] Li C-Z, Pang Y, Li X-G. Proceedings of the 15th Annual Pittsburgh Coal Conference, 1998, in the form of a CD-ROM.

[30] Tan L, Pang Y, Li X-G, Mathews JF, Li C-Z. Proceedings of the Eighth Australian Coal Science Conference, Australian Institute of Energy, 1998. p. 321–6.

[31] Li C-Z, Tan LL. (Part III of this series) Fuel 2000;79:1899.

[32] Zhu Q, Money SL, Russell AE, Thomas KM. Langmuir 1997;13:2149.

[33] Stanczyk K, Dziembaj R, Piwowarska Z, Witkowski S. Carbon 1995;33:1383.

[34] Kapteijn F, Moulijn JA, Matzner S, Boehm H-P. Carbon 1999;37:1143.

[35] Mitchell SC, Snape CE, García R, Ismail K, Bartle KD. Fuel 1994;73:1159.

[36] Ibarra JV, Bonet AJ, Moliner R. Fuel 1994;73:933.

[37] Maes II, Mitchell SC, Yperman J, Franco DV, Marinov SP, Mullens J, Poucke LCV. Fuel 1996;75:1286.

Formation of NO$_x$ and SO$_x$ precursors during the pyrolysis of coal and biomass. Part III. Further discussion on the formation of HCN and NH$_3$ during pyrolysis [☆]

Chun-Zhu Li*, Li Lian Tan

Department of Chemical Engineering, P.O. Box 36, Monash University, Vic. 3800, Australia

Abstract

The formation of HCN and NH$_3$ from the pyrolysis of coal (and biomass) is discussed based on our experimental data as well as the data in the literature, including the pyrolysis of N-containing pyrrolic and pyridinic model compounds reported in the literature. The pyrolysis of the model compounds and the thermal cracking of coal pyrolysis volatiles appear to be in good qualitative agreement in terms of the onset decomposition temperature, the main intermediates and the final N-containing product (HCN). The formation of NH$_3$ requires the presence of condensed phase(s) of carbonaceous materials rich in hydrogen. Direct hydrogenation of the N-sites by the H radicals generated in situ in the pyrolysing solid is the main source of NH$_3$ from the solid. The initiation of the N-containing heteroaromatic ring by radical(s) is the first step for the formation of both HCN and NH$_3$. While the thermally less stable N-containing structures are mainly responsible for the formation of HCN, the thermally more stable N-containing structures may be hydrogenated slowly by the H radicals to NH$_3$. The formation of NH$_3$ and the formation of HCN are controlled by the local availability of radicals, particularly the H radicals, in the pyrolysing solid. The increased yield of NH$_3$ (and HCN) with increasing heating rate can be explained by the rapid generation of the H radicals at high heating rates, favouring the formation of NH$_3$ (and HCN) over the combination of N-containing ring systems within the coal/char matrix. The size of the N-containing heteroaromatic ring systems and the types of substitutional groups also play important roles in the formation of HCN and NH$_3$. © 2000 Elsevier Science Ltd. All rights reserved.

Keywords: HCN; NH$_3$; NO$_x$ precursors; Coal; Biomass; Pyrolysis

1. Introduction

HCN and NH$_3$ are two important gaseous NO$_x$ precursors formed during the pyrolysis of coal and biomass. While the studies on the pyrolysis of simple N-containing compounds in the gas phase provide some basis for the understanding of the formation of HCN in the gas phase, the formation of HCN and NH$_3$ in a pyrolysing solid particle is still poorly understood. There is now little doubt that the majority of the nitrogen in coal exists in N-containing heteroaromatic ring systems [1–8]. Thus the conversion of the nitrogen in coal to NH$_3$ requires hydrogen. There are various hypotheses in literature [3,9,10] about the mechanisms of the formation of NH$_3$ during pyrolysis. However, these hypotheses do not seem to agree well with each other and do not seem to

explain the existing experimental data well [11–13]. The factors governing the formation of HCN instead of NH$_3$, or vice versa, from the pyrolysis of coals, are still unclear. The roles that the free radicals play in the release of the nitrogen in coal during pyrolysis remain imperfectly understood [8,11,12,14]. The purpose of this paper is to further discuss the formation of HCN and NH$_3$ during pyrolysis. We will base our discussion not only on our own experimental data but also on the data in the literature, including the data on the pyrolysis of N-containing model compounds. It should be pointed out that our mention of the literature data in this paper is not in any way meant to be comprehensive.

2. Pyrolysis of N-containing model compounds and the thermal cracking of volatiles from the pyrolysis of coal

Extensive studies [15–24] have been carried out on the pyrolysis of simple N-containing model compounds in order to better understand the reactions taking place during the pyrolysis of such complicated fuels as coal. However, somewhat "contradicting" results are often reported in the

---

[☆] Presented in part in the Open Forum "New Approaches to Modelling in Coal Science" at the 10th International Conference on Coal Science, 12–17 September 1999, Taiyuan, China.

* Corresponding author. Tel.: + 61-3-9905-9623; fax: + 61-3-9905-5686.

E-mail address: chun-zhu.li@eng.monash.edu.au (C.-Z. Li).

Table 1
Pyrolysis of N-containing pyrrolic and pyridinic model compounds (some examples reported in the literature)

Compounds	Authors	Experimental conditions	Comments
Pyrrole Pyridine Quinoline Benzonitrile	Axworthy et al. [17]	About 3% of model compounds injected into a heated quartz capillary.	HCN as a major product with some NH_3 formation. "Carbonaceous residue" formed.
Pyridine	Sugiyama et al. [18]	About 10% (?) in He fed into a tubular flow reactor.	NH_3 as the major product. Formation of soot (about half of the carbon in the feed could not be detected in the gas after reaction).
Pyrrole Pyridine 2-picoline Butenentriles	Mackie et al. [15,16,19–21]	0.06–0.5% of model compounds in inert gas pyrolysed in shock-tube reactors at pressures of 7–20 atm.	HCN and various nitriles as the main N-containing products with little NH_3, if any, detected/reported. Little or no soot formed.
Pyridine 2-picoline	Ikeda and Mackie [22]	Model compounds (2–3.5%) in Ar pyrolysed in a completely stirred fused-silica reactor.	Formation of high molecular mass materials speeds up the decomposition.
Pyrrole	Lifshitz et al. [23]	1% of pyrrole in Ar pyrolysed in a shock-tube reactor	HCN and nitriles as main products.
N-6 (e.g. acridine) N-5 (e.g. carbazole) Others with oxygenated groups	Hämäläinen et al. [24]	Powders (<63 μm) fed into an entrained flow reactor.	Both HCN and NH_3 as the major products.

literature. For example, the pyrolysis of pyrrole and pyridine in shock-tube reactors [15,16] showed that HCN was the main N-containing final product and that NH_3 was absent in the reaction products. On the "contrary", the pyrolysis of pyridine in a quartz capillary reactor [17] and in a tubular reactor [18] showed the presence of NH_3 [17] or even NH_3 as the main N-containing gaseous reaction product [18]. A more detailed comparison is given in Table 1.

Closer examination of the literature such as those shown in Table 1 showed two noticeable points about the pyrolysis of both pyrrolic and pyridinic N-containing model compounds. Firstly, the formation of NH_3 during pyrolysis appears to be related to the formation of "high molecular mass materials", "tary materials" or soot (see the comments in Table 1). The formation of these carbonaceous materials (soot or soot precursors) seems, in addition to the chemical structure of the parent model compound [22], to be more or less related to the use of high concentrations of the model compounds as well as the use of long residence time (see Table 1). For example, when pyrrole, pyridine, 2-picoline and butenenitriles in low concentrations were pyrolysed in shock-tube reactors [15,16,19–21], neither soot/soot precursors were observed, nor NH_3 was reported to be an important N-containing product. However, when pyridine and benzonitrile in relatively high concentrations were pyrolysed in a heated quartz capillary [17], some NH_3 was observed in the product stream together with the formation of carbonaceous residue. This indicates that the presence of condensed phase(s) of carbonaceous materials rich in hydrogen (i.e. soot or soot precursors) is a necessary condition for the formation of NH_3 from the pyrolysis of

these N-containing model compounds. Secondly, once the carbonaceous materials (soot) are present that catalyse the further formation of more carbonaceous materials, NH_3 can be formed at very low temperatures. For example, a significant amount of NH_3 could be formed from the pyrolysis of pyridine at 680°C [18] when there was a lot of "soot" formed (only about 60% of the carbon in the feed could be accounted for in the gas phase [18]).

The gas-phase thermal cracking of the volatiles from the pyrolysis of coal seems to agree at least qualitatively with the pyrolysis of the N-containing model compounds in the gas phase. The first aspect of this agreement is about the onset temperature for the N-containing heteroaromatic rings to decompose. Fig. 1 shows the decomposition of pyrrole and pyridine at 0.5 s residence time predicted by extrapolating the kinetic data obtained for the model compounds at much shorter residence time [15,22]. According to the data in Fig. 1, under conditions similar to those in a fluidised-bed reactor [11] (0.5 s residence time), the onset temperature for the decomposition of the model compounds would be around 800°C. This was indeed seen to be the onset temperature for the N-containing heteroaromatic ring systems in the coal pyrolysis volatiles to decompose. The thermal cracking of volatiles from the pyrolysis of a bituminous coal in the fluidised-bed reactor was studied with X-ray photoelectron spectroscopy (XPS) of the tar samples produced at different temperatures [8]. The XPS results [8] showed that the nitrile groups, not present in the raw coal nor in the tars produced at temperatures lower than 700°C, began to appear in the tars when the temperature was raised to about 700–800°C. The

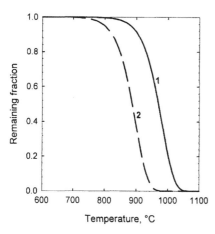

Fig. 1. Thermal decomposition of pyrrole and pyridine at residence time of 0.5 s predicted by extrapolating the kinetic data obtained at much shorter residence time [15,22]. Curve 1, pyridine, kinetic parameters, from Ref. [22]; curve 2, pyrrole, kinetic parameters from Ref. [15].

appearance of the nitrile groups is an indication of the rupture of the N-containing heteroaromatic rings.

The second aspect of the qualitative agreement between the thermal decomposition of model compounds and that of coal pyrolysis volatiles is the observation of nitriles as the intermediates. As was pointed out above, the pyrolysis of model compounds [15–17,19–23], particularly without the formation of soot or soot precursors, all showed the nitriles to be the intermediates in the conversion to HCN. The XPS studies of the tar samples [8] also showed the concentration of nitriles to increase with temperature, confirming the findings on the pyrolysis of model compounds [15–17,19–23]. No convincing evidence on the presence of amines in the tar samples studied was found [8]. Nitriles are the most important intermediates during the conversion of N-containing heteroaromatic ring systems (both in the model compounds and in the coal pyrolysis volatiles) into HCN in the gas phase.

The third aspect of the qualitative agreement between the thermal decomposition of model compounds and that of coal pyrolysis volatiles is about the main final N-containing products. The thermal cracking of the volatiles from the pyrolysis of coal was studied in a fluidised-bed reactor in tandem with a plug-flow reactor [11,25]. The concentration of N-containing species in the gas phase was quite low, at the order of the magnitude of 10^{-5} g-N l^{-1}. The results [11,25] showed that the main final product containing nitrogen in the gas phase was HCN. The formation of HCN from the cracking of the volatiles started at around 700–800°C [11,25], in agreement with the prediction (Fig. 1) from the data on the pyrolysis of model compounds. Only a very small amount of NH_3, probably concurrently with the formation of soot [25], was produced from the thermal cracking of volatiles.

3. Formation of HCN and NH_3 from the pyrolysing solid fuel particles

3.1. General discussion

From the discussion above, it is clear that the formation of a significant amount of NH_3 from the thermal cracking of N-containing pyrrolic or pyridinic heteroaromatic ring systems in the gas phase is unlikely. However, the pyrolysis of coals and biomass does show the formation of considerable amounts of NH_3. The formation of HCN and NH_3 in the pyrolysing solid particles must have gone through very different reaction mechanisms from those during the pyrolysis of the model compounds in the gas phase.

It may be argued that part of the NH_3 observed from the pyrolysis of biomass (such as bagasse [13]) is due to the presence of some amino groups in the biomass. In fact, this does not seem to be the main source of NH_3 even from the pyrolysis of biomass. As was discussed in Part I of this series [13], the pyrolysis of amino acids in the bagasse would likely yield HCN through the formation of N-containing heteroaromatic ring systems [26]. Furthermore, NH_3 from the pyrolysis of coals cannot be mainly due to the presence of amino groups, because the amino groups, if any, exist in coals in negligible concentrations [1,3–8]. Therefore, the direct and/or indirect hydrogenation of the nitrogen in the heteroaromatic structures is the main source of NH_3 from the pyrolysing solid particles. Our experimental results presented in the previous part [27] showed that the indirect routes, either through the hydrogenation of HCN on the char surface [9] or through the hydrogenation of HCN in the gas phase [10], cannot account for the observed NH_3 yields from the pyrolysis of a brown coal, which is rich in hydrogen. For higher rank coals whose chars are depleted with hydrogen, it becomes even less likely for the hydrogenation of HCN on the char surface to be an important route of NH_3 formation. We believe that the direct hydrogenation of the nitrogen in the pyrolysing coal/char particles is the main route for the formation of NH_3.

We believe that the active hydrogen required for the hydrogenation of the nitrogen in heteroaromatic ring systems is generated from the thermal cracking reactions taking place inside the pyrolysing solid coal/char particles. There are several major ways for the generation of active hydrogen in the char. For example, the bulky substitutional groups (particularly abundant in brown coals) may be broken off from the (hydro)aromatic ring systems, possibly through the formation of free radicals. Dehydrogenation of the aliphatic material or dehydrogenation of hydro-aromatic ring systems would also seem to generate free radicals, i.e. the H radicals. The condensation reactions between the aromatic ring systems would also provide a source of active hydrogen. Free radicals, particularly the H radicals, thus generated in situ and possibly *existing in the forms of adspecies on the char surface*, would then

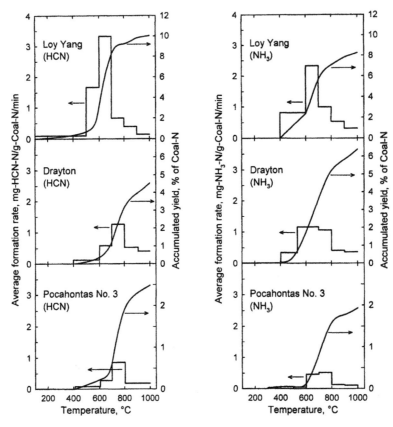

Fig. 2. Average formation rates of HCN and NH₃ from the pyrolysis of the Loy Yang brown coal (C = 68.5 wt% daf), the Drayton bituminous coal (C = 82.1 wt% daf) and the Pocahontas No. 3 low-volatile bituminous coal (C = 91.0 wt% daf). The coal samples were heated at 6.67 K min⁻¹ from room temperature to 1000°C in argon.

attack the N-containing ring systems, resulting in the initial activation of the ring. The H radicals seem to be much more active than other radicals (e.g. CH_3) in initiating the rupture of the heteroaromatic ring systems [8,14]. Free radicals (especially H) would probably then continue to migrate to the N-containing site, resulting in the full hydrogenation of the N site for the release of NH_3 finally. The formation of NH_3 from the thermal cracking of nascent coal/char can be considered as a "self-gasification" process of char by the active hydrogen generated in situ within the char. The reaction mechanisms outlined here differ from those in the gas phase. In the gas phase, under pyrolysis conditions, the chance for an N-containing system to come into collision with 2 or 3 H radicals together or even consecutively to form an NH_3 molecule is very small. On the other hand, the H radicals may exist as adspecies on the char surface.

3.2. Selectivities of HCN and NH₃

The set of rank-ordered coals described in Part I [13] were also pyrolysed in our reactor [13] both at the high and slow heating rate following the experimental procedures described previously [13,27]. Fig. 2 shows the formation rates of HCN and NH_3 when the samples were heated at the slow heating rate of 6.67 K min⁻¹ from room temperature to 1000°C. The trends in HCN and in NH_3 are very similar. In particular, for a given coal, the formation rates of HCN and NH_3 at the slow heating rate reached maxima over the similar temperature range. For example, the formation rates of HCN and NH_3 from the Loy Yang brown coal reached the maxima over the temperature range of 600–700°C. This observation suggests that the formation of HCN and the formation of NH_3 were controlled probably by the same process, viz. the generation of free radicals to activate the N-containing heteroaromatic ring systems. This is the first step for the formation of both HCN and NH_3.

It should be pointed out that the formation of NH_3 would require a larger number of active radicals, particularly H, than the formation of HCN. However, the reasons that some

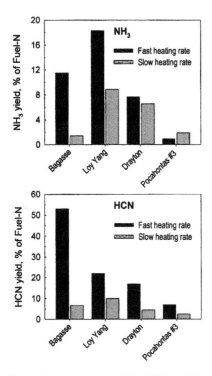

Fig. 3. Effects of heating rate on the yields of HCN and NH₃ from the pyrolysis of a set of rank-ordered solid fuel samples. The final temperature was 800°C for the fast heating rate experiments. During the slow heating rate experiments, the coal/biomass samples were heated at 6.7 K min⁻¹ from room temperature to 1000°C.

free radicals are more and more difficult to generate with increasing rank. Taken together, the rupture of the hetero-aromatic ring systems becomes increasingly difficult with increasing rank.

The data in Fig. 3 also show that the sensitivities in the yields of HCN and NH_3 to the changes in heating rate decrease with increasing rank. As was pointed out in Part II [27], the volatiles were released from coal at low temperatures when the coal sample was heated slowly. The majority of volatiles from the pyrolysis of both brown and bituminous coals could be released at temperatures lower than 700°C [28–31]. From the discussion given above, at this temperature level, the N-containing heteroaromatic ring systems would be little cracked to release HCN. Thus, the contribution of volatile thermal cracking to the observed HCN yield was minimal at the slow heating rate. However, in the case of the yields of NH_3, the reduced extends of volatile cracking reactions at the slow heating rate are not responsible at all for the reduction in the yield of NH_3 with decreasing heating rate. This is because, as was discussed above, the thermal cracking of volatiles does not produce any significant amounts of NH_3. The effect of heating rate on the yield of NH_3 (and HCN) can be explained by the mismatch between the generation of H radicals and the hydrogenation of the N-sites inside the pyrolysing solid particles. Once an N-containing heteroaromatic ring system is activated initially, the ring system could undergo two types of reactions. The first type of the reactions is for the ring system to rupture to form HCN or NH_3. For example, if more H radicals are supplied rapidly, the N-site may be hydrogenated further to form a NH_3. The second type of the reactions is for the N-containing ring system to combine with other ring system(s) to form a larger heteroaromatic ring system. For example, if the H radicals could not be supplied rapidly, the N-site may undergo recombination reactions to form a more stable structure in the pyrolysing solid particle. The new structure would be more difficult to activate/rupture than the original N-containing ring system(s), even at higher temperatures.

The outcome of the competition between the two types of the reactions mentioned above, particularly for the formation of NH_3, is affected to a large extent by the availability of radicals. The more radicals are available in the pyrolysing solid particles, the more NH_3 (and HCN) is formed. As the availability/concentrations of radicals are greatly affected by the heating rate [32], the yields of NH_3 and HCN would be sensitive to changes in heating rate. This is particularly true for the brown coal and the bagasse (Fig. 3), especially for the formation of NH_3 that requires more H than the formation of HCN. Heating the coal/bagasse slowly would result in the generation of H radicals at slow rates but at low temperatures, possibly from the breaking off of the substitutional groups and the dehydrogenation of hydro-aromatic ring systems. This allows for the activated N-containing ring systems to combine before NH_3 can be formed. Therefore, the H radicals generated in this way

are not effectively used for the formation of NH_3. Whereas heating the coal rapidly to high temperatures would result in the generation of the H radicals rapidly, favouring the hydrogenation of the N-site to NH_3 rather than the recombination of the N-sites.

Heating coal rapidly to high temperature also favours the interactions between the volatiles and the pyrolysing coal/char particles. Some volatiles may also react with the nascent char to yield some NH_3. However, this would not be a major source of NH_3 as our experimental data for the Loy Yang brown coal [27] showed that the NH_3 yield did not change very significantly with total gas flow rate (i.e. residence time of volatiles) or coal feeding rate (i.e. the freshness of the char).

3.4. Comparison between pyrolysis in argon and gasification in CO_2

In order to further test our hypothesis mentioned above on the formation of NH_3 and HCN in the pyrolysing solid coal/char particles, some experiments were also carried out in CO_2 under otherwise identical experimental conditions to those in argon outline previously [13,27]. Figs. 4 and 5 compare the formation of NH_3 and HCN in Ar and in CO_2. In making such a comparison, consideration was given to the choice of coal feeding rate. In particular, it was found [27] that the formation of HCN could be suppressed by increasing coal feeding rate. For this reason, a much higher coal feeding rate of 220 mg min^{-1} was used during pyrolysis than the feeding rate of 110 mg min^{-1} during gasification. In other words, the HCN formation during gasification is compared to the lowest HCN formation rate during pyrolysis. The data in Fig. 4 clearly demonstrate that the formation rates of HCN and NH_3 are both suppressed by the use of CO_2 instead of argon. These results may be understood by considering the consumption of the N-sites or the H radicals by CO_2-related species on the char surface. CO_2, being a mild oxidant at 800°C, would be adsorbed on the surface of the coal/char to form surface oxides/complexes. The N-sites, particularly the unstable N-sites with the potential for the formation of NH_3 or HCN, would very likely be the active sites for the adsorption of CO_2. Two possible reactions would decrease the HCN and NH_3 formation rate. The first possibility is the further reaction between the N-site and the adsorbed CO_2, leading to the loss or the gasification of the N-site. The second possibility is that the surface oxides/complexes somehow block the access of the N-site by the H radicals.

In the "not-feeding" periods, the formation rate of HCN in CO_2 dropped rapidly to almost zero while the formation rate of HCN in argon was still measurable up to 30 min. This indicates that the very active N-sites that had the potential to form HCN were consumed rapidly by CO_2 during gasification. This further supports our hypothesis mentioned above that thermally unstable, thus reactive, N-containing

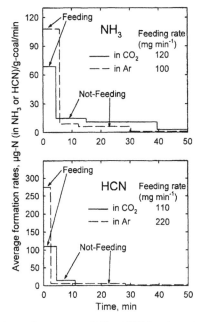

Fig. 4. Average formation rates of NH$_3$ and HCN from the pyrolysis/gasification of Loy Yang brown coal in argon and CO$_2$ at the fast heating rate. Temperature: 800°C.

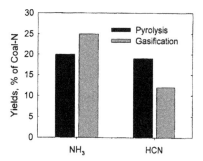

Fig. 5. Yields of NH$_3$ and HCN from the pyrolysis/gasification of Loy Yang brown coal in argon and CO$_2$ at the fast heating rate. Experimental conditions as in Fig. 4.

structures are mainly responsible for the formation of HCN. The overall yield of HCN is lower in gasification than in pyrolysis (Fig. 5).

In the case of NH$_3$ formation in the "not-feeding" periods, the formation rate of NH$_3$ was seen to increase from pyrolysis in argon to gasification in CO$_2$. In fact, NH$_3$ was still seen to form after 2 h in CO$_2$ as the gasification progressed slowly at 800°C. Apparently, as the slow gasification gradually broke down the char structure, new N-sites would be exposed on the surface together with the generation of more H radicals. The H radicals could then migrate to the new N-sites to lead to the formation of NH$_3$. Given the suppression of NH$_3$ formation in the "feeding" period, the slow formation of NH$_3$ in the "not-feeding" periods during gasification makes the overall yield of NH$_3$ higher in gasification than in pyrolysis, as is shown in Fig. 5.

4. Conclusions

1. In the gas phase, the thermal cracking of coal pyrolysis volatiles agree at least qualitatively with the pyrolysis of pyrrolic and pyridinic N-containing compounds in terms of the onset decomposition temperature, the main intermediates and the main final N-containing product (HCN).
2. The formation of NH$_3$ during the pyrolysis of pyrrolic and

pyridinic N-containing heteroaromatic ring systems (as model compounds or in the coal pyrolysis volatiles) requires the presence of condensed phase(s) of carbonaceous materials rich in hydrogen. Direct hydrogenation of the N-site by the H radicals generated in situ from the thermal cracking of the pyrolysing solid is the main source of NH$_3$ in the solid.
3. The formation of HCN and the formation of NH$_3$ in the pyrolysing solid particles are controlled by the same process, i.e. the in situ generation of radicals, particularly the H radicals.
4. The thermally unstable N-containing structures are mainly responsible for the formation of HCN, whereas the thermally more stable N-containing structures can be converted slowly into NH$_3$.
5. The formation of HCN and the formation of NH$_3$ are rank dependent and are sensitive to changes in heating rate, which affect the local availability of the H radicals in the pyrolysing solid.
6. The size of the N-containing heteroaromatic ring systems and the types of substitutional groups also play important roles in the formation of HCN and NH$_3$.
7. The reaction mechanisms proposed for the formation of NH$_3$ and HCN during pyrolysis are also supported by the data obtained in CO$_2$ at 800°C.

Acknowledgements

We gratefully acknowledge the financial support of this study by the Australian Research Council (ARC) and the New Energy and Industrial Technology Development Organisation (NEDO) in Japan. L.L. Tan gratefully acknowledges the scholarships (MGS and OPRS) from Monash University.

References

[1] Bartle KD, Perry DL, Wallace S. Fuel Processing Technology 1987; 15:351.

78

[2] Nelson PF, Buckley AN, Kelly MD. 24th International Symposium on Combustion. The Combustion Institute, Pittsburgh, PA, 1992. p. 1259–67.

[3] Kambara S, Takarada T, Yamamoto Y, Kato K. Energy and Fuels 1993;7:1013.

[4] Buckley AN. Fuel Processing Technology 1994;38:165.

[5] Kelemen SR, Gorbaty ML, Kwiatek PJ. Energy and Fuels 1994; 8:896.

[6] Knicker H, Hatcher PG, Scaroni AW. Energy and Fuels 1995;9:999.

[7] Solum MS, Pugmire RJ, Grant DM, Kelemen SR, Gorbaty ML, Wind RA. Energy and Fuels 1997;11:491.

[8] Li C-Z, Buckley AN, Nelson PF. Fuel 1998;77:157.

[9] Bassilakis R, Zhao Y, Solomon PR, Serio MA. Energy and Fuels 1993;7:710.

[10] Baumann H, Möller P. Erdol und Kohle—Erdgas—Petrochemie vereinigt mit Brennstoff-Chemie 1991;44:29.

[11] Li C-Z, Nelson PF, Ledesma EB, Mackie JC. 26th International Symposium on Combustion. The Combustion Institute, Pittsburgh, PA, 1996. p. 3205–11.

[12] Li C-Z, Buckley AN, Nelson PF. Proceedings of the 14th Annual Pittsburgh Coal Conference, 1997, in the form of a CD-ROM.

[13] Tan LL, Li C-Z. (Part I of this series) Fuel 2000;79;1883.

[14] Li C-Z, Ledesma EB, Buckley AN, Nelson PF. Proceedings of the Seventh Australian Coal Science Conference. Australian Institute of Energy, 1996. p. 171–8.

[15] Mackie JC, Colket III MB, Nelson PF, Esler M. International Journal of Chemical Kinetics 1991;23:733.

[16] Mackie JC, Colket III MB, Nelson PF. The Journal of Physical Chemistry 1990;94:4099.

[17] Axworthy AE, Dayan VH, Martin GB. Fuel 1978;57:29.

[18] Sugiyama S, Arai N, Hasatani M, Kawamura S. Environment Science and Technology 1978;12:175.

[19] Terentis A, Doughty A, Mackie JC. The Journal of Physical Chemistry 1992;96:10334.

[20] Doughty A, Mackie JC. The Journal of Physical Chemistry 1992; 96:10339.

[21] Doughty A, Mackie JC. The Journal of Physical Chemistry 1992; 96:272.

[22] Ikeda E, Mackie JC. Journal of Analytical and Applied Pyrolysis 1995;34:47.

[23] Lifshitz A, Tamburu C, Suslensky A. The Journal of Physical Chemistry 1989;93:5802.

[24] Hämäläinen JP, Aho MJ, Tummavuori JL. Fuel 1994;73:1894.

[25] Ledesma EB, Li C-Z, Nelson PF, Mackie JC. Energy and Fuels 1998;12:536.

[26] Haidar NF, Patterson JM, Moors M, Smith Jr WT. Journal of Agricultural Food Chemistry 1981;29:163.

[27] Tan LL, Li C-Z. (Part II of this series) Fuel 2000;79:1891.

[28] Li C-Z, Wu F, Xu B, Kandiyoti R. Fuel 1995;74:37.

[29] Li C-Z, Nelson PF. Energy and Fuels 1996;10:1083.

[30] Sathe C, Pang Y, Li C-Z. Energy and Fuels 1999;13:748.

[31] Li C-Z, Sathe C, Kershaw JR, Pang Y. Fuel 2000;79:427.

[32] Li C-Z, Madrali ES, Wu F, Xu B, Cai H-Y, Güell AJ, Kandiyoti R. Fuel 1994;73:851.

Emissions Reduction: NO_x/SO_x Suppression
A. Tomita (Editor)

Nitrogen sulphur interactions in coal flames

E. Hampartsoumian*, W. Nimmo, B.M. Gibbs

Department of Fuel and Energy, School of Process, Environmental and Materials Engineering, The University of Leeds, Leeds LS2 9JT, UK

Received 10 September 2000; accepted 2 October 2000

Abstract

The control of NO_x emissions from coal combustion is a major factor in reducing the environmental impact of coal-fired power plants. This paper investigates the effect of sulphur on NO production in pulverised coal flames using an experimental approach. The NO emissions from a 20 kW down-fired combustor were measured with and without SO_2 doping for a range of operating conditions and coal types with total sulphur contents between 0.6 and 2.9%. Some tests were repeated using coals that were acid-washed to remove pyritic sulphur. The NO emissions were statistically correlated against 52 parameters describing the physical and chemical properties of the coals.

The experimental results show that the change in the NO emission due to SO_2 addition is dependent upon the stoichiometry of the flame and that sulphur can interact with the production of NO in a heterogeneous coal flame in a manner similar to that reported previously for gaseous and liquid fuel flames. For fuel-rich conditions, sulphur addition enhanced the production of NO from fuel-bound nitrogen by up to around 20%. The change was larger for coals with a lower inherent fuel-S content (<0.5 wt%). The experimental trends also provide strong evidence that the in-flame sulphur nitrogen interactions are controlled by the availability and production of inorganic (pyritic) sulphur from the coal matrix. This is supported by statistical correlation of the data which has identified specific coal properties, such as pyritic sulphur content and the total volatile yield, as being amongst the most influential parameters in predicting the NO emission for the system employed. These parameters are the key indicators for the release of sulphur at low to intermediate temperatures and would imply that the NO enhancement mechanism under fuel-rich conditions appears to be associated with the early release of pyritic-S into the developing flame to subsequently interact with volatile nitrogeneous species. © 2001 Elsevier Science Ltd. All rights reserved.

Keywords: Sulphur; NO_x; Combustion; Coal; Nitrogen; Emissions; Flames

1. Introduction

Understanding the mechanisms for the formation of nitrogen oxides and the influence of sulphur compounds is of fundamental importance to the design and development of low NO_x combustion systems. The fact that sulphur can influence the production of NO_x is established [1–5]. Most of the earlier reported work was concerned with bench-scale flames and the overall number of studies undertaken so far is very limited, there has been no comparable work concerned with investigating these effects for coal combustion. Previous work [6] using a pilot-scale, oil-fired furnace has shown that under air-staging conditions resulting in a fuel-rich primary flame zone, the effect of sulphur is to enhance fuel-NO formation whilst, under fuel-lean conditions and with negligible fuel-N present, the effect of S is to reduce the formation of thermal-NO. More recently, a comprehensive mechanism has been developed describing C2 hydrocarbon oxidation, NO_x formation and the interaction of

sulphur/nitrogen species in such flames [7]. The implications of S/N interactions in coal flames with respect to the accurate modelling and design of, principally fuel-rich, low NO_x combustion techniques needs to be established. This paper uses an experimental approach to investigate sulphur/nitrogen interactions in pulverised coal flames where the sulphur and nitrogen concentrations in the fuel can be much higher than with other fuel types.

2. Experimental procedure

All tests were performed in a 20 kW thermal, refractory-lined, down-fired pulverised coal furnace shown in Fig. 1. The combustion chamber had a diameter of 100 mm and was 1.5 m long. Gas measurements could be made at 1 m downstream of the burner and also at the exit. Combustion air was split between a swirling secondary air flow and axial flow primary air, which also served as pneumatic transport for the pulverised coal (see Fig. 1). The air/fuel ratio was varied by altering the secondary air flow while maintaining a constant primary airflow. In this way, the coal gun exit

---

* Corresponding author. Tel.: +44-113-233-2497.
  E-mail address: e.hampartsoumian@leeds.ac.uk (E. Hampartsoumian).

Reprinted from *Fuel* 80 (7), 887–897 (2001)

80

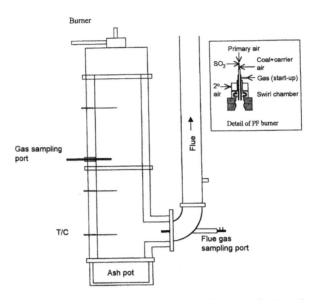

Fig. 1. Schematic of down-fired, 25 kW combustor and pulverised coal swirling burner details.

velocity was held constant along with the transport air required for coal feeding. The bulk of the coal used had a particle size less than 125 μm with about 55–60% less than 63 μm. The SO_2 was added to the primary air supply. Some testing was also undertaken using H_2S to establish the effects with an alternative S additive. This reactor configuration provided a well defined and controllable system which could be related to a practical coal flame, and offered a simple means of varying the total sulphur available without altering other parameters that could also affect NO production. Gas analysis was performed by on-line analysers for NO and NO_x (chemiluminescence), SO_2 and CO (NDIR) and O_2 (paramagnetic). Shielded Pt/PtRh thermocouples were used to obtain an indication of the peak

temperatures in the flame which were typically around 1550–1650 K depending on operating conditions.

A suite of eight bituminous coals, whose properties are shown in Table 1, was chosen with the aim of obtaining as wide a variation of organic to pyritic sulphur ratios as possible. The total elemental sulphur content varied from 0.6 to 2.9 wt%. Coal samples were analysed for pyrite sulphur by a modified version of the chromium reduction method described by Canfield [8] and modified by Newton et al. [9]. The method is known to be specific to sulphide species although it should be noted that it also extracts any elemental sulphur present in the sample. In addition, the process will remove any sulphate species present in the sample. Newton et al. [9] quote a recovery rate and precision of

Table 1
Proximate and ultimate analysis of coals (wt%, dry, ash-free)

	Asfordby	Betts Lane	Hunter Valley	Kaltim Prima	Koorn-fontein	La Jagua	Thoresby	Galatia
VM	37.3	26.8	31.4	42.8	27.1	36.1	30.9	32.1
FC	52.6	56.2	56.3	51	59.1	55.2	52.7	59.5
Ash	10.1	17.0	12.3	5	13.9	8.7	16.4	8.4
Untreated								
C	81.2	86.9	83.0	79.9	82.4	80.2	84.3	86.0
H	6.5	5.4	5.4	6.1	5.1	5.6	4.9	3.8
O	10	3.1	9.2	10.5	9.8	12.1	6.4	7.3
N	1.55	1.7	1.84	1.83	2.02	1.54	1.89	1.8
S (org)	0.49	1.03	0.48	1.36	0.45	0.47	1.33	0.59
S (pyr)	0.26	1.9	0.11	0.27	0.28	0.16	1.07	0.50
Total	0.75	2.93	0.59	1.63	0.73	0.63	1.4	1.09
Treated								
S (org)		1.04		1.36			1.34	
S (pyr)		1.26		0.033			0.35	

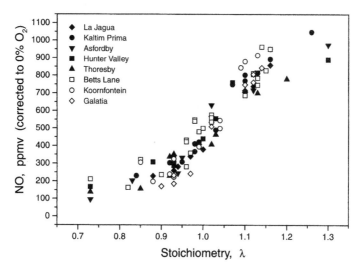

Fig. 2. Variation of NO emission (corrected to 0% O_2) as a function of air/fuel ratio.

94.25 ± 5.24% for pyrite. Chemical desulphurisation of three selected coals was undertaken using essentially a scaled-up version of this process in order to treat coals with different inorganic/organic sulphur ratios as shown in Table 1. One problem encountered was the tendency for excessive effervescence due to the rapid production of H_2S from the fine-grained pyrite. Adding ethanol helped by reducing the surface tension and inhibiting frothing. Coarse-grained pyrite in the sample severely limits the reaction rate and therefore the amount of sulphur removal. An attempt to remedy this was made by first settling the coal in water and removing the dense fraction that settled after 5 min.

The organic sulphur contents were determined by combusting the washed and dried residue from the chromous acid in a Parr Bomb filled with oxygen at 30 bar pressure. The liberated sulphur was precipitated as barium sulphate using barium chloride solution (after removal of metal hydroxides, adjustment to pH 2.5–3 and heating to 70–80°C) and then filtered and dried. The sulphur content of the sample was then determined gravimetrically.

3. Experimental results and discussion

3.1. NO emissions

The variation in flue gas NO concentration for all the coals tested is shown in Fig. 2 over a range of air/fuel equivalence ratios from fuel-rich, at $\lambda = 0.73$, to fuel-lean, at $\lambda = 1.3$. The data are corrected for dilution effects to 0% O_2. There was, as expected, a significant variation in NO production between the various coals. This can partly be accounted for on the basis of experimental error and the

small differences in the nitrogen contents of the coals (Table 1). It is also a consequence of the variations in the pyrolysis and combustion characteristics of the coals. The largest sensitivity to stoichiometry change occurs between $\lambda = 0.95$ and $\lambda = 1.15$. The data in Fig. 2 were used in the statistical analysis described later together with other important measured and computed physical and chemical coal properties. Some typical responses of the NO emission from the reactor to the addition of sulphur as SO_2 are shown in Fig. 3 for tests performed under fuel-rich and fuel-lean conditions. For comparison, Fig. 3 shows the effect on the concentration of, principally, thermal-NO for natural gas combustion, only in the same combustor. When operated in fuel-rich conditions the NO emissions nearly always increased on SO_2 addition, whereas at fuel-lean conditions, the NO emissions decreased. For natural gas combustion, the NO emission was seen to decrease under both fuel-rich and fuel-lean conditions. The maximum amount of SO_2 added (up to 8000 ppmv in the exit flue) was always less than 0.5% of the total flow into the reactor and in all cases, the change in the NO emission was much greater than that accounted for by dilution or mixing effects. There was also no noticeable effect of adding SO_2 on the average temperatures measured in the reactor, so it is likely that the effects are due to chemical rather than physical changes.

The relative change in the NO emission due to sulphur addition as SO_2 is shown in Fig. 4 for all the coals. This data clearly shows that, with the exception of Asfordby, the NO emission increased for fuel-rich flame conditions ($\lambda < 1$, points above the line) whereas at fuel-lean conditions ($\lambda > 1$) the NO emission was reduced. It is apparent that two different chemical mechanisms are present and the operating stoichiometry determines which pathway is dominant. These results for coal are in agreement with those

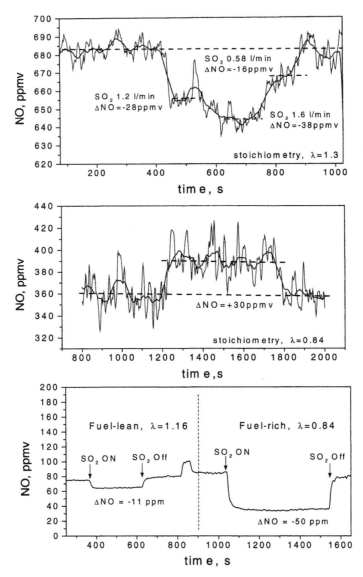

Fig. 3. Experimental measured exit gas NO response to SO₂ addition for (a) Kaltim Prima, fuel-lean ($\lambda = 1.3$); (b) Hunter Valley, fuel-rich ($\lambda = 0.84$); and (c) natural gas only, at fuel-rich and fuel-lean conditions.

reported previously for gaseous [2,3,5] and liquid [6,7] fuel combustion. However, the magnitude of the change is not as great as previously observed for liquid fuels [7] and is probably a consequence, in this work, of achieving less intimate mixing of fuel and sulphur dopant. It can be concluded from these results that the NO enhancement and reduction mechanisms associated with sulphur interaction occur simultaneously in the coal flame depending on local mixing conditions and that the overall NO emission is the net result of these competing mechanisms.

The change in NO emission due to the influence of SO₂ for fuel-lean conditions is shown in Fig. 5. For these conditions, previous work [1] suggests that the reduction of NO concentration is due to a decreased availability of oxidising radicals, O and OH, in the flame radical pool as the presence of a third-body like SO₂ induces catalytic recombination to form H₂O and O₂ through intermediate HSO₂ species. This reduction in the concentration of O and OH has a consequential effect on the rates of formation of thermal-NO from N₂ via the Zeldovich mechanism [10]. It is also possible that

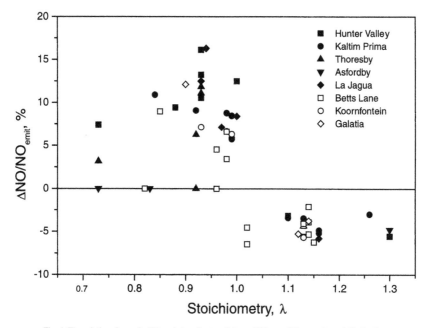

Fig. 4. The relative change in NO emission due to sulphur addition as SO_2 at various air/fuel ratios.

the reduction in NO may result from reduced fuel-N oxidation due to depletion of O and OH radicals which are important in the oxidation of volatile nitrogen species such as NH. Fig. 5 indicates that the relationship is apparently linearly dependent on SO_2 concentration, which may be expected due to the presence of excess O and OH radicals.

Further evidence for NO reduction by SO_2 can be seen in Fig. 3c which show results for natural gas operation only without coal, and hence for this case, the NO emission decreased for both fuel-lean and fuel-rich flame conditions due to the absence of fuel derived nitrogenous species. The principal route to NO formation under these conditions is via N_2 oxidation with increasing amounts of prompt-NO at fuel-rich conditions.

The effect of sulphur addition on NO enhancement for a fuel-rich flame is shown in Fig. 6. The extent of the possible increase in NO emissions under fuel-rich conditions is clearly evident for a heterogeneous coal flame from these

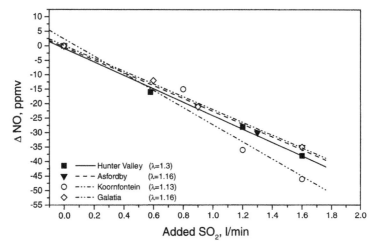

Fig. 5. NO reduction at fuel-lean conditions showing effect of SO_2 addition (air/fuel ratio, λ as shown).

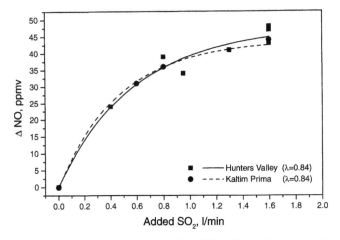

Fig. 6. NO increase at fuel-rich conditions showing effect of SO$_2$ addition (air/fuel ratio, λ as shown).

results and provides strong evidence that an interaction between sulphur and nitrogen chemistry may result in a further chemical pathway to NO formation as reported previously for other fuel types. Sensitivity analysis from a recent kinetic study [7] has shown that the following reaction is an important NO formation reaction under moderately fuel-rich conditions;

$$SO + NH = NO + SH \qquad (I)$$

NH is formed from the inherent nitrogen content of the coal via the initial volatile form of nitrogen, HCN as follows;

$$HCN + OH = NH_2 + CO \qquad (II)$$

$$NH_2 + OH = NH + H_2O \qquad (III)$$

The SO is produced via reaction (IV) by oxidation of fuel-S oxidised early in the flame to form SO$_2$, which subsequently forms SO in the more fuel-rich zones of the flame.

$$SO_2 + H = SO + OH \qquad (IV)$$

Sulphur in the form SH may be further reduced to form H$_2$S or oxidised to form SO$_2$ depending on the local stoichiometry in the flame. It is probable that the formation rate of NO in reaction (I) will be limited by the availability of NH species, hence the observed limit to the change in NO in Fig. 6 as the concentration of sulphur is increased.

The experimental trends observed for coal can also be explained by the above mechanism. This implies that the S–N interactions are probably occurring mostly in the gas phase, subsequent to the release of volatile sulphur and nitrogen species from the coal matrix. The release of nitrogen and sulphur from the coal matrix is a complicated chemical and physical process with interactions within the organic matrix, on the surface and surrounding gas phase. The differences and scatter in the data with respect to the response of each coal to sulphur addition, as shown for

example in Fig. 4, may therefore be a reflection of the different pyrolysis behaviour of the organic and inorganic sulphur components in the coal, as well as the different forms of nitrogen, and this aspect is considered in the next section.

3.2. Effect of coal type

Nitrogen in coal is predominantly in pyrrolic form (55–80%) with pyridinic, quaternary and amino forms. It is generally accepted that NO$_x$ formation cannot be predicted based solely on coal-N functionality. This is due to the complex transformations which the nitrogen compounds undergo when heated and depends on heating rate, final temperature and particle size. All forms of fuel-N which are released as volatiles (including tar-N) form HCN, NH$_3$ with some N$_2$. Fuel-N which is not released as volatile-N remains in the char and this is principally in pyridinic form with lesser amounts of pyrrolic-N [11]. Under fuel-rich conditions char-N will form N$_2$ and NH$_3$ [12]. HCN can also be formed during char pyrolysis and may undergo further conversion to NH$_3$ in the pores of the solid particle. The char can also react with NO to reduce it to N$_2$.

Sulphur is an integral part of coal and can vary typically in content from under 0.5 to 5% by mass. It is present in mineral form or organic form with some elemental sulphur present as a result of weathering. Inorganic sulphur is present principally as pyrite (FeS$_2$) or as sulphates. The organic sulphur content of coals is rank dependent with thiophene sulphur present in greater quantity in higher rank coals, whereas non-thiophene sulphur (poly-aromatics, thiols, sulphides, di-sulphides and mercaptan compounds) is present in lesser amounts. The main gaseous species from the devolatilisation of coal is H$_2$S, with COS, CS$_2$ and smaller amounts of lower mercaptans and thiophenes [13–15]. Sulphur in the tar products from coal pyrolysis is mainly in

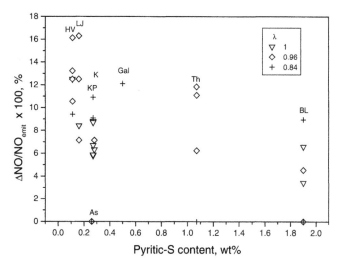

Fig. 7. Change in NO concentration due to the addition of S as SO_2 as a function of initial coal pyritic-S contents. Air/fuel ratio, λ, as shown (HV, Hunter Valley; KP, Kaltim Prima; Gal, Galatia; Th, Thoresby; BL, Betts Lane; K, Koornfontein; As, Asfordby).

the form of thiophenes and aromatic sulphides [16] but the partitioning of sulphur between char, tar and gas is dependent on initial sulphur distribution in coal, rank, heating rate and final temperature [17]. The principal form of organic sulphur in char is the more stable thiophene, which is derived from the transformation of simpler compounds by, for example, cyclisation reactions and possibly from the interaction of elemental-S from FeS_2 pyrolysis with the organic matrix [13,18]. Similarly, sulphate sulphur may form organic sulphur by interaction with the coal matrix [19,20]. It has also been reported that the decomposition temperatures of inorganic sulphur compounds are lowered, when they are in contact with the coal matrix. For instance, the decomposition temperature of pyritic-S is reduced from 500 to about 300°C and that of sulphate-S is lowered to below 500°C. The coals used in the present study contained pyritic-S in the range of 35–84% of total sulphur. Of interest is the identification of any specific effect of the different forms of sulphur in the coal on the production of NO.

Fig. 7 shows the change in NO with SO_2 addition for a range of fuel-rich flame conditions as a function of the coal pyritic sulphur content. Despite the scatter in the data, the underlying trend at each operating condition for all coals suggests that the change in the NO emission due to SO_2 addition is more significant for the coals with lower pyritic sulphur. Since this is the least thermally stable form of sulphur associated with a coal matrix [13], it will, under severe pyrolysis conditions, be quickly released early in the flame. For the coals with a higher pyritic sulphur content, sulphur nitrogen interactions will not be limited by the availability of sulphur and consequently we see that there is less to be gained with respect to changes in the NO emission by artificially increasing the concentration

of S species in the flame. In these cases, the increase in NO concentration will be limited by the availability of NH species rather than S species. In contrast, the availability of sulphur will be more limited for the lower pyritic coals so that the compensating effect of adding SO_2 results in a larger relative change in the NO emission. The important role played by pyritic sulphur is not unexpected since volatile sulphur release and conversion to SO is required prior to interaction with NH species from fuel-N pyrolysis. Obviously, the situation will be complicated by the pyrolysis characteristics and physical and chemical properties of the coals which, while all bituminous in nature, cannot be assumed to be similar and would provide a further explanation for the scatter seen in the experimental data and the fact that some coals followed the trend better than others. Again, these results would support the earlier assumption that the interactions are more likely to occur in the gas phase and are more influenced by the physical behaviour of each coal with respect to how the important reacting species are generated.

3.3. Depyritised coal tests

The aim of these tests was to investigate further the relative influence of the different forms of sulphur (organic and inorganic) by comparing the NO emissions and the effect of adding SO_2 for a coal before and after pyritic sulphur removal. Three coals were chosen with low, intermediate and high levels of pyritic-S, respectively. Table 2 shows the effect of removing pyritic sulphur on the emission of SO_2 for the three coals. The emissions are in line with expectations based on the total elemental sulphur analysis before and after treatment and the differences between the measured and calculated SO_2 levels were well within the ±5% error

Table 2
SO$_2$ emissions at $\lambda = 1.1$ (ppmv, dry). Sulphur (%); pyritic-S in brackets (d.a.f.)

Coal	Total S, raw (%)	SO$_2$ emission, raw coal	Calculated SO$_2$, raw coal	Total S, treated (%)	SO$_2$ emission, treated coal	Calculated SO$_2$, treated coal
Betts Lane	2.93 (1.9)	2000	2064	2.4 (1.26)	1650	1616
Thoresby	2.4 (1.07)	1671	1633	1.69(0.35)	1225	1146
Kaltim Prima	1.63 (0.27)	1300	1219	1.39(0.033)	1025	1045

band quoted for the accuracy of the analysis of pyritic sulphur.

We were not able to establish clearly a conclusive difference in the NO for a particular coal before and after treatment since the small changes in NO were within the scatter of the experimental data obtained for the coal prior to desulphurisation. However, the effect of sulphur addition as SO$_2$ on the NO emission did show significant differences between the untreated and treated coals. This is shown in Fig. 8 which compares the change in NO emission against the parameter $S_i - S_{ds}/S_i$ which is a measure of the amount of pyritic-S removed from the coals on a scale of 0 to 1, where 1 represents complete removal of pyritic-S (Betts Lane $S_i = 1.9$, Kaltim Prima $S_i = 1.1$ and Thoresby $S_i = 0.27$). The change in NO emission due to the addition of SO$_2$ for the untreated (raw) coal is also plotted below the treated coal data points for convenience and ease of comparison (for these points the parameter $S_i - S_{ds}/S_i$ has no meaning). For the coals with the lower initial pyritic-S content, $S_i - S_{ds}/S_i$ is higher indicating that the desulphurisation process was more efficient for these coals. However for these coals, any additional increase in the NO concentration due to SO$_2$ addition was small. In contrast, for the Betts Lane coal, the difference in the NO increase was significant (93 ppmv for the desulphurised coal compared to 27 ppmv for the raw coal). Again, this can be explained by considering the availability of volatile S in the flame. For the two coals with the lower initial pyritic-S content which respond similarly to further S addition, whether treated or untreated, the limiting factor in both cases is the low availability of volatile S from the coal matrix in the first place. The untreated Betts lane coal responds less well to artificial increase in SO$_2$ concentration in the flame since the availability of sulphur is not a limiting factor as this coal, has the highest initial pyritic-S content. After desulphurisation, the NO response to external sulphur addition is significantly greater since volatile sulphur availability from the treated coal is now a limiting factor in comparison to the untreated coal.

3.4. Statistical analysis

The NO emissions from any coal combustion system are strongly dependent on operating conditions. In this study the NO emissions from the flame, within a narrow temperature range from 1550–1650 K, varied from a low of about 150 ppmv (corrected to 0% O$_2$) at $\lambda = 0.73$ to around 900 ppmv at $\lambda = 1.3$ (see Fig. 2). There was also a significant variation for any given stoichiometry between the coals used. In order that this variation may be accounted for, a systematic analysis of the data was performed using the MINITAB statistical analysis package so as to quantify the effects of fuel-S on fuel NO$_x$ formation. Additionally, data on the pyrolysis characteristics for each coal were obtained using the FG-DVC programme [21] and from the DTI collaborative NO$_x$ programme [22] (experimentally

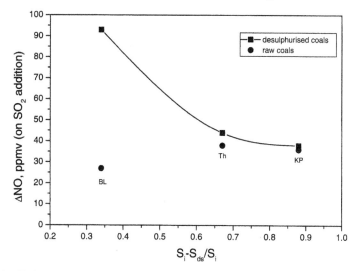

Fig. 8. The effect of initial pyritic-S content on observed increase of NO concentration on the addition of SO_2 for three desulphurised coals, Thoresby (Th), Kaltim Prima (KP) and Betts Lane. S_i, pyritc-S raw coal; S_{ds}, pyritic-S desulphurised coal. (ΔNO for raw coals shown for comparison only.)

measured, high temperature volatiles yields). The FG-DVC code employs a triangular interpolation technique to correlate data on coals which are not included in its database. The statistical analysis which follows was based on six of the coals, Asfordby, Betts Lane, Hunter Valley, Koornfontein, La Jagua and Thoresby for which experimental drop tube devolatilisation data was available at fuel-rich conditions.

Initially, 52 parameters, listed in Table 3, were correlated with measured data on NO emission. This gave a preliminary indication of the parameters to be included in subsequent multiple regression analysis. A stepwise approach was adopted so that a minimum number of variables could be selected for the model by a process of comparison and elimination. Regression models were developed for air/fuel equivalence ratios of $\lambda = 0.85$, 0.97, 1 and 1.33. The NO emission data at each value of λ was derived by interpolation of experimental for each individual coal and represents the measured value at each condition.

The preliminary correlation for the richest condition, $\lambda = 0.85$, suggested that the principal factors were the experimentally determined volatiles yield at 1350°C, the pyritic sulphur content and the rank indicator O/C. The derived regression equation containing these terms was:

$$NO = 225 + 8.22 \, devol1350 - 112 \, Pyr\text{-}S - 4368 \, O/C \quad (1)$$

The R^2 value of 93.1% is high which indicates a good fit between measured and predicted NO concentrations. The multiple regression analysis also gave a good R^2 (adj) value of 82.7%, which indicates a high degree of correlation since this statistical parameter accounts for the number of variables used in the model.

Using the same variables as Eq. (1), the regression equation at $\lambda = 0.97$ was:

$$NO = 640 - 1.51 \, devol1350 - 37.2 \, Pyr\text{-}S - 1510 \, O/C \quad (2)$$

The R^2 value of 87.3% indicates a good fit between measured and predicted NO concentrations. The multiple regression analysis also gave a reasonable R^2 (adj) value of 68.2%. However, the best equation for this condition was obtained by replacing devol1350 by the proximate volatiles yield, which resulted in a regression fit with an R^2 value of 98.4% and an R^2(adj) value of 95.9%. The derived regression equation for these predictors is given by:

$$NO = 682 - 3.82 \, Volatiles - 38.3 \, Pyr\text{-}S - 1304 \, O/C \quad (3)$$

At $\lambda = 1.0$, the best fit equation was obtained by the inclusion of the FG-DVC derived variable for nitrogen released as gas from the coal matrix, gas-N, instead of the rank parameter O/C.

$$NO = 984 - 5.45 \, devol1350 - 37.3 \, Pyr\text{-}S - 16.3 \, gas\text{-}N \quad (4)$$

The R^2 value of 98% and a R^2(adj) of 95.1% for Eq. (4) indicates an extremely good fit at this condition.Finally, for an oxygen rich flame ($\lambda = 1.33$), the balance between NO production and the reduction of NO by volatiles/char is shifted towards NO production. The statistical model shows only moderate predictive ability for this condition using the same variables with an R^2 of 86% and a R^2(adj) of 64.9%.

$$NO = 1379 - 17.5 \, devol1350 + 115 \, Pyr\text{-}S + 2686 \, O/C \quad (5)$$

Table 3
Measured and computationally derived variables (based on experimental and FG-DVC [21] data from this study except where indicated)

NO	Measured NO concentration, ppmv
devol1000[a], devol1350[b], devol1500[a]	Rate of devolatilisation at 1000,1350 and 1500°C
Volatiles	Volatiles content, wt% d.a.f. (proximate analysis)
C, H, O, N	Ultimate analysis, wt% d.a.f.
O/C	Oxygen/carbon ratio, molar
H/C	Hydrogen/carbon ratio, molar
TOT-S	Measured total S in coal, wt% d.a.f.
Pyr-S	Measured pyritic S in coal, wt% d.a.f.
Org-S	Measured organic S in coal, wt% d.a.f.
char-N[a], tar-N[a], gas-N[a]	Nitrogen in char, tar and gas, wt% of total N
HCN[a]	HCN content of gas-N
NH3[a]	NH₃ content of gas-N
char-S, tar-S, gas-S,	Sulphur in char, tar and gas, wt% of total S
TR	Rate of tar release, g/min
TY	Tar yield, wt%
Char + ASH	Combined char and ash yield, wt%
CO2,CO,H2O,CH4,C2H4,H2	Rate of CO_2, CO, H_2O, CH_4, C_2H_4, H_2 release, g/min
COY, H2OY, C2H4Y, H2Y	CO, H_2O, C_2H_4, H_2 yield, wt%
CH4Y	Yield, wt%
CO2Y	CO_2 yield, wt%
Ole + Par	Olefins and paraffins yield, wt%
TVol	Total rate of devolatilisation, g/min
Gas	Volatiles gas yield, wt%
H2S	Rate of H_2S release, g/min
SO2	Rate of SO_2 release, g/min
COS	Rate of COS release, g/min
CS2	Rate of CS_2 release, g/min
NH3	Rate of NH₃ release, g/min
HCN	Rate of HCN release, g/min
NO2	Rate of NO_2 release, g/min
H2SY	H_2S yield, wt%
SO2Y	SO_2 yield, wt%
COSY	COS yield, wt%
CS2Y	CS_2 yield, wt%
NH3Y	NH₃ yield, wt%
HCNY	HCN yield, wt%
TVolY	Total volatiles yield, wt%
vol-N	Volatile N yield (NH3 + HCN)

[a] FG-DVC data [22].
[b] Experimental data [22].

The best fit is obtained using the FG-DVC derived total volatiles yield, resulting in a very good R^2 and R^2(adj) values of 89.8% and 87.3%, respectively.

$$NO = 1855 - 26.4 \text{ TVolY} \qquad (6)$$

As expected, the multiple regression NO emission Eqs. (1)–(6) include parameters that are known to be important factors in the production of NO from coal combustion in oxidising and reducing conditions. These are the experimentally measured high temperature volatiles yield at 1350°C (devol1350), the rank indicator O/C, and the total amount of nitrogenous species (including HCN, NH₃ etc.) released into the gas phase, gas-N (obtained from the FG-DVC model). The absence of coal elemental nitrogen content as an important parameter is not surprising since the experimental NO emission data used in the statistical study are based on a suite of coals with nitrogen contents in a very narrow range from 1.54 to 2.0% (d.a.f.), with three of the coals within 0.1% of the mean of 1.8%.

Of significance to this study is the outcome that the pyritic sulphur content of the coal was found to be a statistically important factor to the prediction of the NO emission. Furthermore, this was valid for all flame conditions with the possible exception of the leaner flames, and would be expected from consideration of the S/N interaction reaction mechanism for lean conditions which is dominated by thermal rather than fuel-NO interactions. Overall, the statistical analysis supports the previous conclusions, based on the experimental observations, that the pyritic sulphur content plays a part in the chemistry of NO production from coal combustion.

4. Conclusions

This study has demonstrated that there is an interaction between S and N in coal flames and that the NO emission is influenced by the S/N ratio in the flame, which is in turn controlled by the coal composition, pyrolysis behaviour and physical properties. Consequently, physical and kinetic models of NO formation from coals, which try to describe the devolatilisation behaviour of coals, must also take into account pyritic sulphur release and reaction kinetics for improved accuracy. For conditions that would be present in a staged fuel-rich flame, the effect of sulphur might be to inhibit NO reduction reactions by up to 20% of the NO produced.

The effect of sulphur addition in the form of SO_2 on the emission of NO has been investigated for coal combustion for a range of flame conditions. The change in NO emission was found to be dependent on the stoichiometry of the flame, with fuel-rich conditions resulting in NO enhancement, and NO reduction under fuel-lean conditions.

The effect of sulphur on NO chemistry for fuel-rich conditions appears to be controlled by the release of volatile pyritic sulphur in the flame. This was substantiated by multivariate correlation of the experimental data which identified the pyritic sulphur content, together with the coal volatile yield, as amongst the more important coal property parameters used to predict the NO emission. Tests with desulphurised coals supported the association of pyritic sulphur with NO production from coal flames.

Acknowledgements

The authors would like to thank the UK EPSERC for a grant to undertake this work. The contributions of S. Bottrell and R. Newton in analysing the coals used and producing desulphurised coals is gratefully acknowledged. We would also like to thank M. Whitehouse, CRE Limited, for providing some of the coals samples and devolatilisation data.

References

[1] De Soete G. Institut Francais du Petrole. Final report no. 23306. Ruiell Malmaison, France, 1975.

[2] Pfefferle LD, Churchill SW. Ind Engng Chem Res 1989;28:1004.

[3] Pfefferle LD, Churchill SW. Combust Sci Technol 1986;49(5–6):235–49.

[4] Wendt JOL, Wootan EC, Corley TL. Combust Flame 1983;49:261.

[5] Tseregounis SI, Smith OI. Combust Sci Technol 1983;30(1–6):231–9.

[6] Hampartsoumian E, Nimmo W. Combust Sci Technol 1995;110–111:487–504.

[7] Nimmo W, Hampartsoumian E, Hughes KJ, Tomlin AS. Proc Combust Inst 1998;27:1419–26.

[8] Canfield DE, Raiswell R, Westrich JT, Reaves CM, Berner RA. Chem Geol 1986;54:149–55.

[9] Newton RJ, Bottrell SH, Dean SP, Hatfield D, Raiswell R. Chem Geol 1995;125:317–20 (Isotope Geoscience Section).

[10] Zeldovich YaB, Sadovnikov PYa, Frank-Kamanetski DA. Academy of Sciences of the USSR. Institute of Chemical Physics, 1947.

[11] Chambrion P, Suzuki, Zhang Z-g, Kyotani T, Tomita A. Energy Fuels 1997;11:681–5.

[12] Wu Z, Ohtsuka Y. Energy Fuels 1997;11:477–82.

[13] Chen HK, Li BQ, Yang JI, Zhang BJ. Fuel 1998;77:487–93.

[14] Garcia-Labiano F, Adanez J, Hampartsoumian E, Williams A. Fuel 1996;75(5):585–90.

[15] Shao D, Hutchison EJ, Heidbrink J, Pan WP, Chou CL. J Anal Appl Pyrolysis 1994;30:91–100.

[16] Schuttte K, Rotzoll K, Schugerl K. Fuel 1989;68:1603–5.

[17] Patrick JW. Fuel 1993;72:281–5.

[18] Griglewicz G, Wilk P, Yperman J, Franco DV, Maes II, Mullens J, van Poucke LC. Fuel 1996;75(13):1499–504.

[19] Ibarra JV, Palacios JM, Moliner R, Bonet AJ. Fuel 1994;73(7):1046–50.

[20] Griglewicz G, Jasienko S. Fuel 1992;71:1225–9.

[21] FG-DVC version 8.1b. Advanced Fuel Research Inc., Connecticut, USA.

[22] Marraffino AM, Richards DG, Whitehouse M. CRE/DTI internal report GT00029 (restricted commercial), 1998.

The influence of calcined limestone on NO_x and N_2O emissions from char combustion in fluidized bed combustors

H. Liu[*], B.M. Gibbs

Department of Fuel & Energy, The University of Leeds, Leeds LS2 9JT, UK

Received 20 September 2000; accepted 5 December 2000

Abstract

Limestone addition to a coal-fired fluidized bed combustor usually results in an increase in NO/NO_x emissions and a decrease in N_2O emissions. This observation has been explained in the literature by the effect of limestone on: (i) the conversion of coal volatile-N to NO/NO_x and N_2O; (ii) the decomposition of N_2O; and (iii) the pool of H, OH and/or O radicals due to sulphur capture. Since char-N is also an important source of NO/NO_x and N_2O under fluidized bed combustion conditions, any effect of calcined limestone on the conversion of char-N to NO/NO_x and N_2O can also contribute to the above observation. The effect of calcined limestone on the conversion of volatile-N species, i.e. NH_3 and HCN has been well documented. However, so far few studies have studied the effect of limestone on the conversion of char-N under fluidized bed combustion conditions. In this study, a series of batch-type char combustion tests was carried out in a bench-scale bubbling fluidized bed combustor with different bed materials. These tests reveal that in addition to reducing N_2O emissions from char combustion, calcined limestone also enhances the conversion of char-N to NO/NO_x similar to the case of volatile-N. Possible reaction mechanisms to explain the experimental results of char combustion tests have been discussed. © 2001 Elsevier Science Ltd. All rights reserved.

Keywords: Char combustion; Fluidized bed combustion; Limestone; NO_x and N_2O emissions

1. Introduction

Because of its very low NO_x emissions and high in situ desulfurisation efficiency, circulating fluidized bed combustion (CFBC) has been widely accepted as one of the most advanced, environmentally benign coal combustion technologies. However, the recognition of nitrous oxide (N_2O) as an extremely long life and strong radiative forcing greenhouse gas and a stratospheric ozone destroyer has imposed concerns of the possible adverse impact on the environment of the wider application of coal-fired circulating fluidized bed combustors that emit high levels of N_2O (potentially 200 ppmv or higher) [1,2]. Fuel-bound nitrogen is the main source for both NO_x and N_2O emissions from a coal-fired fluidized bed combustor. Both the formation and destruction of NO_x and N_2O in a fluidized bed combustor are more or less catalytically affected by bed materials, including char, limestone/calcium oxides, calcium sulphates, ash and sand.

Many studies have investigated the effect of limestone

addition on NO_x and N_2O emissions from coal-fired fluidized bed combustors [3–7]. Due to the importance of volatile-N on NO_x and N_2O emissions, many researchers have also specifically investigated the effect of limestone on the conversion of volatile-N, especially HCN and NH_3, to NO_x and N_2O [8–11]. Limestone addition usually results in a higher NO_x emission and a lower N_2O emission and this is most pronounced in laboratory or bench-scale plants and at high molar ratios of Ca/S. Furthermore, the NO_x emission is more often influenced by limestone addition than is the N_2O emission [3,12]. These observations have been attributed to [4,7,9,11,12]: (i) limestone catalyses the decomposition of N_2O; (ii) the oxidation of HCN, the major N_2O precursor from volatile-N, over limestone has a high selectivity for NO_x formation and a low selectivity for N_2O formation under fluidized bed combustion conditions; (iii) the pool of H, OH and/or O radicals, which can effectively destroy N_2O and are necessary to form NO from NH_3 and HCN related species, increases as a result of reduced SO_2 concentration; (iv) limestone/CaO catalyses the oxidation of CO which can reduce NO effectively when catalysed by char; and (v) limestone/CaO catalyses the conversion of HCN to NH_3. However, char-N is also an important contributor to NO_x and N_2O emissions from

* Corresponding author. Tel.: +44-113-233-2513; fax: +44-113-244-0572.

E-mail address: fue5hl@leeds.ac.uk (H. Liu).

Reprinted from *Fuel* **80 (9)**, 1211-1215 (2001)

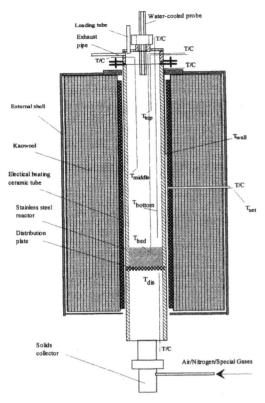

Fig. 1. A schematic diagram of the BFB combustor system.

suspected that the effect of calcined limestone on the conversion of char-N to NO/NO$_x$ and N$_2$O could be responsible partially for this observation [7]. This suspicion has been investigated in this study by conducting batch-type char combustion tests in a bench-scale bubbling fluidized bed (BFB) combustor. Two different chars, one from a high volatile bituminous coal and the other from an anthracite were produced in situ within the BFB combustor before being subjected to batch-type combustion tests. Both silica sand and calcined limestone diluted by silica sand have been used as the bed materials. The comparison of the conversions of char-N to NO/NO$_x$ and N$_2$O with different bed materials will reveal the effect of calcined limestone on the conversion of char-N under fluidized bed conditions.

2. Experimental

Fig. 1 shows the schematic of the bench-scale, electrically heated, BFB combustor system used in the present work. The combustor itself, including its distributor, was made of stainless steel and 65 mm in diameter and 500 mm in length. Both silica sand, within the size range of 212–600 μm, and a UK limestone (Bennite Middle Cut) containing 98.8% calcium carbonate within the size range of 212–355 μm, diluted by sand were used as the bed materials. The inventory of bed materials was 300 g and the static bed height was roughly 60 mm. The flow rate of fluidizing gas was fixed at 1.217×10^{-4} m^3 s^{-1} at STP and the plug flow gas residence time inside the combustor above the distributor was about 3 s at 1073 K. Two chars, one from a UK bituminous coal and the other from an anthracite were produced in situ within the combustor before being subjected to batch-type combustion tests. The analyses of the UK bituminous coal and the anthracite are given in Table 1. The parent coal particles ranged in size from 1.40 to 3.35 mm. All chars were produced at a bed temperature of 1073 K with an initial weight of 1.0 g of a parent coal, whereas char combustion tests were conducted at two different temperatures (1023 and 1073 K). After the combustor reached the set temperature for char production (1073 K) with nitrogen as the fluidizing gas, parent coal particles were loaded into

coal-fired fluidized bed combustors. But, so far few studies have specifically investigated the effect of limestone on the conversion of char-N to NO$_x$ and N$_2$O.

Earlier work [7] by the authors on the effect of limestone addition on NO/NO$_x$ and N$_2$O emissions from a pilot-scale CFB combustor observed that limestone addition above the secondary air injection ports reduced more N$_2$O, but unfortunately increased more NO/NO$_x$ as well, compared with limestone addition at the bottom of the riser. It was

Table 1
Analysis of coals (wt% as analysed)

	Proximate			Ultimate	
	Bituminous coal	Anthracite		Bituminous coal	Anthracite
Moisture	3.7	1.9	C	73.1	88.6
Ash	3.5	2.8	H	4.8	3.4
Volatile matter	37.9	7.6	O[a]	12.3	1.2
Fixed carbon[a]	54.9	87.7	N	1.3	1.3
			S	1.3	0.8
Gross calorific value (MJ/kg)				28.92	34.50

[a] By difference.

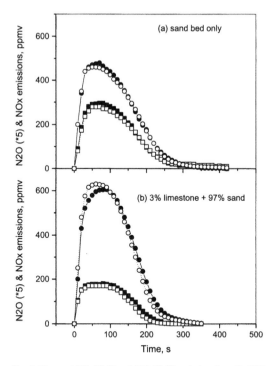

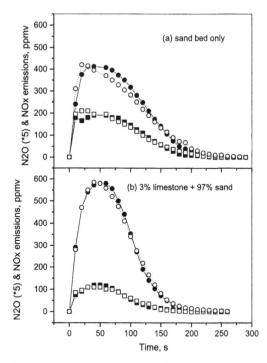

Fig. 2. Measured NO$_x$ (O ●) and N$_2$O (□ ■) emissions from the BFB combustor during anthracite char combustion with different bed materials. $T = 1073$ K. Closed symbols: RUN 1, Open symbols: RUN 2.

Fig. 3. Measured NO$_x$ (O ●) and N$_2$O (□ ■) emissions from the BFB combustor during bituminous coal char combustion with different bed materials. $T = 1073$ K. Closed symbols: RUN 1, Open symbols: RUN 2.

the combustor and pyrolysed for 30 min with the released volatiles being continuously swept away by the fluidizing nitrogen. Subsequently, the combustion tests could be started by switching the fluidizing gas from nitrogen to air. If a char combustion test was to be conducted at a different temperature from that of char production, the reactor would be continuously fluidized by nitrogen and electrically heated or cooled until the desired combustion test temperature had been reached. Due to volatiles and volatile-N released during pyrolysis strongly depending on the pyrolysis temperature but being insensitive to time after being held at the fixed temperature for 30 min, the combustion test temperatures were always set equal to or below the char production temperature ($\sim$1073 K) so that volatiles and volatile-N released or char and char-N left for combustion tests were constant for a particular parent coal.

Concentrations of CO, O$_2$, NO/NO$_x$ and N$_2$O at the exhaust of the combustor were continually measured by on-line analysers and recorded by a chart recorder. All the results of char combustion tests reported were those of the averages from two to three duplicate runs. As will be shown later in Section 3, the repeatability of the char combustion tests was very good and the uncertainty in the amount of NO$_x$ or N$_2$O emitted during the char combustion period was usually less than 5% of the average from duplicate runs.

3. Results and discussion

Char combustion tests with the BFB combustor. Figs. 2 and 3 show the measured NO$_x$ and N$_2$O emissions from the anthracite char and the bituminous coal char combustion tests with two different sets of bed materials at a bed temperature of 1073 K, respectively. Experimental results from two duplicate runs are shown in both Figs. 2 and 3 and obviously the repeatability of the char combustion tests is excellent. Table 2 shows the percentage conversion of char-N to NO$_x$ and N$_2$O, which was estimated by integrating the NO$_x$ and N$_2$O emission profiles shown in Figs. 2 and 3 and assuming that char-N and weight loss during pyrolysis were the same as the parent coal-N and the proximate volatile matter, respectively. Figs. 2 and 3 and Table 2 clearly demonstrate that the peak NO$_x$ emission and the percentage conversion of char-N to NO$_x$ increase, whereas the peak N$_2$O emission and the percentage conversion of char-N to N$_2$O decrease when 3% (by weight) of the silica sand bed is replaced by limestone. Similar results were also obtained for both the anthracite char and the bituminous coal char burned at 1023 K (Table 2). These results indicate that in addition to reducing N$_2$O emissions from char combustion, the calcined limestone also enhances the conversion of char-N to NO$_x$ as in the case of volatile-N conversion to NO$_x$.

Table 2
Percentage conversion of char-N to NO_x and N_2O (%) (char production temperature: 1073 K)

	Combustion temperature (K)	Anthracite char			Bituminous coal char		
		NO_x	N_2O	NO_x+N_2O	NO_x	N_2O	NO_x+N_2O
Sand bed only	1073	51.08	12.18	63.26	42.28	8.34	50.62
3% Limestone+97% sand	1073	62.99	6.98	69.97	51.99	3.93	55.92
Sand bed only	1023	50.36	13.50	63.86	37.62	8.00	45.62
3% Limestone+97% sand	1023	68.20	6.10	74.30	45.47	3.49	48.96

Therefore, the often observed increases in NO_x emissions and decreases in N_2O emissions, due to limestone addition, from coal-fired fluidized bed combustors are also caused by the influence of limestone on the conversions of char-N to NO_x and N_2O. This partially explains the earlier observation [7] that limestone addition above the secondary air injection ports reduced more N_2O but also increased more NO_x than limestone addition at the bottom of the CFB riser.

Table 2 also shows that the total percentage conversion of char-N to NO_x and N_2O increased by up to about 3.3–5.3% (absolute) for the bituminous coal char and about 6.7–10.4% (absolute) for the anthracite char when the bed materials were changed from sand to the mixture of limestone and sand. This indicates that although limestone can reduce N_2O emission it may increase the total emissions of NO_x and N_2O. In addition, when limestone is used as a part of the bed materials in a coal-fired fluidized bed combustor, it will be partially sulphated and its effect on NO_x and N_2O emissions will be decreased. Therefore, one may have to accept that the total N_2O and NO_x emissions from coal-fired fluidized bed combustors will increase as long as limestone addition is used for sulphur capture.

The difference in the conversion of char-N to NO_x and N_2O between the bituminous coal char and the anthracite char is significant as shown in Table 2. The conversion of the anthracite char-N to NO_x or N_2O was always higher than that of the bituminous coal char under the same char combustion conditions and with the same bed materials. The total conversion of char-N to NO_x and N_2O for the anthracite char was about 12.6–14.0% (absolute) at 1073 K and about 18.2–25.3% (absolute) at 1023 K higher than that for the bituminous coal char. These observations agree with the rank dependence of char-N conversion to NO_x and N_2O reported by Pels et al. [13].

It is well-known that there exists a trade-off between NO_x and N_2O in fluidized bed combustion of coals and also of chars [13]. However, in the present study, the effect of temperature on the conversion of char-N to NO_x and N_2O was complicated by NO_x and N_2O decompositions which occurred inside the reactor and were catalysed by bed materials including char particles and calcined limestone particles [14]. As shown in Table 2, the changes in the conversions of char-N to N_2O were negligible for both chars and both sets of bed materials when the combustion temperature increased from 1023 to 1073 K. On the other

hand, for the bituminous coal char, the conversions of char-N to NO_x clearly increased with both sets of bed materials when the combustion temperature increased from 1023 to 1073 K. This trend was also observed for the anthracite char burned with the silica sand bed. These results agree with Pels et al. [13]. However, when the anthracite char was burned with the limestone and sand mixture bed, an increase in bed temperature from 1023 to 1073 K resulted in a decrease in NO_x conversion which contradicts with the trends shown by the bituminous coal char burned with either set of bed materials and the anthracite char burned with the silica sand bed. The reduction of NO_x by char and carbon monoxide, which can be catalysed by calcined limestone and produce N_2O (and N_2), may be one of the reasons for the observation.

Mechanisms of char-N oxidation. Release of nitrogen oxides during char combustion has been reviewed by Wójtowicz et al. [2], Johnsson [15] and Thomas [16]. It has been concluded that NO is the primary nitrogen combustion product and NO is reduced on the surface or in the pores of the char to form N_2 and/or N_2O [16]. De Soete et al. [17] recently proposed the following heterogeneous reactions for the combustion of char-bound nitrogen:

$$O_2 + (-C) + (-CN) \rightarrow (-CO) + (-CNO), \qquad (1)$$

$$(-CNO) \rightarrow NO + (-C), \qquad (2)$$

$$NO + (-C) \rightarrow (1/2)N_2 + (-CO), \qquad (3)$$

$$NO + (-CNO) \rightarrow N_2O + (-CO), \qquad (4)$$

$$N_2O + (-C) \rightarrow N_2 + (-CO), \qquad (5)$$

where species written inside brackets are adsorbed. Reaction (4) was considered to be the most probable heterogeneous way of forming N_2O from char-nitrogen and its rate was determined for two bituminous coal chars. However, some experimental evidence [18–21] clearly suggests that homogeneous reactions are involved in the formation of nitrogen oxides from combustion of char-N. Klein and Rotzoll [18] compared NO and N_2O emissions from char combustion in a sand bed with those in a MgO-particles bed. They found that N_2O emissions decreased to below detection limit and NO emissions increased dramatically when MgO particles were used as the bed material instead of sand. From these results, they postulated that a gas-phase nitrogen intermediate, most possibly HCN, whose further reactions were catalytically affected by MgO particles was involved in the formation of nitrogen oxides from char-N. Jones et al. [19] studied the reactive intermediates produced in the temperature-programmed combustion of carbons derived from the high-pressure carbonisation of polycyclic aromatic compounds containing well-defined nitrogen and sulphur functionalities. Gas sampling ~1 cm above the char detected HCN, $(CN)_2$ and COS in addition to CO, CO_2, NO and SO_2. This indicated that the reactive species were undergoing homogeneous reactions leading to the corresponding oxides. Winter et al. [20,21] applied an iodine addition technique to study the mechanisms of NO and N_2O formation from a bituminous coal char and a petroleum coke char combustion under fluidized bed conditions. It was concluded that during char combustion, NO was mainly heterogeneously formed while N_2O was homogenously formed through HCN which was released from char-N.

The results of the present study shown in Table 2 support the claim that at least to some extent, the formation of NO_x and N_2O from char-N involves homogeneous reactions of a nitrogen intermediate, most possibly HCN. The significant increases in NO_x conversions from char-N when 3 wt% of the sand bed was replaced by limestone indicated that NO_x formation was catalytically influenced by calcined limestone particles. It was well documented that calcined limestone particles could catalytically enhance the formation of NO_x from the oxidation of HCN [3,9,11,22]. Therefore, the effect of calcined limestone particles on the conversions of char-N to NO_x and N_2O observed in the present study could well be explained by assuming that HCN is the nitrogen intermediate which undergoes further homogeneous reactions leading to the formation of NO_x and N_2O during char combustion.

4. Conclusions

The catalytic effect of limestone on the conversion of volatile-N to NO_x is only partially responsible for the often observed increase in NO_x emissions when limestone is added to a fluidized bed combustor as char combustion tests conducted in the present work with the BFB combustor demonstrate that calcined limestone also enhances the conversion of char-N to NO_x. Further analyses of the char combustion test results support the claim that the formation of NO_x and N_2O from char-N involves homogeneous reactions of a nitrogen intermediate, most possibly HCN.

Acknowledgements

UK EPSRC is acknowledged for the financial support of the research work reported in this paper.

References

[1] Mann MD, Collings ME, Botros PE. Prog Energy Combust Sci 1992;18:447.
[2] Wójtowicz MA, Pels JR, Moulijn JA. Fuel Process Technol 1993;34:1.
[3] Jensen A, Johnsson JE, Dam-Johansen K. Twenty-sixth Symposium (International) on Combustion. The Combustion Institute, 1996. p. 3335.
[4] Shimizu T, Tachiyama Y, Fujita D, Kumazawa K, Wakayama O, Ishizu K. Energy Fuels 1992;6:753.
[5] Åmand L-E, Leckner B, Dam-Johansen K. Fuel 1993;72:557.
[6] Collings ME, Mann MD, Young BC. Energy Fuels 1993;7:554.
[7] Liu H, Gibbs BM. Fuel 1998;77:1569.
[8] Iisa K, Salokoski P, Hupa M. In: Anthony EJ, editor. Proceedings of the 11th International Conference on Fluidized Bed Combustion. Montreal, Canada: ASME, 1991. p. 1027.
[9] Jensen A, Johnsson JE, Dam-Johansen K. In: Rubow L, editor. Proceedings of the 12th International Conference on Fluidized Bed Combustion. San Diego, CA: ASME, 1993. p. 447.
[10] Lin W, Johnsson JE, Dam-Johansen KD, van de Bleek CM. Fuel 1994;73:1202.
[11] Hayhurst AN, Lawrence AD. Combust Flame 1996;105:511.
[12] Wójtowicz MA, Pels JR, Moulijn JA. Fuel 1994;73:1416.
[13] Pels JR, Wójtowicz MA, Kapteijn F, Moulijn JA. Energy Fuels 1995;9:743.
[14] Rodriguez-Mirasol J, Ooms AC, Pels JR, Kapteijn F, Moulijn JA. Combust Flame 1994;99:499.
[15] Johnsson JE. Fuel 1994;73:1398.
[16] Thomas KM. Fuel 1997;76:457.
[17] De Soete GG, Croiset E, Richard J-R. Combust Flame 1999;117:140.
[18] Klein M, Rotzoll G. In: Hupa M, Matinlinna J, editors. Proceedings of the 6th International Workshop on Nitrous Oxide Emissions. Turku/Åbo, Finland, 1994. p. 239.
[19] Jones JM, Harding AW, Brown SD, Thomas KM. Carbon 1995;33:833.
[20] Winter F, Wartha C, Löffler G, Hofbauer H. Twenty-sixth Symposium (International) on Combustion. The Combustion Institute, 1996. p. 3325.
[21] Winter F, Wartha C, Löffler G, Hofbauer H, Preto F, Anthony EJ, Can J. Chem Engng 1999;77:275.
[22] Jesen A. PhD. thesis. Department of Chemical Engineering, Technical University of Denmark, 1996. ISBN 87-90142-14-4.

A. Tomita (Editor)

Suppression of nitrogen oxides emission by carbonaceous reductants

Akira Tomita[*]

Institute for Chemical Reaction Science, Tohoku University, 2-1-1 Katahira, Sendai 980-8577, Japan

Abstract

The present status of NO$_x$ emission from power stations and automobiles is first summarized, and the controlling regulations in respective areas are reviewed. In spite of much progress, we have to further reduce the NO$_x$ emission in all the areas. In order to develop more effective technology, the fundamental understanding of the relevant reactions is essential. The heterogeneous reactions, like NO$_x$ and N$_2$O formation from coal char, NO$_x$ and N$_2$O reduction with carbon, and NO$_x$ reduction with hydrocarbon gases over heterogeneous catalysts are not well understood yet. This paper briefly summarizes our recent studies on the heterogeneous reactions of NO$_x$ formation and destruction. The importance of surface nitrogen species is emphasized in all the reaction systems. The presence of such surface species plays a very important role, not only in NO$_x$ destruction on carbon surfaces, but also in the NO$_x$ release during coal char combustion. Finally, future research areas are identified, where we need to understand what actually happens under high-temperature reaction conditions. © 2001 Elsevier Science B.V. All rights reserved.

Keywords: NO$_x$; Coal combustion; Diesel; Carbon; Surface species; Mechanism

1. Introduction

We are now standing at a turning point of centuries. In the last century, the energy situation has changed a lot. A hundred years ago, the main energy sources were coal and wood, and the use of oil had just started. Nobody could anticipate the growth of nuclear energy. The change in the next hundred years will be much faster than the past. It is quite difficult to forecast the energy situation in 2100, but we are obligated to make every effort to leave a better society for our grandchildren. We need to save as much

[*] Tel.: +81-22-217-5625; fax: +81-22-217-5626.
E-mail address: tomita@icrs.tohoku.ac.jp (A. Tomita).

Reprinted from *Fuel Processing Technology* **71** (1-3), 53-70 (2001)

energy as possible and we also have to leave a clean environment. The natural consequence of this argument will be the acceleration of the shift from the conventional energy system that heavily depends on fossil energy to the renewable energy system using solar, wind, biomass energies, etc. If such a shift can be made, our future generations will be free from the energy problem. Unfortunately, this will not be the case except for some limited locations. The absolute available amount of renewable energy will be insufficient to meet the increasing demand. For at least for more than three decades from now, we have no other option but to depend on fossil energy resources. Therefore, it is necessary to develop new technology to use these sources more efficiently and cleanly, because fossil energy resources have an intrinsic shortcoming, that is, the emission of many undesirable by-products. Among them, the NO_x emission is one of the greatest anxiety. NO_x, together with N_2O, causes many hazardous phenomena, such as acid rain, smog formation, global warming, and ozone layer depletion. We have been developing many NO_x-controlling technologies, but these are still not enough to meet the more stringent regulation in future. In this paper, the present state of the art for controlling NO_x emissions from various sources and future research directions will be reviewed. In order to avoid discursiveness, the focus will be put on the chemistry of NO_x destruction reactions, particularly on the reduction with carbonaceous materials, including hydrocarbon gases.

2. State of the art of NO_x reducing technology in Japan

2.1. General trend of NO_x formation

Fossil fuel combustion in industry and transport sectors is a major source of NO_x emission. As shown in Fig. 1, roughly speaking, half of NO_x comes from industrial activities and another half comes from transportation sources [1]. Fossil fuel-fired power plants are one of the most important NO_x producers. Among mobile sources, diesel

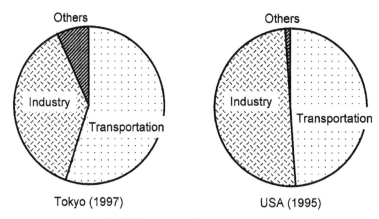

Tokyo (1997)   USA (1995)

Fig. 1. NO_x emission from various sources.

trucks and buses emit approximately three-quarters of NO_x, and gasoline cars emit the remaining one-quarter. In this paper, we will mainly deal with the emission and control of NO_x from pulverized coal combustion systems, but the emission from other sources will also be reviewed for comparison.

In the last several decades, many controls for the reduction of NO_x have been applied [2]. Fig. 2 shows the effect of such primary controls for NO_x reduction from coal-fired power plants in Japan. It is obvious that the emission has been suppressed quite successfully. In addition, many flue gas de-NO_x units have been installed as a secondary measure in the last two decades. The capacity (million m^3/h) has increased from 40 in 1980 to 330 in 1995. In spite of these efforts, the average concentration of atmospheric NO_2 has been constant at 0.02 ppm for the same period of time. This is partly due to the increasing number of cars and the absence of effective NO_x removal technology especially for heavy-duty diesel cars. Therefore, the development of more effective measures is strongly desired in each sector.

2.2. Pulverized coal combustion

Without any measures, pulverized coal-fired power plant may emit more than 700 ppm of NO_x. Thanks to many efforts in the last 25 years, emission has been reduced to a considerable extent. Japanese national standard for NO_x emission from pulverized coal-fired power plant is 200 ppm for these 20 years, but usually local authorities have enforced more stringent emission standards, usually less than 60 ppm. Recently, an emission level as low as 15 ppm has been agreed for a plant under development. As primary measures, low NO_x burners, overfire air (air staging), reburning (fuel staging),

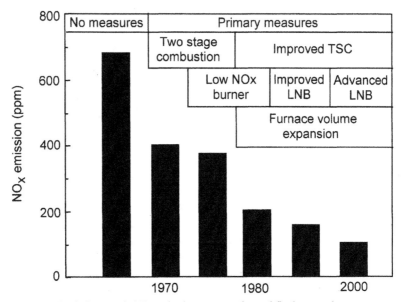

Fig. 2. Progress in NO_x reduction measures for coal-fired power plant.

flue gas recirculation and others are adopted, and selective catalytic reduction or selective non-catalytic reduction are widely used as secondary measures [3]. Table 1 summarizes the reactions and the processes that are related to the NO_x control technology.

The extent of use of these technologies differs from country to country. In Japan, both low NO_x burner and air staging are frequently combined as primary measures, and sometimes even flue gas re-circulation is adopted. Primary measures achieve 30–70% reduction in NO_x emission when applied separately, and a higher reduction, up to 80%, can be achieved when applied in combination. Therefore, the reduction to 150–200 ppm level is easily achieved in advanced coal-fired power stations even without secondary measures. If these primary measures are combined with secondary measures, the emission level can be suppressed to as low as 50–70 ppm. In Japan, actually more than 60% of power plants use a combination of primary and secondary measures. Recently, it was reported that a new burner was able to suppress the NO_x emission level to < 100 ppm, by keeping the unburnt carbon level at 3% [4]. Therefore, it would not be impossible to achieve a severe standard like 10 ppm that will be enforced in the near future. On the contrary, most of the power stations in USA do not use any secondary measures and only either air staging or low NO_x burners are adopted. Therefore, the NO_x emission level is somewhat higher than those of Japanese power stations.

2.3. Fluidized bed combustion

The easiest way to suppress the NO_x emission in coal combustion is to decrease combustion temperature. Thus, fluidized bed combustion (FBC) that is operated at around 800–900°C has attracted much attention. NO_x emissions from commercial FBC plants are 100–150 ppm even without any measures for NO_x control. However, the emission of N_2O is rather high in such a temperature region. Roughly speaking, N_2O emission from FBC is around 10–200 ppm, whereas that from a pulverized coal combustor is only 0.5–2 ppm. There is no regulation for N_2O emission at the present moment, but the necessity of its control is well recognized. The mechanism of the N_2O formation and destruction is not fully elucidated, and therefore the understanding of these reactions is essential for finding the best solution to reduce the N_2O emission.

2.4. NO_x emission from mobile sources

For gasoline-powered automobiles, the so-called three-way catalyst excellently reduces the emission of NO_x together with hydrocarbons and CO. The Japanese emission standard for NO_x from gasoline-powered passenger cars is 0.25 g/km and it will be reduced to 0.08 g/km in the very near future. A recent challenge in this field is the use of the lean-burn engine for improving energy efficiency. For this engine, the conventional three-way catalyst is inapplicable because of the presence of much O_2. The development of a new active catalyst is strongly required, and several promising catalysts are proposed and used in commercial cars [5,6].

The NO_x emission standard in Japan for heavy-duty diesel engines has been decreasing year after year. Nevertheless, the emission is significantly higher than that for

Table 1
NO_x reduction reaction and technology

Reduction mode	Pulverized coal combustion	Fluidized bed combustion	Diesel vehicle	Gasoline vehicle
Gas phase reduction by hydrocarbon gas, etc.	Low NO_x burner, Reburning, Flue gas recirculation	–	Delayed fuel injection, Exhaust gas recycle	Delayed fuel injection, Exhaust gas recycle
Reduction by gas over catalyst	Selective catalytic reduction by NH_3	–	Selective catalytic reduction	Three-way catalyst
Reduction by carbon	Reduction by char	Reduction by char	Simultaneous removal of particulate and NO_x	–

a gasoline car. Effective controls have been explored for a long time, but it has not been easy to find out a good solution. Engine modification and exhaust gas recycle measures have been applied. They were successful to some extent but the removal rate was not enough. Other attempts that use catalysts to reduce NO_x in the exhaust gas have been attempted. A great break-through was made in 1990. Effective catalyst systems for the selective reduction of NO_x with hydrocarbon gases were discovered independently by several groups [7–9]. An addition of a small amount of hydrocarbon gas quantitatively reduces NO_x over a catalyst in the presence of O_2. Extensive catalyst searches have been done since this finding. One of the promising systems is a base-metal catalyst with 1% of diesel oil injection as reductant [10]. However, the NO_x reduction rate is still only 30%, and thus, a further break-through is necessary.

3. Reaction path of NO_x reduction to N_2

3.1. NO_x-related reactions in pulverized coal combustion

Fig. 3 presents a semi-quantitative flow diagram of nitrogen species during combustion with a low NO_x burner. The final emission of NO_x is determined as the balance of NO_x-forming and NO_x-reduction reactions. Simply speaking, the purpose of the NO_x controlling technology is either to decrease NO_x formation by moderating combustion temperature or to enhance the reduction of NO_x that was formed in the previous steps. These technologies became possible only through the fundamental understanding of these reactions. It can be said that the principal mechanisms in the homogeneous phase

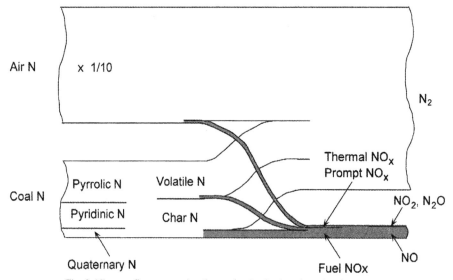

Fig. 3. Nitrogen flow pattern in advanced pulverized coal combustion system.

have almost been identified. The formation of thermal NO_x (Zeldovich NO_x) is initiated by the reaction between O radical and N_2 in air.

$$N_2 + O \rightarrow NO + N \tag{1}$$

$$N + O_2 \rightarrow NO + N \tag{2}$$

Another source of thermal NO_x is known as prompt NO_x, which is formed through the decomposition of N_2 with the aid of hydrocarbon. These are the only NO_x formation mechanisms in natural gas combustion, but are not predominant in pulverized coal combustion.

On the other hand, the contribution of NO_x derived from fuel nitrogen (fuel NO_x) is much more important, and its formation reaction can be described as follows,

$$C(N) \rightarrow I(N) \tag{3}$$

$$I(N) + O\,(or\,O_2, OH) \rightarrow NO + \cdots \tag{4}$$

where C(N) denotes the nitrogen in char while I(N) represents nitrogen-containing intermediate species like CN, HCN, NH and NH_2. Approximately 75–90% of NO_x is from this source when low NO_x burners are used (Fig. 3). The main reaction pathways for fuel nitrogen are illustrated in Fig. 4. The volatile nitrogen and the char nitrogen are oxidized to NO, and the NO is reduced by such reductants as hydrocarbon species and solid char. Among these reactions, the reduction of NO_x over the char surface is quite complex and not yet fully understood. In order to simulate the reactions taking place in the combustor, it is necessary to establish a good model that takes all of these complicated reactions into consideration. Many attempts have been made to simulate the NO_x emission during pulverized coal combustion. However, the agreement with the observed data is not always satisfactory. This is partly because of the lack of reliable experimental data on the heterogeneous reaction [11–13]. Thus, the formation of NO_x from fuel nitrogen and the decomposition of NO_x over the char surface should be elucidated in detail.

3.2. Reduction of NO_x in homogeneous phase

Fig. 5 illustrates four distinct zones in low NO_x burner flame. In NO_x reduction zone, homogeneous reduction of NO by CH, CH_3, CN or NH radical takes place, forming

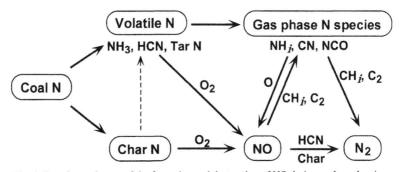

Fig. 4. Reaction pathways of the formation and destruction of NO during coal combustion.

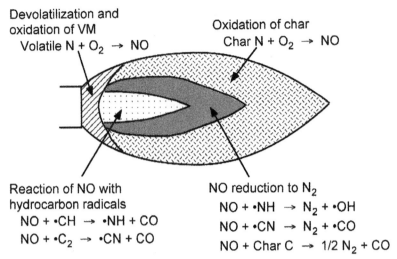

Fig. 5. NO$_x$ formation and destruction near a low NO$_x$ burner.

intermediate species. In some boilers, reburn fuel is injected into the upper part of the boiler. Natural gas and coal are frequently used for this purpose. Some of the main reactions in these zones are as follows.

$$NO + CH \rightarrow HCN + O \tag{5}$$

$$NO + CH_3 \rightarrow HCN + H_2O \tag{6}$$

Tang and Ohtake [14] simulated NO formation in pulverized coal combustion where they took 72 elementary reactions with 25 chemical species. The reaction rates for these homogeneous NO reductions are available and all of them are presented in Arrhenius type expressions. A model to simulate homogeneous reactions in the reburning zone was presented by Chen et al. [15].

3.3. Reduction of NO$_x$ with hydrocarbon gas in the presence of a catalyst

V_2O_5/TiO_2 type catalysts are widely used to reduce NO$_x$ from the flue gas of pulverized coal combustors. The reductant is usually NH$_3$, and this is out of the scope of this paper.

NO$_x$ reduction by hydrocarbons over heterogeneous catalysts has been attempted in connection with the control of NO$_x$ emission in mobile systems. The three-way catalyst is used in gasoline-powered cars, where NO$_x$ is reduced over Pt–Rh–Pd catalysts by hydrocarbon gas and CO. The search for new catalysts for diesel engine cars has been underway for many years. As stated before, several active catalysts that can selectively reduce NO with a small amount of hydrocarbon gas in the presence of excess O$_2$ have been found [7–9]. This reduction takes place only in the presence of O_2, but the role of O_2 is not very clear. General features for this reaction system are as follows.

1. Low temperature activity only with O_2 and hydrocarbon.
2. Starting temperatures for hydrocarbon oxidation and NO reduction are close.
3. At higher temperature, hydrocarbon oxidation becomes predominant.

The reaction scheme proposed by Iwamoto and Mizuno [16] is as follow.

$$NO + O_2 \rightarrow NO_2 \tag{7}$$

$$NO_2 + HC \rightarrow I(CHON) \tag{8}$$

$$HC + O_2 \rightarrow I(CHO) \tag{9}$$

$$NO + I(CHON) \text{ or } I(CHO) \rightarrow N_2 \tag{10}$$

I(CHON) and I(CHO) represent active intermediate species containing respective elements. The formation of these species is essential in this mechanism. The role of hydrocarbon and O_2 is to generate this active species. The last step is still not clear, and no reasonable explanation has been presented on how two nitrogen atoms get together to produce the N_2 molecule. The understanding of these reactions and the design of a good catalyst system are very challenging subjects.

3.4. Reduction of NO_x by solid carbon

As stated above, reliable kinetic data and detailed reaction mechanisms on the formation of NO_x from char nitrogen and its reduction by solid carbon are not sufficient yet, in spite of their importance in determining the overall NO_x emission during pulverized coal combustion with low NO_x burners. A part of NO_x is rapidly reduced by gaseous reductants produced during devolatilization as stated in Section 3.2, but these reductants are consumed very rapidly. Even after this stage, NO_x formation continues to accompany char combustion. The remaining NO_x would be either reduced to N_2 on the surface of char or emitted as it is. Thus, char would function as the main reductant in the later stage. Goel et al. [17] proposed that the heterogeneous reaction with the char is the main NO destruction mechanism at high temperatures. Most of NO reduction on the carbon will occur in the pores where the oxygen concentration is very low. The mechanism is very complicated, because the concentrations of both reactant and product gases vary with the depth of penetration into the pores.

4. Chemistry of NO_x reduction by carbon

4.1. Reaction of carbon with NO

In this section, the chemistry of NO_x reduction by carbon will be discussed. There are many reports and several excellent reviews. For example, Thomas has reviewed nitrogen in coal, nitrogen behavior on pyrolysis, NO reaction with carbon, and influential factors on NO$_x$ formation during coal combustion [18]. Aarna and Suuberg [19] have examined previously reported extensive kinetic data on NO reduction with carbon. Therefore, instead of repeating these aspects, our recent results related to NO$_x$ reduction by carbon will be summarized. We have carried out fundamental studies on the reaction of pure carbon (phenol–formaldehyde resin char) instead of coal char. The use of pure carbon is beneficial from various aspects in understanding the C–NO reaction mechanism; there are no volatile matter, no mineral matter and no nitrogen in pure carbon. The most salient feature of our approach is that we attempted to establish a detailed nitrogen mass balance during the C–NO reaction. Most combustion people are mainly interested in the formation of NO$_x$ from char nitrogen during char combustion, and not in N$_2$ formation. This is evident from the fact that most combustion studies were carried out by using air, that is O$_2$/N$_2$ mixture. Even when O$_2$/inert gas mixture was used instead of O$_2$/N$_2$ mixture, the nitrogen mass balance has been neglected in many studies [20–25], except for a recent study by Ashman et al. [26]. Therefore, we have attempted to examine the detailed mass balance throughout the C–NO reaction, and by doing so we could quantify the amount of nitrogen accumulated on the carbon surface during the reaction. Furthermore, we clarified the role of such surface species in the reaction. The reaction temperature we used was below 1000°C. Therefore, the information obtained cannot be applied to pulverized coal combustion, but it can be used to understand the phenomenon occurring during fluidized bed combustion.

One of the most important results is that the nitrogen accumulation on carbon surface from gas-phase NO was clearly confirmed from an accurate mass balance. One example is shown in Fig. 6 [27]. Many researchers had presumed the presence of such surface nitrogen species, but there had been neither direct evidence nor quantitative data. Also, from the XPS analysis, the presence of surface nitrogen species was confirmed [27,28]. XPS analysis showed the formation of pyridinic, pyrrolic and quaternary nitrogen on the carbon surface. The C(NO) species was minor, if any. Fig. 7 shows one example, but the relative ratio among these species greatly depends on the sample preparation conditions,

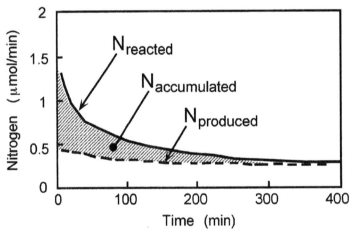

Fig. 6. Nitrogen mass balance during the C–NO reaction at 600°C.

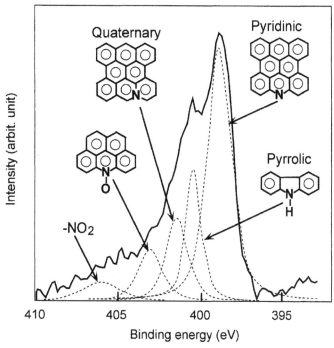

Fig. 7. N_{1s} XPS profile after the C–NO reaction at 900°C.

such as heat-treatment temperature, residence time at high temperature, the presence of oxidizing gas, and other factors. The initial step of C–NO reaction can be described as

$$NO + C( ) \rightarrow C(N) + C(O) \tag{11}$$

where C( ), C(N) and C(O) denote the surface free site, surface nitrogen and oxygen species, respectively. The XPS profiles observed for nitrogen species formed on pure carbon are generally similar to those observed for nitrogen species on coal chars.

Then, we identified the N_2 formation mechanism in the C–NO reaction by step response experiments using isotope gases [29,30]. The reactant gas was switched from $^{14}N^{16}O$ to $^{15}N^{18}O$ during the reaction with carbon, and the product gases were analyzed as in Fig. 8. The appearance of $^{14}N^{15}N$ immediately after the introduction of $^{15}N^{18}O$ strongly suggested that the main route for N_2 formation be as follows.

$$C(N) + NO \rightarrow N_2 + C(O) \tag{12}$$

This holds true in a wide temperature range from 600°C to 1000°C. This study presented the elementary reaction step of N_2 formation for the first time. Before this study, it was thought rather ambiguously that N_2 is formed from two C(N) species [13,18–20,31].

$$2C(N) \rightarrow N_2 + 2C( ) \tag{13}$$

It is not easy for two nitrogen atoms on the carbon surface to combine with each other because of the rather small concentrations. However, our results do not completely

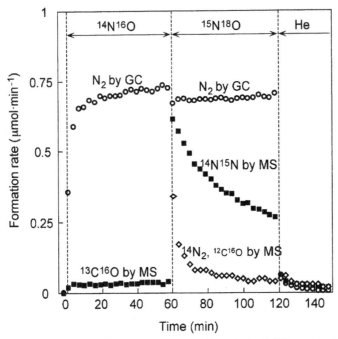

Fig. 8. N_2 evolution profile during a step response experiment in the C–NO reaction at 850°C.

rule out the possibility of this reaction, since a small amount of $^{14}N_2$ was observed in the step response experiment [30]. Perhaps this mechanism will be more important at higher temperatures, where the mobility of these surface nitrogen species would increase and thus the chance for two nitrogen atoms to form a $N{\equiv}N$ bond may become large.

4.2. Effect of O_2 on the reaction of C–NO

In the coal combustion system, the NO reduction by char takes place more or less in the presence of O_2. The influence of O_2 on the C–NO reaction is very remarkable. N_2 formation rate is significantly increased by the presence of O_2, and N_2O formation is also greatly affected by O_2. It is surprising to note that there has been almost no systematic study on this important C–NO–O_2 system. Therefore, we examined the effect of O_2 concentration on the formation of N_2 and N_2O and confirmed the above enhancement effect [27,32,33]. The N_2 and N_2O formation profiles are very complicated ones as shown in Fig. 9. Nitrogen accumulation on the carbon surface was also observed as it was in the absence of O_2 [33]. It was revealed that the main N_2 formation path is similar to Eq. (12). A small amount of N_2O was observed even in the absence of O_2, and this brings a marked contrast with many previous reports [22,23,34,35], where it was concluded that the formation of N_2O during C–NO reaction was only possible in the presence of O_2. From a fundamental point of view, this finding is important to elucidate a detailed reaction mechanism.

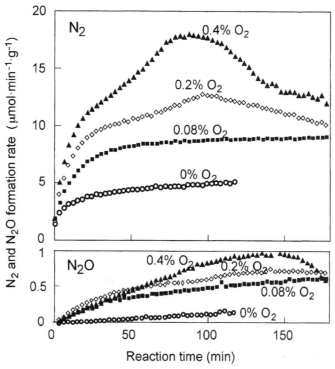

Fig. 9. N_2 and N_2O formation rate during the C–NO reaction in the presence of O_2. NO concentration: 0.05% (4.4 μmol min^{-1}); Temperature: 850°C.

It should be noted that the reaction profile for this reaction has several similarities with that of catalytic NO_x reduction with hydrocarbon gas described in Section 3.3. In both cases, the presence of O_2 is important to activate reductants, either hydrocarbon gas or solid carbon. Furthermore, both reactions are similar in a sense that the reaction is selective in the case of catalytic reduction of NO_x with hydrocarbon, while it is non-selective in NO_x reduction by carbon.

4.3. Reaction of carbon with N_2O

Compared with the reaction with NO, the reduction of N_2O by carbon and the formation of N_2O from coal char have been investigated to only a small extent. These reactions are not so important for high temperature combustion, but are very important in a low temperature process like fluidized bed combustion. As in the case of NO, the formation and destruction of N_2O concurrently take place in the combustor. One of the interesting subjects in this field is to clarify the contribution of heterogeneous reduction of N_2O on the carbon surface.

In the case of the reduction of N_2O by carbon, it has been assumed that N_2 is produced from N_2O without any bond breaking between two nitrogen atoms, although there was no direct evidence. We have carried out a step response experiment, and found the reaction mechanism to be that given in Eq. (14), as was postulated in many studies. There is little N≡N splitting during C–N$_2$O reaction [36].

$$N_2O + C(\ ) \rightarrow N_2 + C(O) \tag{14}$$

4.4. Simulation of C–NO and C–N$_2$O reactions by molecular orbital calculation

The chemisorption of the NO molecule on a carbon surface and the subsequent decomposition of the surface species were simulated by an ab initio molecular orbital theory calculation [37]. For simplicity, a single layer of polyaromatic compounds was employed as a model for carbon. The molecular orbital calculation of the system including both the model carbon and NO molecule was made, and its geometrical parameters were optimized. The adsorption of NO on carbon edge sites resulted in the formation of several types of stable NO containing complexes, C(NO), although this was a minor species when examined by XPS analysis [28]. To elucidate the N$_2$ formation route from the surface nitrogen complexes, one more molecule of NO was put on the N atom of the C(NO) species. The calculation predicted the formation of six-member ring complex including NNO bonding. The bond population analysis predicted a probable route for the dissociation of such a complex. The N–O bond is so weak that C(N≡N) complex will be formed. The C–N bond in the resultant species is much weaker than N–N bond, and it can be easily dissociated to release the N$_2$ molecule, as illustrated in Fig. 10. This reaction scheme is very similar to Eq. (2), which was experimentally implied.

For the C–N$_2$O reaction, the calculation predicted the release of N$_2$ and the formation of C(O) when N$_2$O was put on the carbon edge [37]. This is also in agreement with the experimental observation described in Section 4.3.

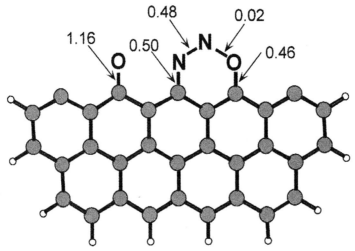

Fig. 10. Stable structure of model carbon with two adsorbed NO molecules. Numbers indicate the parameter of bond strength.

. NO$_x$ reduction on char surface during coal char combustion

Many studies have suggested the importance of secondary reactions for the formation N$_2$ during coal combustion [13,18,24]. However, few researchers have shown an erest in the quantitative analysis of secondary reactions. We have carried out the ction of coal char with O$_2$/He in a packed-bed reactor through complete conversion the char, and established the nitrogen mass balance. Fig. 11 shows the result of thermal reaction at 850°C with different bed heights. Several common features can be luced from these profiles. (1) The major product was N$_2$, followed by NO, with a y small amount of N$_2$O, (2) the ratio of NO/N$_2$ was very small at the beginning but :ame almost unity at the final stage of reaction, and (3) the ratio decreased with :reasing bed height. All of these observations can be explained by the occurrence of :ondary reaction on the char surface. We would like to emphasize again the important e of trapped surface nitrogen. NO is expected to be a primary nitrogen-containing)duct in the char-nitrogen combustion. NO thus produced would react with the char

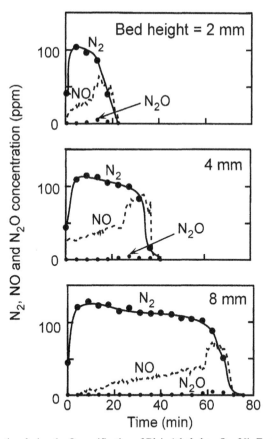

Fig. 11. Gas evolution during the O$_2$-gasification of Blair Athol char. O$_2$: 2%; Temperature: 850°C.

112

downstream to produce C(N), which then react with another NO to produce N_2 as in Eq. (12). As the remaining char is consumed, NO would escape from the bed without being converted to N_2. Thus, the fractional conversion of char nitrogen to NO increased with char burn-off. Therefore, the N_2 formation mechanism during coal char combustion is very similar to that in the C–NO reaction. The primary reaction product is NO, and this reacts with char to form surface nitrogen species as in Eq. (11). Probably, there may be some difference in reactivity between inherent char nitrogen and newly formed char nitrogen, but at present we have no means to distinguish these two species. In any case, N_2 is formed by the reaction between C(N) and NO. Such secondary reactions are not only important in the present packed-bed system, but would hold true even if char particles are separately distributed as in an entrained bed reactor.

So far there have been many arguments on whether the type of surface functionality has some effect on the NO_x formation behavior during combustion [18]. Recently Stanczyk [25] concluded that there is no relationship between nitrogen structure on the original char surface and NO_x evolution. This seems quite reasonable if it is assumed that most of N_2 is produced from NO and newly formed surface nitrogen species via Eq. (12), since the original nitrogen functionality might be of no significance at this stage.

4.6. NO reduction by particulate matter over Pt catalyst

There is a possibility to utilize the C–NO reaction to clean up the exhaust gas from a diesel engine. If the carbonaceous particulate matter can be used as reductant for NO_x, the simultaneous removal of NO_x and particulate would be an ideal method. We attempted to use a Pt catalyst supported on Al_2O_3/cordierite honeycomb. Unfortunately, this is a non-selective reaction. In other words, carbon is consumed more than required for NO_x reduction. Nevertheless, the complete removal of particulate matter by using NO_x as an oxidant is quite attractive. As is shown in Fig. 12, both NO and carbon could be completely removed. It is noteworthy that the carbon consumption and NO_x

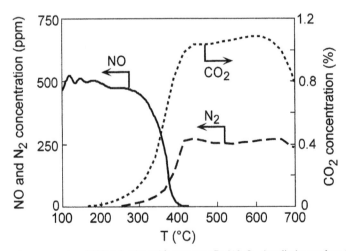

Fig. 12. Temperature programmed NO reduction with soot over Pt/Al_2O_3/cordierite catalyst. NO: 510 ppm; O_2: 1%.

reduction take place in the same temperature range as in the selective catalytic reduction using hydrocarbon gas discussed in Section 3.3. Due to the imbalance between NO_x and particulate matter, more reductant would be necessary in a practical operation.

5. Conclusion

For many years, NO_x controlling measures have been greatly developed, but in the near future we will have to further reduce the NO_x emission both from coal combustion and from automobile exhausts. In order to design more effective technology, the understanding of the relevant reactions is essential. Above all, knowledge of heterogeneous reactions should particularly be improved. For example, NO_x and N_2O formation from coal char, NO_x and N_2O reduction with carbon, and NO_x reduction with hydrocarbon gases over heterogeneous catalysts are not well understood when compared with homogenous reactions. In this paper, recent studies in our laboratory on the heterogeneous reactions of NO formation and destruction were summarized briefly.

With respect to coal char combustion and NO reduction on carbon, much data in the relatively low temperature region has been accumulated. However, data on NO_x emission during high temperature coal combustion and NO_x reduction by carbonaceous materials in such a high temperature region are insufficient. We have to further investigate what actually happens in high temperature combustors. Through these studies, we may be able to simulate NO_x formation behavior and design a better combustor with lower NO_x emissions.

The suppression of NO_x emission from diesel engines is another urgent subject. In this area, the development of high performance catalysis is strongly desired. However, the detailed reaction mechanism of the NO_x reduction over a catalyst is still unclear. The understanding of this mechanism would be helpful for designing a better catalyst. The use of particulate matter as a reductant for NO_x would also be an interesting subject to investigate.

Acknowledgements

This work was partially supported by a Grant-in-Aid for the Scientific Research on from the Ministry of Education, Science, Sports and Culture, Japan (08455369, 08232211, and 09218207).

References

[1] Environmental Agency, Government of Japan, 1999 Environmental White Paper (June 1999); L.R. Raber, Chem. Eng. News (April 14, 1997) 10.
[2] H.N. Soud, K. Fukasawa, Developments in NO_x abatement and control. IEA Coal Res. (1996) 1–125.
[3] S.C. Mitchell, NO_x in pulverized coal combustion. IEA Coal Res. (1998) 1–54.
[4] M. Kimoto, Report of Annual Meeting of Central Research Institute of Electric Power Industry (1999), pp. 71

[5] S. Matsumoto, Shokubai 39 (1997) 210.

[6] K. Komatsu, Shokubai 39 (1997) 216.

[7] M. Iwamoto, N. Mizuno, Shokubai 32 (1990) 462.

[8] W. Held, A. Koenig, T. Richter, L. Puppe, SAE Paper (1990) 900496.

[9] H. Hamada, Y. Kintaichi, M. Sasaki, T. Ito, Appl. Catal. 64 (1990) L1.

[10] M. Horiuchi, Shokubai 39 (1997) 240.

[11] H. Tominaga, M. Harada, T. Ando, Y. Suzuki, Shigen to Kankyo 6 (1997) 333.

[12] D. Förtsch, F. Kluger, U. Schnell, H. Spliethoff, K.R.G. Hein, Twenty-Seventh Symposium (International) on Combustion. The Combustion Institute, Pittsburgh, 1998, p. 3037.

[13] J.M. Jones, P.M. Patterson, M. Pourkashanian, A. Williams, Carbon 37 (1999) 1545.

[14] B. Tang, K. Ohtake, Int. Symp. Coal Combust., Beijing (1988) 199.

[15] W. Chen, L.D. Smoot, T.H. Fletcher, R.D. Boardman, Energy Fuels 10 (1996) 1036;
W. Chen, L.D. Smoot, S.C. Hill, T.H. Fletcher, Energy Fuels 10 (1996) 1046.

[16] M. Iwamoto, N. Mizuno, J. Autom. Eng. 207 (1993) 23.

[17] S.K. Goel, A. Morihara, C.J. Tullin, A.F. Sarofim, Twenty-Fifth Symposium (International) on Combustion. The Combustion Institute, Pittsburgh, 1994, p. 1051.

[18] K.M. Thomas, Fuel 76 (1997) 457.

[19] I. Aarna, E.M. Suuberg, Fuel 76 (1997) 475.

[20] G.G. De Soete, Twenty-Third Symposium (International) on Combustion. The Combustion Institute, Pittsburgh, 1990, p. 1257.

[21] T. Shimizu, Y. Sawada, T. Adschiri, T. Furusawa, Fuel 71 (1992) 361.

[22] C.J. Tullin, S. Goel, A. Morihara, A.F. Sarofim, J.M. Beer, Energy Fuels 7 (1993) 796.

[23] G.F. Kramer, A.F. Sarofim, Combust. Flame 97 (1994) 118.

[24] B. Feng, H. Liu, J. Yuan, Z. Lin, D. Liu, B. Leckner, Energy Fuels 10 (1996) 203.

[25] K. Stanczyk, Energy Fuels 13 (1999) 82.

[26] P.J. Ashman, B.S. Haynes, A.N. Buckley, P.F. Nelson, Twenty-Seventh Symposium (International) on Combustion. The Combustion Institute, Pittsburgh, 1998, p. 3069.

[27] T. Suzuki, T. Kyotani, A. Tomita, Ind. Eng. Chem. Res. 33 (1994) 2840.

[28] Ph. Chambrion, T. Suzuki, Z.-G. Zhang, T. Kyotani, A. Tomita, Energy Fuels 11 (1997) 681.

[29] Ph. Chambrion, H. Orikasa, T. Suzuki, T. Kyotani, A. Tomita, Fuel 76 (1997) 493.

[30] Ph. Chambrion, T. Kyotani, A. Tomita, Energy Fuels 12 (1998) 416.

[31] J.M. Levy, L.K. Chan, A.F. Sarofim, J.M. Beer, Eighteenth Symposium (International) on Combustion. The Combustion Institute, Pittsburgh, 1981, p. 111.

[32] H. Yamashita, H. Yamada, A. Tomita, Appl. Catal. 78 (1991) L1.

[33] Ph. Chambrion, T. Kyotani, A. Tomita, Twenty-Seventh Symposium (International) on Combustion. The Combustion Institute, Pittsburgh, 1998, p. 3053.

[34] M. Horio, M. Mochizuki, J. Koike, Fourth Japan–China Symposium on Coal and C1 Chemistry, Suita. 1993, p. 249.

[35] C.J. Tullin, A.F. Sarofim, J.M. Beer, J. Inst. Energy 66 (1993) 207.

[36] K. Noda, Ph. Chambrion, T. Kyotani, A. Tomita, Energy Fuels 13 (1999) 941.

[37] T. Kyotani, A. Tomita, J. Phys. Chem. 107 (1999) 3434.

Emissions Reduction: NO$_x$/SO$_x$ Suppression
A. Tomita (Editor)
© 2001 Elsevier science Ltd. All rights reserved

Sulfation behavior of limestone under high CO$_2$ concentration in O$_2$/CO$_2$ coal combustion

H. Liu, S. Katagiri, U. Kaneko, K. Okazaki*

Department of Mechanical Engineering and Science, Tokyo Institute of Technology, 2-12-1, O-okayama, Meguro-ku, Tokyo, 152-8552, Japan

Accepted 16 September 1999

Abstract

Coal combustion with O$_2$/CO$_2$ is promising because of its easy CO$_2$ recovery, extremely low NO$_x$ emission and high sulfation efficiency. The mechanism of direct sulfation of limestone under high CO$_2$ concentration, different from that of CaO–SO$_2$ sulfation, is one of the factors to account for its high sulfation efficiency. However, many unknowns about the mechanism of direct sulfation exist. The mechanisms and kinetics of direct sulfation were examined. Both experiment and modeling suggested that sintering is much mitigated during direct sulfation of limestone. The rate of direct sulfation does not decrease so much with sulfation degree as CaO–SO$_2$ sulfation. Direct sulfation is favorable for high sulfation degree. Furthermore, the diffusivity in the product layer demonstrates high temperature dependence and hardly changes with sulfation degree. This result is associated with the fact that the calcium sulfate layer produced is porous owing to CO$_2$ formation. © 2000 Elsevier Science Ltd. All rights reserved.

Keywords: O$_2$/CO$_2$ coal combustion; Direct sulfation; CO$_2$ rich flame

1. Introduction

O$_2$/CO$_2$ coal combustion, essentially different from the conventional flue gas recycling for thermal-NO reduction, is one of the several promising new technologies associated with mitigating the CO$_2$ rise in the atmosphere (Fig. 1). This process uses pure oxygen instead of air and recycles most (around 80%) of the flue gas, but exhausts only a small fraction of the total flue gas. The CO$_2$ concentration in the flue gas may be enriched up to 95% and easy CO$_2$ recovery therefore becomes possible. Furthermore, it has been identified that the conversion ratio from fuel-N to exhausted NO is automatically reduced to less than about one fourth of that with coal combustion in air [1]. The contributions of different factors to the NO decrease were quantitatively identified and it was revealed that the contribution of a reduction of recycled NO in the furnace is dominant and amounts to 50–80% [1]. Besides that, high sulfation efficiency is also possible. Therefore, with this technology, it is possible to realize easy CO$_2$ recovery, extremely low NO$_x$ emission and high sulfation efficiency simultaneously.

There are many possible factors to account for the high sulfation efficiency. The main factor is the system itself. In a system of O$_2$/CO$_2$ coal combustion, SO$_2$ is enriched

in furnace owing to the recirculation of flue gas and consequently a good sulfation condition is provided, whereas only a small fraction of the total gas is exhausted. Therefore, the total SO$_2$ emission can be much lower (the effect and mechanism of the system itself will be discussed in detail in another paper). Moreover, the mechanism of direct sulfation under high CO$_2$ concentration, different from that of CaO–SO$_2$ sulfation, is possibly another important reason for the high sulfation efficiency. In conventional coal combustion, the indirect sulfation of limestone (including calcination and CaO–SO$_2$ sulfation) occurs:

$$CaCO_3 \rightarrow CaO + CO_2 \uparrow \qquad (1)$$

$$CaO + SO_2 + 1/2O_2 \rightarrow CaSO_4 \qquad (2)$$

The sulfation efficiency in conventional coal combustion is usually low because of the sintering of the sorbent. In contrast, in O$_2$/CO$_2$ or pressurized coal combustion (when CO$_2$ partial pressure is higher than the equilibrium pressure), the calcination is inhibited and the limestone is subject to a direct sulfation reaction [2,3]:

$$CaCO_3 + SO_2 + 1/2O_2 \rightarrow CaSO_4 + CO_2 \uparrow \qquad (3)$$

It was concluded that direct sulfation enables higher degrees of sulfation than those observed from CaO–SO$_2$ sulfation, because the counter-diffusion of the CO$_2$ generated during

---

* Corresponding author. Tel.: +81-3-5734-3335; fax: +81-3-5734-2892.
E-mail address: okazakik@mech.titech.ac.jp (K. Okazaki).

Reprinted from *Fuel* **79 (8)**, 945–953 (2000)

116

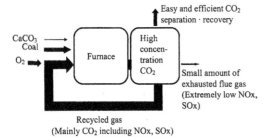

Fig. 1. Schematic of O_2/CO_2 coal combustion.

the direct sulfation reaction forms a porous product layer that offers less diffusional resistance than the essentially nonporous layer formed during the CaO–SO_2 sulfation reaction [4,5].

In O_2/CO_2 coal combustion, CO_2 concentration in the flue gas may be enriched up to 95% owing to the gas recirculation. Under so high CO_2 concentration, the sulfation behavior is not clear. Therefore, it is necessary to study the sulfation behavior under high CO_2 concentration. However, so far no studies aimed at sulfation in O_2/CO_2 coal combustion were reported. Literatures and data concerning direct sulfation are also limited. The kinetics of direct sulfation has been investigated by some researchers [2,4–9] but there are still many unknowns. Among the previous investigations on direct sulfation, most were made with TGA [2,4–6,8–17],

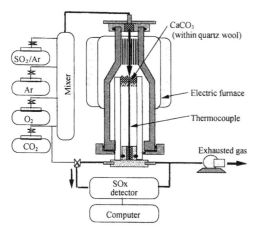

Fig. 2. Fixed-bed reactor.

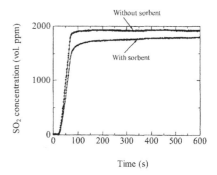

Fig. 3. Typical curves of SO_2 concentration with and without sorbent.

i.e. from weight change, but TGA suffers from many limitations owing to its low gas throughput. The problems are compounded by the dependence on weight change as the sole response normally measured. Any deviation from the assumed reaction mechanism or stoichiometry as conditions are varied can lead to erroneous interpretations of the results. These constraints result in considerable uncertainty. On the contrary, the fixed-bed reactor (differential reactor) enables homogeneous contact between sorbent and reacting gas by dispersing the particles in a quartz wool substrate, through which the gas flows. Besides that, it derives sulfation degree directly from the change in SO_2 concentration. Therefore, experiment with a fixed-bed is needed. Moreover, the control mechanism of direct sulfation is also controversial. Fuertes et al. [8] and Snow et al. [4] argued that direct sulfation is chemically controlled, whereas Iisa and Hupa [14] concluded that direct sulfation is only limited by diffusion. Some other authors concluded that reaction and diffusion are both important [5,17]. Meanwhile, the comparison studies between direct sulfation of limestone and CaO–SO_2 sulfation are also limited.

The main objective of this paper is to investigate the behavior and mechanism of direct sulfation of limestone under high CO_2 concentration as one of the possible mechanisms to realize high sulfation efficiency in addition to easy CO_2 recovery and extremely low NO_x emission in O_2/CO_2 coal combustion. We studied direct sulfation under different environment and with different purpose from others.

2. Experimental

The experiments were performed in a fixed-bed reactor (Fig. 2). A quartz tube, 0.02 m in inner diameter and 0.65 m in length, was used in order to prevent catalytic oxidization of SO_2 at high temperatures. Premixed gases entered the reactor at the top of the quartz tube and were heated as they moved downward. The temperature in the reactor was measured by a thermocouple placed immediately under the sample. Bed temperature and gas phase concentration of SO_2 were measured on-line. The pre-experiment demonstrated that at sample weight of 0.2 g, the gas flow rate had no effect on sulfation when it was above 6.7×10^{-5} m^3/s. In order to minimize mass transfer limitations between bulk gas and sample layer, gas flow rate of 1×10^{-4} m^3/s was selected (at sample weight of 0.2 g). Furthermore, a very thin layer of sorbent dispersed in a quartz wool substrate was used to achieve differential conditions for valid kinetic data (with the difference between inlet and outlet SO_2 concentration within 10%). Prior to an experiment, pure CO_2 was fed into the quartz tube from the bottom to avoid the calcination of the limestone. The reactor was brought to a stationary state under the desired reaction conditions. Fig. 3 demonstrates typical curves of SO_2 concentration with and without sorbent. The induction period is very short comparing with the overall reaction process. The sulfation degree of sorbent X, as a function of time, was calculated by integrating the difference between the SO_2 concentration from the reactor without sorbent and the SO_2 concentration with sorbent (Fig. 3). At the end of an experiment (when the desired time was reached) the sulfation reaction was stopped by switching

Table 1
Constituent of limestone (wt%)

	$CaCO_3$	SiO_2	Al_2O_3	Fe_2O_3	MgO	P_2O_5	Moisture
Limestone A	99.10	0.04	0.10	0.03	0.16	0.09	0.00
Limestone B	99.19	< 0.03	< 0.03	< 0.03	0.59	< 0.03	0.10

118

Table 2
Experimental conditions

Temperature (K)	883–1123
O_2 concentration (vol.%)	0–20
SO_2 concentration (mol/m³)	0–2.56×10^{-2}
at 1123 K)	(0–2360 vol. ppm)
CO_2 concentration (vol.%)	20–80
Ar	As balance
Mean diameter of sorbent	28.3 (limestone A),
particles (μm)	8.4 (limestone A), 54.0
	(limestone B)
Sample weight (g)	0.2
Total gas flow rate (m³/s)	1×10^{-4}
Total pressure (Pa)	1.013×10^5

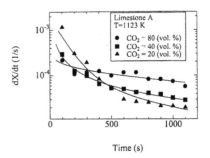

Fig. 5. Effect of CO_2 concentration on sulfation of limestone at 1123 K in an atmosphere of 10% O_2, 2.09×10^{-2} mol/m³ (1920 ppm) SO_2 and Ar as balance (limestone A, particle size: 28.3 μm).

SO_2 into pure CO_2 to avoid calcination of limestone (similar to the curve in Fig. 3, the delay of gas-switching is very short). Following an experiment the sulfated particles were weighed again to make sure the results were correct (with error within 10%). The experiments were performed with two limestones widely used in Japan (mainly with limestone A). The characteristics of the limestones and experimental conditions are given in Tables 1 and 2, respectively. We adopted small particles in order to make sure the difference of SO_2 concentration with and without sorbent was measurable, especially at low temperature, because we calculate sulfation degree from the difference of SO_2 concentration. Furthermore, we intended to minimize mass transfer limitations between bulk gas and particles. The calcine used for comparison experiment (CaO–SO_2 sulfation) was obtained through the calcination of the limestones (Table 1). The pressure was 1.013×10^5 Pa.

The pre-experiment showed that when O_2 concentration was above 5%, it had little effect on sulfation degree, which suggested that the reaction order with respect to O_2 is zero as long as O_2 is much enough. Therefore, an O_2 concentration of 10% was adopted as our experimental condition in order to avoid the effect of O_2 concentration. Furthermore, the order of the reaction with respect to SO_2 was estimated to

be approximately unity in the SO_2 concentration range investigated.

3. Experimental results

3.1. Temperature dependence

The variation of sulfation degree with the reaction time is illustrated in Fig. 4 for different temperatures (limestone A, direct sulfation). Obviously the sulfation degree increased with temperature, suggesting that the $CaSO_4$ formed from direct sulfation is stable and does not decompose in this temperature range. Furthermore, even at sulfation degree of 0.4, the curve in Fig. 4 still keeps strong increasing tendency, implying that direct sulfation is favorable for high sulfation degree.

3.2. Effect of CO_2 concentration on limestone sulfation

In order to clarify that the strong increasing tendency at sulfation degree of 0.4 in Fig. 4 is because of the characteristics of the limestone itself or because of the sulfation mechanism under high CO_2 concentration, experiments were conducted with limestone A under several CO_2 concentrations. We choose 80 and 20% CO_2 to correspond to the sulfation during O_2/CO_2 coal combustion and conventional coal combustion in air, respectively. CO_2 (40%) was selected as the intermediate atmosphere. As shown in Fig. 5, the reaction rate curve for 80% CO_2 is comparatively flat. In contrast, in an atmosphere of 20% CO_2, the reaction rate was high in the beginning but followed by a drastic decrease. Furthermore, the tendency of the curve for 40% CO_2 was the intermediate. These results revealed the sulfation behavior is quite different under different CO_2 concentrations.

In an atmosphere of 20% CO_2, the indirect sulfation (including calcination of limestone and CaO–SO_2 sulfation) took place. In an atmosphere of 80% CO_2, the calcination was inhibited and the limestone was subject to direct sulfation reaction. As CO_2 increases, the sulfation switches from indirect sulfation to direct sulfation of limestone. Therefore,

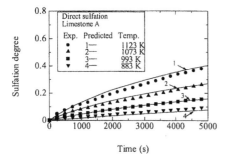

Fig. 4. Variation of sulfation degree with time in an atmosphere of 10% O_2, 80% CO_2, 2.09×10^{-2} mol/m³ (1920 ppm) SO_2 and Ar as balance (limestone A, particle size: 28.3 μm).

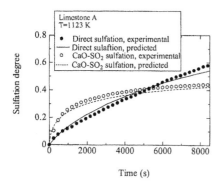

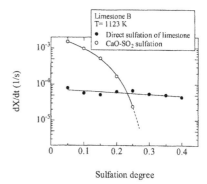

Fig. 6. Sulfation degree at 1123 K in an atmosphere of 10% O_2, 2.09 × 10^{-2} mol/m³ (1920 ppm) SO_2 and Ar as balance (limestone A, particle size: 28.3 μm). CO_2 is 80 and 20% for direct sulfation of limestone and CaO–SO_2 sulfation, respectively.

Fig. 8. Sulfation reaction rate at 1123 K in an atmosphere of 10% O_2, 2.09 × 10^{-2} mol/m³ (1920 ppm) SO_2 and Ar as balance (limestone B, particle size: 54.0 μm). CO_2 is 80 and 20% for direct sulfation of limestone and CaO–SO_2 sulfation, respectively.

the different tendency in Fig. 5 reveals the difference between direct and indirect sulfation of limestone. The result is so because the porous CaO particles have a higher specific surface area than the calcium carbonate particles (our mercury porosimetry measurement showed that the specific surface area of CaO particles and calcium carbonate particles was 24.4 and 1.3 m²/g, respectively). However, pore plugging took place during indirect sulfation and accordingly, the indirect sulfation reaction rate decreased drastically. In an atmosphere of 40% CO_2, the calcination is slower than in an atmosphere of 20% CO_2, and consequently, before complete calcination of limestone, both indirect and direct sulfation of limestone took place simultaneously. Thus the tendency for 40% CO_2 was the intermediate between the two others.

3.3. Comparison with CaO–SO_2 sulfation

We are particularly interested in the comparison between

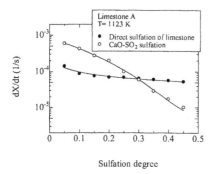

Fig. 7. Sulfation reaction rate at 1123 K in an atmosphere of 10% O_2, 2.09 × 10^{-2} mol/m³ (1920 ppm) SO_2 and Ar as balance (limestone A, particle size: 28.3 μm). CO_2 is 80 and 20% for direct sulfation of limestone and CaO–SO_2 sulfation, respectively.

direct sulfation of limestone and the widely studied CaO–SO_2 sulfation, but the sulfation of limestone in an atmosphere of 20% CO_2 and 40% CO_2 as shown in Fig. 5 includes both calcination and CaO–SO_2 sulfation. In order to compare direct sulfation of limestone with CaO–SO_2 sulfation without the influence of limestone calcination, the experiment on CaO–SO_2 sulfation was also conducted with the calcine prepared through the calcination of limestone A. Fig. 6 compares the sulfation degrees of direct sulfation and CaO–SO_2 sulfation. Before 5000 s of reaction time, the sulfation degree of CaO–SO_2 sulfation was higher, but it was reversed thereafter. It can be seen that direct sulfation is favorable for higher sulfation degree.

Fig. 7 compares reaction rate between direct sulfation of limestone and CaO–SO_2 sulfation at 1123 K. Similar to Fig. 5, the reaction rate for direct sulfation did not decrease so much with sulfation degree as CaO–SO_2 sulfation. Although at low sulfation degree the CaO–SO_2 sulfation was faster than direct sulfation of limestone, at high sulfation degree (after sulfation degree of about 0.3), it was reversed. At sulfation degree of 0.45, the rate of direct sulfation was about six times as high as that of CaO–SO_2 sulfation. It is implied that the sintering is much mitigated during direct sulfation of limestone in O_2/CO_2 coal combustion comparing with CaO–SO_2 sulfation during conventional coal combustion in air.

Experiments on limestone B were also made in order to get more evidence. As shown in Fig. 8, similar to the case of limestone A, the sintering is much mitigated during direct sulfation of limestone comparing with CaO–SO_2 sulfation. CaO–SO_2 sulfation (limestone B) suffered so much from sintering that the sulfation almost stopped at a sulfation degree of 0.25. The difference between direct sulfation and CaO–SO_2 sulfation was pronounced for limestone B. Fuertes also compared direct sulfation of calcium carbonate (particle size: 16 μm) and CaO–SO_2 sulfation and found that at sulfation degree above 0.6, direct sulfation exhibited higher reaction rate than CaO–SO_2 sulfation [8].

120

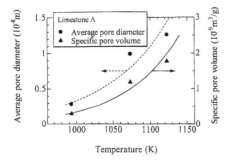

Fig. 9. Specific pore volume and pore size of sulfated product layer versus sulfation temperature in an atmosphere of 10% O_2, 80% CO_2, 2.09 × 10^{-2} mol/m^3 (1920 ppm) SO_2 and Ar as balance (limestone A, particle size: 28.3 μm).

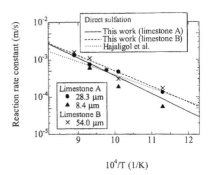

Fig. 10. Arrhenius plot of k_s in an atmosphere of 10% O_2, 80% CO_2, 2.09 × 10^{-2} mol/m^3 (1920 ppm) SO_2 and Ar as balance.

The high sulfation degree obtained for direct sulfation of limestone in Fig. 6, together with the much higher sulfation rate than CaO–SO_2 sulfation at high sulfation degree as shown in Figs. 7 and 8, all suggested that the sintering is much mitigated and it is easier to reach high sulfation degree during direct sulfation of limestone comparing with CaO–SO_2 sulfation. Besides that, these experimental results revealed that the same limestone exhibited much different sulfation behavior under different CO_2 concentrations. It is implied that the mitigation of sintering and the good performance of sorbent during direct sulfation of limestone is attributed to the sulfation mechanism under high CO_2 concentration rather than the characteristics of the limestone itself.

3.4. Pore structure results

So far the pore structure investigation during direct sulfation, especially the investigation on product layer, is very limited. The dependence of average pore radius of calcium oxide on calcination temperature of limestone was studied by some researchers [18], whereas we studied the pore structure characteristics of $CaSO_4$ product layer during direct sulfation. The pore structure characterization for sulfation process is much different from and more difficult than the case of calcination process because the $CaSO_4$ product layer is fragile and liable to be destroyed during analysis. In order to gain better insight into the overall mechanism of the direct sulfation of limestone, pore structure characterization was conducted with gas adsorption (N_2 at 77 K) analyses. Precursor particles (limestone A,

28.3 μm) and particles partially sulfated at different temperatures (sulfation degree = 0.4) were examined.

The specific pore volume and pore size of sulfated product layers were derived from the gas adsorption data of limestone particles and sulfated particles through proper calculations. From Fig. 9 it can be seen that both specific pore volume and average pore diameter increase approximately exponentially with temperature. Fig. 9 clearly shows quantitative changes in the pore structure of the product layers with temperature. From reaction (3) it is known that the sulfation reaction rate, and accordingly CO_2 formation, increase with temperature. Therefore, the increase in pore volume with temperature is probably attributed to the increase in CO_2 formation with temperature.

3.5. Reaction kinetics

At the beginning of the reaction (sulfation degree = 0.03), the product layer of a sorbent particle is so thin that the sulfation process is considered to be chemically controlled. An expression of reaction rate constant, $k_s = 19 \exp(-90,000/RT)$, was obtained for direct sulfation of limestone A from Arrhenius plot in Fig. 10 (k_s is reaction rate constant on apparent area basis). Table 3 compares the frequency factor and activation energy of k_s between our results and that obtained by Hajaligol (particle sizes: 2–3, 5–7 and 10–12 μm) [5]. It was revealed that among the three limestones, the frequency factor is different but the activation energy is close to each other (with about ±10% difference).

Table 3
Chemical kinetics of direct sulfation of limestone

	Frequency factor (m/s)	Activation energy (kJ/mol)
This work (limestone A)	19.0	90
This work (limestone B)	7.6	80
Hajaligol et al. [5]	1.5	69

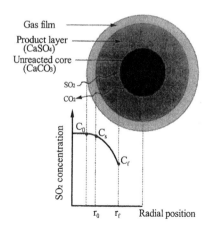

Fig. 11. Schematic illustration of pore diffusion model.

4. Modeling and discussion

4.1. Conservation equations

The sulfation of limestone is a very complicated process. Some parameters such as effective diffusivity are very difficult to obtain directly. Modeling is also necessary to investigate this process. Therefore, a pore diffusion model was developed for direct sulfation of limestone. The scheme of the model is shown in Fig. 11. The simplifying assumptions are as follows: (1) the system is isothermal; (2) the reaction is irreversible, single and first order (with respect to SO_2 concentration at the surface of unreacted core) with an Arrhenius-type dependence on temperature; (3) the diffusivity inside the product layer is uniform; (4) the particle is spherical; (5) the overall particle size does not change during reaction; (6) there is no reaction inside the unreacted core; (7) O_2 has no effect on sulfation when its concentration is above 5% (which is valid according to the experimental result). The unsteady mass conservation equation is as follows (in product layer):

$$\frac{1}{r^2}\frac{\partial}{\partial r}(r^2 D_e \frac{\partial C}{\partial r}) = \frac{\partial C}{\partial t} \quad (4)$$

with boundary conditions given by:

$$D_e \frac{\partial C}{\partial r} = h_D(C_0 - C_s) \quad (5)$$

at $r = r_0$ (D_e : effective diffusivity of SO_2 in particle)

$$D_e \frac{\partial C}{\partial r} = k_s C \quad \text{at } r = r_f \quad (6)$$

with the subscripts s and f refer to particle surface and interface between product layer and unreacted core respectively; $C = SO_2$ molar concentration (mol/m^3); $C_0 = SO_2$ molar concentration in bulk gas (mol/m^3); $D_e =$ effective diffusivity of SO_2 in product layer (m^2/s); $r =$ radius (m);

$r_0 =$ initial radius (m); $h_D =$ mass transfer coefficient of SO_2 in gas film (m/s); $k_s =$ reaction rate constant on area basis (m/s). Furthermore, h_D is obtained from [19]

$$Sh_0 = 2.0 + 0.6Re^{1/2}Sc^{1/3} \quad (7)$$

where $Sh_0 = 2r_0h_D/D$, Sherwood number of SO_2; $Re = 2r_0U/\nu$, Reynolds number; $Sc = \nu/D$, Schmidt number; $U =$ linear velocity of the gas stream flowing past the particle (m/s); $\nu =$ kinematic viscosity (m^2/s); $D =$ diffusivity of SO_2 in gas film (m^2/s). Considering the low Reynolds number in our case, $Sh_0 = 2.0$ was taken to derive h_D.

The total mass balance follows:

$$\frac{dr_f}{dt} = -\frac{k_s C_f}{\rho} \quad (8)$$

where ρ is particle density (mol/m^3). Deriving the above-mentioned equations, the dimensionless unsteady mass conservation equation is given as:

$$\frac{1}{\xi^2}\frac{\partial}{\partial \xi}(\xi^2 \frac{\partial \psi}{\partial \xi}) = \frac{\partial \psi}{\partial \theta} \quad (9)$$

The boundary conditions of Eq. (9) are written as:

$$\frac{\partial \psi}{\partial \xi} = Sh(1 - \psi) \quad \text{at } \xi = 1 \quad (10)$$

$$\frac{\partial \psi}{\partial \xi} = \frac{k_s r_0}{D_e}\psi \quad \text{at } \xi = \xi_f \quad (11)$$

where $\psi = C/C_0$, dimensionless concentration of SO_2; $\xi = r/r_0$, dimensionless radial position; $\theta = D_e t/r_0^2$, dimensionless time; $Sh = r_0 h_D/D_e$, modified Sherwood number of SO_2.

The initial condition is given by:

$$\psi = 0 \quad \text{at } \theta = 0 \quad (12)$$

The above-mentioned pore diffusion model was also adopted to predict the sulfation degree in CaO–SO_2 sulfation. The effective diffusivity in particle for CaO–SO_2 sulfation was supposed to be varying with conversion according to [20]

$$D_e = D_{e0}\frac{\epsilon}{\epsilon_0}\frac{\tau_0}{\tau} \quad (13)$$

and allowance was made for the change of porosity with reaction considering the increase in molar volume from CaO to CaSO$_4$ [20]:

$$\frac{\epsilon}{\epsilon_0} = 1 - \frac{(Z - 1)(1 - \epsilon_0)X}{\epsilon_0} \quad (14)$$

where $\tau = 1/\epsilon$ and $\tau_0 = 1/\epsilon_0$ were taken [21]; $\epsilon_0 =$ initial porosity; $\epsilon =$ porosity; $\tau_0 =$ initial tortuosity; $\tau =$ tortuosity; $X =$ sulfation degree of sorbent (molar fractional conversion of Ca from CaO to CaSO$_4$); $Z =$ ratio of molar volume of solid phase after reaction to that before reaction; $D_{e0} =$ initial effective diffusivity of SO_2 in product layer during CaO–SO_2 sulfation (m^2/s). From Eqs. (13) and

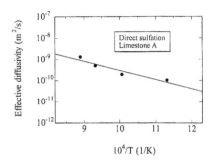

Fig. 12. Temperature dependence of effective diffusivity in particle in an atmosphere of 10% O_2, 80% CO_2, 2.09×10^{-2} mol/m^3 (1920 ppm) SO_2 and Ar as balance (limestone A, particle size: 28.3 μm).

(14) we have:

$$D_e = D_{e0}\left[1 - \frac{(Z-1)(1-\epsilon_0)X}{\epsilon_0} \right]^2 \qquad (15)$$

4.2. Estimation of effective diffusivity

The effective diffusivity of SO_2 in the particle during direct sulfation of limestone was estimated through fitting of the experimental results by using the above-mentioned model. The experimental data before 1000 s in Fig. 4 was used to estimate the effective diffusivity. The estimated effective diffusivity, as shown in Fig. 12, demonstrated high temperature dependence and exhibited approximately exponential relationship with temperature. Considering this tendency and referring to the similar relationship obtained by others [5,12], we supposed an exponential relationship and obtained a regressed formula of $D_e = 6.71 \times 10^{-6} \exp(-10,000/T)$. The diffusion in particle during direct sulfation of limestone is complicated and vague. This formula merely represents a semi-empirical, mathematical relationship of data but has no clear physical meaning.

We supposed an exponential relationship to regress the effective diffusivity because CO_2 formation increases exponentially with temperature according to reaction (3), which may lead to exponential increase in pore volume of the product layer. The exponential increase in specific pore volume with temperature may lead to the exponential dependence of effective diffusivity on temperature, because effective diffusivity in particle is proportional to the porosity, or specific pore volume, of the particle. Our N_2 adsorption measurement results showed that the specific pore volume increases approximately exponentially with temperature. The specific pore volume of the product layer sulfated at 1123 K is about six times as high as that of the product layer sulfated at 993 K. And the effective diffusivity at 1123 K is about seven times as high as at 993 K. According to these results, very likely the high temperature dependence of the effective diffusivity is attrib-

uted to the approximately exponential increase in specific pore volume of the product layer with temperature owing to the increase in CO_2 formation. Nevertheless, there is not enough evidence to reach a conclusion yet. Obviously further research is needed to clarify the physical meaning of the temperature dependence of effective diffusivity and the diffusion mechanism in particle during direct sulfation of limestone.

4.3. Prediction of sulfation degree

The above-mentioned equations were numerically solved with a finite-difference method. The introduction into Eqs. (9)–(12) of the estimated kinetic parameters and diffusivity allows the variation of sulfation degree with time to be predicted. It was identified that the diffusional resistance in gas film is negligible comparing with that inside the particle. The effective diffusivity in particle estimated from the experimental data before 1000 s (Fig. 4) was used to predict the sulfation behavior thereafter. We used a sole value of effective diffusivity D_e to predict sulfation degree for the whole reaction process (at 1123 K) while achieved satisfactory result (Fig. 6), suggesting that the effective diffusivity hardly changes with sulfation degree during direct sulfation of limestone.

In prediction of CaO–SO_2 sulfation, $k_s = 0.058$ m/s (reaction rate constant on apparent area basis, obtained from experiment) and $D_{e0} = 2.3 \times 10^{-9}$ m^2/s derived from data fitting of experimental result before 1000 s at 1123 K were used in the modeling to predict the sulfation behavior thereafter (Fig. 6). We used effective diffusivity which decreases with sulfation degree according to Eq. (15) to predict the degree of CaO–SO_2 sulfation (at sulfation degree of 0.4, D_e is about one tenth of the initial effective diffusivity D_{e0}). It can be seen from Fig. 6 that Eq. (15) can approximately represent the variation tendency of effective diffusivity of SO_2 in particle during CaO–SO_2 sulfation, i.e. the effective diffusivity decreases during CaO–SO_2 sulfation. However, the sulfation degree of CaO–SO_2 sulfation predicted with a sole effective diffusivity D_{e0} is two times as high as the experimental one at $t = 8500$ s, i.e. the prediction with a sole effective diffusivity D_{e0} overestimates the degree of CaO–SO_2 sulfation. In contrast, for direct sulfation of limestone, we used a sole value of effective diffusivity D_e to predict sulfation degree for the whole reaction process (at a given temperature) while achieved satisfactory result. This is another evidence that sintering is much mitigated during direct sulfation of limestone comparing with CaO–SO_2 sulfation.

The sulfation of limestone is a very complicated process, whereas the shape of the curve of sulfation degree, which was much different for direct sulfation and CaO–SO_2 sulfation, was satisfactorily predicted by the model proposed by us. Even though simple, this model could be used to predict the behavior of both direct sulfation of limestone and CaO–SO_2 sulfation, at least under the conditions investigated.

Furthermore, it was identified that a shrinking unreacted core model neglecting either diffusional resistance or chemical resistance overestimates the sulfation degree, suggesting a mechanism of both chemical reaction and diffusion control during direct sulfation of limestone. Szekely et al. [19] proposed an index, (τ_{pld}/τ_{rxn}), to represent the ratio of diffusional to chemical reaction resistance ($\tau_{pld} = \rho r_0^2/(6D_e C_0)$, characteristic time for complete sulfation under total control of diffusion(s); $\tau_{rxn} = \rho r_0/(k_s C_0)$, characteristic time for complete sulfation under total control of chemical reaction(s)). When $0.1 < (\tau_{pld}/\tau_{rxn}) < 10$, the process was considered to be under the control of both chemical reaction and diffusion. It was revealed that the value of (τ_{pld}/τ_{rxn}) is between 2 and 5 for direct sulfation in the temperature range of this work, which is another evidence of both chemical and diffusional control.

5. Conclusions

The high sulfation efficiency in O_2/CO_2 coal combustion is mainly attributed to the system itself. Meanwhile the mechanism of direct sulfation of limestone under high CO_2 concentration, different from that of $CaO–SO_2$ sulfation, is another important factor. The behavior and mechanism of direct sulfation of limestone in O_2/CO_2 coal combustion were examined through fixed-bed reactor experiments, pore structure characterization and modeling. It was found that direct sulfation is controlled by both chemical reaction and diffusion. Both experiment and modeling suggested that sintering is much mitigated during direct sulfation of limestone. The sulfation rate does not decrease so much with sulfation degree as $CaO–SO_2$ sulfation. At sulfation degree of 0.45, the rate of direct sulfation was about six times as high as that of $CaO–SO_2$ sulfation. During direct sulfation of limestone, the diffusivity in the product layer demonstrates high temperature dependence and hardly changes with sulfation degree. This result is associated with the fact that the calcium sulfate layer produced under high CO_2 concentration in O_2/CO_2 coal combustion is porous owing to CO_2 formation. Direct sulfation of limestone is favorable for high sulfation degree and consequently enables better sorbent utilization.

The mechanism of direct sulfation in O_2/CO_2 coal combustion, combined with the effect of the system itself, could realize high efficiency of SO_2 removal. The high sulfation efficiency, together with easy CO_2 recovery and extremely low NO_x emission, makes O_2/CO_2 coal combustion promising for protection of global environment.

Acknowledgements

The authors are grateful to the Electric Power Development Co., Ltd for their support of this study. Help from Idemitsu Kosan Co., Ltd and Ishikawajimaharima Heavy Industry, Co., Ltd is also appreciated.

References

[1] Okazaki K, Ando T. Energy 1997;22(2/3):207.
[2] Tullin C, Ljungstrom E. Energy Fuels 1989;3:284.
[3] Iisa K, Tullin C, Hupa M. Conf (Int) Fluid Bed Combust [Proc] 1991;11:83.
[4] Snow MJH, Longwell JP, Sarofim AF. Ind Engng Chem Res 1988;27:268.
[5] Hajaligol MR, Longwell JP, Sarofim AF. Ind Engng Chem Res 1988;27:2203.
[6] Yrjas P, Iisa K, Hupa M. Fuel 1995;74(3):395.
[7] Illerup JB, Dam-Johansen K, Lunden K. Chem Engng Sci 1993;48(11):2151.
[8] Fuertes AB, Velasco G, Fernandez MJ, Alvarez T. Thermochim Acta 1994;242:161.
[9] Krishnan SV, Sotirchos SV. Can J Chem Engng 1993;71:244.
[10] Tullin C, Nyman G, Ghardashkhani S. Energy Fuels 1993;7:512.
[11] Davini P. Fuel 1995;74(7):995.
[12] Fuertes AB, Velasco G, Fuente E, Alvarez T. Fuel Process Technol 1994;38:181.
[13] Fuertes AB, Velasco G, Fuente E, Parra JB, Alvarez T. Fuel Process Technol 1993;36:65.
[14] Iisa K, Hupa M. Symp (Int) Combust [Proc] 1990;23:943.
[15] Buroni M, Carugati A, Piero GD, Pellegrini L. Fuel 1992;71:919.
[16] Iisa K, Hupa M, Yrjas P. Symp (Int) Combust [Proc] 1992;24:1349.
[17] Iisa K, Hupa M. J Inst Energy 1992;65:201.
[18] Dogu T. Chem Engng J 1981;21:213.
[19] Szekely J, Evans JW, Sohn HY. Gas–solid reactions, New York: Academic Press, 1976. p. 12–77.
[20] Bhatia SK, Perlmutter DD. AICHE J 1981;27(2):226.
[21] Reyes S, Jensen KF. Chem Engng Sci 1985;40(9):1723.

A. Tomita (Editor)

CO$_2$, NO$_x$ and SO$_2$ emissions from the combustion of coal with high oxygen concentration gases

Y. Hu, S. Naito, N. Kobayashi*, M. Hasatani

Department of Energy Engineering and Science, Nagoya University, Furo-cho, Chikusa-ku, Nagoya 464-8603, Japan

Received 15 August 1999; accepted 9 January 2000

Abstract

The emissions of CO$_2$, NO$_x$ and SO$_2$ from the combustion of a high-volatile coal with N$_2$- and CO$_2$-based, high O$_2$ concentration (20, 50, 80, 100%) inlet gases were investigated in an electrically heated up-flow-tube furnace at elevated gas temperatures (1123–1573 K). The fuel equivalence ratio, ϕ, was varied in the range of 0.4–1.6. Results showed that CO$_2$ concentrations in flue gas were higher than 95% for the processes with O$_2$ and CO$_2$-based inlet gases. NO$_x$ emissions increased with ϕ under fuel-lean conditions, then declined dramatically after $\phi = 0.8$, and the peak values increased from about 1000 ppm for the air combustion process and 500 ppm for the O$_2$(20%) + CO$_2$(80%) inlet gas process to about 4500 ppm for the oxygen combustion process. When $\phi > 1.4$ the emissions decreased to the same level for different O$_2$ concentration inlet gas processes. On the other hand, NO$_x$ emission indexes decreased monotonically with ϕ under both fuel-lean and fuel-rich combustion. SO$_2$ emissions increased with ϕ under fuel-lean conditions, then declined slightly after $\phi > 1.2$. Temperature has a large effect on the NO$_x$ emission. Peak values of the NO$_x$ emission increased by 50–70% for the N$_2$-based inlet gas processes and by 30–50% for the CO$_2$-based inlet gas process from 1123 to 1573 K. However, there was only a small effect of temperature on the SO$_2$ emission. © 2000 Elsevier Science Ltd. All rights reserved.

Keywords: Coal combustion; Pollutant emissions; High oxygen concentration gas

1. Introduction

Emissions of greenhouse gases, chiefly CO$_2$, by human way are the main contributors to the anticipated change in the climate. The largest source of CO$_2$ is the combustion of fossil fuels for power generation. On the third Conference of the Parties to the Framework Convention on Climate Change held in Kyoto in December 1997, the so-called Annex-1 countries agreed to reduce greenhouse emissions by 5% in the period 2008–2012 compared with the 1990 levels. As the global demand of energy, greatly dependent on fossil fuel, mainly being coal, is increasing, CO$_2$ discharge will further increase. It is necessary to recover CO$_2$ from flue gas in combustion processes and to use CO$_2$ to make products or for other useful purposes. However, the CO$_2$ concentration of the exhaust gas discharged from the present coal firing power plants is low because the combustion air contains only about 21% of oxygen and the remaining 79% mostly in the form of

nitrogen is discharged as exhaust gas. The recovery of CO$_2$ from this flue gas is very costly.

A significant use of coal has serious consequences to generate large amount of pollutants. Among the pollutants the main ones are SO$_2$, a major contributor to acid rain, NO and NO$_2$, collectively called NO$_x$, which plays an important role for photochemical smog and acid rain. Considerable effort has been made and many techniques [1–6] have been developed to reduce the emission of NO$_x$ and SO$_2$ from the combustion processes. Some of these techniques have been widely used in industry. However, a minimization of the pollutant emissions by these technologies is generally not compatible with high combustion efficiency. New cost-effective methods for further reduction of NO and SO$_2$ are needed to meet stricter requirements in the future.

Recently, the process of coal combustion with oxygen (CCO), which separates nitrogen from the combustion air in advance, has attracted special attention [7–13]. The concentration of CO$_2$ in the exhaust gas of this new process can reach 95% or higher [7] because nitrogen is separated from combustion air in advance. Compared with the conventional process of coal combustion with air (CCA), CCO process possesses the following advantages. (1) The recovery of CO$_2$ from exhaust gas will become easier and

---

* Corresponding author. Tel.: + 81-52-789-3383; fax: + 81-52-789-3842.

E-mail address: hkoba@mhlab.nuce.nagoya-u.ac.jp (N. Kobayashi).

Reprinted from *Fuel* **79** (15), 1925-1932 (2000)

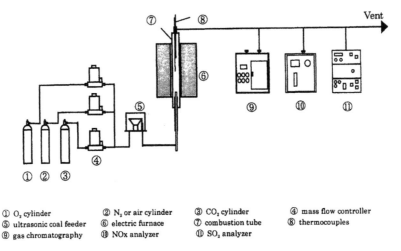

① O₂ cylinder	② N₂ or air cylinder	③ CO₂ cylinder	④ mass flow controller
⑤ ultrasonic coal feeder	⑥ electric furnace	⑦ combustion tube	⑧ thermocouples
⑨ gas chromatography	⑩ NOx analyzer	⑪ SO₂ analyzer	

Fig. 1. The schematic drawing of the experimental apparatus.

less costly, and zero emission may even be realized in this process. (2) Combustion efficiency will be enhanced by higher concentration O_2 combustion and boiler efficiency will increase because the amount of exhaust gas is greatly reduced. (3) The amount of NO_x produced in combustion may be reduced because of the elimination of atmospheric nitrogen fixation to NO. (4) Combustion temperature can be controlled with the recirculation of part of the flue gas (chiefly CO_2), making the process more flexible in operation and coal rank.

The aim of the present study is to make some fundamental experiments on this new process, including the effects of the concentration of oxygen in the inlet gas and other operation conditions on the emissions of CO_2, SO_2, NO_x. In the experiments, besides pure oxygen, $O_2 + N_2$ and $O_2 + CO_2$ with different O_2 concentration were used as inlet gases. $O_2 + N_2$ was selected for comparing with the CCA process and examining the effect of N_2 in inlet gas on the pollutant emissions. $O_2 + CO_2$ was selected for simulating the process of flue gas recirculation. This fundamental research will be helpful to understand and develop this new combustion process.

2. Experimental

A schematic drawing of the experimental, quasi-one-dimensional, electrically heated combustor and its flow system is shown in Fig. 1. The combustion chamber consisted of a cylindrical alumina tube with an inner diameter of 28 mm, which was heated by a SiC element. The heated part was 300 mm long. Four Pt/Pt–13%Rh thermocouples were placed at different positions along the axis of the tube. The feed gases were supplied from gas cylinders and regulated by mass flow controllers. Pulverized coal was supplied continuously by an ultrasonic feeder and sent pneumatically by the feed gas to the combustion zone from the bottom of the tube. The properties of the used coal are given in Table 1. The coal feed rate was up to 180 g/h. The total inlet gas flow rate was controlled between 1.0 and 1.5N l/min in the experiments to maintain the gas resident time of 2 s at different temperatures. The expected concentration was realized by changing the O_2 and base gases (N_2 or CO_2) flow rate. Because of the small size of coal particles, high volatile matter and low ash content (2%), the slip velocity between gas and unburned coal or char was so small that the resident time of the particles was nearly the same as that of the gas. This resident time is similar to those in industrial furnaces.

The exhaust gas composition was determined using an on-line gas chromatography for O_2, CO, CO_2, N_2, a chemiluminescent analyzer for NO and NO_2, and a SO_2 analyzer. These instruments were calibrated periodically by injecting standard gases, respectively.

3. Results and discussion

3.1. Flue gas composition

Fig. 2 shows the concentrations of O_2, CO_2, N_2 and CO in flue gases as functions of fuel equivalence ratio, ϕ, at

Table 1
The properties of coal

Size (μ)	Proximate (wt%, db)			Ultimate (wt%, daf)				
	FC	VM	Ash	C	H	N	S	O
60–125	50.0	48.0	2.0	73.5	5.2	1.4	1.1	18.8

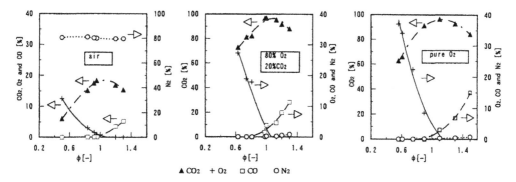

Fig. 2. The variations of O_2, CO_2, N_2 and CO in flue gas with fuel equivalence ratio at 1273 K.

1273 K for the CCO, CCA and O_2 + CO_2 processes. As shown in Fig. 2, in the fuel-lean combustion, CO_2 concentration increased with ϕ, on the other hand, in the fuel-rich combustion, CO_2 concentration decreased with ϕ as CO was produced increasingly. The peak of CO_2 concentration was only about 16% for the CCA process, it was, however, more than 95% for the CCO process and the O_2 + CO_2 process (both are called as N_2-free processes). N_2 concentration in flue gas for the CCA process maintained as high as 80%, but for the N_2-free processes it was only less than 1%, mainly derived from the fuel-nitrogen in coal. High CO_2 concentration can make the recovery of CO_2 from the N_2-free process more efficient than from the traditional CCA process. In addition, the concentration of O_2 decreased

largely as equivalence ratio increased in the fuel-lean combustion, but at the same equivalence ratio O_2 concentration for the CCO process was much higher than that for the CCA and O_2 + CO_2 processes. Experiments at other temperatures of 1123, 1423 and 1573 K showed the same results on flue gas composition.

3.2. NO_x emissions

Fig. 3 shows NO_x emissions versus fuel equivalence ratio at different O_2 concentrations for N_2- and CO_2-based inlet gases at 1273 K, expressed in ppm and mg-N/g-Coal-fed at the upper and lower rows, respectively. In all the cases the emitted NO_x content (ppm) in flue gases increased initially with ϕ in the fuel-lean region, and then declined dramatically as ϕ approached and exceeded the stoichiometric point (fuel-rich region). A similar trend was reported for the air process by other researchers [14–16]. The peaks of NO_x were at an equivalence ratio of around 0.8 for all the kinds of inlet gases used in the experiments. The NO_x concentration at peak decreased as the O_2 concentration in inlet gas decreased, from as high as 4500 ppm for the pure oxygen process to less than 1000 ppm for the air process, and to about 500 ppm for the O_2(20%) + CO_2(80%) inlet gas combustion process. However, NO_x emission indexes (mg-N/g-Coal-fed) monotonically decreased with the increase of ϕ in both fuel-lean and fuel-rich combustion, as shown in the lower row of Fig. 3. A similar trend was reported by Courtemanche et al. [15] in the combustion of coal and man-made organic materials with air. In the fuel-lean combustion, the low ratio of coal to inlet gas led to low NO_x content in flue gas; but due to the oxidizing atmosphere more fuel-N conversion to NO_x, resulting in a high NO_x emission index. The further decline of NO_x emissions may be attributed to the reducing atmospheres that favored the formation of HCN and NH_3 [17,18] and the reduction of NO to molecular nitrogen by homogeneous and heterogeneous reaction [18–26]. Although NH_3 and HCN were not measured in the experiments, the explanation may be

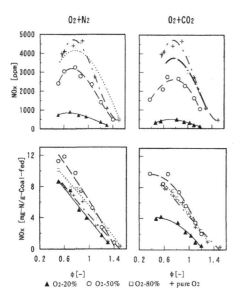

Fig. 3. NO_x emission versus fuel equivalence ratio at different O_2 concentrations for N_2- and CO_2-base inlet gases at 1273 K.

128

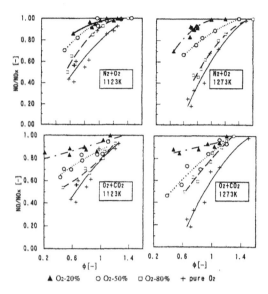

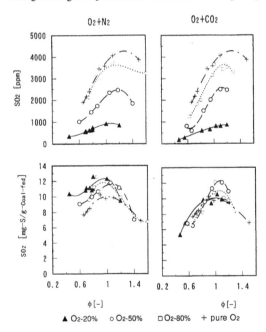

▲ O₂-20%   ○ O₂-50%   □ O₂-80%   + pure O₂

Fig. 4. The effect of fuel equivalence ratio and O₂ concentration on NO/NO$_x$ for N₂- and CO₂-based inlet gases at 1123 and 1273 K.

confirmed by the appearance and the increase of CO and N₂ (for N₂-free inlet gas process), as shown in Fig. 2. Particularly, NO$_x$ concentration in flue gas for the processes with inlet gas of higher O₂ concentration decreased more quickly

O₂+N₂   O₂+CO₂

▲ O₂-20%   ○ O₂-50%   □ O₂-80%   + pure O₂

Fig. 5. SO₂ emission versus fuel equivalence ratio at different O₂ concentrations for N₂- and CO₂-based inlet gases at 1273 K.

after the peak and approached as low as that of the processes with lower O₂ concentration inlet gas when $\phi > 1.4$. This evidence implies that such NO$_x$ emission controlling technology as staged combustion [2,3], which has been widely used in the CCA process in industry, can be introduced to CCO and O₂ + flue gas recirculation processes. The NO$_x$ concentration may reach the same level as the CCA process so that the amount of NO$_x$ emission will be largely reduced.

Comparing the two columns of Fig. 3, NO$_x$ emissions both in ppm and in mg-N/g-Coal-fed for the CO₂-based inlet gases are always lower than those for the N₂-based inlet gases process at each O₂ concentration level. This is due to non-NO$_x$ formation via fixation of atmospheric (molecular) nitrogen in CO₂-based gases, and/or the reduction of NO on char surface through the NO/CO/char reaction [7,23–26]. CO₂ previously existing in inlet gas reacted with coal and/or char in the beginning of combustion to produce CO in the experimental temperature region [27], which promoted the reduction of NO [24].

NO$_x$ concentration in flue gas increased with the increase of O₂ concentration in inlet gases for both N- and CO₂-based inlet gases in the range of experimental equivalence ratio, as shown in the upper row of Fig. 3. However, as shown in the bottom row of Fig. 3, the NO$_x$ emission indexes reached the highest level when O₂ concentration was 50% for both the N₂- and the CO₂-based processes, although the value for the CO₂-based inlet gas processes changed little with the O₂ concentration in inlet gas except for the 20%O₂ + 80%CO₂ process. When O₂ concentration in inlet gases was higher than 50%, NO$_x$ emission indexes decreased, and for the pure oxygen process it went even back to the same level as that of the air process.

In all the experiments in this paper, NO and NO₂ were simultaneously measured by the chemiluminescent analyzer. Fig. 4 shows the ratio of NO/NO$_x$ in flue gas as a function of fuel equivalence ratio at different O₂ concentrations. The ratios for N₂-based inlet gas and CO₂-based inlet gas followed the same trend, as shown in Fig. 4. Under fuel-lean conditions the changes of NO/NO$_x$ with equivalence ratio were quite different for the various O₂ concentration inlet gases. The NO/NO$_x$ for the inlet gases with higher O₂ concentration decreased more quickly than that with lower O₂ concentration as equivalence ratio decreased. On the other hand, under fuel-rich conditions NO/NO$_x$ for all the kinds of inlet gases approached 1.0 as ϕ increased. These results suggest that de-NOx operation for the CCO process should be reconsidered. Since NO/NO₂ ratio affects the consumption rate of de-NO$_x$ reagent (NH₃) in the de-NO$_x$ processes, i.e. consumption rate for NO₂ reduction with selective non-catalytic reduction approach will be 4/3 compared to NO reduction. The chemical reaction formulas are expressed as $4NO + 4NH_3 + O_2 \rightarrow 4N_2 + 6H_2O$ and $6NO_2 + 8NH_3 \rightarrow 7N_2 + 12H_2O$, respectively.

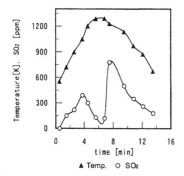

Fig. 6. Changes of temperature and SO₂ emission in batch combustion.

3.3. SO₂ emissions

Plots of the SO₂ emissions in the combustion processes versus equivalence ratio at different O₂ concentrations for N₂- and CO₂-based inlet gases are shown in Fig. 5. For both the N₂- and the CO₂-based inlet gas processes, SO₂ emissions in ppm increased with the equivalence ratio in fuel-lean region, as shown in the upper row of Fig. 5, and then appeared to slightly decrease after $\phi > 1.2$. The decrease of SO₂ emission in the fuel-rich region may partly be the result of the formation of other flurous matters such as H_2S, COS and CS_2 and reduction of SO₂ [28] in the reducing atmo-

sphere. Experiments of coal combustion with oxygen in batch operation showed, as in Fig. 6, that there were two peaks of SO₂ emission in the combustion process. This means that there are two types of S in the coal: some of the S are volatile and others combined strongly with fixed-C in char. From this point, the retention of sulfur in char or unburned coal may be another important reason for the decrease of SO₂ emission in the fuel-rich region. Beer [29] and Chang [30] attributed the decrease of SO₂ emission to aluminosilicate components in ash, which favored retention of sulfur.

3.4. Gas temperature effect

The effects of gas temperature on NO$_x$ emission for the N₂- and the CO₂-based inlet gas processes are shown, respectively, in Figs. 7 and 8. The results showed that NO$_x$ emission increased with gas temperature for all the cases examined, the peak values increased by 50–70% for the N₂-based and 30–50% for the N₂-free processes when gas temperature increased from 1123 to 1573 K. The exception is the process with $20\%O_2 + 80\%CO_2$ inlet gas, in which NO$_x$ emissions were very low at each temperature and increased by almost eight times from 1123 to1573 K. The explanation is that high concentration of CO₂ in inlet gas promoted the reaction of CO₂ with C to CO, and NO$_x$ was reduced by CO even in the fuel-lean region especially at lower temperatures. It may be confirmed by the fact that CO

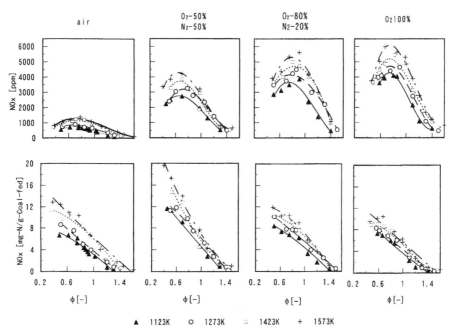

Fig. 7. NO$_x$ emissions versus equivalence ratio and temperature at different O₂ concentration for N₂-based inlet gases.

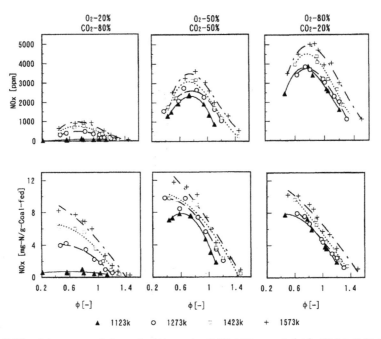

Fig. 8. NO$_x$ emissions versus equivalence ratio and temperature at different O$_2$ concentration for CO$_2$-based inlet gases.

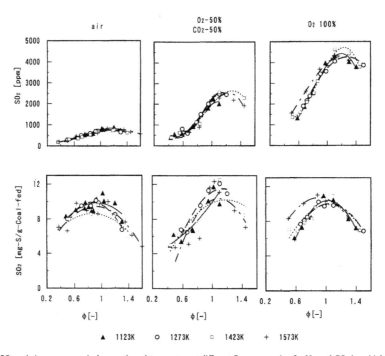

Fig. 9. SO$_2$ emissions versus equivalence ratio and temperature at different O$_2$ concentration for N$_2$- and CO$_2$-based inlet gases.

was detected at any ϕ in the experiments at 1123 K. Many researchers [15,31–35] reported the effects of temperature on NO_x emissions for CCA. Haussann and Kruger [33] reported that the release of fuel-nitrogen from coal increased by 30–40% with temperature in the same range of temperature as this paper. Courtemanche [15] reported NO_x emissions increased by up to 25% when gas temperature increased from 1300 to 1600 K in the combustion experiments of a high volatile bituminous coal with air.

In the fuel-rich region, NO_x emissions in the higher temperature processes decreased to the same level as those in the lower temperature processes at $\phi > 1.4$.

For the processes with the same O_2 concentration in inlet gas, NO_x emissions from the CO_2-based gas process were always lower than those from the process with N_2-based inlet gas for each temperature examined. And it can be further seen that the process with $50\%O_2 + 50\%N_2$ inlet gas emitted the most NO_x per unit of coal at any temperature. This can be explained that this inlet gas possessed high concentration of both O_2 and N_2, which promoted the thermal formation of NO_x from N_2 and the conversion of fuel-N to NO.

Fig. 9 presents the effects of gas temperature on SO_2 emissions for the air, the inlet gas of $50\%O_2 + 50\%CO_2$ and the pure O_2 processes. It is seen that there was a small effect of temperature on the SO_2 emissions for N_2-based, CO_2-based and pure oxygen processes.

4. Conclusion

1. CO_2 concentration in flue gas was only about 16% for the processes with air, but higher than 95% for the processes with N_2-free (O_2, $O_2 + CO_2$) inlet gas. Recovery of CO_2 from the N_2-free processes will be easier and more efficient than the traditional air processes.
2. NO_x emissions increased with ϕ in fuel-lean combustion and declined quickly as ϕ approached and exceeded the stoichiometric point. The peaks were around $\phi = 0.8$ for all the cases in the experiments and the peak values increased from 1000 ppm for the air process and 500 ppm for the $O_2(20\%) + CO_2(80\%)$ inlet gas process to about 4500 ppm for the oxygen combustion process. When $\phi > 1.4$ NO_x emissions will decrease to the same level for all the inlet gas processes in this experiment. NO_x emission indexes decreased monotonically as ϕ increased under both fuel-lean and fuel-rich conditions. CO_2-based inlet gas processes emitted less NO_x than N_2-based inlet gas processes with same O_2 concentration in inlet gases. NO_x emissions reached highest level from the process with N_2-based inlet gas containing $50\%O_2$.
3. SO_2 emissions increased with ϕ in fuel-lean combustion and decreased slightly when $\phi > 1.2$.
4. There was a large effect of temperature on NO_x emis-

sion. NO_x emissions at peak increased by 50–70% for the N_2-based inlet gas processes and by 30–50% for the CO_2-based inlet gas processes from 1123 to 1573 K. NO_x emitted from higher temperature processes reached almost the same level as lower temperature processes when $\phi > 1.4$. On the other hand, temperature had little effect on SO_2 emissions.

Acknowledgements

The authors greatly acknowledge the financial support for this research provided by Japanese Science Promotion Society (JSPS).

References

[1] Bleckner B. In: 26th Symposium (International) on Combustion, The Combustion Institute, Pittsburgh, Pennsylvania, 1996. p. 3231–41.
[2] Yamagishi K, Nozawa M, Yoshie T, et al. In: 15th Symposium (International) on Combustion, The Combustion Institute, Pittsburgh, Pennsylvania, 1974. p. 1157–68.
[3] Wendt JOL, Sterling CV, Matovich MA. In: 14th Symposium (International) on Combustion, The Combustion Institute, Pittsburgh, Pennsylvania, 1973. p. 897–906.
[4] Zhou CQ. In: 26th Symposium (International) on Combustion, The Combustion Institute, Pittsburgh, Pennsylvania, 1996. p. 2091–8.
[5] Koebel M, Elsener M. Chem Engng Sci 1998;53(4):6657.
[6] Nordin A, Eriksson L, Ohman M. Fuel 1995;74(1):128.
[7] Takano S, Kiga T, et al. IHI Engng Rev 1995;28(4):160.
[8] Garay PN. In: Proceeding of Second American Conference on Electric Power 1994. p. 280–3.
[9] Andries J, Becht JGM. Energy Convers Mgmt 1996;37(6-8):855.
[10] Shirakawa K, Noguchi Y. Proceeding of Conference on the Utilizing Technologies of Coal (Japan) 1992;2:235–44.
[11] Shirakawa K, Noguchi Y, et al. In: Proceeding of Conference on the Utilizing Technologies of Coal (Japan) 1993;3:50–60.
[12] Kaneko M, Liu H, et al. In: Proceeding of 36th Symposium on Combustion (Japan) 1998. p. 146–8.
[13] Iino K, Suwa T, Suzuki K. In: Proceeding of 36th Symposium on Combustion (Japan) 1998. p. 302–4.
[14] Spliethoff H, Greul U, et al. Fuel 1996;75(5):560.
[15] Courtemanche B, Levendis YA. Fuel 1998;77(3):183.
[16] Breen BP, Bell AW, Bayard de Volo N. In: 13th Symposium (international) on Combustion, The Combustion Institute, Pittsburgh, Pennsylvania, 1970. p. 391–401.
[17] Nelson PF, Buckley AN, Kelly MD. In: 24th Symposium (international) on Combustion, The Combustion Institute, Pittsburgh, Pennsylvania, 1992. p. 1259–67.
[18] Martin FJ, Dederick P.K. In: 16th Symposium (international) on Combustion, The Combustion Institute, Pittsburgh, Pennsylvania, 1976. p. 191–8.
[19] Kremer H. In: 21st Symposium (international) on Combustion, The Combustion Institute, Pittsburgh, Pennsylvania, 1986. p. 1217–22.
[20] Li YH, Lu GQ, Rudolph V. Chem Engng Sci 1998;53(1):1.
[21] Song YH, Blair DW, et al. In: 18th Symposium (international) on Combustion, The Combustion Institute, Pittsburgh, Pennsylvania, 1981. p. 53–63.
[22] Haynes BS. Combust Flame 1977;28:81.
[23] Li YH, Radovic LR, Lu GQ, Rudolph V. Chem Engng Sci 1999;54:4125.

132

[24] Ohtsuka Y, Wu Z. Fuel 1999;78:521.

[25] Chu X, Schmidt LD. Ind Engng Chem Res 1993;32:1359.

[26] Suzuki T, Kyotani T, Tomita A. Ind Engng Chem Res 1994; 33:2840.

[27] Dulta S, Wen CY. Ind Engng Chem Process Des Dev 1977; 16(1):20.

[28] Bauer SH, Jeffers P, Lifshitz A, Yadava BP. In: 13th Symposium (international) on Combustion, The Combustion Institute, Pittsburgh, Pennsylvania, 1970. p. 417–25.

[29] Beer JM. In: 22nd Symposium (international) on Combustion, The Combustion Institute, Pittsburgh, PA, 1988. p. 1–16.

[30] Chang KK, Flagan RC, Gavalas GR, Sharma PK. Fuel 1986; 65:75.

[31] Spliethoff H, Greul U, et al. Fuel 1996;75(5):560.

[32] Pershing DW, Wendt JOL. Ind Engng Chem Process Des Dev 1979;18(1):60.

[33] Song YH, Pohl JH, Beer JM, Sarofilm AF. Combust Sci Technol 1982;28:31.

[34] Mitchell SC. NO_x in pulverised coal combustion, IEA COAL RESEARCH, April, 1998.

[35] Haussmann GJ, Kruger CH. In: 23rd Symposium (International) on Combustion, The Combustion Institute, Pittsburgh, PA, 1990. p. 1265–71.

Modelling

Modeling of nitrogen oxides formation and destruction in combustion systems

S.C. Hill[1],*, L. Douglas Smoot[2]

Advanced Combustion Engineering Research Center (ACERC), 75A CTB, Brigham Young University, Provo, UT 84602, USA

Received 3 May 1999; revised 10 March 2000; accepted 10 March 2000

Abstract

The formation of nitrogen oxides (NO$_X$) in combustion systems is a significant pollutant source in the environment, and the control of NO$_X$ emissions is a world-wide concern as the utilization of fossil fuels continues to increase. In addition, the use of alternative fuels, which are typically of lower quality, tends to worsen the problem. Advances in the science of NO$_X$ reactions, mathematical modeling, and increased performance of computer systems have made comprehensive modeling of NO$_X$ formation and destruction a valuable tool to provide insights and understanding of the NO$_X$ reaction processes in combustion systems. This technology has the potential to enhance the application of various combustion techniques used to reduce NO$_X$ emissions from practical combustion systems. This paper presents a review of modeling of NO$_X$ reactions in combustion systems, with an emphasis on coal-fired systems, including current NO$_X$ control technologies, NO$_X$ reaction processes, and techniques to calculate chemical kinetics in turbulent flames. Models of NO$_X$ formation in combustion systems are reviewed. Comparisons of measured and predicted values of NO$_X$ concentrations are shown for several full-scale and laboratory-scale systems. Applications of NO$_X$ models for developing technologies, in order to reduce NO$_X$ emissions from combustion systems are also reported, including the use of over-fire air, swirling combustion air streams, fuel type, burner tilt angle, use of reburning fuels, and other methods. © 2000 Elsevier Science Ltd. All rights reserved.

Keywords: Pollutants; Combustion; Modeling; NO$_X$; Emissions

Contents

* Corresponding author. Tel.: +1-810-378-2804; fax: +1-810-378-3831.
  E-mail address: scotty@byu.edu (S.C. Hill).
[1] Combustion Computations Laboratory Manager and ACERC Research Associate.
[2] Professor, Chemical Engineering and ACERC Founding Director.

Reprinted from *Progress in Energy and Combustion Science* **26 (4-6)**, 417-458 (2000)

Nomenclature

A	pre-exponential rate constant (units of k)
A	specific area, m^2/kg
D_{im}	molecular diffusion coefficient of species i in mixture, m^2/s
D_Y	turbulent diffusion coefficient, m^2/s
E	activation energy, kcal/kmol
f	gas mixture fraction
k	reaction rate coefficient, m/s, m^3/kmol,...
K_{eq}	equilibrium constant
M	molecular weight, g/gmol
$P( )$	probability density function
P	pressure, N/m^2
r	radial distance, m
R	universal gas constant, kcal/kgmol K
$S_p^{\eta_1}$	gaseous production rate from volatiles
$S_p^{\eta_2}$	gaseous production rate from char
T	temperature, K
t	time, s
u	axial velocity, m/s
u'	fluctuating velocity component, m/s
v	radial velocity, m/s
w	angular velocity, m/s
w	mass reaction rate, kg/m^3 s
W	mass reaction rate, kg/m^3 s
x	axial distance, m
X	molar fraction
Y	mass fraction
y	fluctuating component of mass fraction

Subscripts and superscripts

0, 1, 2, 3	forward reaction number
$-1, -2, -3$	reverse reaction number
a	from volatiles
b	from char
$-$	mean value
$\sim$	Favre-averaged value
b	exponent
char	char
eq	equilibrium
e	external
f	fully reacted value
H	hydrogen
i	ith species
N	nitrogen
O	oxygen
OH	hydroxide
p	particle
vols	volatiles

Greek symbols

α	mass fraction

η	coal gas mixture fraction
ρ	gas density, kg/m^3
ζ	mass fraction nitrogen, initially converted to $HCN + NH_3$
ω	mass fraction
π_i	ratio $(\tilde{Y}_i/\tilde{Y}_i^f)$
γ	fractional change in external surface area of char due to swelling or fracturing
θ	cylindrical angle, radians

Abbreviations and acronyms

ACERC	Advanced Combustion Engineering Research Center
CFD	Computational Fluid Dynamics
CHEMKIN	Chemical Kinetic Code
CPU	Computer Processing Unit
DNS	Direct Numerical Simulation
EDC	Eddy Dissipation Concept
HP	Hewlett Packard
ILDM	Intrinsic Low-Dimensional Manifold
LEM	Linear Eddy Model
LPC	Lean Premixed Combustion
MB	Megabytes Computer Memory
MW	Megawatt
MW_e	Megawatt-electric
NO_X	Nitrogen Oxides, NO, NO_2, N_2O
OFA	Overfire Air
PCGC-3	Pulverized Coal Gasification and Combustion-3-Dimensional
PDF	Probability Density Function
PSR	Perfectly Stirred Reactor
SCR	Selective Catalytic Reduction
SNCR	Selective Non-catalytic Reduction
SR	Stoichiometric Ratio

1. Introduction

Nitrogen oxides are a significant threat to the environment, and combustion systems are a major source of these pollutants. Nitrogen oxides consist of nitric oxide (NO), nitrogen dioxide (NO_2), and nitrous oxide (N_2O). NO and NO_2 are collectively referred to as NO_X. Combustion systems commonly utilize fossil fuels such as gas, petroleum, and coal. Coal is generally burned in fixed-bed, fluidized-bed, or entrained-flow combustors. Recent work on gaseous combustion has focused on lean-premixed combustion in gas turbines. During the past several decades, work on NO_X formation and control has been substantial, including gas, petroleum-based, and coal NO_X modeling and measurements, demonstration of very low NO_X burner technologies, and other NO_X control technologies. International work in NO_X formation and control also has been extensive during this decade, as noted in reviews by Miller and Bowman[1], Hayhurst and Lawrence[2], Bowman[3], Kramlich and Linak[4], and Smoot et al.[5].

A significant portion of this work has involved computer simulations of the NO_X reaction processes. Computer modeling of the NO_X formation and destruction process in combustion systems provides a tool that can be used to investigate and improve understanding of these complex systems. This modeling work has involved three approaches. In one approach, only the NO_X reaction process is modeled, using hundreds of elementary reactions coupled with a priori or simplified descriptions of the temperature and flow field. A second approach makes use of engineering correlations of measured NO_X effluent data for a variety of test conditions in order to estimate NO_X emissions[6]. A third approach couples a simplified description of the NO_X reaction process with a detailed, 'comprehensive' description of the combustion and flow processes. This third approach for modeling the NO_X reaction process requires a joint solution of detailed CFD equations for turbulent flow, combined with reduced chemical reaction mechanisms, and is referred to as comprehensive modeling, and it is emphasized herein.

Comprehensive modeling of NO_X reaction processes in combustion systems has been ongoing for over two decades. Caretto[7] provided an earlier review of mathematical modeling of pollutant formation, including carbon monoxide, oxides of nitrogen, hydrocarbons, sulfur oxides, and soot. He discusses the approach used with simple flow models and comprehensive models of combustion systems for modeling of pollutant formation. The goal of mathematical models was prediction of trends, and the a priori prediction of actual emissions levels was generally not considered possible. Several solutions of comprehensive models had been reported for gaseous non-reacting and reacting systems. Application of mathematical models to predict pollutant emissions was very limited, and "real" comparisons of theory and experiment were non-existent. Substantial progress has been made in the area of computer modeling of combustion systems since this time, and considerable progress is still being made. Combustion modeling has been used progressively more in the last decade to improve the design and operation of combustion systems, especially in the areas of pollutant control and ash management.

Comprehensive modeling of NO_X reaction processes in combustion systems requires simulation of both the turbulent fluid dynamics and the chemical kinetics in the system being modeled. Hundreds of elementary reactions are required to provide a detailed description of the NO_X reaction process in laminar combustion systems. However, it is not currently feasible to use such generalized reaction mechanisms to model this process in a turbulent, reacting system in which large reaction kinetics schemes are coupled with the turbulent fluid dynamics. Consequently, global reactions or reduced mechanisms are typically used in comprehensive combustion codes to describe the NO_X reaction processes.

This paper presents a review of comprehensive modeling

Table 1
Costs and reduction potential for various NO_X control technologies [15]

Technology	Retrofit cost ($/kW)	% NO_X reduction
Combustion modifications	Range: 0–10	Range: 10–25
	Average: 3.2	Average: 22.5
Overfire air	Range: 5–15	Range: 10–30
	Average: 8.2	Average: 23
Low NO_X burners	Range: 6–40	Range: 20–60
	Average: 16.1	Average: 51
Natural gas reburning	Range: 14–50	Range: 40–60
	Average: 32	Average: 51
Selective non-catalytic reduction	Range: 5–50	Range: 20–70
	Average: 19.6	Average: 43.1
Selective catalytic reduction	Range: 80–180	Range: 60–90
	Average: 123.5	Average: 77.5

of NO_X reactions in combustion systems. Techniques for NO_X control are discussed, followed by a discussion of the salient features of NO_X formation processes in turbulent combustion systems and a survey of NO_X models from the literature. The emphasis of this paper is on coal combustion systems, but NO_X formation in gaseous and oil-fired systems is also discussed. A detailed discussion of the NO_X model developed by the authors and colleagues is included. Finally, several applications of NO_X modeling are shown.

2. NO_X control

Post-combustion clean-up of NO_X is viable, but modifying the combustion process often controls NO_X most economically [8]. Common NO_X control technologies include burner design, air-staging, fuel-staging, over-fire air, flue gas recirculation, reburning, and advanced reburning. These control technologies can often be used in combination. The objective of most of these modifications is to minimize the reaction temperature and/or the contact of nitrogen from the fuel with oxygen in the combustion air, while creating a fuel-rich zone in which NO_X can be reduced to N_2. One way this is achieved is to control the mixing and heat-up such that the ignition zone is very fuel-rich, but still provides enough energy for appreciable reaction to occur. Low-NO_X burners accomplish this with swirling, secondary air-flow combined with tertiary air staging to stabilize a fuel-rich ignition zone prior to significant mixing with the remaining combustion air. Early work by Pershing and Wendt [9] showed this technique to be much less effective with a coal char than with a high volatile coal, which illustrates the strong coupling between burner modifications and fuel characters. Major drawbacks of the modifications to reduce NO_X emissions include reduced coal burnout, increased carbon monoxide, and increased corrosion problems [10].

Staging is used to create a fuel-rich zone near the burners by introducing only a portion of the combustion air into the main-burner region, and introducing the remaining combus-

tion air downstream. This allows most of the coal nitrogen to be released in a fuel-rich region, which favors formation of N_2 over NO_X. Wendt [11] showed that using two staging locations resulted in significantly lower NO_X emissions than with classical single-location staging. Unfortunately, staging can be relatively expensive to implement in existing combustors and can have a negative impact on performance [12].

The two most common post-flame techniques for NO_X control are selective catalytic reduction (SCR) and selective non-catalytic reduction (SNCR). With SCR, ammonia is injected near the combustor exit, and can reduce NO_X emissions by as much as 90% [13]. SCR can be relatively expensive due to the high cost of the catalyst. With SNCR, ammonia is injected farther upstream in the furnace in a higher temperature zone (900–1100°C) [14]. SNCR is only effective over a relative narrow temperature window, which precludes its use in many applications. Table 1 shows a list of the most common combustion modifications used to reduce NO_X emissions in coal combustion systems and the associated costs of these modifications [15]. Although advanced reburning is not included in the table, NO_X reduction of 70–90% has been reported and retrofit costs should be similar to those reported for SNCR [5]. Table 1 shows that the cost of upgrading practical combustion systems can easily exceed two million dollars per 100 MW of generating capacity. However, emission penalties of several million dollars can be incurred annually if NO_X emissions levels exceed allowable limits [15].

The NO_X reaction processes are very complex, and comprehensive modeling of NO_X formation and reduction in turbulent reacting systems is difficult. However, it provides a valuable tool to provide insights and understanding of this process in combustion systems. NO_X models can be used to investigate and optimize operating and design parameters and investigate various combustion techniques used to reduce NO_X emissions from practical combustion systems. These include fuel type, air-staging, low-NO_X burners, product gas recirculation, equivalence ratio, and particle size, all of which aim to reduce NO_X formation through control of the local temperature and fuel/air ratio where the NO_X reactions occur. Reductions of up to five-fold have been demonstrated with these technologies, which are sometimes used in combination [16]. Even if only helping to provide incremental reductions in NO_X emissions, these models can result in considerable cost savings.

3. NO_X formation and destruction processes

3.1. Overview

NO_X formation and destruction processes in combustion systems are very complex. Mitchell [6] provides a recent review of the reaction processes involved in NO_X formation in pulverized coal combustion systems, including work in laboratory-scale, pilot-scale, and full-scale furnaces.

During combustion, nitrogen from the combustion air or fuel is converted to nitrogen-containing pollutants such as NO, NO_2, N_2O, NH_3, and HCN. The pollutant species formed depends principally on the temperature and fuel/oxygen ratio in the combustion zone. The ratio of NO_2 to NO_X is about 0.1 in typical methane/air flames, but it can rise to as high as 0.9 in low-temperature, low-NO_X flames due to the high ratio of H to NO concentrations [17]. The amount of NO_2 formed is very sensitive to the fluid dynamics in the flame zone [18], and unexpectedly high levels of NO_2 have been measured in premixed flames [19]. NO_2 formation from NO tends to occur in regions where rapid cooling takes place, such as the mixing region of hot combustion gases with the inlet air [19]. Measurement of NO_2 is sensitive to the measurement technique since NO_2 formation can occur inside the sampling probe [19–21]. A review paper provides a detailed discussion of NO_2 emissions and measurement techniques [22].

Nitrous oxide (N_2O) contributes to the greenhouse effect in the troposphere and participates in ozone depletion in the stratosphere. Several investigators have shown that N_2O emissions are not typically significant within coal combustion systems [22,23]. N_2O can be formed by a number of reactions in gas and coal reactors, but it rapidly reacts with H and OH radicals to form N_2. Typical reactions involving N_2O in gas and coal combustion systems are shown elsewhere [13]. N_2O survives only in the hotter, fuel-rich, flame regions and is destroyed downstream [24]. Peak N_2O values in coal systems are typically less than 2% of peak NO values, but in the post-flame zone (900–1250°C), N_2O formation is negligible [24,25]. Some N_2O (ca. 50 ppm) does result from fluidized-bed combustion due to their lower operating temperatures [4]. A review of N_2O emissions from combustion and industrial systems is available [4].

NO from combustion systems results from three main processes: thermal NO, prompt NO, and fuel NO. Previously published reviews (e.g. Refs. [3,4,9]), document various details of one or more of these basic processes. The seminal contributions of Zeldovich (thermal NO_X) [26], Fenimore (prompt NO_X) [27], and Pershing and Wendt (fuel NO_X) [9] should also be noted. The reactions involved in these processes are slow relative to the reactions of the major combustion species. Therefore, kinetic rate expressions must be used to model these processes. Thermal NO is formed from oxidation of atmospheric nitrogen, prompt NO is formed by reactions of atmospheric nitrogen with hydrocarbon radicals in fuel-rich regions of flames, and fuel NO is formed from oxidation of nitrogen bound in the fuel. Each of these will be discussed subsequently.

A survey of the literature shows that in recent years, a large amount of work has been done related to nitrogen pollutants from turbulent, combustion systems. Although a large portion of this work has emphasized gas-phase systems, a considerable amount of the work has dealt specifically with coal-fired systems. This is due to the large fuel-nitrogen content of coal (about 1%), which can result in substantial NO_X emissions. Thermal NO is the main source of NO in gaseous combustion systems, and fuel NO is the main source of NO in coal-fired systems.

Comprehensive models of the combustion and NO_X reaction processes are valuable tools, which can be used to study the pollutant formation process. Modeling of NO_X formation in turbulent combustion systems requires a description of numerous physical processes, including the general fluid dynamics, the local mixing process, heat transfer, and chemical kinetics. In addition, these processes occur simultaneously and are intimately coupled, which must generally be accounted for when modeling these systems. A comprehensive model of nitrogen pollutants in practical combustion systems requires coupling of the nitrogen pollutant kinetics with the local turbulent fluid dynamics. The literature on NO_X pollutant formation shows several comprehensive reaction sets used to describe the net rate of NO_X formation in various types of flames [1,28–32]. These reaction sets give insight into the formidable task involved in a detailed kinetic description of NO_X reaction processes in turbulent combustion systems. It has not been feasible to couple a lengthy detailed kinetic mechanism with the turbulent mixing process. A practical kinetic mechanism must currently use a "minimum" set of reactions, often global or overall in nature, to approximate the essential features of the NO_X reaction process. The kinetic mechanism must contain sufficient detail to adequately describe the NO_X reaction process in turbulent combustion systems, but must consist of sufficiently few reactions to allow for coupling with the turbulent mixing process. Coupling with the turbulent mixing process requires calculation of the time-mean reaction rates for the kinetic mechanism. The approach used for calculation of each reaction rate in the kinetic mechanism is dependent on the relative time-scales of the reaction rate and the turbulent fluctuations.

Turbulent flows have been a subject of active research for about a century, but many of the problems related to mathematical descriptions of turbulent systems still exist. This is especially true in heterogeneous combustion systems such as encountered in pulverized-coal flames. Several workers have reviewed the status of turbulence modeling, and discussed the limitations and problems related to turbulence modeling [33–37].

Models of nitrogen pollutants are commonly de-coupled from the generalized combustion model and executed after the flame structure has been predicted. The basis for this assumption is that the formation of trace pollutant species does not affect the flame structure, which is governed by fast fuel-oxidizer reactions. Another advantage of this approach is computational efficiency. Further, solving the pollutant model equations jointly with the combustion model equations is far more complex. The time required to solve the system of equations for the combusting fuel can require many hours of computer time while the pollutant submodels typically converge in a fraction ($\sim$10%) of the time required to converge the combustion case. Thus, NO_X submodel parameters and pollutant formation mechanisms can be more easily investigated by solving the NO_X submodel using restart files for a pre-calculated flame structure.

3.2. Nitrogen release from the fuel

Modeling fuel-NO_X formation from fuels containing nitrogen requires description of the release of the nitrogen from the fuel, including both the rate of nitrogen release and the form or species in which the nitrogen is released. Both are important in developing pollutant control strategies. The form of nitrogen in the coal determines which nitrogen species are released during the devolatilization process. Coal nitrogen is thought to occur primarily in aromatic ring structures [38]. It is postulated that this nitrogen is released mainly in the tar molecules and light gases during high temperature devolatilization [39,40]. In coal particles, this is accomplished during the devolatilization and char oxidation processes. The heating rate and final temperature of the coal particle may influence its devolatilization characteristics, and thus the subsequent amount of nitrogen devolatilized from the coal (fuel-nitrogen). In coal combustion systems, the particle size, heating rate, and final temperature are dependent on the type of combustor. Fluidized-bed combustors have heating rates 10^3–10^4 K/s, reaction temperatures 1100–1200 K, and particle sizes from 1 to 5 mm [41]. Pulverized coal, entrained-flow combustors have heating rates 10^4–10^5 K/s, reaction temperatures 1700–1800 K, and particle sizes typically <100 μm [41]. Several workers have reported that both a higher heating rate and higher final particle temperature produced a greater yield of volatiles [42,43]. However, other workers have shown that the yield of volatiles is independent of the heating rate, and is principally a function of the peak temperature reached in the reactor [44–46]. The time-dependence of the yield of devolatilization products has been observed by most researchers, including results from two different types of

devolatilization experiments [47]. Solomon et al. [48] provide a review of different pyrolysis test methods and discuss effects of various parameters on devolatilization products and yields.

The particle heating rate and final temperature can also influence the extent of nitrogen devolatilization, and the amount of nitrogen remaining in the coal particle [9]. Studies show that nitrogen release increases with increasing heating rate [49], and is a function of final temperature [50,51]. Most experimental studies of coal pyrolysis indicate that the coal nitrogen release is divided between the volatiles and the residual char [52]. Consequently, nitrogen release from both coal pyrolysis and char oxidation must be considered when modeling NO_X reactions in coal combustion systems. Secondary pyrolysis of coal tar and light gases results in the formation of HCN and soot, and is an important part of nitrogen evolution during volatiles combustion [38,48]. Eddings et al. [53] presented a model that differentiates between the volatile nitrogen and char nitrogen. The volatile nitrogen is tracked separately from the char nitrogen and uses a different reaction mechanism. These changes resulted in an improved capability to predict the effects of coal type on NO formation. One limitation of the model is that an a priori knowledge of the volatile and char nitrogen mass fractions is required. Models have been developed recently, which predict the split of coal nitrogen between volatiles and char based on the characteristics of the coal and the combustion environment [54], although, application of these models in comprehensive CFD codes has not been published.

Coal contains approximately 1% nitrogen by weight depending on the coal type, with bituminous coals generally having the highest nitrogen levels and anthracites the lowest nitrogen levels [50,55,56]. NO_X emissions generally increase with the fuel nitrogen content of the coal [51]. Devolatilization studies show that nitrogen leaves the coal at a rate proportional to, but somewhat higher than the rate of coal weight loss [39,40,49,57]. Coal nitrogen loss is less dependent on coal type than is coal mass loss [58]. In addition, nitrogen evolution is more sensitive to temperature than is the coal mass devolatilization. Nitrogen is emitted more slowly than the coal mass at lower temperatures and more rapidly at higher temperatures [55,58–60]. Typically, low-rank coals release a larger fraction of nitrogen in the light gases, and a lower fraction in the tars, compared to medium-rank coals [38].

The timing of nitrogen evolution is important since if the nitrogenous species are evolved into an oxidizing environment, significant fuel-NO_X formation will result. However, if the nitrogenous species are evolved in the absence of oxygen, the fuel-nitrogen can form N_2 rather than NO. Thus, early contact of nitrogenous species with oxygen should be avoided to minimize NO_X formation [57,61]. Consequently, the volatiles content of the coal can have a significant impact on the amount of final NO emissions from a coal system because of its early release [62]. This is further substantiated by Yang et al. [63] who showed that volatile matter is the most important NO_X-forming property for bituminous coals. This is another important reason to correctly model nitrogen release from the coal during coal pyrolysis and char oxidation.

At higher temperatures (above 1400–1500 K), up to 70–90% of the coal nitrogen is devolatilized [49,61,64]. Data show that 0–20% of nitrogen in the coal is evolved in the early volatiles, primarily as HCN and NH_3 [42,49,65,66]. The HCN is formed primarily from the tars and nitrogen bound in aromatic rings [38], and the NH_3 is formed primarily from the amines in the coal [67]. Heterocyclic compounds of pyridine, quinoline, pyrrole, and benzonitrite are mostly pyrolyzed to HCN at typical combustion temperatures [13]. Chen et al. [51] observed that more HCN than NH_3 was formed from bituminous coals, but more NH_3 was formed from subbituminous and lignite coals. This is consistent with the observation that the number of aromatic rings in coal decreases with coal rank, and the number of molecular cyclic structures increases. Consequently, the conversion of fuel nitrogen to NH_3 increases for low-rank coals [68]. However, NH_3 is not generally detected during coal devolatilization in either heated-grid or entrained-flow experiments [38].

Bose et al. [69] observed that HCN appearance in the gas phase from devolatilization precedes NH_3 even for low-rank coals. Tests by Ghani and Wendt [70] showed initial NH_3 concentrations from coal devolatilization are far less than HCN concentrations for Utah bituminous, Beulah lignite, Illinois bituminous, and Texas lignite coals. They concluded that NH_3 is formed from HCN under oxidative pyrolysis conditions. Haussmann and Kruger [71] also observed that HCN appears to be the major light-gas product of primary and secondary pyrolysis for both Montana Rosebud subbituminous coal and Pittsburgh #8 bituminous coal.

The remainder of the nitrogen compounds in the coal is evolved late in the devolatilization process with high molecular hydrocarbons and tars [42,55,57,60,67,72,73]. One theoretical study showed that the higher the heating rate of the particles, the greater the production of HCN during the devolatilization process. Several studies [22,74] show that HCN is not released during the char oxidation phase of coal combustion; consequently, NO_X produced during the char combustion process is not formed from homogeneous oxidation of HCN.

3.3. Homogeneous NO_X reactions

NO_X can be formed or destroyed by at least four separate reaction processes in the gas phase, which are classified as: thermal NO, prompt NO, fuel NO, and NO reburning.

3.3.1. Thermal NO

Thermal NO is formed from oxidation of atmospheric nitrogen at relatively high temperatures in fuel-lean environments, and has a strong temperature-dependence. This process is described by the widely accepted Zeldovich two-step mechanism [64,75]:

$$N_2 + O = NO + N \tag{1}$$

$$N + O_2 = NO + O \tag{2}$$

An additional elementary reaction is often added to the thermal NO mechanism in what is referred to as the modified or extended Zeldovich mechanism, which is applicable in some cases when NO formation from the Zeldovich mechanism is under-estimated [76]:

$$N + OH = NO + H \tag{3}$$

The extended Zeldovich mechanism also considers the effect of oxygen and hydrogen radicals on NO formation. Typically, these radicals are calculated from equilibrium considerations. However, with this assumption the predicted thermal NO concentrations are always lower than the observed concentrations [76–79]. Predicted thermal NO using measured oxygen radical concentrations show good agreement with measured concentrations [76,80]. Table 2 lists the commonly used parameters for these thermal-NO reactions.

The thermal NO reactions take place in a few tens of microseconds, and are highly dependent on temperature, residence time, and atomic oxygen concentration [64,81]. The first reaction is usually accepted as being the rate-determining step due to its high activation energy, and results in the Zeldovich mechanism being temperature-sensitive [56,75,82]. The temperature dependence also results from the temperature sensitivity of the O atom equilibrium concentration [75]. The last reaction shown above is usually negligible except in fuel-rich flames [75,81].

The overall rate for the three reversible, thermal NO reactions is [80]:

$$\frac{d[NO]}{dt} = 2[O]\left\{ \frac{k_1[N_2] - \frac{k_{-1}k_{-2}[NO]^2}{k_2[O_2]}}{1 + \frac{k_{-1}[NO]}{k_2[O_2] + k_3[OH]}} \right\}, \tag{4}$$

$g\ mol\ cm^{-3}\ s^{-1}$

A simplified expression is obtained by assuming initial concentrations of NO and OH are low so that only the forward rates of the Zeldovich mechanism are significant

$$\frac{d[NO]}{dt} = 2k_1[O][N_2], \qquad g\ mol\ cm^{-3}\ s^{-1} \tag{5}$$

Rate equations (4) and (5) are coupled to the fuel oxidation process through competition for the oxygen atom, whose local concentration must be estimated when a comprehensive kinetic scheme is not used to compute the fuel oxidation chemistry. Accurate estimation of the non-equilibrium oxygen concentration is critical for the correct prediction of thermal NO reactions [18]. In fuel-lean, secondary combustion zones, where CO is oxidized to

CO_2, oxygen atoms have been assumed to be in equilibrium with O_2 [68]

$$[O] = \{K_{eq}[O_2]\}^{1/2} \qquad (6)$$

In the flame region where hydrocarbons are consumed, oxygen atom concentrations often significantly exceed the level predicted by Eq. (6). A "partial equilibrium" expression has also been employed for fuel-rich zones where O and OH radical concentrations may be approximated by fast-shuffle oxy-hydrogen reactions and where CO is further oxidized to CO_2 [68]

$$[O] = K_{eq}\frac{[O_2][CO]}{[CO_2]} \qquad (7)$$

Eqs. (6) and (7) yield equivalent O concentrations when CO, CO_2, and O_2 are in chemical equilibrium. Eq. (7) will yield different results when a rate expression is also used to calculate CO oxidation to CO_2. Use of Eq. (6) gave far better results for thermal NO_X concentrations formed in combustion of natural gas and air with equivalence ratios in the range from about 0.9 to 1.0 in a laboratory reactor [13]. Coelho and Carvalho [83] show predictions of thermal NO formation for an oil-fired utility furnace using four different methods to estimate the [O] and [OH] concentrations. Their predictions showed a two-fold difference in predicted NO concentrations for the different models, but there was not sufficient measured data to determine which method was most accurate. Chen et al. [84] show comparisons of the equilibrium [O] and the [O] predicted from CHEMKIN [85] simulations of C_2H_4/O_2 flames. In this comparison, the equilibrium [O] is much higher than the predicted [O]; however, a near-linear relationship exists between the [O_2] and the predicted [O]. It is clear that reliable general estimation of the atomic oxygen concentration is far from resolved.

At mean temperatures below 1600–1800 K, thermal NO formation by the Zeldovich mechanism is significantly reduced [56,64] Thermal NO formation is also much less prevalent in fuel-rich environments because of the low concentration of oxygen atoms. However, it may be important to account for the "super equilibrium" concentration of oxygen atoms when calculating thermal NO [82].

As a result of the lower temperatures (<1800 K) and the locally fuel-rich nature of most practical pulverized-coal flames (which lack oxygen atoms), the Zeldovich mechanism does not appear to be a major source of NO in these flames [30,50,57,81,86,87]. This was further substantiated by tests with a premixed coal flame in which argon was substituted for N_2 without significant changes in the amount of NO measured [88]. Similar tests showed that thermal NO contributed less than 15% of the total NO at a stoichiometric ratio of 1.2, but the contribution of thermal NO approached 0% at a stoichiometric ratio of 1.0 [89].

3.3.2. Prompt NO

Prompt NO is formed by the reaction of atmospheric nitrogen with hydrocarbon radicals in fuel-rich regions of flames, which is subsequently oxidized to form NO. This was first reported by Fenimore [27] who studied NO formation in Meker-type burners from gaseous pyridine and ammonia. He observed a higher NO formation rate in the fuel-rich regions of hydrocarbon flames. The prompt NO mechanism has been further confirmed by several researchers [1,25,69,76,80,90,91].

The main reactions proposed to describe this process in hydrocarbon flames are [92]:

$$N_2 + CH_X = HCN + N + \cdots \qquad (8)$$

$$N_2 + C_2 = 2CN \qquad (9)$$

$$N + OH = NO + H \qquad (10)$$

The first reaction is postulated as the dominant reaction. It is estimated that HCN is involved in approximately 90% of the prompt NO formed [81,92]. Once the HCN is formed, it follows the same reaction pathway as HCN formed from coal devolatilization [81,93].

Prompt NO formation occurs in fuel-rich regions where hydrocarbon radicals increase the formation of HCN through the following reactions [94,95]:

$$CH + N_2 = HCN + N \qquad (11)$$

$$CH_2 + N_2 = HCN + NH \qquad (12)$$

Since the prompt NO mechanism requires a hydrocarbon to initiate the reaction with nitrogen, this mechanism is much more prevalent in fuel-rich flames than in fuel-lean hydrocarbon flames [75,96]. Prompt NO reactions are neglected in many NO models due to the increased complexity of the nitrogen chemistry and also due to intimate coupling of these reactions with the fuel oxidation steps. In addition, prompt NO is only significant in very fuel-rich systems, and is a small portion of the total NO formed in most combustion systems. In practical combustion systems which are usually operated fuel-lean or very close to stoichiometric, the contribution of prompt NO to the total NO formed is likely to be small [81]. However, a study was performed of a large-scale oxy-natural gas furnace in which prompt NO was predicted to account for 14–17% of the total exit NO depending on the firing configuration [97]. Staged coal combustion systems are usually operated with a fuel-rich stage that will increase prompt NO formation. Also, in coal diffusion flames, the reaction zone is often slightly fuel-rich which can promote prompt NO formation [81]. However, tests in which N_2 in the combustion air was replaced by argon showed no change in NO concentrations [81]. Thus, prompt NO is most prevalent in cooler, hydrocarbon flames of "clean" fuels (those containing no fuel-nitrogen), and is less prevalent in hot, fuel-lean systems with substantial amounts of fuel-nitrogen [81].

3.3.3. Fuel NO

Fuel NO is formed from nitrogen bound in the fuel, and is

Table 2
Reaction rate parameters for extended Zeldovich mechanism [68,76,80]

Reaction expression[a]	A (cm^3 $gmol^{-1}$ s^{-1} K^{-1})	β	E (J/g mol)
$O + N_2 \rightarrow NO + N$	1.36×10^{14}	0	315,900
$N + O_2 \rightarrow NO + O$	6.40×10^9	1.0	26,300
$N + OH \rightarrow NO + H$	3.28×10^{13}	0	No temperature dependence

[a] $k = AT^{\beta} \exp(-E/RT)$, cm^3 g mol^{-1} s^{-1} K^{-1}.

usually assumed to proceed through formation of HCN and/ or NH_3 which are oxidized to NO while being competitively reduced to N_2 according to the overall reactions [98]:

$$HCN/NH_3 + O_2 \rightarrow NO + \cdots \qquad (13)$$

$$NO + HCN/NH_3 \rightarrow N_2 + \cdots \qquad (14)$$

Nitrogen bound in the coal is released during the devolatilization process. A fraction of the nitrogen (α) is rapidly converted to HCN, and the remaining portion of the fuel nitrogen reacts to form NH_3. These two species react to form either NO or N_2 depending on the local conditions. The NO formed can be reduced by heterogeneous reaction with char particles. This is schematically illustrated in Fig. 1. Various global reaction rates have been proposed by investigators, such as DeSoete [98], Bose et al. [69], and Mitchell and Tarbell [99]. In addition, the global reaction rates are typically measured under controlled conditions in the temperature range of interest, and apply to both fuel-rich and fuel-lean conditions. Table 3 lists the rate parameters for each of these global reaction sets.

When modeling the fuel NO reaction process, fuel nitrogen is assumed to evolve from the fuel as HCN or NH_3. These species react to form either NO or N_2 by Eqs. (13) and (14) depending on the local environment. In fuel-rich regions, these nitrogen-containing species will typically be reduced to N_2, and in fuel-lean regions they are generally oxidized to form NO. Controlling the local environment in which nitrogen is released from the fuel is a primary means of controlling NO emissions.

Fuel NO is the dominant NO formation mechanism in flames which contain nitrogen in the fuel, and typically accounts for more than 80% of the NO formed in these systems [100]. Fuel NO is formed more readily than thermal NO because the N–H and N–C bonds most common in fuel-bound nitrogen are much weaker than the triple bond in molecular nitrogen which must be broken for thermal NO formation [50]. The conversion of nitrogen species to NO is only weakly dependent on the specific nitrogen compound [9].

Experimental evidence indicates that HCN is normally the first measurable volatile nitrogen species [70,101]. The appearance of fuel nitrogen as NH_3 in the gas phase may be important during char oxidation or during high temperature gasification of low-rank coals as indicated by the experimental data of several investigators [102,103]. HCN may consist of only a small portion of the nitrogen compounds initially evolved from the coal, but the remainder of these nitrogen compounds is rapidly converted to HCN in the absence of oxygen [55,104]. In fuel-rich gas flames doped with nitrogen-containing compounds, HCN is the major product of the fuel-nitrogen/hydrocarbon interactions, and fuel-nitrogen exists mainly as HCN just downstream of the reaction zone [31,32,49,92,93,105–107]. This agrees with the finding that HCN is the most stable nitrogen product in the primary reaction zone of high temperature, premixed hydrocarbon flames [27].

In rich, coal dust/oxidizer flames, HCN is one of the predominant intermediate species [104,108–110]. The HCN formed may subsequently decay to NH_i which further

Table 3
Reaction rate parameters for fuel NO mechanisms

Reaction expression[a]	Investigator	A	E (J/g mol)
$HCN + O_2 \rightarrow NO + \cdots$	DeSoete [98]	1.0×10^{10}	280,300
$HCN + NO \rightarrow N_2 + \cdots$	DeSoete [98]	3.00×10^{12}	251,000
$HCN \rightarrow NH_3 + \cdots$	Bose et al. [69]	Parameters are coal dependent	
	Mitchell and Tarbell [99]	1.94×10^{15}	328,500
$NH_3 + O_2 \rightarrow NO + \cdots$	DeSoete [98]	4.00×10^6	133,900
	Mitchell and Tarbell [99]	$\dfrac{3.48 \times 10^{20}}{1 + 7 \times 10^{-6} \exp(2110/T)}$	50,300
$NH_3 + NO \rightarrow N_2 + \cdots$	DeSoete [98]	1.80×10^8	113,000
	Bose [69]	1.92×10^4	94,100
	Mitchell and Tarbell [99]	6.22×10^{14}	230,100

[a] $k = A \exp(-E/RT)$.

144

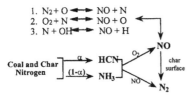

Fig. 1. Schematic representation of the fuel NO formation and reduction process [188].

decays to NO or N_2 by the reaction sequence shown previously in Fig. 1 [31,93,98,106,110,111]. High initial concentrations of HCN decay rapidly to low values at reaction times on the order of 100 ms for systems with stoichiometric ratios >0.70 [93].

In systems with fuel-nitrogen present and stoichiometric ratios <0.70, hydrocarbons recycle NO back to HCN by reactions such as [87,107]:

$$NO + C_X H_Y = HCN + \cdots \qquad (15)$$

This is the same reaction involved in the NO reburning process, which is discussed in the next section. The HCN reacts further to form NO or N_2. The relative yields of NO and N_2 depend mainly on the local stoichiometric ratio [112]. For sufficient residence times in a very fuel-rich gas, the fuel-nitrogen converts to relatively small amounts of NO and large amounts of N_2. The yield of NO from the coal nitrogen will increase for higher stoichiometric ratios [110–112]. For pulverized-coal flames with normal levels of excess air, approximately 20–40% of nitrogen devolatilized from the coal is converted to NO with the remainder presumed to be converted to N_2 [49,110,112].

NO can also be reduced to N_2 by both heterogeneous and homogeneous reactions, but the homogeneous reactions seem to dominate the NO reduction process [113]. This is especially true with low-NO_X burners, which are designed to create a fuel-rich region to promote homogeneous NO reduction. The dominant reaction resulting in NO_2 formation involves NO reduction, and homogeneous NO reduction usually involves several nitrogen-containing species [98,114]. Homogeneous NO reduction partially explains the sensitivity of the NO concentration to the local stoichiometric ratio. At fuel-rich conditions, less oxygen is available for NO formation, and more nitrogen-containing species are available for homogeneous reduction of NO to N_2. Homogeneous NO reduction is also the principle used in the Thermal De-NO_X process [14] which can result in a substantial reduction of the NO concentration. In this application, NH_3 is added to combustion systems to provide a nitrogen species for homogeneous NO reduction.

For fuel-lean systems, NO is the main nitrogen pollutant formed, and low amounts of HCN and NH_3 are observed. As the system becomes more fuel-rich, less NO is produced, but higher amounts of HCN and NH_3 are observed [31,59,104,106,108,110]. The relative amounts of HCN and NH_3 formed appear to be functions of coal type with

bituminous coals producing higher quantities of HCN and lignite coals producing higher quantities of NH_3 [51,108,115]. Lower temperatures also result in decreased NO emissions, but increased HCN and NH_3 emissions [31,87,92,93]. Preheating of inlet air up to 300 K above ambient temperatures in a turbulent diffusion gas flame has little effect on fuel NO levels [87].

Fuel NO has been shown to have a weak temperature dependence in turbulent, diffusion-type, pulverized-coal flames [55,57,112,116,117]. This could be attributed to the offsetting effects of volatiles production and stoichiometric ratio on NO formation at different temperatures. At lower temperatures, fewer volatiles are produced which reduces nitrogen evolution from the coal. However, this fuel-nitrogen is evolved into a relatively fuel-lean environment that increases the proportion of NO formed from the devolatilized nitrogen [112]. At higher temperatures, more volatiles are produced which increases nitrogen evolution, but the nitrogen is evolved into a relatively fuel-rich environment, which results in a lower percent conversion of nitrogen to NO.

NO reduction is also influenced by the presence of CO, which tends to create additional radicals (OH, H, and O) by the branching CO oxidation which then participate in the reduction process. A shift to lower temperatures for NO reduction has been observed when CO is present [118,119].

Homogeneous fuel NO formation is also sensitive to burner design and various techniques for fuel and air contacting [91,117,120]. Wendt [121] gives a summary of the techniques used to reduce NO emissions in practical combustion systems. In general, near-stoichiometric premixed flames result in higher NO emissions than diffusion flames. However, this effect is also related to coal type, and some coals seem insensitive to burner type [104]. This dependence on coal type could be attributed to differences in the split of fuel nitrogen between the volatiles and the char for different coals [61]. A higher rate of mixing is a factor, which can increase the yield of NO from NH_3 [116]. The higher yield of NO could be due to the greater production of radicals caused by the intense mixing, but is primarily the result of increased contact of the nitrogen volatiles and oxidizer [31,120]. The effect of fuel/air mixing tends to decrease as the mixture becomes more fuel-rich [117].

3.3.4. Reburning

Reburning is a process used to reduce NO emissions by the addition of a hydrocarbon-containing fuel into the downstream combustion zone, as shown previously in Eq. (15). The additional fuel creates a fuel-rich region, which provides CH_i radicals that react with NO_X to form HCN, which can then be reduced to N_2 through reaction shown previously for fuel-NO. This reduction reaction can also be a sink for NO. The reburn-zone is usually operated at lower temperatures and with higher hydrocarbon concentrations than occur with air-staging [22]. The effectiveness of the reburning process is dependent on numerous factors including the local mixing and stoichiometry in the reburning region [122,123]. A major drawback of reburning is the necessity of creating a fuel-rich region to be effective. This is accomplished by the addition of a hydrocarbon fuel, which then requires additional reactor size for complete burnout of the reburning fuel. The fuel-rich zone also creates a reducing environment, which can lead to furnace deterioration. Recent work [11,124,125] has shown that reburning can be performed in overall oxidizing conditions by exploiting the locally fuel-rich regions that occurs in diffusion flames. Overall fuel-lean reburning can be most effective in full-scale systems because of the locally fuel-rich regions resulting from unmixedness in these large systems [125]. Data for fuel-lean reburning indicates that slower mixing and lower-temperatures in the reburning zone enhance NO reduction [125].

Because of the relatively high cost of natural gas, a range of fuels have been used for reburning, including coal [126–128], wood [129,130], heavy fuel oil [131], coke oven gas [131], as well as natural gas [132–134]. Smart et al. [131] showed in a semi-industrial-scale furnace (2.5 MW) that the amount of NO reduction is independent of reburning fuel type and up to 89% reduction can be achieved. Wendt [11] suggested that natural gas produces more hydrocarbon species in the NO reduction zone. However, he showed that with coal reburning, HCN was destroyed more rapidly, which is a key intermediate to reduce NO to N_2 in fuel-rich regions. Coal can also offer economic advantages over gas as a reburning fuel, and experiments show NO reduction efficiencies comparable to those of natural gas reburning [135], with low-rank coals. Wendt [11] also indicated that bound-nitrogen in the reburning fuel is not an issue, regardless of the nitrogen content in different reburning fuels. Based on the investigations into the reduction efficiency of different reburning fuels, Kicherer, et al. [136] concluded that a high NO-reduction level can be achieved under the following conditions: high volatile matter of reburning fuel, long residence time, optimized mixing conditions, and very fine grinding if solid reburning fuels are used. They indicated that homogeneous reduction mechanisms are more effective than heterogeneous reactions. However, Chen and Ma [137] observed that lignite as a reburning fuel has a higher efficiency for NO reduction than CH_4 or other coals. They noted that heterogeneous reactions contributed to higher levels of NO reduction compared with homogeneous reactions under certain conditions. Moyeda et al. [138] also confirmed experimentally that the low-rank coals generally perform in an equivalent manner to natural gas, while high-rank coals are generally less effective as reburning fuels. The reaction mechanisms, which dominate the reburning process, will vary depending on the type of fuel used for reburning, and the amount of hydrocarbons present to participate in the reburning process [22].

The rate parameters for the global reburning reaction were determined by Chen et al. [84]:

$$CH_i + NO \rightarrow HCN + \cdots \tag{16}$$

The reaction rate parameters were determined from simulations with the CHEMKIN code [85] using 254 elementary reactions of NO in premixed, laminar hydrocarbon-containing flames.

Research is also underway to develop and test *advanced reburning* concepts that provide higher amounts of NO reduction. A recent review of NO reburning and advanced reburning was published by Smoot et al. [5]. In this approach, ammonia is injected downstream of the main reburning zone in a slightly fuel-rich zone, followed by additional combustion air. Reductions of up to 85–90% have been reported with this technique [8]. Work was also underway to provide detailed test data for reburning and advanced reburning technologies in a pilot-scale reactor, including species profiles within the reactor [139] and to develop theoretical models of advanced reburning for use in comprehensive combustion codes [140].

3.4. Heterogeneous NO_X formation and destruction

In combustion systems containing solid fuel such as coal, NO_X can be both formed and reduced by reactions with the fuel [57]:

$$char + O_2 \rightarrow NO \tag{17}$$

$$char + NO \rightarrow N_2 + \cdots \tag{18}$$

These heterogeneous reactions occur with the solid carbonaceous substance remaining after devolatilization (e.g. coal char). The net amount of NO_X formed from heterogeneous reactions is strongly dependent on the intrinsic reactivity and internal surface area of the char [141]. These characteristics of the char will vary depending on several factors, including fuel-type, air-staging, and burner injection and the conditions that led to the formation of the char from the parent coal [142]. Shimizu et al. [141] found the reactivity of char to NO per unit internal surface area to be very similar for nine different chars. This indicates that the difference in NO formation for various chars is due to the amount of internal surface area in the different chars. Typically, the overall contribution of NO from char combustion is less than from combustion of the coal volatiles [22,112]. However, heterogeneous NO_X reactions are affected less by combustion modifications than homogeneous NO_X reactions [143]. This makes control of NO_X formed from heterogeneous reactions more difficult, and increases the importance of correctly modeling these reactions. Heterogeneous NO_X reduction is generally dominated by homogeneous NO_X reduction [113].

The time-scale for char oxidation is typically much longer than for coal devolatilization, being in the order of 30–1000 ms for pulverized-coal flames [39,116]. As much as half of the nitrogen still remains in the coal after devolatilization, and the amount of nitrogen retained appears to be a function of the stoichiometry [112,144]. The amount of nitrogen retained in the char also tends to increase with

146

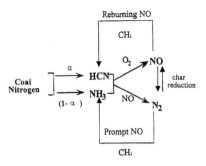

Fig. 2. Schematic representation of the complete NO reaction process in coal systems.

particle size, but does not seem to be strongly influenced by the heating rate [145]. The nitrogen retained can be heterogeneously oxidized to form NO. Fuel NO_X formed in this manner typically accounts for 20–30% of the total NO_X formed [116,146,147]. However, with higher devolatilization temperatures, less nitrogen remains in the char, and the amount of NO_X formed from heterogeneous oxidation of char may be reduced [57]. NO_X formation from char is insensitive to temperature, but is strongly dependent on char particle size and the oxygen content of the surrounding gases.

The percent conversion of char nitrogen to NO_X is much lower than the percent conversion of devolatilized nitrogen [49,57,116]. However, char nitrogen conversion of up to 40% has been measured [57]. One experimental study showed 12–16% conversion of fuel-nitrogen to NO_X for pulverized char combustion compared to 28% conversion for a pulverized-coal with the same nitrogen content [116]. More reactive chars have been shown to have a lower char nitrogen to NO conversion [52,148]. Thus, with char, the conversion of nitrogen to NO_X is about half the conversion expected from coal combustion.

NO can also be reduced by heterogeneous reactions with the char in coal combustion systems, and heterogeneous reactions can result in substantial reduction of the NO_X concentration, particularly under fuel-rich conditions [55,57,93,149,150]. The rate of NO_X reduction is much higher in coal than in liquid fuel flames, and the amount of NO_X reduction is strongly dependent on temperature

and char availability [91,93,150]. The rate of heterogeneous NO_X reduction has been reported by several investigators [91,93,113,150–152]. In theory, higher levels of NO_X reduction could be achieved by increasing the char loading and/or increasing the residence time of the char in the reactor [93].

Recent theoretical work [123] suggests that heterogeneous reactions with soot particles can be a significant form of NO reduction, and can even be the most important mechanism for NO removal with bituminous coals. Additional work is needed to better understand the heterogeneous reactions of NO and soot, and the effects of soot on the gas phase reactions which involve NO.

The various NO reaction processes are represented schematically in Fig. 2. This includes nitrogen release from the coal, fuel-NO reaction, prompt NO, thermal NO, heterogeneous NO and NO reduction by reburning. Fiveland and Wessel [17] reported the predicted amounts of NO formation and reduction that resulted from most of the various steps in the above mechanism. Two simulations of a 1.76 MW wall-fired furnace were reported. In the first simulation, the furnace was fitted with standard burners, and in the second simulation, the furnace was fitted with low-NO$_X$ burners. The amounts of NO attributed to the various steps in the mechanism are shown in Table 4. This table shows that the predicted NO concentrations are dominated by the fuel-NO reactions, and that thermal-NO formation and heterogeneous-NO reduction were relatively small for this case. These reactions could be more significant in full-scale furnaces, because of higher temperatures. Comparison of the values for the standard and low-NO$_X$ burners shows that the significant difference is an increase in homogeneous NO reduction. This results from the fuel-rich region created by the low-NO$_X$ burners.

4. Chemical kinetics computations in turbulent flames

The purpose for development of the NO_X pollutant model is to provide a tool for the prediction of NO_X concentrations in practical combustion systems. A comprehensive model of the NO reaction process involves solution of species conservation equations for the various pollutant concentrations. The combustion process typically takes place in a turbulent environment, which requires special considerations when obtaining a time-mean reaction rate, and when solving the

Table 4
Increase (+) and decrease (−) of NO concentrations (ppm) attributed to formation/reduction by various mechanisms [17]

Mechanism	Process	Standard burners	Low-NO$_X$ burners
$HCN \xrightarrow{O_2} NO$	Fuel-NO$_X$ form	+2493	+2595
$NO \xrightarrow{HCN} N_2$	Fuel-NO$_X$ reduction	−1607	−2297
$NO \xrightarrow{char} N_2$	Heterogeneous reduction	−5	−4
$N_2 \xrightarrow{O_2} NO$	Thermal NO$_X$	+17	+14
Total		+898	+292

species continuity equation. The importance of accounting for chemistry–turbulence interactions when predicting NO_X concentrations was illustrated by Smith et al. [153]. There is a significant difference between the NO_X profiles predicted from mean temperature and mean composition profiles and those obtained when the kinetic rates are integrated with respect to fluctuating temperatures and concentrations [153]. The species conservation equation is usually time-averaged to account for the effects of the turbulent fluctuations. Unfortunately, in most cases, the time-mean reaction rates cannot be calculated from the time-mean values of temperature and concentration because the reaction rates are highly non-linear functions of these values. Numerous approaches have been used to account for effects of the turbulent fluctuations on the reaction rates. These approaches will be discussed briefly in order of increasing complexity. More detailed reviews of these approaches are given by McMurtry and Queiroz [37] and Brewster et al. [154].

4.1. Turbulence models

At the lowest level of complexity, the time-averaged Navier–Stokes equations are solved for the time-mean field variables, and these values are used directly to solve for the time-mean reaction rates. This approach simply ignores the effects of the turbulence on the chemical reactions and can introduce significant errors for turbulent, reacting flows [37].

At the second level of complexity are the moment closure methods. These methods include both the classical approach of time-averaging the Navier–Stokes equations, as well as the more recent approach of conditional-averaging. With moment closure methods, a turbulence model is used to provide closure for the Reynolds stresses in the momentum equations, and a combustion model is used to obtain the time averaged reaction rates in the species continuity and energy equations. The $k-\epsilon$ turbulence model is the most widely used model to obtain Reynold's stresses [155]. In this model, two transport equations similar to the other equations of motion are solved, one for the turbulent kinetic energy (k), and another for the rate of dissipation (ϵ) of the kinetic energy. The values of k and ϵ are used with several proportionality constants to calculate the local values of the eddy viscosity [156]. Modifications to the $k-\epsilon$ model which include multiple-time scale effects [157–159] and non-isotropic effects [160] are also used.

Several different combustion models are used with the moment closure methods, including: the Eddy Dissipation Concept (EDC) model [152], the assumed PDF method [100,161,162], and the Linear Eddy Model (LEM) [37,163–166]. The EDC model assumes that combustion occurs at the molecular level when mixing occurs, and the rate is proportional to the inverse of the turbulence time scale, k/ϵ. The EDC model is based on the original Eddy Breakup Model proposed by Spalding [167]. The most

significant difference between these two models is that the EDC model does not allow reactions to occur unless both the fuel and oxidizer mix on the molecular level at sufficiently high temperatures.

The assumed Probability Density Function (PDF) method [161,162] uses one or more "progress variables" to track the mixing in the reactor, which is subsequently used to calculate the chemical reactions, based on the local composition and energy level. These progress variables are allowed to fluctuate proportionally to the turbulence level. A shape is assumed for the PDF of the progress variables, and transport equations are solved for the first and second moments (the mean and variance, respectively) of the progress variables. The time-averaged values of the conserved scalars are calculated by integrating the instantaneous values over the PDF. This method is most applicable to diffusion-limited combustion where the chemical reactions can be assumed to be mixing limited. The Linear Eddy Model [163–166] also focuses on the small-scale mixing effects, but it explicitly distinguishes between turbulent mixing, molecular diffusion, and chemical reactions [168]. This model is in the early stages of development and implementation in CFD codes.

At the third level of complexity are the Monte Carlo PDF methods [169,170]. These methods are capable of incorporating complex chemistry and accurately modeling the interactions of the chemistry and turbulence. These methods are also applicable to both pre-mixed and non-pre-mixed combustion. In addition, Monte Carlo methods treat the chemistry exactly, and no modeling of the reaction rates is required. In these methods, a PDF transport equation is solved using a large number of fluid particles to represent the statistics of the turbulent flow field.

Monte Carlo PDF methods are classified according to the random variables considered, and the type of Monte Carlo method used to solve the PDF transport equation. The random variables, which have been considered in modeling turbulent combustion, include the following: composition, velocity–composition, and velocity–composition–dissipation PDFs. Both Eulerian and Lagrangian Monte Carlo methods have been used to solve the composition PDF transport equation, but only Lagrangian methods have been used with PDFs that treat velocity as the random variable [169]. Monte Carlo PDF methods are computationally expensive compared to other PDF methods, and are only feasible for reduced chemistry mechanisms [154]. Additional discussion of Monte Carlo PDF methods is provided elsewhere [154]. This topic is discussed further in Section 4.3.

At the fourth level of complexity are Large Eddy Simulation methods [37,171], in which the large scales of the turbulent flow-field are resolved explicitly and the small scales are modeled. These methods resolve the large-scale motion of the fluid by solving the instantaneous motion equations on a coarse mesh; although, sub-grid turbulence and combustion models are required for closure. These methods have not reached a level of maturity for application to practical combustion systems.

At the highest level of complexity are the direct numerical simulation (DNS) methods [37,171,172], in which the motion equations are solved directly on a fine grid, and the instantaneous, small-scale structure of the flame is resolved. This method is very computationally intensive, and it is not currently feasible for application to practical combustion systems. Its main role has been as a research tool to provide fundamental insight into turbulent flows and for development of improved statistical submodels.

4.2. Moment closure methods

The moment closure method is used in many of the NO models found in comprehensive combustion codes, and its implementation will be discussed in more detail. Additional details regarding the implementation of the other methods outlined above are provided by references in the appropriate section. In the moment closure method, the procedures used to obtain a time-mean reaction rate depend on the relative time-scales of the kinetics and the turbulent fluctuations, and these procedures can be classified as: (1) kinetics in low turbulence; (2) slow kinetics; (3) fast kinetics; and (4) intermediate kinetics. Use of the moment closure model for pollutant concentrations requires solution of the species conservation equation [173]. This equation can be solved to calculate the mass fractions of the NO species and N-containing pollutants (e.g. NO, HCN, and NH_3 are typically considered) throughout the flow field. The species continuity equation in three dimensions for cylindrical coordinates is [173]:

$$\frac{\partial \rho Y_i}{\partial t} + \frac{\partial \rho u Y_i}{\partial x} + \frac{\partial \rho v Y_i}{\partial r} + \frac{\partial \rho w Y_i}{\partial \theta} - \frac{\partial}{\partial x}\left(D_{im}\rho \frac{\partial Y_i}{\partial x}\right)$$

$$- \frac{1}{r}\frac{\partial}{\partial r}\left(\rho r D_{im} \frac{\partial Y_i}{\partial r}\right) - \frac{\partial}{\partial \theta}\left(\frac{\rho}{r}D_{im}\frac{\partial Y_i}{\partial \theta}\right) = W_i \quad (19)$$

Eq. (19) is valid for laminar and turbulent systems. The combustion process usually takes place in a turbulent environment where all of the dependent variables fluctuate rapidly with time and position as a result of the turbulence. Thus, in turbulent reacting systems, the species conservation equation is usually time- or Favre-averaged to obtain the time-mean mass fraction.

The time-averaged species conservation equation is obtained by decomposing the instantaneous variables into their average and fluctuating components such as:

$$\bar{u} = u + u' \quad (20)$$

Similar expressions for v, w, ρ, Y_i, and W_i are substituted into Eq. (19), and the entire equation is time-averaged. This results in a number of additional terms, which include the fluctuating components. Turbulence closure is invoked by using the conventional gradient diffusion approximation to obtain the appropriate turbulent diffusion coefficient [174].

The final Favre-averaged species continuity equation becomes:

$$\bar{W}_i = \bar{\rho}\tilde{u}\left(\frac{\partial \tilde{Y}_i}{\partial x}\right) + \frac{\bar{\rho}\tilde{v}}{r}\left(\frac{\partial \tilde{Y}_i}{\partial r}\right) + \bar{\rho}\tilde{w}\left(\frac{\partial \tilde{Y}_i}{\partial \theta}\right)$$

$$- \left(\frac{\partial}{\partial x}\right)\left(\tilde{D}_Y \frac{\partial \tilde{Y}_i}{\partial x}\right) - \frac{1}{r}\left(\frac{\partial}{\partial r}\right)\left(r\tilde{D}_Y \frac{\partial \tilde{Y}_i}{\partial r}\right) \quad (21)$$

$$- \left(\frac{\partial}{\partial \theta}\right)\left(\frac{\tilde{D}_Y}{r}\frac{\partial \tilde{Y}_i}{\partial \theta}\right)$$

The $\bar{W}_i$ term in the Favre-averaged species continuity equation represents the time-mean net rate of formation of species i, and is the net difference between the rate of formation and the rate of depletion of species i from chemical reaction. One of the largest problems in solving the species conservation equation in turbulent systems is to obtain a value for $\bar{W}_i$. Several techniques have been developed to obtain the net time-mean reaction rate ($\bar{W}_i$) in turbulent systems; however, each method is limited in application to certain types of reaction mechanisms. Several of these methods are discussed below.

If the turbulent fluctuations are small, they will not affect the kinetic reaction rate. In this case, the local, time-mean reaction rate is equivalent to the local instantaneous reaction rate. This means that time-mean values for temperature and composition can be used to calculate the reaction rate without introducing significant error. This technique may be applicable to regions in particle-laden combustion systems with very low turbulence levels. Since most coal combustors have high turbulence levels, this technique is generally not applicable to pollutant formation kinetics in combustors.

Some chemical reactions are very slow compared to the characteristic time-scale of the local turbulence. One example of such reactions is the heterogeneous char oxidation reaction, which occurs during coal combustion. For a slow chemical reaction, unless the turbulent fluctuations are very large, the various fluctuations do not affect the reaction rate since the response times of the reactions are much longer than the time-scale of the turbulent fluctuations. For this type of reaction, the time-mean reaction rate can be calculated using the time-mean properties. Some of the pollutant formation reactions may fit into this category. One example is the heterogeneous decay rate of NO with char, which may be slow relative to the local turbulence time-scales. However, it should be noted that turbulence time-scales may vary significantly for different coal combustors.

Some chemical reactions are very fast compared to the characteristic time-scale of the local turbulence. Some gas phase combustion reactions are an example of reactions that can be approximated by fast kinetics under certain conditions. For fast kinetic reactions, the assumption is made that the mixing process is the rate-limiting step, and the reaction is completed as soon as the reactants are mixed. The assumption that the reactions are *mixing-limited* allows the mean gas phase species composition to be calculated from statistical considerations [175,176].

In diffusion flames, the fuel and oxidizer are initially in separate streams, and it is necessary for the fuel and oxidizer to mix on a molecular level before any reaction can occur. A similar phenomenon occurs in coal flames, since the coal must devolatilize and the resultant volatiles must mix locally with the oxidizer before any reaction can occur. If it is assumed that the gas phase combustion processes in these systems can be treated as fast kinetic reactions, the reaction rates may be only functions of the local extent of gas mixing and/or the local extent of coal reaction. In this procedure, the local extent of mixing of the inlet gas streams is defined by the gas phase mixture fraction (f) [173]:

f = (mass of gas from primary stream)/(total mass of

gas from primary and secondary streams) (22)

If a single mixture fraction is used to track the off-gas from the coal, the local mass fraction of gas originating in the coal (extent of coal reaction) is defined by the coal gas mixture fraction (η) [173]:

η = (mass of gas originating from the coal)/(total mass of

gas from coal, primary and secondary streams) (23)

Once the extent of mixing and/or the extent of coal reaction (f and η) are known locally throughout the reactor, the local elemental composition and enthalpy (for adiabatic systems) level can be calculated. These quantities can be used with the Gibbs free energy scheme to calculate equilibrium species composition and gas temperatures. The effects of the turbulence are incorporated as shown previously by decomposing f and η into time-mean and fluctuating quantities [177].

If all the variables in the rate expression are only functions of f and/or η, then the time-mean reaction rate can be obtained by convolution over a joint probability density function (PDF) for f and η. In most cases of pulverized-coal combustion, the stoichiometry will be independent of f since the primary and secondary streams will have the same gas composition (air). Even in the few cases where f is required, the amount of coal off-gas at any point will be only weakly dependent on the amount of primary gas; thus, f and η can be considered statistically independent [177]. This allows for separation of the joint PDF into separate PDFs for f and η. In essence then, the mean chemical composition can be calculated by using f and η, and the calculation of the time-mean reaction rate is avoided since instantaneous composition is based on equilibrium considerations. This convolution technique using joint PDFs in f and η is currently being used to obtain time-mean equilibrium concentrations of major species [178]. This approach also follows closely that proposed by Williams and Libby [179].

Intermediate kinetics result when the time-scale of the

reaction is of the same order as the time-scale of the turbulence. In this case, neither the mixing process, nor the reaction kinetics may be considered rate-limiting, and the turbulent fluctuations must be accounted for directly. The instantaneous and mean species concentrations resulting from intermediate kinetics are different from their equilibrium values, and the appropriate mean reaction rates must be used in calculation of the time-mean mass fraction.

One technique used with intermediate kinetics to determine the time-mean mass fraction is the calculation of the deviation of the local species concentration from its equilibrium or fully reacted value [180,181]. This is accomplished by considering the time-mean mass fraction of species i ($\bar{Y}_i$) to consist of a time-mean equilibrium or fully reacted value ($\bar{Y}_i^f$) and time-mean perturbation from the equilibrium or fully reacted value ($\bar{y}_i$):

$$\bar{Y}_i = \bar{Y}_i^f + \bar{y}_i \quad (24)$$

The fully reacted value ($\bar{Y}_i^f$) is the mass fraction of species i that would result if all the original reactants formed only species i (e.g. if i = HCN, then $\bar{Y}_i$ is the mass fraction of HCN formed if all the fuel nitrogen formed only HCN). This approach does not require the perturbation to be small or large as in small perturbation theory [181]. The reason for separating the mass fraction into these two terms is that the equilibrium or fully reacted value ($\bar{Y}_i^f$) is a function of only f and η, and thus the effects of the turbulent fluctuations can be incorporated directly by the method outlined for fast kinetics.

The time-mean perturbation from the fully reacted value ($\bar{y}_i$) is calculated using a transport equation, similar to the species conservation equation. The transport equation contains both a source term and a sink term for the time-mean perturbation. The sink term is the local, time-mean reaction rate ($\bar{W}_i$) which must be expressible in terms of known time-mean values. This transport equation is solved to obtain the time-mean perturbation from the fully reacted value throughout the reaction field. The local time-mean mass fraction is then found using Eq. (24).

Another technique which can be used with intermediate kinetics for calculation of the time-mean mass fraction requires that the reaction rate be a function of only the local stoichiometry (f and/or η). In this approach, the time-mean reaction rate is obtained by convolution over the probability density functions for f and η as discussed for fast kinetics.

To illustrate this procedure, consider the elementary reaction:

$$O + H \rightarrow OH \quad (25)$$

The instantaneous forward rate is:

$$W_{OH} = \rho^2 Y_O Y_H A \exp(-E/RT) \quad (26)$$

If all the variables in the rate expression are only functions of f and/or η, then the time-mean reaction rate is:

$$\bar{W}_{OH} = \int_f \int_\eta W_{OH}(f, \eta) P(f) P(\eta) \, df \, d\eta \qquad (27)$$

$P(f)$ and $P(\eta)$ are the Favre-averaged probability density functions (PDF). The PDF is the statistical distribution of the variables at any local point in space due to the fluctuations of the variable in time. Once the time-mean reaction rate for a species has been found, the species continuity equation can be solved to find the local time-mean concentration of that species [173].

This approach for intermediate kinetics is similar to the approach outlined previously for fast kinetics except that a kinetic expression is used to describe the reaction process rather than assuming the reaction takes place to thermochemical equilibrium. In addition, the species conservation equation is solved to obtain the mass fraction once the time mean reaction rate is calculated. The major limitation of this approach is that the local instantaneous reaction rate must be a function of only f and/or η. Referring to the example above, this will be true if Y_O, Y_H, and T are only functions of f and/or η. This approach will be too restrictive for many chemical kinetic mechanisms.

Bilger [182] determined experimentally that, for a broad region in heptane/air diffusion flame, the molecular species composition was *only* a function of the mixture fraction. The reaction rates in this flame are also kinetically limited, similar to the pollutant reactions. Fenimore and Fraenkel [183] also reported the same finding for thermal NO concentrations in laminar, gaseous, diffusion flames. Coelho and Carvalho [83] showed that predicted NO concentrations using this approach were independent of the shape of the assumed PDF.

4.3. Monte Carlo PDF methods

As mentioned previously, another method, which has been developed recently for incorporating complex chemistry in turbulent flow systems is the Monte Carlo PDF method [169,170,184]. These methods are very general, and can be applied to pre-mixed and non-pre-mixed combustion. Monte Carlo PDF methods treat complex chemistry exactly, and require no modeling terms for the reaction processes. These methods solve a PDF transport equation using a large number of particles to represent the statistics of the turbulent flow-field. Monte Carlo PDF methods are classified according to the variables that are treated as random variables, and the type of Monte Carlo Method used to solve the PDF transport equation. The methods that have been used in combustion modeling include composition, velocity–composition, and velocity–composition–dissipation PDFs.

Node-based and fluid particle tracking composition PDF methods treat all of the scalars that describe enthalpy and composition as random variables. The probability of having certain gas composition at each location in the reactor can be determined from the composition PDF. In the composition PDF methods, a conventional CFD solver is required to calculate the velocity and turbulence fields, and the PDF code calculates the density field. The advantage of the composition PDF methods is their relative simplicity, and their disadvantages are that they assume gradient diffusion and ignore the effects of density fluctuations on the turbulence. Velocity–composition PDF methods treat the components of velocity as random variables along with the species mass fractions and enthalpy. Here also, the conventional CFD solver is used to solve for pressure and turbulence dissipation fields, and the PDF code calculates the density field. The major advantages of this method are that turbulent diffusion is now treated exactly, and the effects of variable density are included [154].

Velocity–composition–dissipation PDF methods also solve for the turbulence dissipation as a random variable along with the velocity components and species mass fraction and enthalpy. Since this method now contains time- and length-scale information, it does not require an external CFD solver or turbulence model [154].

5. NO$_X$ submodels

A review of NO$_X$ submodels published in the literature is provided in this section. The first NO submodel was apparently that of Zeldovich [26] for the fixation of atmospheric nitrogen to NO, called thermal NO. This submodel is applicable to the high-temperature, fuel-lean regions of combustion systems, and was discussed previously in more detail. This section reviews NO submodels commonly used in comprehensive modeling of NO formation. These submodels require a comprehensive CFD/combustion code to provide the local properties of the flow field such as temperature and composition. The salient features of these comprehensive combustion codes will be discussed first. This will be followed by a detailed discussion of the various NO submodels published in the literature, including that of the authors and co-workers. A discussion of these submodels is followed by a discussion of a recently published NO$_X$ submodel used for gaseous combustion systems, including lean, premixed flames. This submodel was developed specifically for gas turbine systems, and is based on a Monte Carlo PDF approach.

5.1. Comprehensive combustion models

A generalized NO$_X$ submodel requires information of the instantaneous gas phase temperatures and density resulting from the combustion process. Comprehensive combustion models are used to provide this information for the NO$_X$ submodel. Such a comprehensive model must adequately predict the fuel heat-up and ignition and provide an adequate description of nitrogen release from the fuel, including reacting coal and char particles or oil droplets. This involves

Table 5
Components in typical comprehensive combustion models (e.g. PCGC-3)

Process	Description	References
Gas fluid dynamics	Motion equations	[100,209]
	Newtonian fluid	
	Eulerian equations	
	Favre-averaged	
Gas phase chemistry	Fast chemistry, mixing limited	[210]
	Chemical equilibrium calculated by minimization of Gibbs free energy	[211,212]
	Coupled with turbulence using probability density function (PDF), and mixture fractions	
Turbulence	Prandtl mixing length	[155]
	Standard $k-\epsilon$ model	[213]
	Non-linear $k-\epsilon$ model, adjusted for effects of particles	[213]
Radiation	Discrete-ordinates approach S4 (24 directions)	[214,215]
Devolatilization	Two-step model	[44]
	Model constants	[216]
	CPD model	[217]
	Particle swelling allowed	
Char oxidation	First order reactions	[100]
	CO/CO_2 are primary products	
Particle mechanics	PSI-CELL approach	[218]
	Langrangian equations	
	Particle dispersion based on particle drag and gas turbulence	[213]
Solution technique	SIMPLER, SIMPLEC, SIMPLEST	[219]
	TEACH method for central and upwind finite-differencing	
NO_X formation	Two fuel-NO mechanisms	[68,188]
	Thermal NO	
	Global gas phase reactions	
	Heterogeneous NO reduction	

adequately predicting the general flow and heat transfer characteristics of the furnace. Comprehensive combustion models include various submodels of the physical processes occurring in combustion systems, including gaseous fluid dynamics, homogeneous gas phase reactions, radiative and convective heat transfer, devolatilization, heterogeneous reactions, and particle motion. A review by Niksa [185] on coal combustion modeling provides a detailed description of the submodels used in comprehensive modeling and a summary of the submodels used in 15 different comprehensive combustion models.

A three-dimensional model, PCGC-3 (Pulverized Coal Gasification/Combustion-3 Dimensions) [186] has been used in conjunction with the NO_X model described in Section 5.2. PCGC-3 is a generalized code developed by the authors and co-workers for simulating steady-state, reacting and non-reacting gaseous or particle-laden systems, with an emphasis on turbulent, pulverized-coal combustion systems. The model uses conventional numerical methods and a differencing scheme for a completely arbitrary mesh, and can solve the large computational meshes required for simulating practical furnaces. PCGC-3 has been evaluated through extensive comparisons of code predictions with local, experimental profile data from several gas and coal combustion facilities. A more detailed description of PCGC-3 and its evaluation is given elsewhere [186,187].

The various components used in typical comprehensive codes such as PCGC-3 are summarized in Table 5, which lists the process or submodel, a brief description, and several key references. These comprehensive combustion codes and submodels provide the required information for the various NO_X submodels.

5.2. NO_X submodel for gaseous and coal combustion

5.2.1. Fuel nitrogen release

Hill et al. [188] apparently published the first NO_X submodel for coal combustion and it has been expanded and generalized substantially over the past decade or more [68,84,140,189]. This submodel treats only nitric oxide, NO. Two basic assumptions were made related to nitrogen release from the coal during coal devolatilization. The first assumption is that nitrogen is devolatilized at a rate equal to the rate of coal weight loss [39]. This is approximately true, and if an error does exist in this assumption it should be small, because of the rapidity of the devolatilization process [39]. Since coal devolatilization in a typical furnace takes place so rapidly (5–100 ms), a large portion of the nitrogen is devolatilized at approximately the same time and location in the reactor. This would be true whether the rate of nitrogen release was exactly equal to, or slightly greater than the rate of coal weight loss. This assumption allows the amount of nitrogen released from the coal to be calculated directly from the source of particle mass added to the gas phase.

The second assumption made related to coal devolatilization was that most of the nitrogen devolatilized from the coal is evolved as HCN or NH₃ or is rapidly converted to one of these species in the gas phase [109,110]. This assumption is justified by the studies mentioned previously, and simplifies the prediction of NO$_X$ formation from devolatilized nitrogen to a mechanism that proceeds by initial decay of HCN or NH₃ in the gas phase. In this NO model, separate mixture fractions are used to track the off-gas formed from coal devolatilization (η_1) and char oxidation (η_2) [189]. The rate of HCN or NH₃ formation from devolatilized nitrogen is calculated from the equation:

$$w_{\text{vols}} = \zeta_{\text{vols}} \omega_N^{\text{vols}} S_p^{\eta_1} / M_N \tag{28}$$

In this equation, ω_N^{vols} is the mass fraction of nitrogen in the coal, $S_p^{\eta_1}$ the local source of particle mass given to the gas by coal devolatilization, and M_N molecular weight of nitrogen. The symbol ζ_{vols} is a constant used to specify the fraction of coal nitrogen initially converted to HCN and NH₃ during the devolatilization process. The split of this percentage of coal nitrogen between HCN and NH₃ is a function of coal type, and is specified by the user with a separate constant. Unless information is available for the particular coal being used, it is recommended that all of the coal nitrogen is assumed to convert to HCN.

The nitrogen from the char is assumed to form NO and N₂. The rate of nitrogen conversion from char reactions is calculated from the equation:

$$w_{\text{char}} = \zeta_{\text{char}} \omega_N^{\text{char}} S_p^{\eta_2} / M_N \tag{29}$$

In this equation ω_N^{char} is the mass fraction of nitrogen in the char, $S_p^{\eta_2}$ the local source of particle mass given to the gas from char reactions, and M_N is the molecular weight of the nitrogen. The symbol ζ_{char} is a constant used to specify the fraction of char nitrogen converted to NO during the char reaction process. A typical value for ζ_{char} in this equation is 0.15, which assumes that 15% of the nitrogen released during char oxidation is instantaneously converted to NO in the gas phase, with the remainder converted to N₂ [189].

5.2.2. Homogeneous NO$_X$ reactions

Once the coal particle begins to devolatilize, the gaseous constituents react homogeneously. In the gas phase, NO can be formed by three separate reaction processes discussed previously: thermal NO, prompt NO, and fuel NO. The reactions of these processes have a typical time-scale on the order of 1 ms. The integral time-scale of turbulence is around 10 ms. Since these time-scales are nearly the same order of magnitude, the effects of the turbulent fluctuations on the reaction rates cannot be ignored. These reactions come under the category of intermediate kinetics discussed previously.

The calculation of the time-mean reaction rate for the NO

reaction processes requires formulation of the reaction rate as only a function of the local stoichiometry. It was mentioned previously that an experimental observation for a broad region of a gas phase diffusion flame, showed that the molecular species composition was only a function of the local stoichiometry. A similar finding was discussed for thermal NO$_X$ concentrations in laminar, gaseous diffusion flames. Thus, the pollutant species reaction rates may be only functions of the local stoichiometry [180,181].

The stoichiometry in a gaseous or pulverized-coal furnace can be calculated from the mixture fraction (f) and/or the local amount of coal off-gas (η_1) from coal devolatilization and/or the local amount of char off-gas (η_2). If the NO reaction rates can be written as only functions of f, η_1 and η_2, they can then be obtained by convolution over the respective PDFs. This approximation does not require the functional relationship between the mass fraction and local stoichiometry be unique for the entire flow field; it requires only that it be time-independent with respect to the turbulent time-scales. In other words, the reaction time might be different at different points throughout the reactor.

The perturbation analysis outlined previously for intermediate kinetics was used initially in this NO submodel to obtain the time-mean reaction rate. The local mass fractions of HCN and NO$_X$ were written as a fully reacted value plus a perturbation from this fully reacted value. The effects of the turbulent fluctuations on the fully reacted value could be included since this value is only a function of the local stoichiometry, but the turbulent fluctuations in the perturbation were ignored. Often, the perturbation from the fully reacted value is not small, and ignoring the turbulent fluctuations in the perturbation resulted in unrealistic mass fractions.

The problems encountered with the perturbation analysis necessitated formulation of another approach for calculation of the time-mean reaction rate. The basis for this method was the experimental observation mentioned above that the reaction rate is only a function of the local stoichiometry (f, η_1 and η_2) throughout the reactor. For this approach, the local instantaneous mass fraction is taken to be a linear function of its fully reacted mass fraction:

$$Y_i = \pi_i Y_i^f \tag{30}$$

where Y_i^f, the fully reacted mass faction of species i, is only a function of the local stoichiometry, and thus can be calculated from f, η_1 and η_2. The variable π_i is a linearization constant for species i, which represents a fractional conversion or extent of reaction for species i. The variable π_i must be independent of the turbulent fluctuations, although it may be spatially variable. In other words, the reaction time or fractional conversion might be different at different points throughout the reactor, and thus have different Y_i values versus mixture fraction curves. The value of π_i is obtained from the time-mean mass fraction field [173]:

$$\pi_i = (\tilde{Y}_i / \tilde{Y}_i^f) \tag{31}$$

In this equation, π_i is the fractional deviation of the Favre-mean mass fraction i ($\tilde{Y}_i$) from the fully reacted Favre-mean value ($\tilde{Y}_i^f$). The assumed relationship between the mixture fractions and the original amount of reactant, implies that the histories of the fluid elements at a given location in the reactor are not important. Because of the assumed time-independence of π_i, fluid elements at a given location with mixture fractions f, η_1 and η_2 have the same local species mass fractions Y_i. This assumption is based on the experimental observations that the local composition is a unique function of the stoichiometry. The assumed relationship still allows for the history effects of the fluid elements at different locations in the reactor. This results because each π_i is dependent on the local mean velocity, density, etc., through the Favre-mean mass fraction. Thus, fluid elements with the same mixture fractions f, η_1 and η_2 may have different local mass fractions Y_i at different locations. Using the definition of π_i in Eq. (31), the instantaneous Y_i's in Eq. (30) are only functions of the local stoichiometry and hence only functions of f, η_1 and η_2. The time-mean reaction rate can then be obtained for each pollutant species by convolution of the homogeneous pollutant reaction rates shown below (Eqs. (32) and (33)) over the PDF as shown previously in Eq. (27).

The global reaction rates of the major nitrogenous species measured by DeSoete [151] are shown below:

$$w_1 = A_1 X_{HCN} X_{O_2}^b \exp(-E_1/RT) \tag{32}$$

$$w_2 = A_2 X_{HCN} X_{NO} \exp(-E_2/RT) \tag{33}$$

The order of reaction 1, as shown in Eq. (32), varies according to the O_2 mole fraction. The order is zero for O_2 mole fractions greater than 18,000 ppm and 1.0 for O_2 mole fractions less than 2500 ppm. These global reactions are the recommended option to describe the homogeneous NO_X formation and decay processes in the NO pollutant sub-model. Alternate reaction rates proposed by Bose et al. [69], and Mitchell and Tarbell [99] were also incorporated into this NO submodel.

A similar approach used by Nelson et al. [15] took the full set of single-step reactions from Bowman in a perfectly stirred reactor (PSR) to calculate the NO reaction rates for different stoichiometries. The effects of the turbulence were incorporated in the same manner as described here. Their simulation of a full-scale, wall-fired boiler showed good agreement with measured results.

5.2.3. Heterogeneous NO_X reactions

Approximately half of the nitrogen typically remains in the coal after devolatilization, and this nitrogen can be heterogeneously oxidized to form NO. Fuel NO_X formed in this manner accounts for as much as 20–30% of the total NO_X formed. In the coal combustion submodel, the char particles continue to release mass at a slow rate, even after coal devolatilization has taken place. In this NO submodel, heterogeneous NO formation is handled analogously

to homogeneous NO formation. As the char particles release mass, nitrogen is released to the gas phase, and this nitrogen is instantaneously converted to HCN. This HCN then reacts homogeneously by the same pathway as the HCN formed by coal devolatilization. An alternative would be to assume the fuel-N in the char is converted directly to NO through surface oxidation.

NO can also be reduced by interaction with the char particles. The rate of heterogeneous NO reduction as measured by Levy et al. [150] is shown below:

$$w_3 = A_3 \exp(-E_3/RT) A_E P_{NO} \gamma \tag{34}$$

A_E is the external surface area of the char in m^2/kg, P_{NO} the partial pressure of NO in atmospheres, and T the particle temperature. γ is a factor used to include the change in external area of the char particles due to swelling and fracturing during the reaction process. This rate was measured in an electrically heated laminar flow furnace for the temperature range of 1250–1750 K in a $He/CO/H_2$ atmosphere. Size-graded, pulverized Montana lignite char particles were used to measure the amount of heterogeneous NO reduction, but these rates usually describe both the heterogeneous NO formation and reduction. In addition, the selected rate was measured under controlled conditions in the temperature range of interest. This equation has been used in the NO_X pollutant model to describe the heterogeneous reduction of NO.

The heterogeneous NO reactions are shown to be much slower than the homogeneous reactions. A time-scale for the heterogeneous reduction of NO from a typical maximum to minimum is on the order of 1 s [190]. Since the heterogeneous reactions are slow compared to the turbulence time-scale (10 ms), calculation of the heterogeneous time-mean reaction rates is greatly simplified. In regions of intense turbulence, the large fluctuations in gas properties, especially the temperature, can have a significant effect on the mean reaction rate, even for very slow reactions. However, turbulence levels are generally lower in reactors and furnaces where heterogeneous reactions predominate. Thus the time-mean reaction rate for heterogeneous NO decay can be calculated from Eq. (34) using time-mean properties, neglecting the effects of the turbulent fluctuations on T, A_E, and P_{NO}.

5.2.4. Reaction sequences and rates

The fuel NO portion of this NO submodel for coal combustion systems reaction sequence can be depicted as:

$$W_0 = coal - N \rightarrow HCN \tag{35}$$

$$W_1 = HCN + O_2 \rightarrow NO + \cdots \tag{36}$$

$$W_2 = HCN + NO \rightarrow N_2 + \cdots \tag{37}$$

$$W_3 = NO + char \rightarrow CO + \cdots \tag{38}$$

This sequence of reactions is represented schematically in

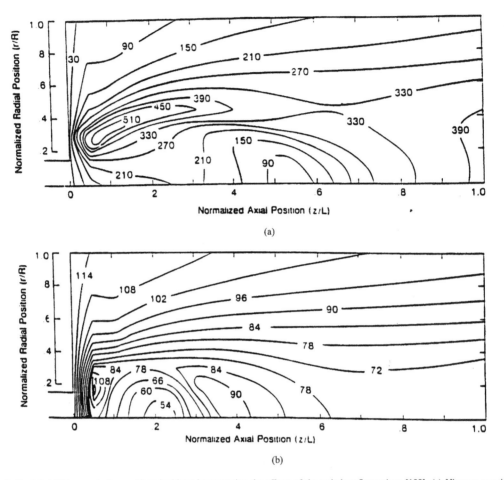

Fig. 3. Predicted NO concentrations with and without incorporating the effects of the turbulent fluctuations [100]. (a) Nitrogen reaction pollutant reaction rates calculated using mean values of properties. (b) Nitrogen pollutant reaction rates calculated using values that include turbulent fluctuations.

Eq. (39):

$$
\text{Coal} - \text{N} \rightarrow \text{HCN}
\begin{cases}
\overset{O_2}{\underset{W_1}{\rightarrow}} \text{NO} \overset{\text{char}}{\underset{W_3}{\rightarrow}} \text{CO} + \text{N}_2 + \cdots \\
\overset{W_2}{\underset{\text{NO}}{\rightarrow}} \text{N}_2
\end{cases}
\tag{39}
$$

The kinetic parameters used for the homogenous reactions in this NO submodel are summarized in Table 3.

The time-mean concentrations for the nitrogen pollutant species are obtained by solution of the species continuity equation. This solution requires calculation of the time-mean net formation rate for the individual species. The time-mean net formation rates for NO and HCN are the difference between the time-mean formation rate and the time-mean depletion rate for these species. The time-mean net formation rates for NO and HCN are:

$$
\bar{W}_{\text{NO}} = (\bar{W}_1 - \bar{W}_2 - \bar{W}_3)M_{\text{NO}} \tag{40}
$$

$$
\bar{W}_{\text{HCN}} = (\bar{W}_0 - \bar{W}_1 - \bar{W}_2)M_{\text{HCN}} \tag{41}
$$

The $\bar{W}_i$s are the time-mean formation and depletion rates shown previously.

The equations for the pollutant species are decoupled from the comprehensive combustion model since the small pollutant concentrations will not significantly affect the major species mass fractions, velocity, density, or temperature fields. However, the equations for the pollutant species are coupled since the pollutant reaction rates are functions of the other pollutant concentrations. The equations are solved following convergence of the combustion model.

5.2.5. Effects of turbulence

The development of this NO submodel has emphasized the effects of turbulent fluctuations on the various properties (e.g. temperature, density, and velocities) on the NO_X formation process. Although the comprehensive analysis and evaluation of the NO submodel is discussed in a later section, a prediction from the NO model is shown here to illustrate the effects of the turbulence. Fig. 3 shows the predicted NO concentrations in a pulverized-coal reactor [191]. The two contour maps show the predicted NO levels for the same case without and with inclusion of the effects of the turbulent fluctuations, respectively. All other input conditions were identical. In Fig. 3(a), the pollutant reaction rates were calculated incorporating the temperature and density fluctuations caused by the turbulence, as discussed previously. In Fig. 3(b), the pollutant reaction rates were calculated using the mean temperatures and densities. This comparison illustrates that the turbulent fluctuations can have a significant impact on the predicted NO_X concentrations in coal systems. Incorporating the effects of the turbulent fluctuations on NO_X formation typically results in lower NO_X concentrations at most reactor locations. The reduction in NO_X probably results primarily from the dominant effect of the lower instantaneous temperatures, oxygen and pollutant concentrations caused by the turbulent fluctuations [173].

5.2.6. Solution approach

This NO submodel [188] is initialized using gaseous concentrations obtained from the combustion code solution. Typically, NO concentrations are significantly less than the equilibrium concentrations. Therefore, the concentrations of HCN, NH_3, and NO are initially set equal to an arbitrary fraction of the concentrations that would occur if the volatile fuel nitrogen at each location were all converted to the respective species. The solution algorithm is iterative as follows: First, source terms are calculated by convoluting the instantaneous reaction rates (dependent on the gas phase concentrations) over the statistical distribution of mixture fractions. Second, species continuity equations are solved for HCN, NH_3, and NO using the source terms. Third, the O_2 mass fraction is updated from a mass balance, based on the computed NO concentrations. Fourth, Steps 1–3 are repeated until convergence on the source terms is achieved. Normally, this NO submodel reaches convergence within 100 iterations for two- and three-dimensional cases. Under-relaxation factors and other numerical parameters can be specified to aid or speed convergence. When joint thermal and fuel NO predictions are made, the model first predicts the amount of thermal-NO and fuel-NO that can be formed. This value is added to the maximum fuel-NO level to provide a reference for the maximum amount of NO that can be formed.

5.3. Other NO_X submodels

Numerous submodels of nitrogen-containing pollutants have been developed by various investigators, as summar-

ized in Table 6. These models share similar features and characteristics to the NO submodel discussed in the previous section [13,50,68,140,173,188]. These submodels are presented in light of the differences with the NO submodel discussed in the previous section. The summary of Table 6 shows the investigators, a brief summary of the submodel, the simulations performed, and comments. Carvalho et al. [192] and Bonvini et al. [193] report the modeling of thermal NO formation in gas- and oil-fired boilers based on the Zeldovich mechanism. Coelho and Carvalho [83], Coimbra et al. [194], Epple et al. [195], and Fiveland and Wessel [17] report NO submodels and the calculation of NO formation in coal-fired boilers. The approach used to model homogeneous and heterogeneous reactions is essentially the same as outlined for the NO submodel discussed previously [68,173,188]. The main difference in the approach used by the different investigators is the estimation of the [O] and [OH] concentration in the calculation of thermal NO. Hedley et al. [97] report the modeling of a large-scale oxy-fuel test furnace to predict thermal and prompt NO formation. De Michele et al. [196] report modeling of gas- and oil-fired boilers using a reaction mechanism based on 27 chemical reactions involving 17 species, but ignore the effects of the turbulent fluctuations on the pollutant reaction processes.

5.4. NO_X submodel for lean, premixed combustion

Most of the NO_X submodels discussed previously are for gaseous and coal combustion in diffusion flames and treat only nitric oxide (NO). A submodel of NO formation in premixed combustion systems, including fuel-lean systems has also been recently reported [154]. This submodel was developed specifically for lean, pre-mixed combustion (LPC) of natural gas in gas turbines but has broader potential application. This newer model is based on a Monte Carlo velocity–composition probability density function method coupled with an unstructured grid flow solver.

Gas turbines operating in this mode minimize thermal NO formation, which typically forms in high temperature, near-stoichiometric regions that occur near the interface of fuel and air mixing. With LPC, the combustion zone is operated with excess air (equivalence ratios of 0.4–0.6) to reduce the flame temperature, and stoichiometric interfaces are eliminated since the fuel and air are premixed. The cooler flame temperature (<1800 K) results in very little thermal NO_X formation. Prompt NO is also negligible at these lean conditions, and essentially all of the NO_X formed results from the nitrous oxide mechanism [154]. Approaches used to couple the turbulence and chemistry in diffusion flames are not applicable to pre-mixed combustion. Various methods and codes used for modeling turbulent, pre-mixed combustion were reviewed by Brewster et al. [154]. The approach selected for use in this premixed combustion submodel is the Monte Carlo velocity–composition PDF method, which was discussed previously. PDF methods treat the chemical

Table 6
NO_X submodels for combustion systems

No	Author(s)/year	Features	Simulation/evaluation	Comments
1	Hill et al. [188]/1984	(1) De Soete [98] and Mitchell and Tarbell [99] global fuel NO mechanisms	Laboratory and utility furnaces	(1) NO, NH_3, and HCN predicted, no N_2O or NO_2 predictions
	Boardman [68]/1988	(2) Zeldovich and extended Zeldovich thermal-NO included [76]	Predictions give general trends near flame region	(2) Extensive predictions and comparisons for different conditions and furnaces
	Smoot et al. [62]/1993	(3) Different NO schemes integrated into a comprehensive code	NO predictions within 5–30% in extensive comparisons	(3) No SO_X /NO_X interactions
		(4) PDF used for turbulent flow		
2	Schnell et al./1991	(1) Fuel NO (De Soete [98]) model, and prompt NO (De Soete [98]) model	150 and 490 MW wall-fired furnace; laboratory drop-tube reactor	(1) Model similar to Hill (1983)
	Epple et al./1992	(2) PDF used for turbulent flow	NO predictions within 20% agreement (limited data available)	(2) Thermal-NO_X in later work
	Epple et al. [195]/1995			(3) No N_2O or NO_2, and no SO_X/NO_X interactions
3	Fiveland and Wessel [17]/1991	(1) Follows Hill et al. (1984) [188] work on NO formation	A 1.76 MW furnace with three configurations	(1) Predicted NO lower than measurements
		2) Only fuel NO included	NO prediction within 20%	2) Thermal-NO, N_2O and NO_2 not predicted; no SO_X/NO_X interactions
		(3) De Soete [98] homogeneous mechanism	Applying model to NO reduction burners	
4	Abbas et al. [24]/1991	(1) De Soete [98] mechanisms for both fuel NO and char NO	A down-fired laboratory furnace with 0.6 m internal diameter and 0.3 m height	(1) Poor predictions of NO near flame region
	Lockwood and Yehia/1992	(2) NH_3 and N_2O measured, but not predicted	Predictions of NO profiles show similar trend of the measurements	(2) No thermal-NO_X
				(3) No SO_X/NO_X interactions
5	Murrells et al./1992	(1) Thermal NO by the extended Zeldovich mechanism [76]	Laboratory-scale furnaces	(1) No N_2O and NO_2 predictions
		(2) Fuel NO by De-Soete [98] mechanism	Predictions are approximately within 20% of the experiments	(2) No SO_X/NO_X interactions
6	Coelho and Carvalho [83]/1995	(1) Thermal NO [76] by extended Zeldovich mechanism; several methods used to estimate [O] and [OH] investigated	250 MW 12 burner, wall-fired, fuel-oil boiler; qualitative agreement except near burner	(1) Predicted NO much lower than measurements
		(2) Fuel NO by De Soete [98] mechanism; investigate different treatments of fuel-N release		(2) N_2O, NO_2 not predicted; no SO_X/NO_X interactions
7	Brewster et al. [154]/1999	(1) Monte Carlo PDF for chemistry	Laboratory-scale furnace and small-scale gas turbines	(1) Good agreement with measured results, ppm levels
	Chen et al. [197]/1997	(2) Pre-mixed and non-premixed gaseous methane combustion		(2) Monte Carlo PDF requires very long calculation time
		(3) Reduced mechanism for chemistry, includes thermal NO and prompt NO and N_2O		(3) Only gaseous combustion
		(4) Coupled chemistry		
		(5) Applicable to lean flames		

reactions exactly, and are very computationally intensive. This results in a practical limit of 3–5 reaction scalars with current computers and solution methods. Consequently, reduced reaction mechanisms or intrinsic low-dimensional manifolds (ILDM) must be implemented. Reduced mechanisms can be developed empirically or from skeletal mechanisms containing hundreds of elementary reactions.

A new five-step reduced mechanism for methane combustion [197] is used in the LPC NO submodel:

$$3H_2 + O_2 + CO_2 \leftrightarrow 3H_2O + CO \tag{42}$$

$$H_2 + 2OH \leftrightarrow 2H_2O \tag{43}$$

$$3H_2 + CO \leftrightarrow H_2O + CH_4 \tag{44}$$

$$H_2 + CO_2 \leftrightarrow H_2O + CO \tag{45}$$

$$2H_2 + CO_2 + 2NO \leftrightarrow 3H_2O + CO + N_2 \tag{46}$$

Reaction (46) models NO formed from thermal, prompt and N_2O reactions. This mechanism was evaluated in a perfectly stirred reactor (PSR) model, and NO predictions at different pressures compared very favorably with predicted values from the full mechanism. Brewster et al. [154] report good agreement of predicted and measured results for lean, pre-mixed, natural gas combustion in an atmospheric pressure, laboratory-scale gas turbine combustor.

6. Model trend comparisons

Model trend comparisons require the use of a comprehensive combustion code in conjunction with a NO_X submodel since the flow properties must be predicted as input to the NO_X submodel. Table 6 provided a brief summary of the features and the evaluation of several published NO_X submodels. Most of the commercial CFD codes available for simulating combustion systems contain some form of pollutant model for simulating nitrogen pollutant formation. These models are similar in their approach to the pollutant model discussed herein, but for the most part do not model the reburning or advanced reburning processes. Published results from these codes are available, although they not as readily available in the open literature due to the sometimes proprietary nature of the simulations. This section contains extensive evaluation results of the NO_X submodel of the authors and co-workers, while other comparisons are also presented.

Evaluation of the NO_X submodels has been conducted for a variety of experimental, two- and three-dimensional, coal reactor configurations and conditions, including swirling diffusion flames, external air-staging, and oxygen-blown gasification [153,188]. The capability of predicting trends associated with variations in operating parameters such as equivalence ratio, air-flow, swirl number, air-staging location, coal type, particle size, and reactor pressure has been demonstrated for bituminous and sub-bituminous coals. The

thermal NO mechanism has also been tested for predicting trends with variation in equivalence ratio and swirl number for a natural gas diffusion flame in a large, laboratory-scale combustor [173]. Several model predictions have also been compared to detailed local flame data. These include coal combustion and coal gasification cases as well as a natural gas flame. Table 7 summarizes the results of several of these evaluations by the authors and co-workers. The table illustrates the effects of flame conditions and combustor operating parameters such as inlet stream swirl number, stoichiometry, and fuel type. The sensitivity of an early version of the NO submodel to kinetic parameters was evaluated by Smith et al. [173]. The sensitivity of the predictions to the form of the fuel NO mechanism was discussed [68] but has not been fully determined.

One method used to evaluate the NO submodel involved comparisons of the predicted and measured observed trends of key variables on nitrogen pollutant formation. These comparisons indicate to what extent the NO_X submodel predicts the effect of various operating parameters on NO_X formation in pulverized-coal systems. The selected cases and key parameters studied are shown in Table 8. The results of some measurements were extrapolated to slightly different conditions to illustrate several points. Several of these cases are discussed below.

6.1. Swirl number

The swirl number is a measure of the angular momentum imparted to the inlet streams, and is defined as:

$$S_t = \text{(flux of angular momentum)/flux of axial momentum}$$
$$\times \text{burner radius} \tag{47}$$

The magnitude of S_t, in essence, indicates the mixing rate of the inlet streams. A higher swirl number indicates a greater flux of angular momentum imparted to the inlet streams. In coal systems, swirl is often imparted to only the secondary inlet stream, which is the situation with the measurements and predictions presented below. The increased angular momentum of the secondary stream generally results in faster mixing of the coal stream and air. Increased mixing enhances ignition, and often causes attachment of the flame to the coal burner and can increase NO concentration. Swirl of the secondary stream is one technique of fuel/air contacting which causes significant changes in NO_X formation. In some cases, increased swirl number increases NO emissions [116,117,120]. This results from increased mixing of secondary air with volatiles products prior to complete combustion. The additional oxygen from the secondary stream mixes with the volatiles, and reacts with the fuel nitrogen to form NO. In other cases, increased swirl initially decreases NO_X emissions [109,115]. The observations of the effect of swirl number on NO emissions are varied and indicate the influence of additional parameters.

Asay et al. [115] and Harding et al. [109] show for a

Table 7
NO$_X$ submodel evaluation results: comparisons between predicted and measured local profile concentrations

Flame type [Ref]	Flame conditions	Remarks
Natural gas–air [68] cylindrical reactor diffusion burner	150 kW; stoich. ratio = 0.67; secondary air swirl no. = 1.5	Thermal NO prediction only. Trends of all radial profiles accurately predicted with the exit profile being 33% below the measured concentrations. Evidence of prompt NO which was not predicted
Utah bituminous [188] coal combustion diffusion burner	300 kW, 1 bar; stoich. ratio = 1.0; secondary air swirl no. = 1.4,2.0	Correct prediction of radial NO profiles in both near and aft regions; observed values predicted within 20% for 2.0 swirl number case and within 30% for 1.4 swirl number case. Thermal NO contributions calculated to be small
Wyoming subbit. [188] coal combustion diffusion burner	300 kW, 1 bar; stoich. ratio = 1.0; secondary air swirl no. = 1.4,2.0	Correct prediction of radial NO profile trends, location of peak NO concentrations differed slightly; observed values predicted within 15% for 2.0 swirl number case and within 30% for 1.4 swirl number case. Thermal NO contributions calculated to be small
Utah bituminous [8,68] coal gasification by O$_2$ in premixed burner; sight-window purge; N$_2$ also included	800 kW, 1 bar; no swirl; stoich. ratio = 0.4	Location and magnitude of peak NO concentrations and general location of NO isopleths correctly predicted. Thermal NO contributions calculated to be important in the near-burner region. Joint thermal and fuel NO predictions shown to improve agreement
Utah bituminous [68] pressurized gasification by O$_2$ in premixed burner	800 kW, 2 and 5 bar; no swirl; stoich. ratio = 0.4	Approximate location and magnitude of peak NO concentrations predicted. Rapid decay of NO due to fuel-rich homogeneous reactions correctly predicted with exit NO concentrations being matched
North Dakota lignite [68] gasification by premixed O$_2$	800 kW, 1 bar; no swirl; stoich. ratio = 0.4	Incorrect NO isopleth predictions. Predictions improved somewhat by allowing nitrogen release as NH$_3$ and by delaying nitrogen release
German brown coal combustion, pilot plant, diffusion burner	6000 kW, 1 bar, no swirl; stoich. ratio ≥ 1.0	Approximate radial and axial profile-trends predicted. Predicted HCN exceeded measured values initially; ~27% average error between measured and predicted NO

subbituminous and bituminous coals respectively, that swirling of the secondary stream first decreases and then increases NO formation. This is the same trend predicted by the NO submodel as shown by the comparisons in Figs. 4 and 5. Fig. 4 compares the predicted and measured NO concentrations as a function of the swirl number for a subbituminous coal. The observed NO concentrations are the average of integrated outlet

values measured at SR = 0.87 and SR = 1.17 and adjusted to stoichiometric conditions. Fig. 5 compares the predicted and measured centerline NO concentrations as a function of the swirl number for a bituminous coal. The predicted and observed profiles have a similar shape for both coal types, and the predicted NO values are lower at most swirl numbers. For low swirl levels, NO emissions decreased as the swirl of the secondary stream was

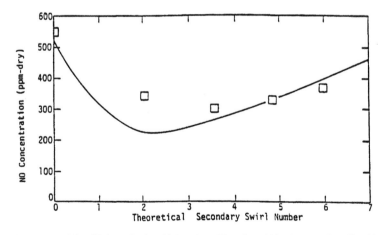

Fig. 4. Effect of secondary stream swirl on NO formation for a high-moisture Wyoming subbituminous coal: predicted (—) and measured (□) effect [100].

Table 8
Summary of NO_X submodel trend validation results

Variable (references)	Flame conditions	Remarks
Coal moisture [188]	Wet and dry Wyoming subbit.	Predictions match observed increase of NO with increasing moisture; ~12% difference between exit values
Particle size [10,188]	165 and 35 μm Utah bituminous	Predicted observed trend of decreasing NO with increasing particle size; ~16% difference between exit values
Swirl number [153] (coal flame)	$0 \leq$ swirl no. ≤ 10 subbit. and bitum.	Observed local minimum in NO concentrations predicted and explained; ~8% difference between exit values
Swirl number [8] (natural gas flame)	$0 \leq$ swirl no. ≤ 5 different primary-tube diameters	Observed trends predicted with only partial success by thermal NO mechanism alone; differences attributed to prompt NO and possibly inaccurate prediction of mixing
Secondary velocity [10]	High-volatile bit., 40–55 m/s	Observed NO emission increased; ~16% postulated to increased mixing of air and coal
Tertiary velocity [10]	High-volatile bit., 40–80 m/s	Observed NO emission decreased; ~15% postulated to delayed mixing of air and coal particles and longer residence-rich environment
CO_2 diluent [68]	$0 \leq CO_2/O_2 \leq 1.5$	Temperature insensitivity of fuel-NO predicted; ~15% difference between exit values
Stoichiometery		
(a) Coal–air diffusion [188]	$0.5 \leq$ st. ratio ≤ 1.2	Trend for swirling and non-swirling conditions matched; ~19% difference for exit values
(b) Coal–air premixed [68]	$0.8 \leq$ st. ratio ≤ 1.2	Observed trend matched; ~12% difference for exit values
(c) Natural gas diffusion [8]	$0.9 \leq$ st. ratio ≤ 1.2	Observed trend matched with the thermal NO mechanism using partial equilibrium expression to estimate O radical concentrations
Air stage location [68]	Primary-zone S.R. and residence time; Kentucky bitum.	Observed trend of reduced NO concentrations correctly predicted for (1) different stoichiometric ratios in the primary zone and (2) increased residence time in the primary zone
Gasification pressure [68]	1 and 5 bar; Utah bituminous	Observed trend of decreasing NO concentrations with increasing pressure correctly predicted; ~30% difference between exit values (2 ppm for 2 bar case)
Char oxidation	Fuel-rich, Utah coal char	Peak NO concentrations correctly predicted but exit values differed significantly (>400%)

increased. For high swirl levels, NO emissions increased as the swirl was increased. The minimum NO concentrations occur at a swirl number of 2.0–3.0 depending on the coal type.

The variations in NO emissions with swirl undoubtedly result from differences in the mixing pattern. The mixing of the primary and secondary streams controls the volatile nitrogen conversion to NO [9]. Lifted flames result in increased NO concentration, and the reduction in NO emissions can be partially attributed to flame stabilization at the burner [42,110]. At $S_t = 0$, the flame was detached from the burner, but as secondary swirl was increased the flame became attached to the coal burner. A delay in ignition allows for increased mixing of the coal and oxidizer prior to ignition and volatiles combustion. This increases the secondary air available during volatiles combustion and increases NO formation.

The rate of nitrogen volatiles release increases substantially when the flame attaches to the burner, and this is probably due to changes in the particle heating rate and peak temperature [9,198] Higher levels of swirl result in more rapid particle heatup by increasing both the mixing rate of the incoming stream with hot recirculating gases

and the furnace residence time in the hot combustion regions [198].

When the flame first attaches to the burner (at lower levels of swirl), the swirling secondary stream results in sufficient mixing to attach the flame to the burner. This stream forms an annulus around the primary stream, which creates a fuel-rich central core. Then oxidizer from the secondary stream is entrained at the outer edge of this core, which ignites the volatiles and attaches the flame to the burner. Heat transfer by radiation from the surrounding flame causes nitrogen devolatilization in a fuel-rich environment that reduces NO formation. As the mixing process is completed, much of the fuel-nitrogen is devolatilized and reacted under fuel-rich conditions, which favor N_2 formation over NO_X formation. As the swirl level is increased, the increased mixing tends to break down the fuel-rich central core. This increases the mixing of the volatiles and oxidizer, which increases NO formation.

6.2. Stoichiometric ratio

The stoichiometric ratio (SR) is a measure of the overall air to fuel ratio in a combustion system. The stoichiometric

160

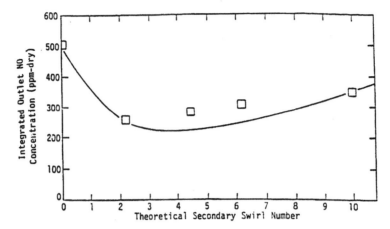

Fig. 5. Effect of secondary stream swirl on NO formation for a bituminous coal: predicted (—) and measured (□) effect [100].

ratio (SR), which is the reciprocal of the equivalence ratio, is defined as:

$$SR = (oxidizer/fuel)/(oxidizer/fuel)_{stoichiometric} \qquad (48)$$

Increased stoichiometric ratio increases NO formation in coal systems [61,109,110,115,117,199]. The local stoichiometry affects NO formation and reduction in various ways, but a primary effect is the amount of oxygen available for NO formation. A fuel-rich environment reduces NO formation from volatile nitrogen.

Rees et al. [110] measured the effect of stoichiometric ratio on NO formation for a Utah bituminous coal in the laboratory-scale combustor. This was a parallel injection case, and the stoichiometric ratio was varied from 0.40 to 1.15. Integrated outlet NO concentrations are shown on a dry basis at 0% excess air. The comparison of measured and predicted NO concentrations is shown in Fig. 6. The effect of the stoichiometric ratio (SR) on NO emissions was measured at numerous SR values as shown, and the model predictions were made at stoichiometric ratios of 0.70, 1.00, and 1.15.

Asay et al. [115] also measured the effect of stoichiometric ratio on NO formation for a high-moisture subbituminous coal in a laboratory-scale combustor. An experimental swirl number of 1.4 was used, and the stoichiometric ratio was varied from 0.57 to 1.17. Outlet NO values were reported on a dry basis at 0% excess air. The effect of stoichiometric ratio on both the measured and predicted outlet NO values is also shown in Fig. 6, for the subbituminous coal.

The lower stoichiometric ratios result in less oxygen available for oxidation of the volatile nitrogen. Nitrogen is devolatilized into a fuel-rich zone that both reduces NO formation and enhances reduction of the NO formed. Particle temperatures increase at higher stoichiometric ratios

which results in greater nitrogen devolatilization, and increased NO formation [9,110].

For both the swirl and non-swirl diffusion flames, the measured and predicted values indicate that the exit NO concentrations increase with stoichiometric ratio. However, the measured effect of stoichiometric ratio is greater than the predicted effect at higher SR values. This could partially result from some assumptions in the generalized combustion code, but it probably results primarily from the underprediction of NO emissions. Thermal NO was not considered in these simulations, and at higher stoichiometric ratios, near unity with higher temperatures, the thermal NO contribution increases. However, even at higher stoichiometric ratios, thermal NO is a relatively small fraction of the total NO_x formed in flames with fuel NO. In addition, coal burnout increases at these values, but is usually underpredicted which results in lower NO concentrations. The discrepancies are also largest for the non-swirl case, which probably results because of differences in the observed and predicted ignition points. The ignition point is dependent on many variables, and accurate prediction of the ignition point is difficult. Swirl usually causes both the predicted and observed ignition to occur at the burner. Thus, NO_x model predictions are usually more reliable for attached flames.

6.3. Particle size

Variations in coal particle size can affect NO_x emissions. Particle size partially determines the rate of particle heat-up and peak temperature [65,200]. These, in turn, influence the extent of coal and nitrogen devolatilization, and the location of the ignition point. Smaller particles experience more rapid heat-up, achieve higher temperatures, and have a larger extent of coal devolatilization. These factors usually

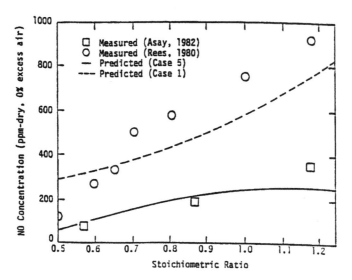

Fig. 6. Comparison of measured (○) [110], (□) [115]; and predicted (- - -) [100] (—) [100] NO concentrations as a function of stoichoimetric ratio for a Utah bituminous coal (case 5) and a high-moisture Wyoming subbituminous coal (case 1).

result in increased nitrogen devolatilization, and will generally increase NO_x formation.

Lee et al. [61] report the effect of particle size on NO concentration for a Utah bituminous coal in a laboratory-scale premixed combustor. The furnace included a large quarl, which required several approximations in the coal combustion code. Particles in two narrow size distributions were obtained by mechanical sieving. The mean particle diameter of the larger particles was 165 μm. From a plot of the particle distribution, the mass mean particle diameter was approximately 35 μm. These narrow, particle size distributions were approximated by discrete particle sizes of 165 and 35 μm, respectively.

The experimental measurements show that the larger particles result in lower NO emissions than the smaller particles. This trend was measured at several stoichiometric ratios, but model predictions were only made at stoichiometric conditions (SR = 1.0). The NO_x model predictions

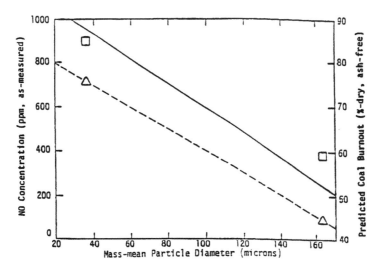

Fig. 7. Comparison of measured (□) [61] and predicted (—) [188] NO concentrations as a function of particle size for a Utah bituminous coal. Predicted (- - △ - -) coal burnout shown as a function of particle size.

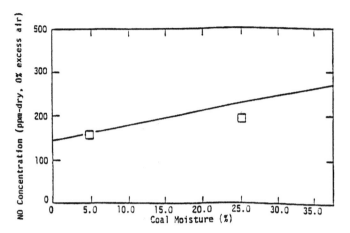

Fig. 8. Comparison of measured (□) [115] and predicted (—) [100] NO concentrations as a function of coal moisture for a high-moisture Wyoming subbituminous coal.

showed the same trend and a similar magnitude of effluent NO concentration as observed experimentally in Fig. 7. Again, a range of values is given for the measured NO concentration, since the lowest NO measurements were made at a stoichiometric ratio of 1.01 for the large particles and 1.02 for the smaller particles. This required extrapolation from the measured curve to obtain the NO values at stoichiometric conditions.

The effect of particle size on NO emissions is the result of differences in particle heating and peak temperatures, which subsequently affects the devolatilization rate. The smaller particles heat at a faster rate and attain a higher peak temperature that results in more coal devolatilization and higher coal burnout [200]. Both of these factors produce more fuel nitrogen, which often results in more NO formation. This is supported by the predicted coal burnout in Fig. 7, which shows that the smaller particles react to a greater extent, resulting in more nitrogen from the coal. In addition, the gases devolatilized from small particles mix and react more rapidly with the surrounding oxidizer than gases devolatilized from large particles [200]. This results in greater oxygen availability for the devolatilized nitrogen, and subsequently more NO formation. Unfortunately, no local measurements were available for these cases, which limits understanding of the physical processes that are occurring.

6.4. Coal moisture

The amount of moisture in the coal particles has an impact on NO_X emissions [115]. Increases in the amount of coal moisture can surprisingly result in a subsequent increase in the NO_X emissions, and the effect of coal moisture on NO_X formation diminishes at higher levels of swirl. The changes in NO_X emissions with coal moisture content

result primarily from differences in the coal ignition point, which is delayed by higher coal moisture levels.

The effect of coal moisture was reported by Asay et al. [115] for a high-moisture Wyoming subbituminous coal. This coal had an inherent moisture content of 25%, part of which was dried to 4.5% moisture to measure the effect of coal moisture. An experimental swirl number of 1.4 was used at a stoichiometric ratio of 0.87. Integrated outlet NO concentrations were reported on a dry basis at 0% excess air. The measured and predicted effects of coal moisture on NO concentration are shown in Fig. 8. As the coal moisture content increases, both the predicted and observed NO emissions increase.

The predicted effect is somewhat greater than the observed effect, especially at higher values of coal moisture. The primary reason postulated for these differences is the difference in the predicted and observed ignition point at the various coal moisture contents. For the low moisture coal, the predicted ignition point occurs at the burner, and is likely the location of the ignition point. However, for the high moisture coal the predicted moisture point is delayed somewhat (about 5 cm). The observed ignition point is probably also delayed, but may not correspond to the predicted ignition point which would lead to differences in predicted and measured NO concentrations.

7. Model applications

Several applications of the comprehensive NO submodel are presented. The first applications discussed are for laboratory-scale and pilot plants. Only simulation results, which are compared with measured data, are included. Many of these applications were for the purpose of identifying design and operating parameters to minimize predicted

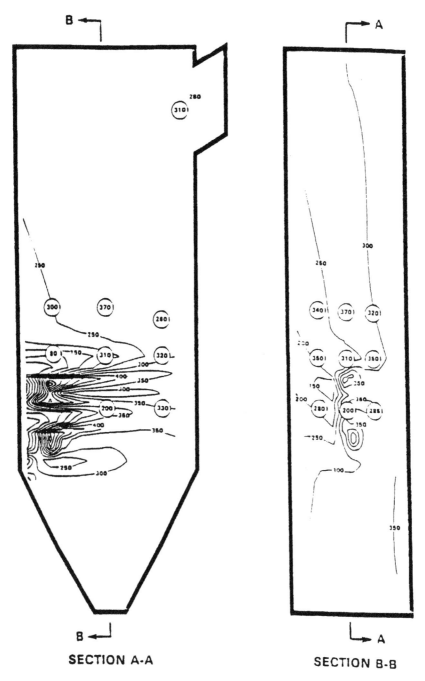

Fig. 9. Comparisons of measured and predicted NO concentrations (ppm) in a 1.76 MW, wall-fired, pulverized coal furnace [17].

Table 9
Operating conditions and coal characteristics for tangentially fired utility furnace 1991 tests [201–206]

	Test 4—Coal 1		Test 5—Coal 2	
Operating conditions				
Gross power (MW$_e$)	85		82	
Excess air (%)	4.5		4.4	
Air-flow rate (kg/h)	29,100		28,350	
Measurements ports	3, 5a, 7, 64		3, 5a, 7, 64	
Basis (wt%)	As received %	Dry	As received %	Dry
Moisture	5.16	–	5.82	–
Volatile matter	35.03	36.93	20.01	21.24
Fixed carbon	52.32	55.09	64.09	68.0
Ash	7.57	7.98	10.08	10.70
Sulfur	2.07	2.18	1.47	1.56
Carbon	76.71	80.87	73.59	78.14
Hydrogen	3.29	3.47	4.30	4.56
Nitrogen	1.14	1.20	1.01	1.07
Oxygen	4.06	4.30	3.73	3.97
Heating value (KJ/kg)	30,690	32,350	30,290	32,160
Mass mean diameter (μm)	31		18	
Hargrove grindability index	59		90	

NO$_X$ emissions. The second set of applications is for full-scale utility boilers. Simulations of these facilities are compared with measured in situ and effluent pollutant concentrations. The final application discussed is for trend analysis of various operating conditions on effluent NO emissions from a full-scale utility boiler.

7.1. Laboratory/pilot plant applications

In addition to the sensitivity studies discussed earlier, model simulations have been made for several laboratory and pilot-plant scale combustors for which local major species, temperature and pollutant concentrations have been measured. Applications include coal combustion, coal gasification and natural gas combustion. Table 7 summarizes the results of several comparisons between the predicted and the measured local NO profile concentrations for these cases. These results for coal reactors demonstrate that the NO submodel can be used to characterize local regions of interest. Such understanding can be helpful to design effective NO$_X$ abatement burners and furnaces and to establish optimum operating conditions. The contributions of thermal NO can be determined using the model, although for natural gas flames, the thermal NO mechanism alone, without prompt NO, may not be sufficient to reliably predict NO$_X$ trends.

Simulations of a 1.76-MW, wall-fired, pulverized coal furnace were reported by Fiveland and Wessel [17]. This small furnace is geometrically similar to many commercial boilers, and is insulated to maintain furnace gas temperatures that are typical of commercial-scale furnaces. NO formation was predicted for three firing configurations: a standard-cell burner, a low-NO$_X$ cell burner, and staged-

combustion using overfire air injection. An NO submodel [17] similar to the NO submodel described in Section 5.2 was used in their simulations of NO formation. Fig. 9 shows a comparison of measured and predicted NO concentrations for the low-NO$_X$ cell burner at the two center-line planes through the height of the furnace. The predicted concentrations show fair agreement with the measured values at most locations in the planes shown. The predicted values tend to over-predict the measured values by 10–50% at most measurement locations, with an average deviation of approximately 30%.

7.2. Utility boiler applications

Several simulations of nitrogen pollutant formation in pulverized, coal-fired utility boilers are presented. These simulations were selected because they include comparisons of local measured quantities with predicted values.

7.2.1. Tangentially fired 85 MW$_e$ utility furnace

The first simulation shown is for the Goudey Station power plant in Johnson City, New York, which is operated by New York State Electric and Gas Co. The furnace is a tangentially fired, forced-recirculation, pulverized coal unit with an 85 MW$_e$ generating capacity. A schematic drawing of the furnace is shown in Fig. 10, which illustrates the different regions of the furnace as well as the dimensions. The boiler has seven access levels, six of which have ports available for data acquisition. The furnace has 16 burners, with four burners in each corner, all of which were operational during the test series.

The predicted quantities include gas temperature and velocities, major species, and nitrogen pollutant concentrations

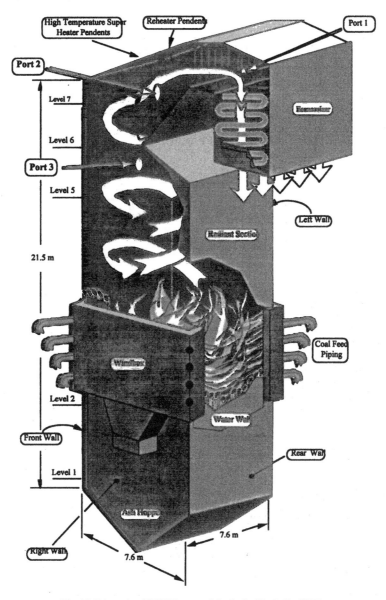

Fig. 10. Schematic of 85 MW$_e$ tangentially fired utility boiler [201].

[62,186]. Only comparisons of nitrogen pollutant concentrations will be presented and discussed here. In situ measurements from this boiler were made for use in evaluating the combustion model and for interpreting combustion processes [201–206]. Measured quantities were spatially resolved gas velocities, temperatures, and species concentrations, particle size distributions, particle velocities, particle number densities, radiative heat fluxes, and NO$_X$ concentrations. For the comparisons shown herein,

measured NO$_X$ concentrations were taken to be principally NO. Test variables were coal type, furnace load, burner tilt, and percent excess air. Access ports were available at levels 2, 3, 5, 6 and 7 through 5 cm × 25 cm corner ports, 10 cm circular ports, or 10 cm × 25 cm burner ports.

The operating conditions and characteristics of the coal for the two sets of 1991 measurements that were simulated are also shown in Table 9. The coal was assumed to have been fed into the furnace in equal amounts through the 16

166

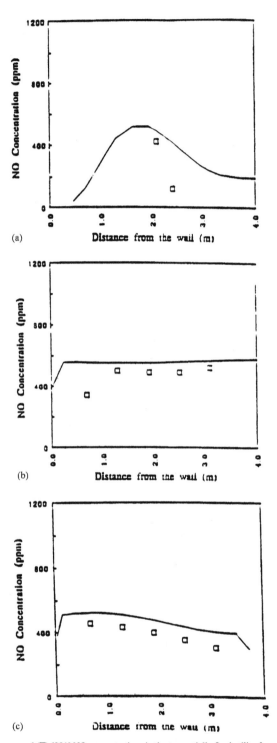

(a)

(b)

(c)

Fig. 11. Predicted (—) [62] and measured (□) [201] NO concentrations in the tangentially fired utility furnace firing Coal 1 for Test 4 at three locations: (a) port 7; (b) port 5; and (c) port 3. (a) at port 7, between upper two burners. (b) at port 5 halfway between top row of burners and top of furnace. (c) at port 3, near furnace top.

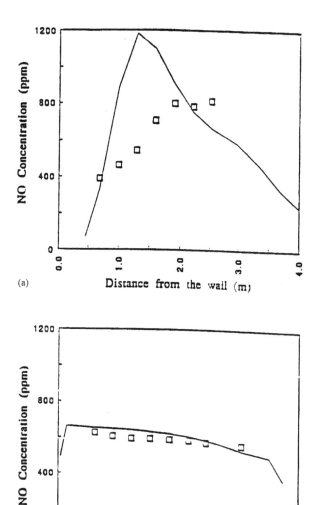

(a)

(b)

Fig. 12. Predicted (—) [62] and measured (□) [201] NO concentrations in the tangentially fired furnace for Test 5 firing Coal 2 at two locations: (a) port 7; and (b) port 3. (a) port 7, between two burners. (b) port 3, near furnace top.

coal burners. The flow rates shown are total flow rates into the furnace for all 16 burners. The coal burners were surrounded by secondary air-flow streams, which carried the remaining air required for combustion into the furnace. There was no coal in the secondary inlet streams. The operating conditions for the two cases were very similar, and the main difference between the tests was the coal type used. Figs. 11 and 12 show comparisons of predictions from PCGC-3 using the NO submodel described in Section 5.2

with measured values of nitrogen pollutants for the two utility boiler test cases discussed above. These plots show nitrogen oxide (NO) pollutant concentrations in ppm versus distance measured from the furnace wall. Reaction rate parameters used for the predictions are those based on the NO submodel discussed previously.

Fig. 11 shows comparisons of measured and predicted NO_x concentrations at several port locations for Test 4. Fig. 11(a) compares NO concentrations at port 7, which is

168

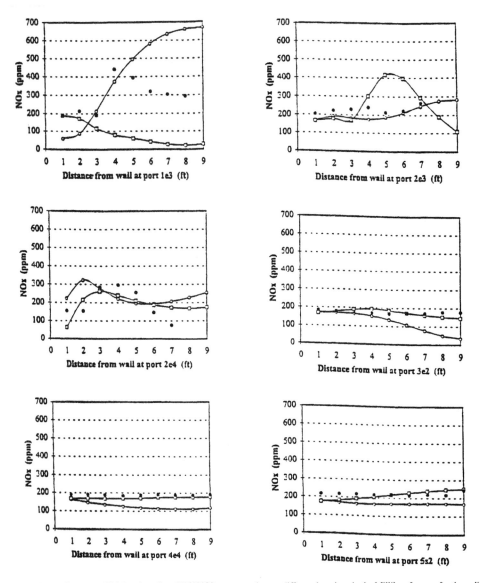

Fig. 13. Comparisons of measured [221] and predicted [207] NO concentrations at different locations in the Milliken furnace for three different computational grids: ○, 192 K; □, 337 K; △, 65 K; ●, Test #3.

located between the upper two burners (there are four burners in each corner) on the front wall, approximately 1.4 m from the right furnace wall. These values are in the near-burner region of the combustion zone. Although the number of measured values is limited, measured and predicted values suggest similar magnitudes and trends. Fig. 11(b) compares NO concentrations at port 5a, which is located on the front wall of the furnace about half way between the top of the

burners and the top of the furnace. Agreement between measured and predicted values is reasonable in this region of the furnace, with the largest discrepancies occurring near the furnace wall. This is consistent with results reported elsewhere which showed higher predicted values of temperatures and oxygen concentrations near the wall for this test [186]. Fig. 11(c) compares NO concentrations at port 3, which is located on the right wall near the top of the furnace in the super-heater pendants. The predicted values show the same trend as the measured values, and are within about 10% of measured values.

Fig. 12 shows comparisons of measured and predicted NO concentrations at two port locations for Test 5. Fig. 12(a) compares NO concentrations at port 7, which is in the near-burner region and corresponds to Fig. 11(a) for Test 4. The predicted values are generally higher than the measured values, and show a sharper profile and a considerably higher peak value. This comparison suggests that the model does not correctly predict the exact location of the ignition point. This is further verified with results reported by Hill and Smoot [186] for temperature and oxygen and CO_2 concentrations. Fig. 12(b) compares NO concentrations at port 3, which corresponds to Fig. 11(c) for Test 4.

Comparison of Figs. 11 and 12 reveals higher measured and higher predicted NO values for Test 5, with Coal 2. Yet, coal 1 (Test 4) has a somewhat higher nitrogen percentage (dry basis) (1.20 vs. 1.07, Table 9); even so, the model properly predicted the higher NO levels for Test 5. The higher NO concentrations likely result from differences in the ignition characteristics of the two coals. The coal used in Test 4 has a higher volatiles content (35 vs. 20%), commonly leading to earlier ignition [186]. This results in more of the nitrogen species being released from the coal during devolatilization and reacting under more fuel-rich conditions to form N_2 rather than NO. Further, the mass mean diameter of the coal used in Test 5 is smaller, which would result in more rapid char burnout and less NO reduction through heterogeneous reactions.

7.2.2. Tangentially fired 160 MW_e utility furnaces

A simulation was also performed for the Milliken furnace located on the Finger Lakes near Lansing, New York. The Milliken furnace is a tangentially fired, forced-recirculation, pulverized coal unit with a rectangular cross-section. The furnace is corner-fired with four levels of burners in each corner and 160 MW_e gross generating capacity, also operated by the New York State Electric and Gas Co. Simulations of the Milliken furnace were completed with three different computational grids of 65, 192, and 337 K cells to determine grid independence [207]. Fig. 13 shows a comparison of measured and predicted values of NO concentrations at several locations in the furnace for the three computational grids. The location shown in Fig. 13(a) is between the top two levels of coal burners, near the close-coupled overfire air ports. In these figures, the measured values are represented by the open symbols, and the predicted values are represented by the closed symbols connected by lines. Fig. 13(b) compares measured and predicted values between levels 3 and 4, which is about 13 feet above the separated overfire air ports. Fig. 13(c) compares measured and predicted values between levels 5 and 6, near the outlet of the furnace. Fig. 13(d) compares measured and predicted values at different ports along the height of the furnace. Most of the difference in the measured

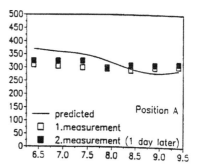

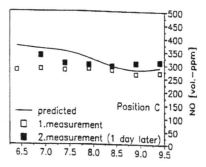

Fig. 14. Comparisons of measured and predicted NO concentrations near the top of a 490 MW wall-fired coal furnace [195].

and predicted NO concentrations likely result from discrepancies in the observed and measured ignition location. This is especially apparent from Fig. 13(d), which shows a significantly higher difference in the measured and predicted values at port 1e3 (just above the second level of burners) than the other ports. Agreement is quite good in the upper regions of the furnace.

A large-scale simulation was also reported for a 490 MW_e, wall-fired, coal furnace [195]. This furnace has eight swirl burners, and was operated using a bituminous coal. Simulations of the furnace were performed for both a staged and unstaged operation. Fig. 14 shows comparisons of measured and predicted NO concentrations profiles for the unstaged case. The values measured were between the front and rear walls near the top of the furnace. The predicted values show good agreement between the measured and predicted NO concentrations at this location in the furnace.

7.3. Trends analysis

The NO submodel of Hill et al. [173,188] has been applied as part of a comprehensive combustion code (i.e. PCGC-3) to a small-scale utility boiler to demonstrate application for prediction of observed trends. This NO model was evaluated by comparing model predictions with effluent

Table 10
1993 85 MW$_e$ utility boiler NO$_X$ effluent concentration tests [220]

Test #	OFA[a]	Burner tilt angle (deg.%)[b]	Excess air
5, Baseline	No	0	20
1, OFA	Yes	0	20
2, +1% O$_2$	Yes	0	30
3, −1% O$_2$	Yes	0	10
8, Tilts up	Yes	+22	20
9, Tilts down	Yes	−20	20

[a] OFA = over-fire air.
[b] Angle in degrees from horizontal.

measurements made in the pulverized, coal-fired utility boiler discussed in the previous section [208].

In one experimental program, 10 tests were performed by the New York State Electric and Gas (NYSEG) Company to measure NO effluent concentrations under various operating conditions. They were designed to evaluate variations in furnace operation and simulate control techniques known to affect NO$_X$ concentrations. The operating variables considered herein were over-fire air, burner tilt, and percent excess air. The tests were performed in October and November of 1993.

One test was performed to establish a baseline, with the top level of burners out of service, and no air injection through this level of burners. The burners were operated with no tilt. In the remaining tests, coal was still injected in only the three lower levels of burners, and the top level was used for injection of over-fire air. The tests that were computer-simulated are shown in Table 10. Test 1 was conducted to determine the effects of over-fire air on NO formation. In this test, the air and coal flow rates were the same as the baseline test, but only over-fire air was injected in the top level of burners, and all burners were operated with no tilt. Tests 2 and 3 were conducted to determine the

Table 11
Operating conditions and coal characteristics for the baseline 1993 Goudey furnace NO$_X$ pollutant tests [220]

Gross power (MW$_e$)	67	
Air-flow rate (kg/h)	384,100	
Coal feed rate (kg/h)	501	
	As received %	Dry %
Moisture	6.14	–
Volatile matter	35.77	38.11
Fixed carbon	50.87	54.20
Ash	7.22	7.69
Sulfur	2.28	2.43
Carbon	71.09	75.74
Hydrogen	4.66	4.96
Nitrogen	1.39	1.48
Oxygen	7.22	7.69
Heating value (kJ/kg)	30,530	32,530
Mass mean diameter (μm)	27.4	

Table 12
PCGC-3 input conditions for baseline simulation of 1993 Goudey utility boiler tests [220]

Primary stream	
Total flowrate (kg/s)	16.7
Temperature (K)	360
Mass fractions	
O$_2$	0.196
N$_2$	0.740
H$_2$O	0.061
Coal loading (kg coal/kg gas)	0.58
Secondary stream	
Gas flowrate (kg/s)	90.0
Temperature (K)	540
Swirl no.	0.0
Mass fractions	
O$_2$	0.209
N$_2$	0.791
Coal particle density (kg/m^3)	1340
Particle size distribution	10.0 μm, 26.6%
	20.0 μm, 30.0%
	31.0 μm, 16.4%
	45.0 μm, 19.4%
	65.0 μm, 7.6%
Mass mean particle diameter	27.4 μm

effect of percent excess air. As with previous tests, all burners were operated with no tilt. In Test 2, the air-flows were increased to obtain approximately 1% increase in the exit O$_2$ concentration. In Test 3, the air-flows were decreased to obtain approximately 1% decrease in the exit O$_2$ concentration. Tests 8 and 9 were conducted to determine the effect of burner tilt on NO formation. In Test 8, the fuel nozzles were tilted upward 22%, and in Test 9 the fuel nozzles were tilted downward 20%.

Simulation of practical furnaces requires a large computational mesh to identify the processes occurring inside of these systems. Since only measured effluent values of NO concentration were available for this study, a smaller grid was used ($48 \times 28 \times 28$) to predict the trends observed with changes in various operating conditions in the boiler.

The operating conditions and coal characteristics used for the baseline test are summarized in Table 11. The characteristics of the coals used for these tests varied only slightly. Consequently, characteristics of the coal were assumed to be the same for all of these simulations. The input conditions used for the simulation of the baseline tests are shown in Table 12. The flow rates shown are total flow rates into the furnace. No information on wall temperatures in the Goudey furnace was available for these tests, and a constant wall temperature of 900 K was assumed.

Comparisons of measured and predicted trends are shown in Figs. 15 and 16. All NO concentrations have been adjusted to 3% excess air. Fig. 15 shows comparisons of the measured and predicted average effluent NO

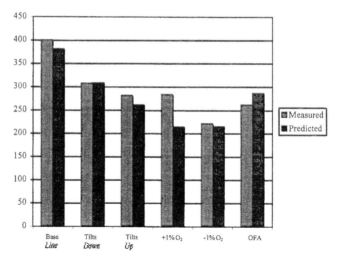

Fig. 15. Comparisons of measured [220] and predicted [208] average effluent NO concentrations (ppm) for several operating conditions in the Goudey 67 MW$_e$ utility boiler (1993 tests).

concentrations for the different tests. Both the measured and predicted values show the same general effects of the various operating conditions. The submodel slightly under predicts the measured NO concentration in all of the tests, except in Test 1 with over-fire air. The largest discrepancy of about 60 ppm (about 70%) between the measured and predicted values occurred with the test

where the air-flow rate was increased to raise the O$_2$ concentration by 1%.

The measured and predicted values both indicate that the major effect on the effluent NO concentrations results from operation with over-fire air. This is shown by comparing the difference between the base-line test and the test where the only change was the use of over-fire air. This change

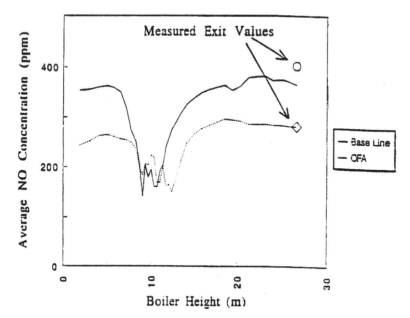

Fig. 16. Average predicted NO concentration (ppm) as a function of burner height for the baseline and over-fire air tests in the Goudey boiler [208].

resulted in a reduction in effluent NO emissions of 36%. The average change in effluent NO concentrations of the other tests was only 12%. The measured values showed that three of the four changes in operating conditions resulted in an increase in effluent NO concentration relative to the test with only over-fire air. Only a decrease in the excess air (-1% decrease in exit $\%O_2$) resulted in a further NO reduction. The predicted values show comparable trends. However, the differences in NO concentrations between these tests are relatively small, which makes it difficult to draw conclusions from the measured and predicted trends.

Since operation with over-fire air had the largest effect on effluent NO emissions, and the NO submodel predicted this trend, Fig. 16 is shown to illustrate some of the differences between these two cases. Fig. 16 shows the predicted average NO concentration as a function of burner height for the baseline test and the test with over-fire air. The measured effluent concentrations are also shown for the two cases. This figure shows that in the case with over-fire air, NO formation is delayed by 1–2 m. Once NO formation begins, it occurs at approximately the same rate, but does not achieve the same peak value as the baseline case.

The bottom of the burners is located at a height of approximately 8.5 m, and the top of the third level of burners occurs at approximately 11.5 m. The top of the fourth level of burners occurs at approximately 12.5 m. These burner locations correlate directly with the different locations of the rapid increase in temperature and NO concentrations. For the case with over-fire air, the region below the top level of burners (below 11.5 m) will be more fuel-rich than in the base-line case. The major factor contributing to the lower rate of NO formation in the case with over-fire air is most likely that the nitrogen is released from the coal in a more fuel-rich, lower temperature region. This environment favors reduction of the devolatilized nitrogen to N_2, and also results in the reaction of any existing NO to N_2 through homogeneous gas phase reduction reactions. Both of these factors contribute to less NO production.

8. Conclusions

This review paper has provided an overview of the NO formation and destruction processes, and the calculational procedures used in the comprehensive simulation of nitrogen pollutants in turbulent, combusting systems. Submodels of nitrogen pollutant formation and destruction used in comprehensive combustion codes were also discussed. Evaluation of these models by comparison with laboratory-scale and full-scale data and applications of these models to practical combustion systems is reviewed. Results demonstrate that these integrated models can be useful to predict, at least qualitatively, NO_X emissions. Comprehensive combustion modeling of NO_X reaction processes has been on-going for over two decades, and considerable progress has been made in the capabilities to model nitrogen-containing pollutant formation and destruction in practical combustion systems. Advances in mathematical modeling and increased performance of computer systems have made comprehensive combustion modeling, including NO_X formation and destruction a valuable tool in understanding NO_X emissions from combustion systems. These models can be used to investigate, design, and optimize various combustion techniques used to reduce NO_X emissions from practical combustion systems.

Caretto [7], in his earlier review of mathematical modeling, concluded that comprehensive modeling of pollutants had great potential, but that considerable work was necessary before this potential could be realized. He emphasized that efforts should focus on the mechanics of the turbulence and its coupling with the chemical reaction process. Considerable progress has been made in these areas since that time. Numerous solutions of comprehensive models have been reported for both gaseous and particle-laden combustion systems. Several of these simulations report comparisons of model predictions with measured profile and effluent data from practical combustion systems. Several simulations of nitrogen pollutant reactions were discussed previously, some of which report comparisons of predictions with measured data from practical combustion systems. Mathematical models have evolved to become an essential part of the design and optimization of combustion systems. Limitations still exist today in understanding and mathematically describing the physical processes occurring in turbulent reaction systems. These limitations, coupled with limitations of computer hardware, restrict our ability to model practical combustion systems. Additional research is still needed to improve both the physical understanding of NO_X reaction processes and mathematical methods to model these reaction processes in practical combustion systems. Several approaches to modeling chemical reactions in turbulent systems, which have been developed over the last several years, offer promise, especially as computer hardware becomes more powerful. Improved numerical efficiency of these methods would accelerate their applications to practical combustion systems.

Acknowledgements

Some of the research on which this paper is based, and some of the preparation of this review paper were sponsored by the Advanced Combustion Engineering Research Center at Brigham Young University. Funds for this center have been received from the National Science Foundation (Engineering Education and Centers Division), The State of Utah (Centers of Excellence), over 50 industrial participants, and five federal agencies. Substantial support from the US Department of Energy's Morgantown Energy Technology Center was particularly helpful in much of the earliest research on this topic at this Center. The authors express their gratitude to all of the above, particularly the earlier substantial contributions of Professor Philip J. Smith, then at Brigham Young University and currently at the University of Utah. The earlier technical work of numerous students at Brigham Young University who worked in the area of NO_X pollutant modeling is also acknowledged. The authors also express their gratitude to Mrs Haley Morales, Mrs Cheryl Paoule, and Mrs Marilyn Olson for their work in preparation of this manuscript.

References

[1] Miller JA, Bowman CT. Progress in Energy and Combustion Science 1989;15:287.

[2] Hayhurst AN, Vince IM. Progress in Energy and Combustion Science 1980;6:35.

[3] Bowman CT. Twenty-fourth Symposium (International) on Combustion, The Combustion Institute, Pittsburgh, PA, 1993. p. 859.

[4] Kramlich JC, Linak WP. Progress in Energy and Combustion Science 1994;20:149.

[5] Smoot LD, Hill SC, Xu H. Progress in Energy and Combustion Science 1998;24:385.

[6] Mitchell SC. IEA Coal Research, London, England, 1998.

[7] Caretto LS. Progress in Energy and Combustion Science 1976;1:47.

[8] Boardman RP, Eatough CN, Germane GJ, Smoot LD. Combustion Science Technology 1993;93:193.

[9] Pershing DW, Wendt JOL. Industrial Engineering Chemical Process Design and Development 1979;18:60–7.

[10] Maier H, Spliethoff H, Kicherer A, Fingerle A, Hein KRG. Fuel 1994;73:1447.

[11] Wendt JOL. Combustion Science and Technology 1995;108:323.

[12] Smart JP, Knill KJ, Visser BM, Weber R. Twenty-second Symposium (International) on Combustion, Pittsburgh, PA, 1988. p. 1117.

[13] Boardman RD, Smoot LD. In: Smoot LD, editor. Fundamentals of coal combustion, Amsterdam: Elsevier, 1993. p. 433.

[14] Lyon RK. Environmental Science Technology 1987;21:231.

[15] Nelson RK, Franklin JD, Scherer B. Power-Gen Americas'94, Livonia, MI, 1994.

[16] Lissausskas RA, Snodgrass RJ, Johnson SA, Eskinaze D. Joint Symposium on Stationary Combustion NO_x Control, Boston, MA, 1985.

[17] Fiveland WA, Wessel RA. Journal of the Institute of Energy 1991;64:41.

[18] Driscoll JF, Chen RH, Yoon Y. Combustion and Flame 1992;88:37.

[19] Hori M. Twenty-second Symposium (International) on Combustion, The Combustion Institute, Pittsburgh, PA, 1988. p. 1175.

[20] Johnson GM, Smith MY, Mulcahy MFR. Seventeenth Symposium (International) on Combustion, Pittsburgh, PA, 1979. p. 647.

[21] Kramlich JC, Malte PC. Combustion Science and Technology 1978;18:91.

[22] Mereb JB, Wendt JOL. Fuel 1994;73:1020.

[23] Linak WP, et al. Journal of Geophysical Research 1990;95:7533.

[24] Abbas T, Costen P, Lockwood FC. Combustion and Flame 1991;92:346.

[25] Kramlich JC, Cole JA, McCarthy JM, Lanier WS, McSarley JA. Combustion and Flame 1989;77:375.

[26] Zeldovich YB. Acta Physicochim, U.S.S.R. 1946;21:577.

[27] Fenimore CP. Combustion and Flame 1976;26:249.

[28] Heap MP, Tyson TJ, Cichanowicz GR, Kau CJ. Sixteenth Symposium (International) on Combustion, Pittsburgh, PA, 1977. p. 535.

[29] Salimian S, Hanson RK. Combustion Science and Technology 1980;23:225–30.

[30] Caretto LS. Western States Section/Combustion Institute, Salt Lake City, UT, 1976.

[31] Takagi T, Tatsumi T. Combustion and Flame 1979;35:17–25.

[32] Levy JM. Fifth EPA Fundamental Combustion Research Workshop, Redondo Beach, CA, 1980.

[33] Mellor GL, Herring HJ. AIAA Journal 1973;11:590.

[34] Spalding DB. In: Murthy SNB, editor. Turbulence modeling: solved and unsolved problems, New York: Plenum, 1975.

[35] Launder BE. Stress transport closures—into the third generation. New York: Plenum, 1975.

[36] Liepmann HW. American Scientist 1979;67:221–8.

[37] McMurty PA, Queiroz M. Turbulent reacting flows. Amsterdam: Elsevier, 1993.

[38] Solomon PR, Fletcher TH. Twenty-fifth Symposium (International) on Combustion, The Combustion Institute, Pittsburgh, PA, 1994. p. 463.

[39] Smoot LD. Eighteenth Symposium (International) on Combustion, The Combustion Institute, Pittsburgh, PA, 1981. p. 1185.

[40] Niksa S, Cho S. Energy and Fuels 1996;10:463.

[41] Radulovic PT, Smoot LD. In: Smoot LD, editor. Fundamentals of coal combustion, Amsterdam: Elsevier, 1993. p. 1 1.

[42] Wendt JOL, Pershing DW. Combustion Science and Technology 1977;16:111.

[43] Loison R, Chauvin F. Chemistry & Industry (Paris) 1964;91:269.

[44] Koyabashi H, Howard JB, Sarofim AF. Sixteenth Symposium (International) on Combustion, The Combustion Institute, Pittsburgh, PA. 1977. p. 411.

[45] Suuberg EM, Peters WA, Howard JB. Seventeenth Symposium (International) on Combustion, The Combustion Institute, 1979. p. 177.

[46] Anthony DB, Howard JB, Hottel HC, Meissner HP. Fifteenth Symposium (International) on Combustion, The Combustion Institute, Pittsburgh, PA, 1975. p. 1303.

[47] Horton MD. In: Smoot LD, Pratt DT, editors. Pulverized coal combustion and gasification, New York: Plenum, 1979. p. 133.

[48] Solomon PR, Serio MA, Suuberg EM. Progress in Energy and Combustion Science 1992;18:133.

[49] Pohl JH, Sarofim AF. Sixteenth Symposium (International) on Combustion, Pittsburgh, PA, 1977. p. 491.

[50] Palmer HB, Seery DJ. Annual Review of Physical Chemistry 1973;24:235–62.

[51] Chen SL, Heap MP, Pershing DW, Martin GB. Ninteenth Symposium (International) on Combustion, Pittsburgh, PA, 1983. p. 1271.

174

[52] Wang W, Brown SD, Hindmarsh CJ, Thomas KM. Fuel 1994;73:1381.

[53] Eddings EB, Smith PJ, Heap MP, Pershing DW, Sarofim AF. ASME Conference, Snowbird, UT, 1994. p. 169.

[54] Smith KL, Smoot LD, Fletcher TH, Pugmire RJ. The structure and reaction processes of coal. New York: Plenum, 1994.

[55] Heap MP, Corley TL, Kau CJ, Tyson TJ. Energy and Environmental Research Corp., Salt Lake City, UT, 1978.

[56] Malte PC, Rees DP. In: Smoot LD, Pratt DT, editors. Mechanisms and kinetics of pollutant formation during reaction of pulverized coal, New York: Plenum, 1979. p. 186.

[57] Wendt JOL. Progress in Energy and Combustion Science 1980;6:201.

[58] Blair DW, Wendt JOL. Exxon Research and Engineering Company, New Jersey, 1981.

[59] Song YH, Beer JM, Sarofim AF. Second Annual Symposium on Stationary Source Combustion, Electric Power Research Institute, New Orleans, LA, 1977.

[60] Solomon PR, Colket MB. Fuel 1978;57:749.

[61] Lee JW, Chen SL, Pershing DW, Heap MP. Western States Section/Combustion Institute, Brigham Young University, Provo, UT, 1979.

[62] Smoot LD, Boardman RD, Brewster BS, Hill SC, Foli AK. Energy and Fuels 1993;7:786.

[63] Yang YB, Naja TA, Gibbs BM, Hampartsourmian E. Journal of the Institute of Energy 1997;70:9.

[64] Malte PC, Pratt DT. Combustion Science and Technology 1974;9:221.

[65] Visona SP, Stanmore BR. Combustion and Flame 1996;105:92.

[66] Morely C. Eighteenth Symposium (International) on Combustion, The Combustion Insitute, Pittsburgh, PA, 1981. p. 23.

[67] Krevelen DWV, Schuger A. Coal science. Amsterdam: Elsevier, 1957.

[68] Boardman RD, Smoot LD. AIChE Journal 1988;34:1573.

[69] Bose AC, Dannecker KM, Wendt JOL. Energy and Fuels 1988;2:301.

[70] Ghani MU, Wendt JOL. Twenty-third Symposium (International) on Combustion, Pittsburgh, PA, 1990. p. 1281.

[71] Haussmann GJ, Kruger CH. Western States Section/The Combustion Institute, Livermore, CA, 1989.

[72] Blair DW, Wendt JOL, Bartok W. Sixteenth Symposium (International) on Combustion, Pittsburgh, PA, 1977. p. 475.

[73] Solomon PR. United Technologies Research Center, CN, 1977.

[74] Goel S, Zhang B, Sarofim AF. Combustion and Flame 1996;104:213.

[75] Sawyer RF. Eighteenth Symposium (International) on Combustion, The Combustion Institute, Pittsburgh, PA, 1981. p. 1.

[76] Bowman CT. Fourteenth Symposium (International) on Combustion, The Combustion Institute, Pittsburgh, PA, 1973. p. 270.

[77] Bowman CT. Western States Section of the Combustion Institute, Salt Lake City, UT, 1991.

[78] Thompson D, Brown TP, Baer JM. Combustion and Flame 1972;19:69.

[79] Drake MC, Correa SM, Pitz RW, Shyy W, Fenimore CP. Combustion and Flame 1987;69:347.

[80] Bowman CT. Progress in Energy and Combustion Science 1975;1:33.

[81] Hayhurst AN, Vince IM. Progress in Energy and Combustion Science 1980;6:35.

[82] Edelman RB, Harsha PT. Progress in Energy and Combustion Science 1978;4:1.

[83] Coelho PF, Carvalho MG. Combustion Science and Technology 1995;108:363.

[84] Chen W, Smoot LD, Hill SC, Fletcher TH. Energy and Fuels 1996;10:1046.

[85] Kee RJ, Rupley FM, Miller JA. SANDIA Report, Livermore, CA, 1991.

[86] Vogt RA, Laurendear NM. Central States Section, The Combustion Institute, Columbus, OH, 1976.

[87] Corlett RC, Monteith LE, Halgre CA, Malte PC. Combustion Science and Technology 1979;19:95.

[88] Milne TA, Beachy JE. Combustion Science and Technology 1977;16:139.

[89] Pershing DW, Martin GB, Berkau EE. AIChE Symposium Series 19, 1975.

[90] Sarofim AF, Williams GC, Modell M, Slater SM. AIChE Symposium Series 148, 1975.

[91] Wendt JOL, Pershing DW, Lee JW, Glass JW. Seventeenth Symposium (International) on Combustion, The Combustion Institute, Pittsburgh, PA, 1979. p. 77.

[92] Morely C. Combustion and Flame 1976;27:189.

[93] Song YH, Blair DW, Siminski VJ, Bartok W. Eighteenth Symposium (International) on Combustion, The Combustion Institute, Pittsburgh, PA, 1981. p. 53.

[94] Sanders WA, Lin CY, Lin MC. Combustion Science and Technology 1987;51:103.

[95] Bowman CT. Chemistry of gaseous pollutant formation and destruction. New York: Wiley, 1991.

[96] Chen GT, Shang JY, Wen CY. DOE Report, DOE/MC/ 11284-166, 1981.

[97] Hedley JT, Pourkashanian M, Williams A. Combustion Science and Technology 1995;108:311.

[98] DeSoete GG. Fifteenth Symposium (International) on Combustion, The Combustion Institute, Pittsburgh, PA, 1975. p. 1093.

[99] Mitchell JW, Tarbell JM. AIChE Journal 1982;28:302.

[100] Smoot LD, Smith PJ. Coal combustion and gasification. New York: Plenum, 1985.

[101] Freihaut JD, Proscia WM. First International Conference on Combustion Technologies for a Clean Environment, Vilamoura (Algarve), Portugal, 1991.

[102] Peck RE, Glarborg P, Johnson JE. Combustion Science and Technology 1991;76:81.

[103] Brown BW, Smoot LD, Hedman PO. Fuel 1986;65:673.

[104] Chen SL, Heap MP, Pershing DW, Nihart RK, Rees DP. WSS/CI, University of California, Irvine, CA, 1980.

[105] Haynes BS. Combustion and Flame 1977;28:113.

[106] Sadakata M, Fujioka Y, Kunii D. Eighteenth Symposium (International) on Combustion, The Combustion Institute, 1981. p. 65.

[107] Malte PC et al. AIAA Eighteenth Aerospace Sciences Meeting, Pasadena, CA, 1980.

[108] Altenkirch RA, Peck RE, Chen SL. Combustion Science and Technology 1979;20:49.

[109] Harding Jr. NS, Smoot LD, Hedman PO. AIChE Journal 1982;28:573.

[110] Rees DP, Smoot LD, Hedman PO. Eighteenth Symposium (International) on Combustion, Pittsburgh, PA, 1981. p. 1305.

[111] Fenimore CP. Seventeenth Symposium (International) on Combustion, The Combustion Institute, Pittsburgh, PA, 1979. p. 661.

[112] Song YH, Pohl JH, Beer JM, Sarofim AF. Combustion Science and Technology 1982;28:31.

[113] Glass JW, Wendt JOL. Ninteenth Symposium (International) on Combustion, The Combustion Institute, Pittsburgh, PA, 1982. p. 1263.

[114] Song YH, Bartok W. Ninteenth Symposium (International) on Combustion, The Combustion Institute, Pittsburgh, PA, 1983. p. 1291.

[115] Asay BW, Smoot LD, Hedman PO. Combustion Science and Technology 1983;35:15.

[116] Pershing DW, Wendt JOL. Sixteenth Symposium (International) on Combustion, The Combustion Institute, Pittsburgh, PA, 1977. p. 389.

[117] Brown RA, Mason HB, Pershing DW, Wendt JOL. American Institute of Chemical Engineers 84th National Meeting, Houston, TX, 1977.

[118] Hulgaard T, Dam-Johnson K. AIChE Journal 1993;39:1342.

[119] Glarborg P, Kristensen PG, Dam-Johansen K. The Journal of Physical Chemistry 1997;101:3741.

[120] Heap MP, Lowes TM, Walmsley R, Bartelds H. EPA Coal Combustion Seminar, Research Triangle Park, NC, 1973.

[121] Wendt JOL. Process modifications for NOₓ control in pulverized coal combustion systems (January 4, Personal Communication to V.J. Kothari, 1983).

[122] Kolb T, Jansohn P, Leuckel W. Twenth-second Symposium (International) on Combustion, The Combustion Institute, Pittsburgh, PA, 1988. p. 1193.

[123] Ostberg M. et al. Twenty-seventh Symposium (International) on Combustion, The Combustion Institute, Pittsburgh, PA, 1998.

[124] Lin WC. University of Arizona, Tuscon, AZ, 1994.

[125] Miller CA, Touati AD, Becker J, Wendt JOL. Twenty-seventh Symposium (International) on Combustion, The Combustion Institute, Pittsburgh, PA, 1998. p. 3189.

[126] DOE. US Department of Energy, Washington DC, 1995.

[127] Wilcox B. Babcock and Wilcox Company, 1990.

[128] TVA. Tennessee Valley Authority, 1992.

[129] Brouwer J. et al. Western States Section of the Combustion Institute, Reno, NV, 1994.

[130] Syverud T, Thomassen A, Gantestad T. World Cement 1994;25:39.

[131] Smart JP, Morgan DJ. Fuel 1994;73:1437.

[132] Folsom BA, Sommer TM, Payne R. AFRE–JFRC International Conference on Environmental Control of Combustion Processes, Honolulu, HI, 1991.

[133] EAER. Congress Clean Coal Technology, E.A.E.R. Corporation, 1990.

[134] Folsom BA. et al. Electric Power Research Institute/Environmental Protection Agency 1995 Joint Symposium on Stationary Combustion NOₓ Control, Kansas City, MO, 1995.

[135] Chen SL. et al. Twenty-first Symposium (International) on Combustion, The Combustion Institute, Pittsburgh, PA, 1986. p. 1159.

[136] Kicherer A, Spliethoff H, Maier H, Hein KRG. Fuel 1994;73:1443.

[137] Chen JY, Ma L. Third Symposium (International) on Coal Combustion, Beijing, China, 1995. p. 594.

[138] Moyeda DK, Li B, Maly P, Payne R. Pittsburgh Coal Conference, Pittsburgh, PA, 1995. p. 1119.

[139] Tree DR, Black DL, Rigby JR, McQuay MQ, Webb BW. Progress in Energy and Combustion Science 1998;24:355.

[140] Xu H. Brigham Young University, Provo, UT, 1999.

[141] Shimizu T, Sazawa Y, Adschiri T. Fuel 1992;71:361.

[142] Thomas KM, Grant K, Tate K. Fuel 1993;72:941.

[143] Visona SP, Stanmore BR. Combustion and Flame 1996;106:207.

[144] Glass JW, Wendt JOL. Western States Section/The Combustion Institute, University of Utah, Salt Lake City, UT, 1982.

[145] Wang W, Thomas KM. Fuel 1992;71:871.

[146] Pohl JH, Sarofim AF. EPA Symposium on Stationary Source Combustion, Atlanta, GA, 1975.

[147] Heap MP, Lowes TM, Walmsley R, Bartelds H, LaVaguerese P. USEFTS Pub. No. EPA-600/2-76-061a, 1976.

[148] Brown SD, Thomas KM. Fuel 1993;72:359.

[149] Snow GC, Grosshandler WL, Malte PC. WSS/CI, Brigham Young University, Provo, UT, 1979.

[150] Levy JM, Chen LK, Sarofim AF, Baer JM. Eighteenth Symposium (International) on Combustion, The Combustion Institute, Pittsburgh, PA, 1981. p. 111.

[151] DeSoete GG. Twenty-second Symposium (International) on Combustion, The Combustion Institute, Pittsburgh, PA, 1988. p. 1117.

[152] Chan LK, Sarofim AF, Beer JM. Combustion and Flame 1983;52:37.

[153] Smith PJ, Smoot LD, Hill SC. AIChE Journal 1986;32:1917.

[154] Brewster BS, Cannon SM, Farmer JR, Meng F. Progress in Energy and Combustion Science 1999:353.

[155] Launder BE, Spalding B. Mathematical models of turbulence. New York: Academic Press, 1972.

[156] Eaton AM, Smoot LD, Hill SC, Eatough CN. Progress in Energy and Combustion Science 1999;25:387.

[157] Hanjalic K, Launder BE, Schiestel R. In: Bradbury LJS, Durst F, Launder BE, Schmidt FW, Whitelaw JH, editors. Turbulent shear flows, vol. 2. New York: Springer, 1980. p. 36.

[158] Fabris G, Harsha PT, Edelman RB. NASA CR-3433, 1981.

[159] Kim SW, Chen CP. Numerical Heat Transfer 1988;16:193.

[160] Speziale CG. Journal of Fluid Mechanics 1987;178:459.

[161] Bilger RW. In: Libby PA, Williams FA, editors. Topics in applied physics: turbulent reacting flows, New York: Springer, 1980. p. 65.

[162] O'Brien EE. In: Libby PA, Williams FA, editors. Topics in applied physics: turbulent reacting flows, New York: Springer, 1980. p. 185.

[163] Kerstein AR. Combustion Science and Technology 1988;60:391.

[164] Kerstein AR. Combustion and Flame 1989;75:397.

[165] Kerstein AR. Journal of Fluid Mechanics 1990;216:411.

[166] Kerstein AR. Combustion Science and Technology 1992;81:72.

[167] Spalding DB. Thirteenth Symposium (International) on Combustion, The Combustion Institute, Pittsburgh, PA, 1971. p. 649.

[168] McMurty PA, Manon S, Kerstein AR. Energy and Fuels 1993;7:817.

[169] Pope SB. Progress in Energy and Combustion Science 1985;11:119.

176

[170] Pope SB. Combustion Science and Technology 1981;25:159.

[171] Schumann U, Friedrich R. Notes on Numerical Fluid Mechanics 1985:15.

[172] McMurtry PA. Mechanical Engineering Department, University of Washington, Seattle, WA, 1987.

[173] Smith PJ, Hill SC, Smoot LD. Nineteenth Symposium (International) on Combustion, Pittsburgh, PA, 1982. p. 1263.

[174] Smoot LD, Smith PJ. Coal combustion and gasification. New York: Plenum, 1985. p. 245.

[175] Smith PJ, Fletcher TH, Smoot LD. Eighteenth Symposium (International) on Combustion, The Combustion Institute, Pittsburgh, PA, 1981. p. 1305.

[176] Bilger RW. Advances in Geophysics 1974;188:349.

[177] Smoot LD, Smith PJ. Coal combustion and gasification. New York: Plenum, 1985. p. 321.

[178] Smith PJ, Smoot LD. Combustion and Flame 1981;42:277.

[179] Williams FA, Libby PA. AIAA Eighteenth Aerospace Sciences Meeting, New York, NY, 1980.

[180] Bilger RW. Combustion Science and Technology 1979;19:89.

[181] Bilger RW. Combustion Science and Technology 1980;22:251.

[182] Bilger RW. Combustion and Flame 1978;30:277.

[183] Fenimore CP, Frankel HA. Eighteenth Symposium (International) on Combustion, The Combustion Insitute, Pittsburgh, PA, 1981. p. 143.

[184] Cannon SM, Brewster BS, Smoot LD. Combustion and Flame 1998;113:135.

[185] Niksa S. Coal Combustion Modeling, IEA Coal Research, London, England, 1996.

[186] Hill SC, Smoot LD. Energy and Fuels 1993;7:874.

[187] Brewster BS, Hill SC, Radulovic PT, Smoot LD. In: Smoot LD, editor. Fundamentals of coal combustion for clean and efficient use, Amsterdam: Elsevier, 1993. p. 567.

[188] Hill SC, Smoot LD, Smith PJ. Twentieth Symposium (International) on Combustion, The Combustion Institute, Pittsburgh, PA, 1984. p. 1391.

[189] Flores DV, Fletcher TH. Combustion Science and Technology 2000;150:1.

[190] Smith IW. Eighteenth Symposium (International) on Combustion, The Combustion Institute, Pittsburgh, PA, 1981. p. 1213.

[191] Smoot LD, Hedman PO, Smith PJ. Progress in Energy and Combustion Science 1984;10:359.

[192] Corvalho MG, Coelho PJ, Costa F. Second European Conference on Industrial Funaces and Boilers, Vilamoura, Portugal, 1991.

[193] Bonvini M, Piana C, Vigevano L. Second European Conference on Industrial Furnaces and Boilers, Vilamoura, Portugal, 1991.

[194] Coimbra C, Azevado J, Carvalho MG. Fuel 1993;73:1128.

[195] Epple B, Schneider U, Schnell U, Hein K. Combustion Science and Technology 1995;108:383.

[196] DeMichele G, Pasini S, Tozzi A. Second European Conference on Industrial Furnaces and Boilers, Vilamoura, Portugal, 1991.

[197] Mallampalli H, Fletcher TH, Chen JY. Journal of Engineering for Gas Turbines and Power 1998;120:703.

[198] Heap MP, Lowes TM, Walmsley R. AFRC/EPA American Flame Days, Chicago, IL, 1972.

[199] Burch TE, et al. Energy and Fuels 1991;5:231.

[200] Seeker WR, Samuelsen GS, Heap MP, Trolinger JD. Eighteenth Symposium (International) on Combustion, The Combustion Institute, Pittsburgh, PA, 1981. p. 1213.

[201] Bonin MP, Queiroz M. Combustion and Flame 1990;85:121.

[202] Butler BW, Webb BW. Heat Transfer in Combustion Systems, ASME/HTD 1990;142:49.

[203] Butler BW, Wilson T, Webb BW. Twenty-fourth Symposium (International) on Combustion, The Combustion Institute, Pittsburgh, PA, 1993. p. 1333.

[204] Huntsman LK. Brigham Young University, Provo, UT, 1990.

[205] Cannon JN, Webb BW, Queiroz M. Fossil Fuel Combustion, ASME 1991;33:49.

[206] Oetlli MC. Brigham Young University, Provo, UT, 1993.

[207] Groberg CJ. MS Thesis, Brigham Young University, Provo, UT, 1996.

[208] Hill SC, Cannon JN. Joint AFRC/JFRC Pacific Rim International Conference on Environmental Control of Combustion Processess, Maui, HI, 1994.

[209] Crowe C, Smoot LD. In: Smoot LD, Pratt DT, editors. Pulverized coal combustion and gasification, New York: Plenum, 1979. p. 15.

[210] Pratt DT. In: Smoot LD, Pratt DT, editors. Pulverized coal combustion and gasification, New York: Plenum, 1979. p. 3.

[211] Kent JH, Bilger RW. Sixteenth Symposium (International) on Combustion, The Combustion Institute, Pittsburgh, PA, 1977. p. 1643.

[212] Lockwood FC, Naguib AS. Combustion and Flame 1975;24:109.

[213] Melville EK, Bray NC. International Journal of Heat and Mass Transfer 1979;22:647.

[214] Fiveland WA. Journal of Heat Transfer, ASME 1987;109:809.

[215] Jamaluddin AS, Smith PJ. Combustion Science and Technology 1988;59:321.

[216] Ubhayaker SK, Stickler DB, von Rosenberg CW, Gannon RE. Sixteenth Symposium (International) on Combustion, The Combustion Institute, Pittsburgh, PA, 1977. p. 427.

[217] Fletcher TH, Kerstein AR, Pugmire RJ, Solum MS, Grant DM. Energy and Fuels 1992;6:414.

[218] Crowe CT, Sharma MP, Stock DE. Fluids Engineering 1977;99:325.

[219] Gosman AD, Pun WM. Imperial College, London, 1973.

[220] NYSEG. New York State Electric and Gas, Binghamton, NY, 1993.

[221] Cannon JN, Webb BW, McQuay MQ. Brigham Young University, Provo, UT, 1997.

Emissions Reduction: NO$_x$/SO$_x$ Suppression
A. Tomita (Editor)

Modelling of the combustion process and NO$_x$ emission in a utility boiler

M. Xu[a,*], J.L.T. Azevedo[b], M.G. Carvalho[b]

[a]*National Laboratory of Coal Combustion, Huazhong University of Science and Technology, Wuhan 430074, People's Republic of China*
[b]*Instituto Superior Técnico, Av. Rovisco Pais, 1096 Lisbon, Portugal*

Accepted 19 January 2000

Abstract

This paper presents numerical simulation of the flow and combustion process in the furnace of a pulverized coal fired utility boiler of 350 MWe with 24 swirl burners installed at the furnace front wall. Five different cases with 100, 95, 85, 70 and 50% boiler full load are simulated. The comparison between the simulation and the plant data is stressed in this study. The heat flux to furnace walls between the measured values and the calculation results is compared. It is found that increasing the load leads to consistent variations in the properties presented and the exception is observed for the full load case where the predicted exit gas temperature is lower than the 95% one and the total heat to the boiler walls is smaller. This might be due to the fact of considering a linear scaling of the input parameters between the 70% and 100% load. The increase of the air flow rate led as expected to a reduction of the furnace outlet temperature and to a small decrease in NO$_x$ emissions. It shows that the NO$_x$ model used shows a higher sensitivity to temperature than to oxygen level in the furnace. The model used considered the De Soete mechanism for the nitrogen from volatiles and the contribution of char was considered in a similar way. The agreement for all cases except the one of 50% boiler load between the calculation results with the plant data validates the models and algorithm employed in the computation. The furnace performance under different boiler loads is predicted and compared in order to meet the requirements of NO$_x$ abatement and avoiding some negative side effects on the furnace. © 2000 Elsevier Science Ltd. All rights reserved.

Keywords: Modelling; NO$_x$ emission; Utility boiler

1. Introduction

The efficient use of pulverized coal is crucial to the utility industry. To achieve a higher combustion efficiency, the major affecting factors such as the particle size distribution, gas and particle temperatures, local heat release, local oxygen concentration, kinetic parameters for coal devolatilization and char oxidation, and coal properties should be understood thoroughly.

In the past two decades, the use of CFD codes for modelling utility boilers is becoming a useful tool to predict the performance of boilers among the scientific and industrial communities [1–8]. It helps engineers to optimize the operating conditions, reduce pollutant emissions, investigate malfunctions in the equipment, evaluate different corrective measures and also improve the design of new boilers. Submodels for simulating the in-furnace processes such as mixing, radiative heat transfer, and chemical kinetics have been developed. The development of the models depends on the availability of accurate and approximate experimental

data for comparison. However, because of the expensive price of measurements of the combustion and heat transfer characteristics and the limitation by the geometry, time, and number of instruments and skills required, assessment of these models is still limited to laboratory-scale [9,10]. Only a few detailed works reported in the literature are concerned with power plants typically below 80 MWe [1,11–13]. This paper addresses the comparison between the predictions and measurements acquired in a pulverized coal-fired boiler of 350 MWe of the Spanish Empresa National de Electricidad, S.A. (ENDESA) with 24 swirl burner installed at the furnace front wall.

The improvement of boiler operation and the development of tools to assess these are continuously pursued. Primary measurements for NO$_x$ reduction have been successfully implemented in power plants allowing to meet the stringent NO$_x$ emission limits. However, in some cases, there is still the need to improve burner or boiler design with the objective of producing NO$_x$ abatement avoiding some negative side effects such as the increase of unburned carbon in the fly ash and modifications in the fouling and slagging in the furnace and superheater surfaces. This paper also concentrates on the predicting of furnace performance for different boiler operating conditions.

* Corresponding author. Tel.: + 86-27-8754-2417; fax: + 86-27-8754-5526.
E-mail address: mhxu@mail.hust.edu.cn (M. Xu).

Reprinted from *Fuel* **79** (13), 1611-1619 (2000)

The boiler geometry and operating conditions is described in Section 2, and is followed by the description of the mathematical model in Section 3. Then the results are presented and discussed in Section 4. Finally, the paper ends with a summary of the main conclusions in Section 5.

2. Boiler geometry and cases description

The data of boiler geometry were taken in one of the boilers of ENDESA, as shown in Fig. 1. The front wall pulverized coal fired boiler is 48.7 m high, 17.13 m wide, 10.67 m deep and with an installed capacity of 350 MWe. Twenty four burners are arranged in an array of four burners disposed in six different levels. The boiler operating conditions considered in this study include full load (100%), 95, 85, 70 and 50% boiler full load corresponding to cases with all burners in service (100, 95 and 85%), 1–4 and 6 rows (from the bottom) of burners in service, 1 and 3–5 rows of burners in service, respectively. All the boiler operating conditions are listed in Table 1. Among them, cases 100, 70 and 50% are the raw cases from the power plant and cases 95 and 85% are interpolated from cases 100 and 70%.

Besides the above simulation for different boiler loads,

corresponding cases of higher excess oxygen levels (more secondary air flow rate) were also calculated to predict the performance of the boiler and validate NO$_x$ models applied in this study.

3. Mathematical models

The mathematical model is based on an Eulerian description for the continuum phase and a stochastic Lagrangean description for the coal particles while the following assumptions are made. (1) The mass sources are provided by the three-dimensional flow at levels corresponding to burner positions. (2) Burner mixing is characterized by the turbulence intensity of the burner. The production of turbulence and its dissipation are modelled with $k-\varepsilon$ model. (3) The polydisperse distribution of coal is segmented into discrete particle groups, assuming Rosin–Rammler distribution. (4) A simple chemically reacting system (SCRS) is adopted where the reaction rates are very fast compared with the mixing rates. (5) Char oxidizes to CO at the particle surface; CO oxidizes to CO$_2$ in the bulk gas.

The well-known $k-\varepsilon$ eddy viscosity/diffusivity model is used to quantify turbulent mixing in the furnace. The representative coal particles are tracked in the combustion chamber using simulated instantaneous gas velocities. The energy balance to the coal particles is used to calculate the particle temperature with time and to describe coal evolution. Char combustion is modelled using a parallel process of surface kinetics and oxygen diffusion. The balance of radiative heat transfer is calculated using the discrete transfer method [14] which is based on the direct solution of the radiation intensity transport equation. The radiative properties of gas are computed using the wide band model. Detailed description of the models is presented in Refs. [15–18].

The velocity, the temperature and the mixture fraction are prescribed at the inlet, whereas the kinetic energy of the turbulence and its dissipation rate are estimated (see, for example, Carvalho, et al. [19]). At the walls, the laws of the wall [20] are employed. Although this approach may be questionable in respect of heat transfer in complex recirculating flows [21], it does not present a serious problem for the present application. In fact, heat transfer to the walls in utility boilers is mainly due to radiation and the convective heat transfer has only a minor contribution. The temperature and emissivity of the walls are specified. At the exit, a zero gradient normal to the boundary is assumed for the dependent variables. The vertical velocity is then corrected to ensure mass conservation.

A post processor is employed for NO$_x$ simulation based on an extension of De Soete's mechanism ([22]) schematically illustrated in Fig. 2. The model is developed on the basis of the solution of balance equations involving NO and its precursors (HCN and NH$_3$) assuming $\alpha = 0.9$. The mechanism for volatiles is extended to include thermal NO$_x$, reburning and NO$_x$ formation from coal char. The

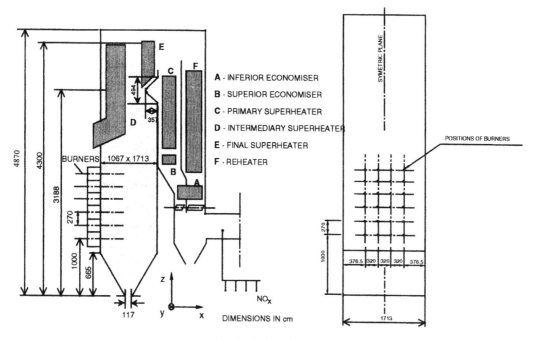

Fig. 1. Sketch of the utility boiler.

nitrogen in coal is assumed to be uniformly divided by the char and volatile matter. For the reburning reaction, the kinetic rate presented by Chen [23] is used together with the eddy break-up model. The kinetic rate derived represents the global reaction rate between light hydrocarbons and NO_x.

The volatile composition can be determined from the parallel volatilization model [24,25] described below which indicates the fraction of CH_4 to account for 10% of the volatile mass release. The parallel volatilization model represents the overall devolatilization process by two mutually competing first order reactions. The rate of weight loss of the coal (d.a.f.) is given by

$$\frac{dm_c}{dt} = -(k_1 + k_2)m_c \tag{1}$$

The rate of devolatilization at any time is

$$\frac{dV}{dt} = (\alpha_1 k_1 + \alpha_2 k_2)m_c \tag{2}$$

and the extent of devolatilization at time t is obtained as

$$V(t) = m_c(0)\int_0^t (\alpha_1 k_1 + \alpha_2 k_2) \exp\left[-\int_0^{t'} (k_1 + k_2)\, dt'\right] dt \tag{3}$$

In the above set of equations, α_1 and α_2 are mass stoichiometric factors denoting the extents of devolatilization via reactions 1 and 2, respectively. The rate constants k_1 and k_2 have Arrhenius form, and are such that reaction-1 has a lower activation energy than reaction-2, with the effect

Table 1
Cases description (*Note*: VM—volatile matter; PA—primary air; SA—secondary air)

Case no.	1	2	3	4	5	6	7	8	9	10
Boiler load (%)	100	95	85	70	50	100	95	85	70	50
Rows of burner in service	1–6	1–6	1–6	1–4,6	1,3–5	1–6	1–6	1–6	1–4,6	1,3–5
N mass fraction in coal (%)	0.54	0.56	0.625	0.66	0.67	0.54	0.56	0.625	0.66	0.67
VM in coal (%)	25.76	25.795	25.865	25.97	24.22	25.76	25.795	25.865	25.97	24.22
Total coal flow rate (t/h)	205	198.6	185.6	166.2	119.2	205	198.6	185.6	166.2	119.2
Total PA flow rate (t/h)	352.8	335.8	301.7	250.7	200.0	352.8	335.8	301.7	250.7	200.0
Total SA flow rate (t/h)	930.4	921.5	903.7	877.0	734.7	1022.7	1008.6	980.4	938.2	826.9

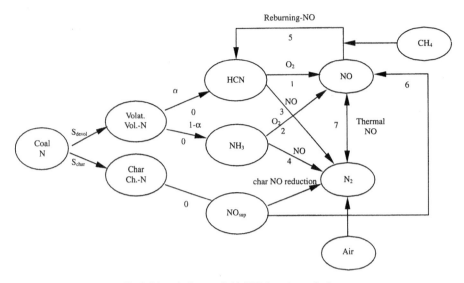

Fig. 2. Schematic diagram of global NO formation mechanism.

that the second reaction becomes operational only at higher temperatures to effect volatile yields in excess of α_1.

Most of the volatile matter is released as tar which will decompose later into lighter hydrocarbons and their radicals but no attempt is made here to simulate these. The reburning reaction rate taken from Chen [23] is considered between NO and methane from volatile and the natural gas. The oxidation rate of methane is higher than that of other volatile species and therefore its concentration can be lower than the fraction of volatile matter considered.

Methane is competitively consumed by combustion and by the reburning reaction. The reaction rate of the latter is slower than the oxidation and therefore the calculations are performed considering that for the reburning reaction methane would be left if complete oxidation occurs. The reburning reaction is considered in the fuel rich zones together with Eqs. (5b) and (7b). These equations are a modification of the reaction rates from De Soete's model while Eqs. (5a) and (7a) are employed for fuel lean conditions with the constant in Eq. (5a) increased by 3.5 compared with the original value proposed by Lockwood and Romo-Millares [26]. The calculation of NO formation from char is considered by the balance of concentration at the particle surface. All char nitrogen is assumed to produce NO at the particle surface proportionally to the char consumption rate (Eq. (10b)). Simultaneous NO reduction is calculated based on its concentration close to the particle surface (Eq. (10c)). The mass source of NO can be obtained from the difference of the two rates and the calculation of NO concentration at the char surface (Eq. (10a)) is obtained equating the source of diffusion flux from the particle to the surroundings (Eq. (10)). It should be noted that NO concentration at char surface can be larger than or smaller than the remote value which is calculated from the mass balance. Mean particle diameters are used in each control volume for the calculation of $S_{\text{Dif,NO}}$

$$S_{0,\text{HCN}} = 2\alpha f_N S_{\text{devol}}(M_{\text{HCN}}/M_{N_2}) \tag{4a}$$

$$S_{0,\text{NH}_3} = 2(1 - \alpha) f_N S_{\text{devol}}(M_{\text{NH}_3}/M_{N_2}) \tag{4b}$$

$$S_{1,\text{NO}} = \rho \times 3.5 \times 10^{10} X_{\text{HCN}} X_{\text{O}_2}^b \exp\left(- \frac{2.805 \times 10^8}{RT}\right) \frac{M_{\text{NO}}}{M_{\text{gas}}} \tag{5a}$$

$$S_{1,\text{NO}} = \rho \times 1.5 \times 10^{10} X_{\text{HCN}} X_{\text{O}_2}^b \exp\left(- \frac{2.646 \times 10^8}{RT}\right) \frac{M_{\text{NO}}}{M_{\text{gas}}} \tag{5b}$$

$$S_{2,\text{NO}} = \rho \times 4.0 \times 10^6 X_{\text{NH}_3} X_{\text{O}_2}^b \exp\left(- \frac{1.340 \times 10^8}{RT}\right) \frac{M_{\text{NO}}}{M_{\text{gas}}} \tag{6}$$

$$S_{3,\text{NO}} = \rho \times 3.0 \times 10^{12} X_{\text{HCN}} X_{\text{NO}} \exp\left(- \frac{2.512 \times 10^8}{RT}\right) \frac{M_{\text{NO}}}{M_{\text{gas}}} \tag{7a}$$

$$S_{3,\text{NO}} = \rho \times 1.1 \times 10^{12} X_{\text{HCN}} X_{\text{NO}} \exp\left(- \frac{2.470 \times 10^8}{RT}\right) \frac{M_{\text{NO}}}{M_{\text{gas}}} \tag{7b}$$

Table 2
Coal particle trajectory data

Number of particle start locations	4
Number of track paths per start locations	10
Maximum number of steps in each trajectory	500

$$S_{4,NO} = \rho \times 1.8 \times 10^8 X_{NH_3} X_{NO} \exp\left(-\frac{1.131 \times 10^8}{RT}\right)\frac{M_{NO}}{M_{gas}} \tag{8}$$

$$S_{5,NO} = \rho \times 2.76 \times 10^6 X_{HC} X_{NO} \exp\left(-\frac{0.7872 \times 10^8}{RT}\right)\frac{M_{NO}}{M_{gas}} \tag{9}$$

$$X_{HC} = X_{CH_4} - X_{O_2}/SR_{CH_4} \tag{9a}$$

$$S_{6,NO} = S_{Dif,NO}(X_{NO} - X_{NO_{sup}}) = S_{0P,NO} - S_{0R,NO}X_{NO_{sup}} \tag{10}$$

$$X_{NO_{sup}} = \frac{S_{0P,NO} + S_{Dif,NO}X_{NO}}{S_{0R,NO} + S_{Dif,NO}} \tag{10a}$$

$$S_{0P,NO} = 2f_N S_{char}(M_{NO}/M_{N_2}) \tag{10b}$$

$$S_{0R,NO} = 4.24 \times 10^4 A_E \frac{m_{char}}{VOL}PX_{NO_{sup}}$$
$$\times \exp\left(-\frac{1.465 \times 10^8}{RT}\right)M_{NO} \tag{10c}$$

$$S_{Dif,NO} = A_E m_{char}\frac{Sh \times M_{NO}D_{NO,gas}}{RTd_p} \tag{10d}$$

$$S_{Thermal,NO} = \rho\frac{2.0X_O(k_3 k_4 X_{O_2} X_{N_2} - k_{-3}k_{-4}X_{NO}^2)}{k_4 X_{O_2} + k_{-3}X_{NO}}\frac{M_{NO}}{M_{gas}} \tag{11}$$

All the Eulerian partial differential equations governing conservation of mass, momentum and energy can be written in the following general form:

$$\frac{\partial(\rho u_j \phi)}{\partial x_j} = \frac{\partial}{\partial x_j}(\Gamma_\phi\frac{\partial\phi}{\partial x_j}) + S_\phi \tag{12}$$

where ϕ stands for the three momentum components, the turbulent kinetic energy k and its dissipation ε, the enthalpy, and the mass fraction of gas species (mixture fraction). Γ_ϕ is the diffusion coefficient of the transported variable ϕ. For

Table 3
Coal particle size distribution

d (μm)	200	130	90	70	50
Mass fraction (%)	0.5	3.0	5.0	10.5	81.0

Table 4
Coal combustion parameters

(a) *Coal devolatilization data*		(b) *Coal char combustion data*	
α_1 (s^{-1})	3.70×10^5	A (kg/(m^2 s Pa$^{0.5}$))	0.052
α_2 (s^{-1})	1.50×10^{13}	k (kJ/kmol)	6.10×10^4
k_1 (kJ kmol^{-1})	7.40×10^4	Mechanism factor	2
k_2 (kJ kmol^{-1})	2.50×10^5	Note: A and k may change for different coal types	

the particular case of the mass conservation equation, variable ϕ is set to unity and the right-hand side of the equation is zero.

Typical parameters and wall boundary conditions employed in the simulation are listed in Tables 2–5.

The governing equations are discretized over a staggered grid using the finite difference method and integrated over each control volume in the computational domain. A non-uniform spacing of control volumes was used, placing more of them in regions near the burners and using a sufficient number of control volumes to obtain grid independent solutions. Each of the equations described in the above general form is tridiagonal and can be solved using TDMA solvers. The equations are solved by performing iterations until the solution satisfies a preset accuracy and SIMPLER algorithm [27] of pressure correction is applied in the iteration.

4. Model results

The computational domain simulates from the bottom up to the roof of the furnace. In this study, the computation domain was discretized using $17 \times 71 \times 90$ grid nodes while the radiative heat transfer equation was solved on a coarser grid with $14 \times 10 \times 34$ control volumes.

Figs. 3–5 show the temperature, oxygen and NO$_x$ distribution in a vertical plane crossing the burners for Case 1 (100% boiler full load) and Case 4 (70% boiler full load), respectively. Comparison of the measured data (available for Cases 1, 4 and 5) with the calculated values are listed in Table 6. Tendency and agreement is found for all the cases except Case 5 (50% boiler full load). It is analysed that the models, especially the submodel of char combustion should be improved for an approved prediction of partial loads. The calculated results using the current models and

Table 5
Furnace wall boundary conditions

Wall	Side walls	Furnace bottom	Furnace exit	Furnace hopper	Platen superheaters
Emissivity	0.6	1	1	1	0.6
Wall temperature (K)	620	350	1000	350	620
Wall resistance	0	0	0	0	0

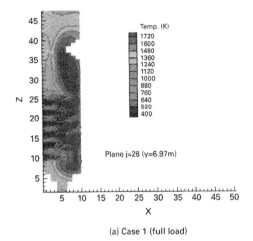

(a) Case 1 (full load)

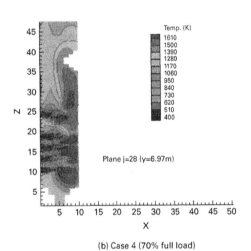

(b) Case 4 (70% full load)

Fig. 3. Temperature distribution of a section crossing burners.

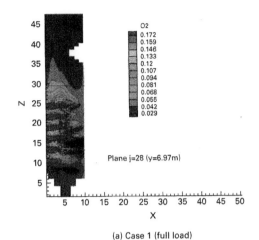

(a) Case 1 (full load)

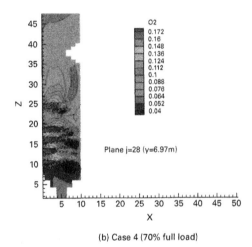

(b) Case 4 (70% full load)

Fig. 4. Oxygen distribution of a section crossing burners.

algorithm for higher loads are reasonable and can be applied.

A comparison of the calculated values for 100 and 70% boiler full load, without or with air leakage, is also listed in Table 6. It is shown that the furnace outlet temperature and total energy to walls are decreased with air leakage existing, while NO_x at the furnace outlet increased a little as expected. Better agreement between the measured values and the calculation results with air leakage than those without air leakage is found and this suggests that air leakage should be considered during boiler simulation. The calculated oxygen content is close to the experimental value. The calculated NO_x emission for full load

decreases while from 50 to 70% load increases as observed. The results for full load presented some peculiarities where the predicted exit gas temperature is lower than the 95% case and the total heat to the boiler walls is smaller. This might be due to the fact of considering a linear scaling of the input parameters between the 70 and 100% load.

The increase of the air flow rate led as expected to a reduction of the furnace outlet temperature and to an increase in NO_x emissions. This shows that the NO_x model used in the current study can predict correctly NO_x changes with different oxygen levels in the furnace.

The maximum observed heat fluxes were compared with the predicted incident heat fluxes. The incident heat fluxes

Table 6
Measured data/calculated results for different cases

Case no.	1	4	5	6	9
Boiler load (%)	100	70	50	100	70
Furnace outlet temperature (K)	1530/1308	1290/1210	1208/1112	1530/1292	1290/1205
Furnace outlet O_2 (%)	2.29/1.65	3.22/2.32	4.40/4.94	2.29/2.80	3.22/3.18
Furnace outlet NO_x (ppm)	419.6/335.6	426.3/399.2	303.6/375.6	419.6/353.5	426.3/397.7
Total energy to walls (kW)	305000/346134	259000/312048	162000/242999	305000/341619	259000/306372

calculated based on clean wall conditions are in agreement with the maximum values observed. The maximum values are hard to define and the average wall conditions do not correspond to clean walls. Therefore based on the comparison of the measured values with the maximum heat fluxes a

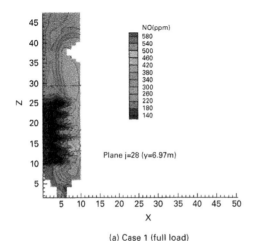

(a) Case 1 (full load)

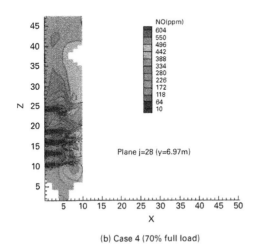

(b) Case 4 (70% full load)

Fig. 5. NO_x distribution of a section crossing burners.

heat transfer resistance at the wall was estimated as about 10 m^2 K/KW. Considering this value uniformly at the walls leads to incident heat fluxes that are in general higher than the measured maximum values. The incident heat flux increases as expected when increasing the heat transfer resistance and therefore the gas temperature. The difference is probably due to the fact that the maximum values observed never correspond to a completely clean wall condition.

The comparison of the absorbed heat fluxes is presented in Fig. 6 showing an acceptable agreement with the heat flux readings. The correlation of the measurements with the calculations is close. The calculated values show smaller variation with position but the deviations due to the experimental data is well within the experimental uncertainty for these data.

The comparison of the heat flux data allow the conclusion that the information from the heat transfer calculations is representative of the reality and therefore the interest on the use of the model results to the support of heat transfer degradation calculations.

5. Conclusions

Three-dimensional simulation of the flow and combustion process in the furnace of a front wall pulverized coal-fired utility boiler was presented in this study. The tendency and agreement of the measured furnace outlet temperature, oxygen, NO_x, total heat energy transferred to walls and superheaters with calculated values was found for all the cases except the one of very low boiler load. Comparisons between measured data with calculated results, with and without air leakage to boiler furnace, suggest that the air leakage should be considered during boiler simulation.

Acknowledgements

This work was financed by the Commission of the European Community through subprogram ACORDE of the BRITE/EURAM Program under Contract No. BRPR-CT96-0198. The authors would like to express their gratitude to ENDESA for the contribution to the experimental data. Partial support from PRAXIS XXI/BPD/16323/98(No. 439.01) of FCT (Fundação para a Ciência e a Tecnologia) of the Ministry of Science and Technology of Portugal to Dr Minghou Xu is gratefully appreciated.

184

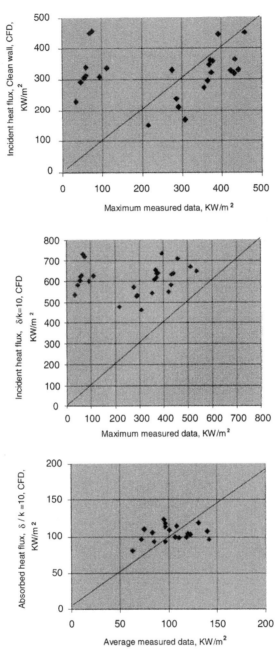

Fig. 6. Comparison of calculated heat flux with plant data.

References

[1] Boyd RK, Kent JH. Twenty-first Symposium (International) on Combustion. The Combustion Institute, Pittsburgh, 1986. p. 265–74.

[2] Robinson GF. J Inst Energy 1985;116–50.

[3] Görner K, Zinser W. ASME 107th Annual Meeting. Anaheim, California, USA, 1986.

[4] Lockwood FC, Papadopoulos C, Abbas AS. Combust Sci Technol 1988;58:5–24.

[5] Fiveland WA, Wessel RA. ASME J Engng Gas Turbines and Power 1988;110:117–26.

[6] Carvalho MG, Coelho P. Engng Comput 1991;7:227–324.

[7] Azevedo JLT, Coelho LMR, Carvalho MG. Combustion Related Organizations—Common and Unified Symposium. Salsomaggiore Terme, Italy, 1994. p. 11–7.

[8] Xu M, Yuan J, Ding S, Cao H. Comput Methods Appl Mech Engng 1998;155:369–80.

[9] Barlow SM. In: Durão et al, editors. Proceedings of International Symposium on Applied LDA to Fluid Mechanics. Lisbon, Portugal, 1982. p. 84.

[10] Costa M, Costen P, Lookwood FC. Combust Sci Technol 1991;75:129–54.

[11] Wall TF, Stewart I, McC. J Inst Fuel 1975;May:235–40.

[12] Bonin MP, Queiroz M. Combust Flame 1991;85:121–33.

[13] Butler BW, Webb BW. Fuel 1991;70:1457–64.

[14] Lockwood FC, Shah NG. Eighteenth Symposium (International) on Combustion. The Combustion Institute, Pittsburgh, 1981. p. 1405–14.

[15] Azevedo JLT, Carvalho MG. Second Int Conf on Combustion Technologies for a Clean Environment. Lisbon, Portugal, 1993.

[16] Coimbra CFM, Azevedo JLT, Carvalho MG. Fuel 1994;73:1128–34.

[17] Yuan JW, Xu MH, Ding SF, Cao HD. Proc Third Int Symposium on Combustion. Beijing, China, 1995. p. 234–41.

[18] Xu MH, Yuan JW, Ding SF, Cao HD. Proc Chinese Soc Electl Engng 1996;16:266–70.

[19] Carvalho MG, Durão DFG, Pereira JCF. Engng Comput—Int J Computer Aided Engng Software 1987;4:23–34.

[20] Launder BE, Spalding DB. Comput Methods Appl Mech Engng 1974;3:269–89.

[21] Coelho PJ, Carvalho MG. Int J Numer Methods Engng 1993;36:3401–19.

[22] De Soete GG. Fifteenth Symposium (International) on Combustion. The Combustion Institute, Pittsburgh, 1975. p. 1093–102.

[23] Chen W. A global reaction rate for nitric oxide reburning. PhD dissertation. Bringham Young University, 1994.

[24] Ubhayakar SK, Stickler DB, Rosenberg CWV, Gannon RE. Sixteenth Symposium (International) on Combustion. The Combustion Institute, Pittsburgh, 1977. p. 427–36.

[25] Kobayashi H, Howard JB, Sarofim AF. Sixteenth Symposium (International) on Combustion. The Combustion Institute, Pittsburgh, 1977. p. 411–24.

[26] Lockwood FC, Romo-Millares CA. J Institute Energy 1992;65:144–52.

[27] Patankar SV. Numerical heat transfer and fluid flow. New York: Hemisphere Publishing Corporation, 1980.

Emissions Reduction: NO_X/SO_X Suppression
A. Tomita (Editor)
© 2001 Elsevier science Ltd. All rights reserved

Mathematical modeling of fluidized bed combustion. 4: N_2O and NO_X emissions from the combustion of char

Z. Chen, Mu Lin, J. Ignowski, B. Kelly, T.M. Linjewile, P.K. Agarwal*

Department of Chemical and Petroleum Engineering, University of Wyoming, Laramie, WY 82071-3295, USA

Received 30 June 2000; accepted 4 December 2000

Abstract

Batch experiments were conducted to investigate the emissions of carbon oxides and nitrogen oxides from the combustion of char prepared from a commercial coal in a bubbling fluidized bed. Combustion gases escaping from the surface of the bed were sampled and analyzed for N_2O, NO, NO_2, CO and CO_2 as a function of time by means of a Fourier Transform Infrared Spectrometer. The experimental variables include char size and loading, inlet oxygen concentration, inlet CO concentration and bed temperature.

A single particle reaction–diffusion model was developed first. The detailed chemistry of NO and N_2O formation destruction is complex. Homogeneous and heterogeneous reactions considered most pertinent were included in an extension of the single particle char combustion model reported earlier. This single porous char particle model was integrated into a three-phase hydrodynamic description of the fluidized bed reactor. This system model for fluidized bed combustion was developed in response to the observation that most previous system models, including those that incorporate details of NO_X emissions, are based on the two-phase theory of fluidization. However, two-phase models are unable to predict the gas back-mixing and the recycle peak in solids-mixing. The non-isothermality of the bed resulting from the gas-phase reactions was taken into account through inclusion of an energy balance for the bubble phase. The effect of the variation in superficial gas velocity on bubble properties and cross-flow was included through an overall mass balance. Calculations from the system model, including details of homogeneous NO_X reactions far from the char particles, compared well with data on the emissions of CO, CO_2, NO and N_2O for various experimental conditions. The validated model was then used to investigate the influence of operating conditions on the conversions of char-nitrogen to NO and N_2O from the simultaneous combustion of char and propane in a fluidized bed with an air/propane mixture as the fluidizing medium. The most significant factors were found to be bed particle size and char diameter. © 2001 Elsevier Science Ltd. All rights reserved.

Keywords: Fluidized bed; Char combustion; NO_X and N_2O emissions

1. Introduction and background

Fluidized bed combustion (FBC) has emerged as an environmentally attractive method for burning coal, primarily due to the low operating temperature employed, usually between 1025 and 1200 K. The low operating temperature results in low NO_X and optimum conditions for SO_X removal with limestone or dolomite. Both SO_X and NO_X are of major environmental concern due to their contribution to acid rain. Moreover, NO_X is also known to play a role in the formation of photochemical smog. The low combustion temperature, however, enhances formation of N_2O ranging from 15 to 200 ppm in comparison to levels observed in pulverized coal combustion boilers at 5 ppm. It is estimated [1] that about 50% of US coal-derived power generation stems from FBC units. This raises concern over increased emission

levels of N_2O — a potent greenhouse gas and stratospheric ozone layer depletion agent.

During combustion of coal, oxides of nitrogen are formed mainly from the nitrogen present in the air (thermal NO_X) and that occurring in the coal. Formation of thermal NO_X is highly temperature dependent and is considered to be insignificant at FBC temperatures, where both NO and N_2O are derived almost entirely from coal nitrogen [2]. Combustion of coal involves a primary step of devolatilization, where the organically bound coal nitrogen is partitioned into volatile-nitrogen and char-nitrogen (char-N). In the volatiles stream, nitrogen is known to exist as NH_3, HCN and tar-nitrogen [3,4]. In the char, nitrogen is bound in aromatic structures [5]. The subsequent step of oxidation results in the conversion of volatile- and char-N into NO, N_2O and N_2.

The factors affecting the production of N_2O and NO in FBC can be broadly categorized into those originating from the fuel type, and those resulting from the combustor design

---

* Corresponding author. Tel.: +1-307-766-4259; fax: +1-307-766-6777.
E-mail address: pagarwal@uwyo.edu (P.K. Agarwal).

Reprinted from *Fuel* **80 (9)**, 1259-1272 (2001)

Nomenclature

a	particle radius (m)
b	outer film radius (m)
C	gas concentration (kmol m^{-3})
d	diameter (m)
D	diffusivity (m^2 s^{-1})
F_P	specific surface area (m^{-1})
h	height above the distributor (m)
K	rate constant (s^{-1})
N/C	initial nitrogen to carbon ratio in the char
r	radial position (m)
R	reaction rate (kmol m^{-3} s^{-1})
T	temperature (K)
U_0	gas velocity (m s^{-1})
U_{mf}	minimum fluidization velocity (m s^{-1})
W_{Char}	nominal char loading (g)
x	char fractional conversion

Greek symbols

α/β	primary CO/CO$_2$ product ratio
γ	adjustable parameter
υ, ζ	stoichiometric coefficients

Subscripts

aver	average
A–J	reaction index
Bulk	bulk phase
HET	heterogeneous reaction
HOM	homogeneous reaction
i	ith gas component
inlet	at the inlet
P	particle

and operating conditions. A review of the literature [1,2,6–10] identifies these factors as the bed temperature, fuel type, excess air, limestone feed, pressure, air staging and boiler load. Except for bed temperature, the effects of these operating conditions on N$_2$O and NO$_x$ emissions in FBC are often contradictory. The contradictions are believed to be due to the different FBC technologies employed — bubbling, circulating, pressurized, as well as combustor size (laboratory to commercial scales) and the complex chemistry of formation and destruction of N$_2$O and NO.

Bed temperature is undoubtedly the single most important parameter affecting the rates of formation and destruction of N$_2$O and NO. Most studies [11–14] report that N$_2$O decreases with an increase in bed temperature, while NO exhibits an inverse effect. On the other hand, Gulyurtlu et al. [15] found a distinct maximum in N$_2$O emissions as the bed temperature was increased from 1000 to 1200 K; the excess air levels as well as the char preparation temperature were found to influence the peak emissions of N$_2$O. It is known that during FBC of coal, the char particles burn at temperatures well above the bed temperature by up to 400 K [16,17]. The actual particle temperature, more relevant in understanding temperature-dependent processes close to the burning char particle, depends on operating parameters such as the sizes of bed material and char, mass of char added (or mass flow rate of char) and the superficial gas velocity (excess air) [17]. Clearly, there is need to interpret and relate the measurements of N$_2$O/NO emissions in terms of these parameters.

The release of NO and N$_2$O also depends on the properties of the coal used; most results in the literature relate emissions with coal rank. N$_2$O emissions are reported to increase with an increase in coal rank from brown coal to anthracite [9]. Char plays a role in the heterogeneous decomposition of both N$_2$O and NO, and low-rank coal chars lead to higher levels of N$_2$O decomposition. This is a consequence, perhaps, of the presence of mineral matter (which catalyzes the decomposition reaction), the higher intrinsic reactivity of the char, or the higher volatile and moisture fractions, which may participate in the homogeneous reduction of N$_2$O. Volatile matter also produces hydrocarbon radicals, which may take part in the homogeneous reduction of N$_2$O. A very significant influence of coal type is its nitrogen content and the manner in which it is partitioned into volatile- and char-N. Two experimental methods have been employed to resolve the conflicting accounts of the contributions of volatile- and char-N to emissions.

Continuous feeding of coal/char in a fluidized bed and observing the difference in concentrations of NO and N$_2$O obtained from the two combustion experiments [18,19]. In this method, it is difficult to account for interaction between volatile species and char. The presence of char particles may result in secondary reactions with N$_2$O and NO produced homogeneously. Secondly, the char prepared separately by pyrolysis may retain different levels of nitrogen to that produced during combustion of coal. Char produced during combustion of coal will experience higher devolatilization temperatures due to the combustion of volatiles in its vicinity.

Combustion of a small batch or a single coal particle while the concentrations of NO and N$_2$O are monitored continuously [13,20]. In these methods, the problem of variability in nitrogen retention in the chars is eliminated. However, the problem of interaction between oxides of nitrogen and the resultant chars remains. In addition, during the devolatilization stage, coal particles tend to float and burn at the bed surface [8], where heat and mass transfer conditions are entirely different from inside the bed. This behavior, which can also be experienced by unrestrained single particles, will definitely influence the formation and destruction of NO and N$_2$O.

In both methods, the temperature of the burning coal and char particles, which would certainly influence reaction rates, is difficult to determine. These issues, perhaps, are responsible for the discrepancies reported in the literature. Tullin et al. [13] report that the contribution to the formation of NO and N$_2$O from char-N is higher. Other researchers [18–20] indicate a greater contribution from volatile-nitrogen.

In this paper, the results of an experimental and modeling investigation on the emissions of carbon oxides and nitrogen oxides from the combustion of char in a bubbling fluidized bed are reported. Batch combustion experiments were conducted using char prepared from a commercial coal in a laboratory-scale bubbling fluidized bed. The experimental variables include char particle size and loading, inlet oxygen concentration, inlet CO concentration and bed temperature. The concentrations of off-gases were measured as a function of time. Admittedly, this approach does not shed any light on the relative contributions of volatile- and char-N to the formation of NO and N_2O during coal combustion. Conversely, operating variables which influence particle temperature — and hence, the rates of formation and destruction of NO and N_2O — are examined in what appears to be greater detail than that reported in the literature.

In terms of mathematical modeling, a single particle reaction–diffusion model was developed first. The detailed chemistry of NO and N_2O formation/destruction is complex [10]. Homogeneous and heterogeneous reactions considered most pertinent were included in an extension of the single particle char combustion model reported earlier [21]. The approach is similar to that adopted by Goel et al. [22–26]. However, there are several significant differences. The homogeneous reactions of CO in the film surrounding the char particle — oxidation, reactions with NO and N_2O — are included. The heterogeneous reactions considered — besides (a) oxidation of carbon to produce CO and CO_2; (b) oxidation of char-N to produce NO and N_2O; and (c) the reduction of NO and N_2O on the char (internal) surface — include the gasification of carbon with CO_2. The inclusion of the gasification reaction is considered important for the larger particle sizes of interest in FBC [21,27]. Also, inclusion of the reactions for the formation (and destruction through oxidation) of CO can exert a significant influence on the predicted NO and N_2O emissions [28]. Finally, linearization of the film CO oxidation reaction rate permits analytical solution for the concentration profiles of various gaseous species. This approach, unlike conventional boundary layer models for char combustion, is able to predict [21] the two extrema in CO/CO_2 ratio as a function of particle temperature observed experimentally [17,29].

This single porous char particle model was integrated into a three-phase hydrodynamic description of the fluidized bed reactor. This system model for FBC was developed [30–32] in response to the observation that most previous system models [16,33,34], including those that incorporate details of NO_x emissions [28], are based on the two-phase theory of fluidization. However, two-phase models are unable to predict the gas back-mixing [35] and the recycle peak in solids-mixing [36]. In previous papers of this series [30–32], calculations from this FBC model have been compared with experimental data on the combustion of (a) char, (b) propane, and (c) simultaneous combustion of char and propane. The agreement, in all cases, was found to be very good. Particularly encouraging was the agreement with data [37] on the variation of CO, CO_2 and O_2 concentrations in the freeboard with time resulting from the batch combustion of char in sand fluidized by air/2.5% (volume) propane. This agreement suggested that the model provided a good representation of the dynamics of the combustion of char as well as combustible gases. In this paper, the calculations from the system model, including details of homogeneous NO_x reactions far from the char particles, are compared with data on the emissions of CO, CO_2, NO and N_2O for various experimental conditions. Note that of all FBC models that incorporate NO_x chemistry, very few [23,38] address N_2O emissions [7]. Finally, the system model is used to explore the influence of operating conditions and the addition of hydrocarbon gases in the fluidizing stream on the abatement of NO and N_2O emissions.

2. Experimental

Batch combustion experiments were conducted in a bubbling fluidized bed. The experimental set up consisted of a fluidized bed combustor fabricated from a stainless steel cylinder, 127 mm ID. The bed was heated by means of two semi-cylindrical electric heaters rated at 7.4 kW. Bed temperature was maintained at a constant value by a PID temperature controller using a type K thermocouple located at 3 cm above the distributor plate. Heat loss was kept to a minimum by wrapping a thick layer of ceramic blanket insulation around the FBC column. The fluidized bed vessel was loaded with 2 kg of sand. Metered fluidizing gas flowed through a pre-heater and a calming section before its introduction into the FBC through a distributor grid. In order to eliminate the potential for formation of thermal NO_x, the fluidizing gas consisted of a mixture of oxygen in argon.

A sub-bituminous coal from the Skull coal mine in Kemmerer, Wyoming was used in this investigation. Analysis of the coal is presented in Table 1. Sample preparation for the experiments involved crushing the coal and sieving to collect the desired size cuts. The crushed coal was then pyrolysed at 1123 K for 30 min in helium flow.

A probe was used for sampling gases in the experiments;

Table 1
Analysis of the coal sample

Skull coal (FMC)	Coal	Char
Proximate analysis (Wt%; as received)		
Moisture	4.9	1.0
Volatile matter	39.5	5.3
Fixed carbon	53.2	89.3
Ash	2.4	4.4
Ultimate analysis (Wt%; dry basis)		
Carbon	75.7	91.9
Hydrogen	5.1	0.1
Oxygen	14.3	1.0
Nitrogen	1.3	1.7
Sulfur	1.2	0.8

Table 2
Experimental conditions

Coal type	Sub-bituminous
Sand size (μm)	200
Char size (mm)	3.0, 4.5
Fluidization velocity (m s^{-1})	0.071–0.095
Bed temperature (K)	873–1173
Oxygen concentration (%)	10, 21

the probe tip was located at the bed surface and the gas velocities were maintained at values greater than minimum fluidization. The probe exit end was connected to an ice cooler to condense out water vapor present in the gas sample. Gas samples, drawn from the combustor by a vacuum pump, were collected from the discharge end in small stainless steel cylinders for off-line analysis using a Fourier Transform Infrared Spectrometer with a narrow band MCT detector. The instrument was calibrated for N_2O, NO, NO_2, CO and CO_2. The detailed experimental protocol may be found elsewhere [39]; a summary of the experimental conditions is presented in Table 2.

3. Mathematical modeling

3.1. Single particle model

Consider a porous spherical char particle exposed to a combustion environment. This particle is assumed to be isothermal; the boundary layer surrounding the particle is also isothermal and at the same temperature as the particle. The complex chemistry of the formation and destruction of NO and N_2O, despite considerable research efforts [1,2,7,8,40], is not completely understood. Simplifications are necessary to keep the problem tractable. These simplifications — relating to the heterogeneous reactions of char with O_2, CO_2, NO and N_2O within the porous particle, and the homogeneous reactions involving CO, O_2, NO and N_2O in the boundary layer — are described in the following sections.

3.1.1. Heterogeneous reactions

3.1.1.1. Char oxidation. CO and CO_2 are the primary products of the char combustion reaction according to

$$\frac{2(\alpha + \beta)}{\alpha + 2\beta} C_{Char} + O_2 \rightarrow \frac{2\alpha}{\alpha + 2\beta} CO + \frac{2\beta}{\alpha + 2\beta} CO_2$$

(Reaction A)

where α/β is the primary CO/CO_2 product ratio given by

$$\alpha/\beta = A_1 \exp(-A_2/T_p). \quad (1)$$

A_1 and A_2 are constants that may depend on the type of char, and T_P is the temperature of the particle. The oxidation

reaction, which can take place within the porous particle, is assumed to be first order with respect to concentration of oxygen. The volumetric (per unit particle volume) rate of oxygen consumption, adapted from Field et al. [41], is written as

$$R_A = K_A C_{O_2} \quad \text{with} \quad K_A = 250 T_P \exp(-19\,000/T_P) F_P, \quad (2)$$

where F_P is the specific internal surface area of the char.

3.1.1.2. Char gasification. The char gasification reaction is assumed to be first order with respect to concentration of carbon dioxide. The volumetric rate of CO_2 consumption, adapting the rate expression of Dutta et al. [42], is taken as

$$C_{Char} + CO_2 \rightarrow 2CO \quad \text{(Reaction B)}$$

$$R_B = K_B C_{CO_2} \quad \text{with} \quad K_B = 4364 \exp(-29\,844/T_P) F_P. \quad (3)$$

3.1.1.3. Formation of NO. As proposed by De Soete et al. [43], char bound nitrogen is first oxidized to NO by a reaction which is first order with respect to oxygen concentration

$$C_{Char}-N_\gamma + \frac{\gamma + 1}{2} O_2 \rightarrow \gamma NO + C_{Char}-O \quad \text{(Reaction C)}$$

The rate of this reaction is assumed to be proportional to the rate of char combustion with the constant of proportionality being defined as γ. Thus the rate of NO formation is

$$R_C = K_C C_{O_2} \quad \text{with} \quad K_C = \frac{2(\alpha + \beta)}{\alpha + 2\beta} \gamma K_A. \quad (4a)$$

It can be shown that γ is essentially the same as the instantaneous N/C ratio of the product gases (Appendix A). On an average, every char-N atom is expected to be linked to one carbon atom suggesting that $\gamma \approx 1$. Croiset et al. [44], assuming that the (N/C) ratio remains constant during combustion, calculated $\gamma \approx 0.984$ for Westerholt char. On the other hand, recent experimental data show that (N/C) changes as a function of temperature, char conversion [45] and the type of char [46]. In particular, for a high-volatile bituminous coal, the instantaneous N/C ratio in the gas phase was very similar to the initial (N/C) ratio in the particles at 1173 K; however, at 873 K, the (N/C) ratio of the product gases increased with the fractional conversion of char [45].

A tentative functionality for γ is proposed here

$$\gamma = (cx + d)(N/C)_{initial}. \quad (4b)$$

c and d are functions of the particle temperature. Also, mass balance for char-N requires that

$$\int_0^1 \gamma \, dx = (N/C)_{initial}. \quad (4c)$$

Thus, $d = 1 - c/2$. Limiting values of c and d can be determined. For low temperatures — 873 K in the case of a bituminous coal used in Ref. [45] — $d = 0$ and $c = 2$; for high temperatures, $d = 1$ and $c = 0$. It is expected that the values of c and d should depend on the type of char. For the present calculations, it was assumed that $\gamma = (N/C)_{\text{initial}}$. The effect of γ on the shapes of NO and N_2O emissions is shown in Appendix A.

3.1.1.4. Formation of N_2O. NO further reacts with bound nitrogen to produce N_2O according to the following heterogeneous reaction [43,44,47]:

$$C_{\text{Char}}\text{-N} + NO \rightarrow N_2O + C_{\text{Char}} \qquad \text{(Reaction D)}.$$

The rate of N_2O formation is also proportional to γ. The modified first order rate expression is

$$R_D = K_D C_{\text{NO}} \qquad \text{with} \qquad K_D = \gamma k_0 F_P \exp(-9000/T_P). \tag{5}$$

k_0 is an adjustable parameter; $k_0 = 3.0$ was found to provide good agreement between the measured and calculated N_2O concentrations in the present work. It should be noted that N_2O formation mechanism remains a matter of debate. Contrary to the approach adopted here, Winter et al. [48,49] believe that HCN is the primary product of the combustion of char-N. HCN, then, oxidizes and reacts homogeneously with NO to form N_2O.

3.1.1.5. Reduction of NO by char. A number of mechanisms have been proposed for the reduction of NO on the surface of char [10,43,50]. The mechanism proposed by Chan et al. [50] assumes reaction between NO and a single carbon site to form chemisorbed oxygen and the release of nitrogen. Johnsson [10] introduced an intermediate step; the intermediate form decomposed to the final products. De Soete et al. [43] proposed a similar mechanism but two adjacent carbon sites were involved. These mechanisms result in release of N_2 and the chemisorbed oxygen as CO.

$$NO + C_{\text{Char}} \rightarrow \tfrac{1}{2}N_2 + CO \qquad \text{(Reaction E)}.$$

Extensive reviews on the NO–carbon reaction are available in the literature [51,52]. Assuming first order reaction with respect to NO concentration, the activation energy for the NO–carbon reaction was found ranging from 45 to 245 kJ/mol [51]. The presence of CO in the gas phase is known to enhance the rate of NO reduction [53]. This may occur through reaction of CO with surface-bound oxygen atoms, leading to regeneration of surface sites for reaction with NO [50,52]. Alternatively, this reaction has also been considered as a part of the heterogeneous reaction between NO and carbon in the presence of CO — thus, CO acts as an oxygen scavenger from the surface of carbon leaving behind a free active carbon site. This reaction can be written as

$$NO + CO \rightarrow \tfrac{1}{2}N_2 + CO_2 \qquad \text{(Reaction F)}.$$

In the present work, two approaches were tested. In the first approach, the reactions of NO with char and with CO were considered separately. The NO–CO reaction was assumed to be first order with respect to NO concentration,

and zero order with respect to CO concentration [53]. Li et al. [54] found that the activation energy for the NO–carbon reaction in the absence of CO was 52 and 118 kJ/mol for the low-temperature regime (773–973 K) and for the high-temperature regime (973–1173 K), respectively. These activation energies were adopted in the simulations. The pre-exponential factors for this reaction were adjusted in order to find a best fit between the predicted NO concentration with experiment, assuming the reaction rate at 973 K with two activation energies was the same. This results in the following volumetric reaction rate for the NO–char reaction:

$$R_E = K_E C_{\text{NO}} \qquad \text{with}$$

$$K_E = \begin{cases} 0.159 F_P \exp(-6255/T_P) & T_P \leq 973 \text{ K}, \\ 555.6 F_P \exp(-14\,193/T_P) & T_P > 973 \text{ K}. \end{cases} \tag{6}$$

The kinetic parameters for the NO–CO reaction in the presence of char particles were taken from Ref. [53],

$$R_F = K_F C_{\text{NO}} \quad \text{with} \quad K_F = 5.67 \times 10^3 T \exp(-13952/T). \tag{7}$$

Simulations using Eqs. (6) and (7) showed that for higher bed temperatures this approach provided good agreement between the calculations and experimental data; for lower bed temperatures, however, the predicted NO concentration was significantly higher than the measurements. This may be because the NO–char reaction is not first order in the low-temperature regime [53].

In the second approach, an overall reaction scheme was adopted; a zeroth reaction order with respect to CO concentration was still taken. This is based on the notion that the two-regime character for the NO–char reaction vanishes [53] in the presence of CO. Activation energy of 60 kJ/mol was used in the calculations. The overall volumetric reaction rate is written as

$$R_E = K_E C_{\text{NO}} \qquad \text{with}$$

$$K_E = 1.0 \times 10^{-3} F_P T_P \exp(-7200/T_P). \tag{8}$$

It will be seen in the following that this approach provides good agreement between the simulations and experimental data for different bed temperatures. It should be pointed out that for higher temperatures, Eq. (8) is in the range of the reaction rates reported in Ref. [52]. For lower temperatures, the reaction rates are considerably higher than those reported in the literature. This may be because the reaction rate is dependent on the type of char — different chars have significantly different NO–carbon reaction rates [52]. On the other hand, the kinetic parameters for the NO–carbon reaction in the presence of CO reported in the literature were often obtained by pre-mixing NO with CO. Reaction occurs with these gaseous reactants diffusing into the char. This is very different from the scenario encountered in this work, where NO and CO are generated within the char particle. Clearly, more research efforts are necessary to further understand the NO–carbon reaction in the presence of CO.

3.1.1.6. Reduction of N_2O by char. This reaction, observed to be faster than that between char and NO, involves several steps. N_2O reacts with carbon to form N_2 and a carbon–oxygen surface complex. This surface complex can, subsequently, react with N_2O to form nitrogen and carbon dioxide [55]. Alternatively, the chemisorbed oxygen may be released as CO. In principle, then, CO, CO_2 and N_2 are the products of reaction with production of CO being favored at higher temperatures [51]. For the present model, the following global reaction has been adopted:

$$N_2O + C_{Char} \rightarrow N_2 + CO \qquad \text{(Reaction G)}$$

The reaction rate, assumed to be first order with respect to N_2O concentration, is written as

$$R_G = K_G C_{N_2O} \qquad \text{with}$$
$$K_G = 13.36 F_P \exp(-16\,677/T_P). \tag{9}$$

3.1.2. Homogeneous reactions

The very many homogeneous reactions (involving gaseous species and radicals) that can occur have been discussed in several reviews [7,10,40,56]. In this work, only the overall reactions between O_2, CO, NO and N_2O have been included. Further, these reactions are assumed to take place only in the film and not in the pores of the char.

3.1.2.1. Homogenous oxidation of carbon monoxide. The reaction rate has been found to depend on the concentrations of moisture, oxygen and carbon monoxide. In order to keep the model solution analytical, the reaction rate was assumed to be first order with respect to concentration of carbon monoxide. The volumetric rate of carbon monoxide consumption, modifying the rate expression of Hautman et al. [57], is expressed as

$$CO + \tfrac{1}{2}O_2 \rightarrow CO_2 \qquad \text{(Reaction H)}$$

$$R_H = K_H C_{CO} \qquad \text{with}$$
$$K_H = 2.23 \times 10^{12} \phi (C_{O_2,Bulk}/2)^{0.25} C_{H_2O,Bulk}^{0.5} \exp(-20130/T), \tag{10}$$

where $C_{O_2,Bulk}$ and $C_{H_2O,Bulk}$ are the bulk concentrations of oxygen and moisture, respectively; ϕ is an adjustable parameter defined as $\phi = 7.93 \exp(-2.48\phi_0)$, where ϕ_0 is the initial equivalence ratio, taken as the ratio of the volumetrically averaged concentrations of CO and oxygen in the film. In the bubble phase, ϕ_0 is simply the ratio of CO to oxygen concentrations.

3.1.2.2. Homogeneous destruction of N_2O. Homogeneous

destruction of N_2O in conditions of interest in FBC can occur through reactions involving

(a) H and OH radicals [56];
(b) oxygen radicals;
(c) carbon monoxide in the presence of water [58,59].

According to Glarborg et al. [60], the rate of reduction by OH radicals has been overestimated by an order of magnitude. For experiments based on the combustion of char, reduction of N_2O by CO in the presence of water appears more likely [38,58–60]. The overall reaction is written as

$$N_2O + CO \rightarrow N_2 + CO_2 \qquad \text{(Reaction I)}$$

Note that concentration of CO is expected to be comparatively high, and that N_2O is likely to be the rate limiting species. The rate of this reaction is reported by Loirat et al. [61]to be

$$R_I = K_I C_{N_2O} \qquad \text{with}$$
$$K_I = 2.51 \times 10^{11} \exp(-23\,180/T) C_{CO,aver}, \tag{11}$$

where $C_{CO,aver}$ is the volumetrically averaged concentration of CO in the film; for the bubble phase, it is CO concentration.

3.1.2.3. Thermal decomposition of N_2O. The final reaction considered is the thermal decomposition of N_2O

$$N_2O \rightarrow N_2 + \tfrac{1}{2}O_2 \qquad \text{(Reaction J)}$$

The following reaction rate reported by Bonn et al. [62]is used:

$$R_J = K_J C_{N_2O} \quad \text{with} \quad K_J = 1.75 \times 10^8 \exp(-23\,800/T). \tag{12}$$

3.1.3. Reaction–diffusion model formulation

The pseudo-steady-state species balances for the gases — O_2, CO, CO_2, NO, N_2O — following the reaction–diffusion approach can be written in a generalized form as

$$\frac{D_P}{r^2} \frac{d}{dr}\left(r^2 \frac{dC_{i,P}}{dr} \right) = \Sigma v_i R_{i,HET} \quad \text{for } 0 \leq r \leq a, \tag{13a}$$

$$\frac{D_{Film}}{r^2} \frac{d}{dr}\left(r^2 \frac{dC_{i,Film}}{dr} \right) = \Sigma \zeta_i R_{i,HOM} \quad \text{for } a \leq r \leq b. \tag{13b}$$

Subscript P refers to the particle. D_P is the effective diffusivity, assumed to be the same for all gases, within the particle and D_{Film} is the molecular diffusivity of gases in the film. v_i and ζ_i are stoichiometric coefficients which convert the volumetric rates of consumption/formation to the rate of consumption for the ith component. The relevant boundary conditions are

$$\frac{dC_{i,P}}{dr} = 0 \text{ at } r = 0 \text{ and } C_{i,\text{Film}} = C_{i,B} \text{ at } r = b, \quad (14a)$$

$$C_{i,P} = C_{i,\text{Film}} \text{ and } D_P \frac{dC_{i,P}}{dr} = D_{\text{Film}} \frac{dC_{i,\text{Film}}}{dr} \text{ at } r = a.$$
$$(14b)$$

Eqs. (13a) and (13b) can be solved analytically if $R_{i,\text{HET}}$ and $R_{i,\text{HOM}}$ are first order with respect to $C_{i,P}$ and $C_{i,\text{Film}}$, respectively. There are ten conservation equations, one for each species in the film and within the porous particle. Integration leads to twenty unknowns, which must be evaluated from the boundary conditions. Ten unknowns were found explicitly using the boundary conditions at the particle center and at the outer edge of the boundary layer. Determination of the remaining ten unknowns required simultaneous solution of the equations using the boundary conditions at the surface of the particle. Analytical solutions were obtained for the concentration profiles of the gas species within the particle and in the films; detailed procedure is given in Appendix B.

3.2. Fluidized bed reactor model

The hydrodynamic description of the bubbling fluidized bed, as in our previous work [30–32], is based on the three-phase model. The non-isothermality of the bed resulting from the gas-phase combustion reactions is taken into account through inclusion of an energy balance for the bubble phase. The effect of the variation in superficial gas velocity on bubble properties and cross-flow is included through an overall mass balance.

The model has two specific features, which make it especially suitable to investigate the operating conditions, which minimize N_2O emissions. Three possible methods have been discussed in the literature [7] for reduction of N_2O emissions from FBC:

1. Operating temperature: N_2O emissions are reduced significantly as the bed temperature is increased to about 1173 K. However, such temperatures lead to a decrease in the efficiency of SO_2 removal and an increase in the NO emission. In addition, transformations of inorganic species may lead to severe cyclone fouling, agglomeration of bed material and defluidization.
2. Fuel gas afterburning [63,64]: Injection of secondary fuel and its combustion increase the temperature and decrease emissions of N_2O.
3. Selective catalytic or non-catalytic reduction: Injection of additives, such as ammonia, with or without catalysts can be used to convert NO_X to nitrogen.

In this FBC model, unlike many others reported in the literature, energy balance for the char particles is included. Thus, the influence of operating conditions on particle temperature is calculated. As discussed earlier, particle temperature is considered to be more significant than bed temperature for heterogeneous destruction reactions. The second feature of importance, in the context of evaluating the efficacy of fuel gas injection as an N_2O control strategy, is the ability to handle gas phase combustion reactions in the bed and in the freeboard.

4. Results and discussion

In this section, results from the measurements and the model predictions are presented first. Symbols are used to indicate the measured concentrations; curves are used to represent the model calculations. The simulations were performed using the amount of char calculated from the total emissions of CO and CO_2 from the fluidized bed; this treatment then takes into account unburned char (ash) remaining in the bed and char loss due to elutriation of fine char particles from the bed.

4.1. Carbon combustion

The kinetic parameters for carbon combustion were determined first through comparison between model predictions and experimental measurements for CO_2 and CO emissions; these parameters are listed in Section 3.1. Besides the kinetic parameters listed above, there are still some other parameters, which require specification. The primary CO/CO_2 product ratio also varies with the type of char; it was calculated using $\alpha/\beta = 750 \, exp(-7200/T_P)$, which is substantially in agreement with that recommended by Rajan and Wen [65]. The specific internal surface area of char was found to depend on the type of char [66], ranging from 3.0×10^5 m^{-1} to 5.0×10^8 m^{-1}; it was taken as 5.0×10^6 m^{-1} in the calculations. The initial porosity of char is expected [66] to be in the range 0.12–0.21 and was assumed to be 0.20 in the simulations.

Fig. 1 shows comparison between calculations and experimental data for CO_2 emissions at different operating conditions. It is clear that the agreement for all the conditions is favorable. Comparison between data and calculations for CO emissions must be considered with greater care. The oxidation of carbon monoxide, found to take place only in the bubble phase and in the freeboard [31], is favored at higher temperatures. As a result, CO concentration can decrease rapidly as the sampling position moves from the fluidized bed to the freeboard at higher bed temperatures. Small variations in the sampling position, as well as bed surface fluctuations, may lead to significant variations in the measured CO concentration. This is borne out by the results shown in Fig. 2. At a bed temperature of 873 K, CO oxidation is virtually negligible; model predictions, Fig. 2(a), are in very good agreement with the measurements. Fig. 2(b,c) shows the results at higher bed temperatures; calculations at three positions (below, at and above the bed surface) are included. It can be seen that, for higher bed temperatures, CO concentrations vary significantly with location. The measured CO concentrations are

194

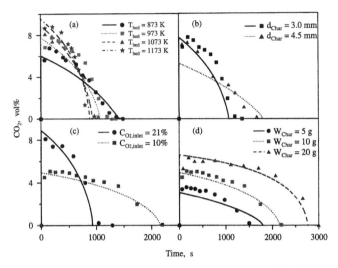

Fig. 1. Comparison of measured and predicted CO_2 concentrations for different conditions: (a) $U_0 = 0.071$, 0.079, 0.087 and 0.095 m s^{-1} (for different T_{bed}), $C_{O_2,inlet} = 21\%$, $d_{Char} = 3.0$ mm, $W_{Char} = 10$ g; (b) $T_{bed} = 973$ K, $U_0 = 0.079$m s^{-1}, $C_{O_2,inlet} = 21\%$, $W_{Char} = 10$ g; (c) $T_{bed} = 1073$ K, $U_0 = 0.087$ m s^{-1}, $d_{Char} = 3.0$ mm, $W_{Char} = 10$g; (d) $T_{bed} = 1073$ K, $U_0 = 0.087$ m s^{-1}, $C_{O_2,inlet} = 10\%$, $d_{Char} = 3.0$ mm.

in the range defined by calculations just below and just above the bed surface.

4.2. N_2O and NO emissions

4.2.1. Effect of char particle size

Measurements of concentrations as functions of time for two char particle sizes — 3.0 and 4.5 mm — are presented in Fig. 3. The 3.0 mm particles took about 1200 s to burn off, whereas the larger, 4.5 mm, particles took about 1700 s. The maximum concentration of N_2O in the sample gas was approximately 15 ppm for the 3.0 mm char particles, while the larger char particles (4.5 mm) produced a peak emission of only 9 ppm. A similar pattern of emission behavior was observed for NO. In this case, a maximum of 170 ppm was observed for the 3.0 mm char particles in comparison to the peak concentration of approximately 120 ppm for the 4.5 mm char particles. Comparison of these results with these for CO_2 emission shown in Fig. 1 establishes the strong link of NO_X emission with char combustion behavior. For the experimental conditions under consideration, combustion is under external mass-transfer controlled conditions. Experimental data from Prins et al. [67]indicate that the mass transfer coefficient decreases with an increase in active (char) particle size and the effect is more pronounced for larger active particle sizes. Based on this argument, it is expected that the peaks in NO_X emission will be increasingly more pronounced as char size decreases for coarse bed particles at bed temperatures of practical interest in FBC.

4.2.2. Effect of bed temperature

The effect of bed temperature on the emissions of N_2O and NO_X from FBC of coal is well documented [7,9]. A range of bed temperatures — 873, 973, 1073 and 1173 K — was investigated. Model calculations are in reasonable agreement with the measurements, Fig. 4. The experimental concentration–time data were integrated to determine total species production as a function of the bed temperature. Total N_2O emission passes through a maximum at an intermediate bed temperature, while total NO increases monotonically as the bed temperature increases. Clearly, at higher bed temperatures, the destruction of N_2O outweighs its formation. From the model calculations, it was found that heterogeneous reactions are most important in the destruction of NO and N_2O in the range of temperature under consideration. Homogeneous reactions become increasingly significant as the bed temperature increases.

4.2.3. Effect of inlet oxygen concentration

Measurements and model predictions for two different inlet oxygen concentrations — 10 and 21% — are presented in Fig. 5. It can be seen that the model predictions are in excellent agreement with the experiment for both cases. Tullin et al. [13]found that below 1073 K, fractional conversion of char-N to NO increased and that to N_2O decreased with increasing temperature. However, an increase in oxygen concentration, with an accompanying increase in particle temperature, leads to higher N_2O and lower NO fractions. In our experiments, the total N_2O emission increases with increase in oxygen concentration (1.875% of char-N for 21% oxygen concentration and

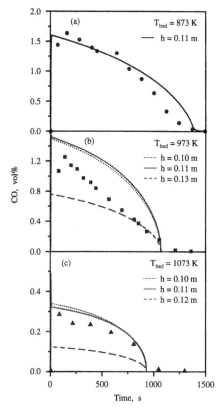

Fig. 2. Comparison of measured and predicted CO concentrations: $U_0 = 0.071, 0.079$ and 0.087 m s^{-1} (for different bed temperatures), $C_{O_2,inlet} = 21\%$, $d_{Char} = 3.0$ mm, $W_{Char} = 10$ g.

1.14% of char-N for 10% oxygen concentration). The conversion of char-N to NO, at about 15%, remains relatively unaffected by oxygen concentration. The higher conversion to N_2O is thought to be consequence of higher instantaneous NO concentration in accord with Reaction D in Section 3.1.

4.2.4. Effect of injection of CO in the inlet gas

CO has been considered as one of the important factors in the homogeneous destruction of both N_2O and NO as discussed in Section 3.1. Experiments were conducted with 500 ppm of CO injected into the bed at 873 K. The low bed temperature was chosen to prevent oxidation of CO thereby increasing the possibility of reaction with oxides of nitrogen. For comparison, experiment without CO injection was also performed at otherwise identical conditions. Results, not shown here, indicate that the effect on the emissions of NO and N_2O is virtually negligible; the model predicts essentially the same results. Experiments were also conducted at a bed temperature of 973 K. Once again, the effect on NO$_X$ emissions was negligible. Clearly,

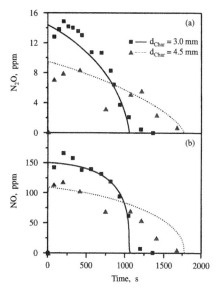

Fig. 3. Influence of char particle size on N_2O and NO emissions: $T_{bed} = 973$ K, $U_0 = 0.079$ m s^{-1}, $C_{O_2,inlet} = 21\%$, $W_{Char} = 10$ g.

additional experiments are required to establish the importance of the homogeneous destruction of NO$_X$. It is possible that the CO injected may not have been sufficient in quantity to cause an appreciable effect. Hayhurst and Lawrence [68] found that the presence of a small amount of iron enhanced the reduction of nitrogen oxides with CO. A balance

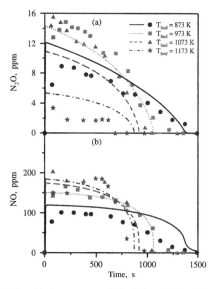

Fig. 4. Effect of bed temperature on N_2O and NO emissions: $U_0 = 0.071, 0.079, 0.087$ and 0.095 m s^{-1} (for different T_{bed}), $C_{O_2,inlet} = 21\%$, $d_{Char} = 3.0$ mm, $W_{Char} = 10$ g.

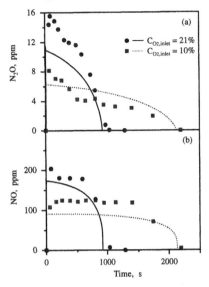

Fig. 5. Influence of O_2 concentration on N_2O and NO emissions: $T_{bed} = 1073$ K, $U_0 = 0.087$ m s^{-1}, $d_{Char} = 3.0$ mm, $W_{Char} = 10$ g.

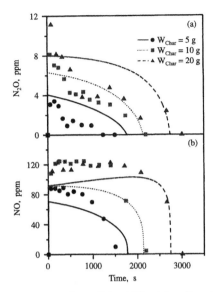

Fig. 6. Effect of char loading on N_2O and NO emissions: $T_{bed} = 1073$ K, $U_0 = 0.087$ m s^{-1}, $C_{O_2,inlet} = 10\%$, $d_{Char} = 3.0$ mm.

between moisture content (which catalyses oxidation of CO), bed temperature, catalyst and quantity of CO injected may be necessary for optimum results.

4.2.5. Effect of char loading

Results of experiments carried out using various nominal char loadings — 5, 10 and 20 g — are shown in Fig. 6. The model calculations are, again, in good agreement with the measurements. The total conversion of char-N to NO$_X$ was also calculated. Conversion to N_2O is about 1% of char-N; conversion to NO is about 15% and decreases slightly for the highest char loading.

4.3. Parametric investigation on the emissions of NO and N_2O

In the previous sections, the model structure was validated through comparison with experiment for the emissions of CO, CO_2, NO and N_2O from char combustion in a fluidized bed. In this section, the model was used to investigate the influence of various operating parameters on the total emissions of NO and N_2O from the simultaneous combustion of char and propane in a bed fluidized with an air/propane mixture. The presence of propane represents an attempt to include the effect of hydrocarbon volatile combustion within the bed [20]. The base case used in the simulations is $T_{bed} = 1073$ K, $d_{Char} = 3.0$ mm, $d_P = 200$ μm, $W_{Char} = 10$g, $C_{O_2,inlet} = 21$vol%, $C_{propane,inlet} = 2$vol% and $U_0 = 2U_{mf}$. The simulated results are presented in terms of the variation of NO and N_2O yields — the ratio

of the total emissions of NO or N_2O to the char-N loading — with increase in the specific parameter.

Fig. 7 shows the effects of char diameter and char loading on the emissions of NO and N_2O. In the range of variables investigated, a maximum of about 10% char-N is converted to NO, however, only 1% char-N is consumed to form N_2O. These conversions are similar to those measured experimentally [20]. The conversion of char-N to N_2O is relatively unaffected with increase in both char size and char loading. The conversion to NO increases significantly with an increase in char diameter and decreases slightly as char loading is increased.

Bed particle size exerts the most significant influence on the conversions of char-N to nitrogen oxides, Fig. 8(a). Increase in the particle diameter leads to a dramatic increase in the conversion of char-N to NO, since the particle temperature increases as the bed particle size is increased; the conversion of char-N to N_2O increases rapidly first, then levels off. As much as 20% of the char-N is converted to NO for larger bed particles. Fig. 8(b) shows the effect of superficial gas velocity on the NO and N_2O yields. Higher gas velocities imply that more oxygen is introduced into the fluidized bed, thus the formation rate of NO increases, leading to an increase in the formation rate of N_2O.

Bed temperature, too, has a significant effect on the N_2O yield, Fig. 9(a). Lower bed temperatures lead to remarkably higher N_2O yields. NO yield exhibits a minimum at about 1000 K as the bed temperature is increased; this, of course, must be a consequence of the balance between the formation and destruction reaction rates. Propane concentration has the least effect on the conversions of char-N to nitrogen

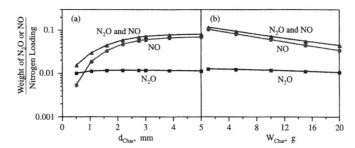

Fig. 7. Influence of char diameter and char loading on N_2O and NO emissions.

oxides, Fig. 9(b). It is clear that both NO and N_2O yields decrease slightly as the inlet propane concentration is increased from 1 to 5% at a bed temperature of 1073 K. The propane concentration may be too small, leading to only small decrease in the oxygen concentration in the bed. Calculations using a bed temperature of 1223 K were also performed. Since propane combustion is enhanced at higher bed temperatures [31] increasing the oxygen consumption, the influence of propane concentration in the inlet gas on NO and N_2O yields becomes more obvious. Note that NO yield decreases only marginally as the bed temperature is increased, however, N_2O yield decreases by an order of magnitude. Certainly, lower NO concentration leads to a lower N_2O formation rate as indicated in Eq. (5). On the other hand, at higher bed temperatures, the destruction rate of N_2O is also enhanced significantly resulting in a decrease in the N_2O yield.

5. Conclusions

Experiments were conducted to investigate the effects of operating conditions — char particle size, bed temperature, oxygen concentration, CO injection and char loading — on the emissions of carbon oxides and nitrogen oxides from char combustion in a bubbling fluidized bed. A single particle reaction–diffusion model has been developed and integrated into a fluidized bed model to simulate the combustion of char particles in a fluidized bed. The hydrodynamic description of the fluidized bed was based on the three-phase model. The non-isothermality of the bed resulting from the gas-phase reactions was taken into account through inclusion of an energy balance for the bubble phase. The effect of the variation in superficial gas velocity on bubble properties and cross-flow was included through an overall mass balance. The following conclusions from the measurements can be drawn:

- Char particle size influences peak NO_X emission. Further, it is expected that the peaks in NO_X emission will be increasingly more pronounced as char size decreases for coarse bed particles.
- Total N_2O emission exhibits a maximum at an intermediate bed temperature. NO emission, on the other hand, increases monotonically with bed temperature.
- Higher oxygen concentrations lead to higher conversion of char-N to N_2O, while conversion to NO remains relatively unaffected.
- Injection of CO in the inlet gas did not influence N_2O and NO emissions at bed temperatures of 873 and 973 K.
- Char loading did not affect the conversions of char-N to NO and N_2O.

Model calculations compare favorably with experimental data for all the operating conditions. The validated model was then used to investigate parametrically the effects of operating conditions on the conversions of char-N to NO and N_2O from the simultaneous combustion of char and propane in a bed fluidized with an air/propane mixture. Bed particle and char diameters were found to have the

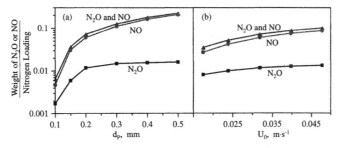

Fig. 8. Effects of bed particle size and gas velocity on N_2O and NO emissions.

198

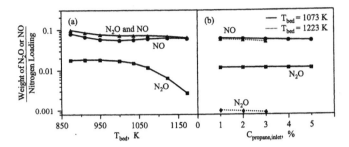

Fig. 9. Effects of bed temperature and propane concentration on N_2O and NO emissions.

most significant effects on NO and N_2O yields. Increase in both bed particle size and char diameter leads to a remarkable increase in the conversion of char-N to NO. Bed temperature also exerts substantial influence on NO and N_2O yields. Other parameters have only minor effects on the conversions of char-N to NO and N_2O.

Acknowledgements

The authors thank the National Science Foundation (Grant No. OSR-9550477) for financial support of the project of which this work is a part. Participation of Jolene Ignowski and Brad Kelly in the project was made possible through funding from the US Department of Energy (Grant No. DE-FC02-91ER75680). The authors also thank Sivakumar Kulasekaran for his contribution to the project.

Appendix A. Derivation of γ expression

Assume that the volumetric formation rate of NO, R_C, is proportional to the volumetric rate (R_{Char}) of char consumption as in Eq. (4a).

Now,

$$R_C = -\frac{1}{V_P}\frac{dN_P}{dt} \quad \text{and} \quad R_{Char} = \frac{1}{V_P}\frac{dC_P}{dt}. \quad (A1)$$

N_P and C_P are nitrogen and char contents (kmol) in the char particles, respectively; V_P is the char volume. Substitution of Eq. (A1) into Eq. (4a) leads to

$$\frac{dN_P}{dC_P} = \gamma. \quad (A2)$$

In addition, during a time period of Δt, the change of nitrogen in the particles, ΔN_P, is given by $N_g V_g$, where N_g is the instantaneous total nitrogen concentration in the gas phase; V_g is the volume of the gas phase. Similarly, $\Delta C_P = C_g V_g$. Thus,

$$\frac{dN_P}{dC_P} = \frac{\Delta N_P}{\Delta C_P} = \frac{N_g}{C_g} = \left(\frac{N}{C}\right)_g. \quad (A3)$$

Combination of Eqs. (A2) and (A3) results in

$$\gamma = \left(\frac{N}{C}\right)_g. \quad (A4)$$

Eq. (A4) indicates that γ is the same as the (N/C) ratio in the product gases instantaneously produced from combustion of char.

We have assumed, Section 3.1, that $\gamma = (cx + d)(N/C)_{initial}$ with $d = 1 - c/2$. It is expected that $0 \le c \le 2$. To examine the effect of parameter c on the shapes of NO and N_2O emissions, calculations were performed and the results are presented in Fig. 10. It is clear that the parameter c has significant influence on both profiles of NO and N_2O instantaneous emissions. For the experimental condition considered, $c \approx 0.3$. Insufficiency of experimental data, at this stage, restricts the derivation of the relation between γ and the particle temperature.

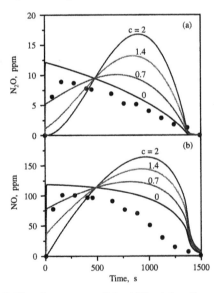

Fig. 10. Effect of parameter c on N_2O and NO emissions: $T_{bed} = 873$ K, $U_0 = 0.071$ m s^{-1}, $C_{O_2,inlet} = 21\%$, $d_{Char} = 3.0$ mm, $W_{Char} = 10$ g.

Appendix B. Single particle model — solutions of Eq. (13a,b)

Since both $R_{i,\text{HET}}$ and $R_{i,\text{HOM}}$ are first order with respect to $C_{i,\text{P}}$ and $C_{i,\text{Film}}$, respectively, Eq. (13a,b) can be solved analytically. To facilitate the solution, the following transformation is used: $\phi_i = C_i r$. Substitution into Eq. (13a) leads to

$$\frac{d^2\phi_{i,\text{P}}}{dr^2} = \frac{\sum v_i R_{i,\text{HET}}}{D_\text{P} r}. \tag{B1}$$

This equation can readily be solved for $\phi_{i,\text{P}}$. Define the following variables

$$\theta_\text{A}^2 = \frac{K_\text{A}}{D_\text{P}}, \qquad \theta_\text{B}^2 = \frac{K_\text{B}}{D_\text{P}}, \qquad \theta_\text{C}^2 = \frac{K_\text{C}}{D_\text{P}}, \qquad \theta_\text{D}^2 = \frac{K_\text{D}}{D_\text{P}},$$

$$\theta_\text{E}^2 = \frac{K_\text{D} + K_\text{E}}{D_\text{P}}, \qquad \theta_\text{F}^2 = \frac{K_\text{G}}{D_\text{P}}, \qquad \theta_\text{G}^2 = \frac{K_\text{H}}{D_\text{F}},$$

$$\theta_\text{H}^2 = \frac{K_\text{I} + K_\text{J}}{D_\text{F}}. \tag{B2}$$

The general solutions for NO and N_2O concentrations can be obtained as follows:

$$\phi_{\text{NO,P}} = A_1 e^{\theta_\text{E} r} + A_2 e^{-\theta_\text{E} r} + \frac{2B_1 \theta_\text{C}^2}{\theta_\text{E}^2 - \theta_\text{A}^2} \sinh(\theta_\text{A} r), \tag{B3}$$

$$\phi_{\text{N}_2\text{O,P}} = A_3 e^{\theta_\text{F} r} + A_4 e^{-\theta_\text{F} r} + \lambda_1 e^{\theta_\text{E} r} + \lambda_2 e^{-\theta_\text{E} r}$$
$$+ 2\lambda_3 \sinh(\theta_\text{A} r), \tag{B4}$$

$$\phi_{\text{NO,Film}} = A_5 r + A_6, \tag{B5}$$

$$\phi_{\text{N}_2\text{O,Film}} = A_7 e^{\theta_\text{H} r} + A_8 e^{-\theta_\text{H} r}, \tag{B6}$$

where

$$\lambda_1 = \frac{A_1 \theta_\text{D}^2}{\theta_\text{F}^2 - \theta_\text{E}^2}, \quad \lambda_2 = \frac{A_2 \theta_\text{D}^2}{\theta_\text{F}^2 - \theta_\text{E}^2}, \quad \lambda_3 = \frac{B_1 \theta_\text{C}^2 \theta_\text{D}^2}{(\theta_\text{A}^2 - \theta_\text{E}^2)(\theta_\text{A}^2 - \theta_\text{F}^2)}. \tag{B7}$$

B_1 is a coefficient numerically evaluated from boundary conditions [21]. To solve for A_1–A_8, the appropriate boundary conditions must be employed. Due to symmetry, the flux of any species at the center of the particle must be zero $(dC_{i,\text{P}}/dr = 0)$. Applying this boundary condition to Eqs. (B5) and (B6) leads to

$$\phi_{\text{NO,P}} = 2A_1 \sinh(\theta_\text{E} r) + \frac{2B_1 \theta_\text{D}^2}{\theta_\text{E}^2 - \theta_\text{A}^2} \sinh(\theta_\text{A} r), \tag{B8}$$

$$\phi_{\text{N}_2\text{O,P}} = 2A_3 \sinh(\theta_\text{F} r) + 2\lambda_1 \sinh(\theta_\text{E} r) + 2\lambda_3 \sinh(\theta_\text{A} r). \tag{B9}$$

Applications of other boundary conditions, Eq. (14a,b), permit evaluation of the remaining constants in Eqs. (B5), (B6), (B8) and (B9). Solutions of equations for species balances of CO, CO_2 and O_2 have been presented elsewhere [21].

References

[1] Mann MD, Collings ME, Botros PE. Prog Ener Combust Sci 1992;18:447.
[2] Thomas KM. Fuel 1997;76:467.
[3] Blair DW, Wendt JOL, Bartok W. 16th Symposium (International) on Combustion. The Combustion Institute, 1977;p 475.
[4] Nelson PF, Buckley AN, Kelly MD. 24th Symposium (International) on Combustion. The Combustion Institute, 1992, p 1259.
[5] Pohl JH, Sarofim AF. 16th Symposium (International) on Combustion. The Combustion Institute, 1977;p 491.
[6] Bonn B, Richter E. Fuel Process Tech 1990;24:319.
[7] Wojtowicz MA, Pels JR, Moulijn JA. Fuel Process Tech 1993;34:1.
[8] Hayhurst AN, Lawrence AD. Prog Ener Combust Sci 1992;18:529.
[9] Takeshita M, Sloss LL, Smith IM. N_2O emissions from coal use. IEA Coal Research, 1993.
[10] Johnsson JE. Fuel 1994;73:1398.
[11] Zhao J, Grace JR, Lim CJ, Brereton CMH, Legros R. Fuel 1994;73:1650.
[12] Collings ME, Mann MD, Young BC. Ener Fuels 1993;7:554.
[13] Tullin CJ, Sarofim AF, Beer JM. J Inst Ener 1993;66:207.
[14] Amand L-E, Leckner B. Combust Flame 1991;84:181.
[15] Gulyurtlu I, Esparteiro H, Cabrita I. Fuel 1994;73:1098.
[16] La Nauze RD. Chem Engng Res Des 1985;63:3.
[17] Linjewile TM. Temperature of carbonaceous particles in a fluidized bed. Ph.D. Thesis, University of Adelaide, Australia, 1993.
[18] Wojtowicz MA, Oude Luhuis JA, Tromp PJJ, Moulijn JA. 11th International Conference on Fluidized Bed Combustion. ASME, 1991;p 1013.
[19] Bramer EA, Valk M. 11th International Conference on Fluidized Bed Combustion. ASME, 1991;p 701.
[20] Hayhurst AN, Lawrence AD. Combust Flame 1996;105:341.
[21] Kulasekaran S, Linjewile TM, Agarwal PK, Biggs MJ. Fuel 1998;77:1549.
[22] Goel SK, Lee CH, Longwell JP, Sarofim AF. Ener Fuels 1996;10:1091.
[23] Goel SK, Beer JM, Sarofim AF. J Inst Ener 1996;69:201.
[24] Goel SK, Zhang B, Sarofim AF. Combust Flame 1996;104:213.
[25] Goel SK, Morihara A, Tullin CJ, Sarofim AF. 25th Symposium on Combustion. The Combustion Institute, 1994;p 1051.
[26] Goel SK, Sarofim, AF, Kilpinen P, Hupa M. 26th Symposium on Combustion. The Combustion Institute, vol. 2, 1996;p 3317
[27] Biggs MJ, Agarwal PK. Chem Engng Sci 1997;52:941.
[28] Jensen A, Johnsson JE. Chem Engng Sci 1997;52:1715.
[29] Takahashi M, Kotaka M, Sekimoto H. J Nucl Sci Technol 1994;31:1275.
[30] Sriramulu S, Sane S, Agarwal PK, Mathews T. Fuel 1996;75:1351.
[31] Srinivasan R, Sriramulu S, Kulasekaran S, Agarwal PK. Fuel 1998;77:1033.
[32] Kulasekaran S, Linjewile TM, Agarwal PK. Fuel 1999;78:403.
[33] Adanez J, Abanades JC. Ind Engng Chem Res 1992;31:2286.
[34] Park D. Fuel 1989;68:1320.
[35] Latham R, Potter OE. Chem Engng J 1970;1:152.
[36] Lim KS, Gururajan VS, Agarwal PK. Chem Engng Sci 1993;48:2251.
[37] Hesketh RP, Davidson JF. Chem Engng Sci 1991;46:3101.
[38] Johnsson JE, Amand L-E, Dam-Johansen K, Leckner B. Ener Fuels 1996;10:970.
[39] Lin M. Emissions of N_2O and NO_x from fluidized combustion of coal. M.S. thesis, University of Wyoming, 1998.
[40] Miller JA, Bowman CT. Prog Ener Combust Sci 1989;15:287.
[41] Field MA, Gill DW, Morgan BB, Hawksley PGW. Combustion of pulverized coal. Leatherhead: BCURA, 1967.

200

[42] Dutta S, Wen CY, Belt RJ. Ind Engng Chem Process Des Dev 1977;16:20.

[43] De Soete GG, Croiset E, Richard J-R. Combust Flame 1999;117:140.

[44] Croiset E, Heurtebise C, Rouan J-P, Richard J-R. Combust Flame 1998;112:33.

[45] Ashman PJ, Haynes BS, Buckley AN, Nelson PF. 27th Symposium (International) on Combustion, The Combustion Institute,1998; p 3069.

[46] Harding AW, Brown SD, Thomas KM. Combust Flame 1996;107:336.

[47] Krammer GF, Sarofim AF. Combust Flame 1994;97:118.

[48] Winter F, Wartha C, Loffler G, Hofbauer H. 26th Symposium (International) on Combustion. The Combustion Institute, 1996;p 3325.

[49] Winter F, Loffler G, Wartha C, Hofbauer H, Preto F, Anthony EJ, Can. J Chem Engng 1999;77:275.

[50] Chan LK, Sarofim AF, Beer JM. Combust Flame 1983;52:37.

[51] Li YH, Lu GQ, Rudolph V. Chem Engng Sci 1998;53:1.

[52] Aarna I, Suuberg EM. Fuel 1997;76:475.

[53] Aarna I, Suuberg EM. Ener Fuels 1999;13:1145.

[54] Li YH, Radovic LR, Lu GQ, Rudolph V. Chem Engng Sci 1999;54:4125.

[55] Amand L-E, Andersson S. In: Proceedings of the 10th international Conference on FBC. ASME, 1989;p 49.

[56] Kilpinen P, Hupa M. Combust Flame 1991;85:94.

[57] Hautman DJ, Dryer FL, Schug KP, Glassman I. Combust Sci Technol 1981;25:219.

[58] Aho MJ, Rantanen JT, Linna VL. Fuel 1990;69:957.

[59] Allen MT, Yetter RA, Dryer FL. Combust Flame 1997;109:449.

[60] Glarborg P, Johnsson JE, Dam-Johansen K. Combust Flame 1994;99:523.

[61] Loirat H, Caralp F, Destriau M, Lesclaux R. J Phys Chem 1987;91:6538.

[62] Bonn B, Pelz G, Baumann H. Fuel 1995;74:165.

[63] Gustavsson L, Leckner B. In: Anthony EJ, editor. Proceedings of the 11th International Conference on FBC, New York: ASME, 1991. p. 1677.

[64] Leckner B, Gustavsson L. J Inst Ener 1991;64:176.

[65] Rajan RR, Wen CY. AIChE J 1980;26:642.

[66] Turnbull E, Kossakowski ER, Davidson JF, Hopes RB, Blackshaw HW, Goodyer PTY. Chem Engng Res Des 1984;62:233.

[67] Prins W, Casteleijn TP, Draijer W, Van Swaaij WPM. Chem Engng Sci 1985;40:481.

[68] Hayhurst AN, Lawrence AD. Combust Flame 1997;110:351.

Research and Development in Commercial Scale Facilities

Emissions Reduction: NO$_X$/SO$_X$ Suppression
A. Tomita (Editor)
© 2001 Elsevier science Ltd. All rights reserved

Combustion technology developments in power generation in response to environmental challenges

J.M. Beér*

Department of Chemical Engineering, Room 66-548, Massachusetts Institute of Technology, Cambridge, MA 02139-4307, USA

Received 27 October 1999; revised 28 March 2000; accepted 28 March 2000

Abstract

Combustion system development in power generation is discussed ranging from the pre-environmental era in which the objectives were complete combustion with a minimum of excess air and the capability of scale up to increased boiler unit performances, through the environmental era (1970–), in which reduction of combustion generated pollution was gaining increasing importance, to the present and near future in which a combination of clean combustion and high thermodynamic efficiency is considered to be necessary to satisfy demands for CO_2 emissions mitigation.

From the 1970s on, attention has increasingly turned towards emission control technologies for the reduction of oxides of nitrogen and sulfur, the so-called acid rain precursors. By a better understanding of the NO_x formation and destruction mechanisms in flames, it has become possible to reduce significantly their emissions via combustion process modifications, e.g. by maintaining sequentially fuel-rich and fuel-lean combustion zones in a burner flame or in the combustion chamber, or by injecting a hydrocarbon rich fuel into the NO_x bearing combustion products of a primary fuel such as coal.

Sulfur capture in the combustion process proved to be more difficult because calcium sulfate, the reaction product of SO_2 and additive lime, is unstable at the high temperature of pulverized coal combustion. It is possible to retain sulfur by the application of fluidized combustion in which coal burns at much reduced combustion temperatures. Fluidized bed combustion is, however, primarily intended for the utilization of low grade, low volatile coals in smaller capacity units, which leaves the task of sulfur capture for the majority of coal fired boilers to flue gas desulfurization.

During the last decade, several new factors emerged which influenced the development of combustion for power generation. CO_2 emission control is gaining increasing acceptance as a result of the international greenhouse gas debate. This is adding the task of raising the thermodynamic efficiency of the power generating cycle to the existing demands for reduced pollutant emission. Reassessments of the long-term availability of natural gas, and the development of low NO_x and highly efficient gas turbine–steam combined cycles made this mode of power generation greatly attractive also for base load operation.

However, the real prize and challenge of power generation R&D remains to be the development of highly efficient and clean coal-fired systems. The most promising of these include pulverized coal combustion in a supercritical steam boiler, pressurized fluid bed combustion without or with topping combustion, air heater gas turbine-steam combined cycle, and integrated gasification combined cycle. In the longer term, catalytic combustion in gas turbines and coal gasification-fuel cell systems hold out promise for even lower emissions and higher thermodynamic cycle efficiency. The present state of these advanced power-generating cycles together with their potential for application in the near future is discussed, and the key role of combustion science and technology as a guide in their continuing development highlighted. © 2000 Elsevier Science Ltd. All rights reserved.

Keywords: Combustion; Power generation; Combined cycles; Air pollution control

* Tel.: +1-617-253-6661; fax: +1-617-253-3122.
  E-mail address: jmbeer@mit.edu (J.M. Beér).

Reprinted from *Progress in Energy and Combustion Science* **26 (4-6)**, 301-327 (2000)

Contents

1. Background

Combustion is the prevailing mode of fossil energy utilization, and coal is the principal fossil fuel of electric power generation. In places where coal is not readily available, heavy fuel oil or natural gas is used in power station boilers.

Natural gas is a premium fuel and, because of its relatively high price, it was used in the past primarily for "peak-shaving" by gas turbine plants. Recent reassessment of the long-term availability of natural gas, the development of combined cycle gas turbine–steam plants with near 60% efficiency, environmental pressures for "decarbonization" of the fuel supply, and deregulation of the electric utility industry in the OECD countries resulted in the acceptance of natural gas as a fuel for base load power generation.

Coal, due to its low cost and broad availability, can be expected to remain in essential supply well into the twenty-first century, barring strong future evidence for high rate, anthropogenic global warming. Notwithstanding such evidence, however, it is prudent public policy to aim at the development and early application of clean coal utilization technology in high efficiency power cycles.

Coal combustion systems for power generation have to satisfy the following demands:

- high degree of burnout of the coal with a minimum of excess air;
- scale up to 500 MWe or larger boiler unit sizes without excessive slagging in the combustion chamber;
- operation with easily removed friable ash deposits, low NO$_x$ emissions obtained by combustion process modifications;
- sulfur capture in the combustion process by additive sorbents; and
- acceptance of coal quality variation, without significant

reduction of combustion efficiency and boiler plant availability.

For combustion turbines the requirements are:

- complete combustion of gaseous or distillate fuel oil;
- low NO$_x$ and low CO emissions;
- oscillation-free operation;
- low pressure drop (2–5%); and
- high temperature durability combustor.

Added to these conditions is the recent requirement for the combustion system to be amenable to CO$_2$ sequestration (by the production of high pressure and/or high CO$_2$ flue gas concentration).

The *travelling grate stoker* was the early coal combustion system for power generation. Travelling grate stokers are capable of burning coals of a wide range of coal rank (from anthracite to lignites). Early experimental studies by Werkmeister [1] and the better understanding of the mechanism of combustion in a fuel bed by Thring [2] helped the development of the travelling grate with controlled air distribution along the grate. Stoker firing, however, remained sensitive to coal fines (<3 mm); it could not be scaled up to beyond about 25 MWe unit capacity, and the boiler efficiency was suppressed by the high excess air ($\approx 40\%$) which had to be used for acceptable coal burnout. Also, high air preheat (above 400 K) was not compatible with the need for cooling the grate by the combustion air. Low air preheat, on the other hand, restricted the application of regenerative feed water heating (using steam bled from the turbine) and hence the raising of the thermodynamic efficiency of the electric power generating cycle.

In an effort to reduce the sensitivity to coal fines and to improve the boiler efficiency, travelling grate stokers were retrofitted with topping pulverized coal combustion (TPC) [3]. The arrangement or TPC allowed for flash drying the coal with flue gas withdrawn from the boiler, and for

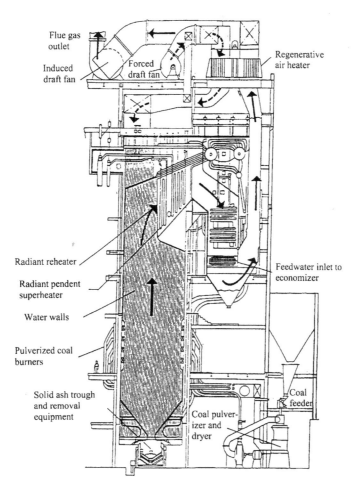

Fig. 1. Schematic illustration of a pulverized coal fired utility boiler.

controlled separation of fines from the larger coal particles. The lumps of coal were fed to the grate and the fines (<0.3 mm) were carried pneumatically to a small grinding mill, ground to pulverized coal fineness, and injected through burners into the combustion chamber above the grate. TPC was successful in improving the boiler efficiency and raising the steaming capacity, but it required on retrofit some additional screen tubes in the combustion chamber and improved flue gas cleaning for the capture of fly ash particles [3].

As the boiler unit capacity has grown and boiler combustion chambers with completely cooled walls were developed, *pulverized coal combustion* (PCC) has become the generally accepted combustion system for power generation. PCC is still the combustion system of choice for coal fired utility boilers. The science and technology of pulverized coal combustion is discussed in several books and reviews (Field et al. [4]; Dolezal [5]; Essenhigh [6]; Beér, Chomiak and Smoot [7]; Smoot and Smith [8]; Wall [9]).

In PCC combustion, the coal is dried and ground to specified fineness; the latter depending mainly on the coal rank and hence the reactivity of the coal. The system of coal preparation: feeding, drying, grinding of the coal and the pneumatic transport of the pulverized coal to the burners is fully integrated with the boiler. For lower reactivity coals, the fineness of grind is increased to create a larger specific surface area of the coal so as to improve conditions for ignition and combustion. In the past, according to rule of thumb, the percentage of pulverized coal residue on the sieve with 76 μm hole sizes was not supposed to exceed the percentage volatile matter in the dry coal. Presently the requirements are more stringent—even high volatile bituminous coals are ground to give less than 10% residue on the 76 μm hole size sieve.

The powdered coal is pneumatically transported to the burners and injected in the form of particle-laden jets into the combustion chamber. The transport air that carries the

206

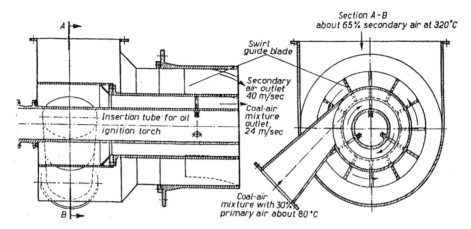

Fig. 2. Circular pulverized coal burner.

coal from the mill to the burners is a small fraction of the total combustion air mainly because its temperature is limited to about 100°C for reasons of safety against ignition and explosion in the mill and the transport pipeline between the mill and the burners. Upon injection into the combustion chamber, the coal particle-laden jet entrains hot combustion products, which raises its temperature and assists the ignition of the cloud of coal particles. The rest of the combustion air, which can be more strongly preheated, is injected separately and admixed with the burning fuel jet in the combustion chamber. A schematic illustration of a PCC boiler is shown in Fig. 1. The combustion chamber is typically of parallelepiped shape; the dimensions of a 300 MW coal-fired boiler would approximately be $15 \times 15 \text{ m}^2$ of cross-sectional area and 45–50 m in height. The combustion chamber walls are completely cooled by steam generating tubes. The particles burn in a mode in which both external diffusion of oxygen to the particle surface and chemisorption of the oxygen at the particle surface and in the pores of the solid char play roles in determining the progress of combustion, with diffusion controlling the burning rate of larger particles at the higher temperatures, and chemical kinetics controlling the burning rate of the small particles as the char burns out in the tail end of the flame. A comprehensive review of the fundamentals of pulverized coal combustion is presented by Essenhigh [6].

Burners are mixing devices designed to ensure ignition, flame stability, and complete burnout of the coal along its path in the combustion chamber. As the pulverized coal particles burn, the flame transfers heat, mainly by thermal radiation, to the steam cooled tube-walls of the boiler. The last few percentages of the residual carbon in the char burns in an environment of depleted O_2 concentration and reduced temperature before the fly ash leaves the combustion chamber and enters the pass of convective heat exchangers. The design of the combustion chamber has to provide for suffi-

cient residence time of the burning particle to complete combustion, and for the cooling of the fly ash to below its "softening temperature" to prevent the build up of ash deposits on heat exchanger surfaces.

While there is a great variety of burner types, the most widespread are circular burners (Fig. 2) and vertical nozzle arrays (Fig. 3). Circular burners are usually positioned perpendicularly to the combustion chamber walls, while the vertical nozzle arrays are in the corners, firing tangentially to the circumference of an imaginary cylinder in the middle of the combustion chamber. The design of circular burners is more concerned with the tailoring of the individual burner-flame while those of vertical nozzle arrays in tangentially fired furnaces rely more on the bulk of the furnace volume for the mixing of the fuel and air streams injected through the nozzle arrays.

In the majority of cases, most of the fly ash formed in pulverized coal combustion is removed from the flue gas in the form of dry particulate matter, with a small proportion (~10%) of the coal ash falling off the tube walls as semi-molten agglomerated ash which is collected from the bottom hopper of the combustion chamber ("bottom ash"). Because of the difficulty of handling and disposing of dry fly ash, an alternative combustion system, *slagging combustion* [5,10], was developed in the 1930s. In slagging combustion, the boiler tubes in the lower part of the furnace are covered by refractory to reduce heat extraction and to allow the combustion temperature to rise to beyond the melting point of the ash. The temperature has to be sufficiently high for the viscosity of the slag to be reduced to about 150 poise, necessary for removal in liquid form. The most notable application of slagging combustion technology in the USA was the Cyclone Furnace in which about 85% of the coal ash could be removed in molten form in a single pass without ash recirculation. Because of the high temperature and the oxidizing atmosphere, slagging furnaces

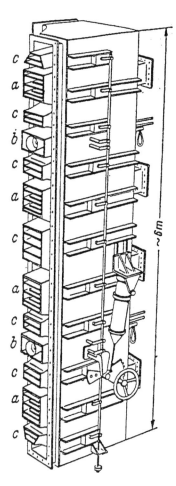

Fig. 3. Vertical nozzle array for tangentially fired PC boilers: (a) pulverized coal mixture; (b) ignition torches; (c) secondary-, top-, and bottom-air.

produced very high NO$_x$ emissions and they fell in disfavor in the 1970s. Their application, however, reemerged in the form of staged combustion, a fuel rich slagging furnace with liquid ash removal followed by the completion of combustion in a fully cooled fuel lean combustion environment. Slagging combustion features presently in the US DOE's Clean Coal Technology Program's Low Emissions Boiler System (Beittel [11]), and also in the High Performance Power Generating System (Seery et al. [12]).

Schematic slagging furnace designs and their primary ash retention is shown in Fig. 4 [5]. The furnace designs aim at maintaining high temperature in the lower part of the furnace, especially above the slag tap, but extracting heat further downstream to cool the fly ash particles, which escaped removal in liquid form, to avoid their deposition

on heat exchange surfaces. The solutions include the covering of the steam generating tubes by refractory in the lower furnace, the narrowing of the cross-section area above the slagging chamber to reduce radiation losses, and to separate the high temperature furnace from the fully cooled section of the combustion chamber by a tube screen. Fig. 4 illustrates the effect of the design and the use of swirl flow upon the first pass ash retention in molten form. Because of the viscous high temperature gas, strong swirl has to be used to drive the particles to the slagging chamber wall where they are captured by a molten slag layer moving down towards the slag tap. Strong swirl, however, demands high airside pressure drop with the associated energy losses. As can be seen in Fig. 4, several of the designs use screen tubes to increase molten ash retention without excessive pressure drop. Screen tubes, however, may cause problems because of the difficult demand upon their operation, that the ash be molten upstream of the slag screen, but be dry and non sticky downstream of it. This requirement leads to restrictions concerning the choice of candidate coals. Also, the tube screen is exposed to high temperature corrosion especially in sulfur bearing, fuel-rich environments (Reid) [13] which may require special measures (e.g. plasma sprayed protective layers) for its protection.

In fluidized bed combustion (FBC) (Fig. 5), crushed coal of 5–10 mm is burned in a hot fluidized bed of 0.5–3.0 mm size inert solids. Less than 2% of the bed material is coal; the rest are coal ash and limestone, or dolomite, which are added to capture sulfur in the course of combustion. The bed is cooled by steam generating tubes immersed in the bed to a temperature in the range of 1050–1170 K. This prevents the softening of the coal ash and the decomposition of CaSO$_4$, the product of sulfur capture.

Because of the relatively low combustion temperature, the overall rate of coal particle oxidation is strongly dependent upon the rate of reaction at the particle surface and in the pores (Borghi et al. [14]). Kerstein and Niksa [15] developed a model of char oxidation: "percolative fragmentation" caused by changes in pore structure of the reacting solid. Experimental observations by Beér et al. [16], Chirone et al. [17], and Sundback et al. [18] supported the model. Fragmentation turned out to be an important extension of the coal combustion model; it explained how unburned carbon in the form of fine fragments finds its way into the fly ash. Small (<200 μm) char fragments are carried over from the freeboard of a FBC or the riser tube of a circulating fluid bed (CFBC), and the residence time of about 3 s at 1100 K is, in the case of low reactivity chars, insufficient for complete combustion.

FBC and CFBC are mature technologies primarily applied to the use of low grade fuels in smaller unit sizes. There are 185 boilers installed in the USA with a total capacity of 6 GW. The largest FBC is a 350 MWe unit in Japan. An impediment to scale up is the FBC coal feed system: a feed point is needed for about every

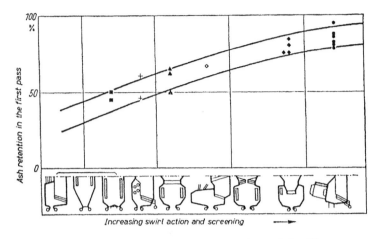

Fig. 4. Primary ash retention in various slagging furnace designs [5].

3 MWe throughput. The advantage of the CFBC over the FBC is that it requires fewer feed points due to the smaller cross-sectional area of the bed, and hence it is more easily scaleable to higher outputs. Also, for a given percentage sulfur capture, CFBC requires a lower Ca/S feed ratio.

2. Pollutant emission control

In the 1970s, applied combustion research has taken a turn from high output, high intensity combustion towards combustion process modifications for reduced pollutant emissions. The combustion generated pollutants of concern

REACTIONS OCCURRING IN FLUIDIZED BED COMBUSTOR

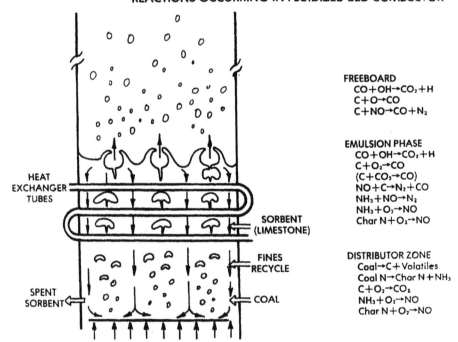

FREEBOARD
$CO+OH \rightarrow CO_2+H$
$C+O \rightarrow CO$
$C+NO \rightarrow CO+N_2$

EMULSION PHASE
$CO+OH \rightarrow CO_2+H$
$C+O_2 \rightarrow CO$
$(C+CO_2 \rightarrow CO)$
$NO+C \rightarrow N_2+CO$
$NH_3+NO \rightarrow N_2$
$NH_3+O_2 \rightarrow NO$
$Char\ N+O_2 \rightarrow NO$

DISTRIBUTOR ZONE
$Coal \rightarrow C+Volatiles$
$Coal\ N \rightarrow Char\ N+NH_3$
$C+O_2 \rightarrow CO_2$
$NH_3+O_2 \rightarrow NO$
$Char\ N+O_2 \rightarrow NO$

HEAT
EXCHANGER
TUBES

SORBENT
(LIMESTONE)

FINES
RECYCLE

SPENT
SORBENT

COAL

Fig. 5. Reactions in fluidized coal combustion [19].

COMBUSTION GENERATED POLLUTANT EMISSIONS

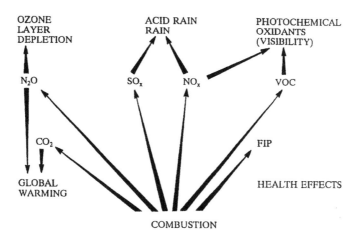

Fig. 6. Combustion generated pollutant emissions.

FORMATION AND REDUCTION OF NITROGEN OXIDES IN COMBUSTION; MECHANISTIC PATHWAYS

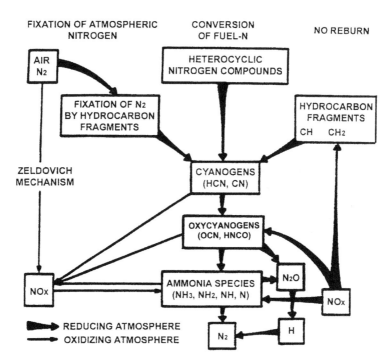

Fig. 7. Chemical pathways of NO$_x$ formation and destruction [20].

Table 1

Capacities of coal-fired units fitted with primary combustion measures for NO$_x$ reduction currently in use world wide (after IEA coal research, 1995)

Country	MWe
Austria	1675
Belgium	555
Canada	9065
China	70
Denmark	4585
Finland	1765
Germany	30,332
Hong Kong	1030
Ireland	305
Italy	3495
Japan	15,320
Koreas(s)	1060
Malaysia	600
Netherlands	4340
Sweden	715
Taiwan	5825
Ukraine	300
UK	12,600
USA	94,483
Total	188.665 GWe

were oxides of sulfur, nitrogen and carbon, and fine organic and inorganic particulates (Fig. 6).

Sulfur in the coal will oxidize to SO$_2$ with a small fraction of the SO$_2$ (about 1–2%) oxidizing further to SO$_3$. The formation of SO$_3$ can be mitigated by very low excess O$_2$ combustion (<0.5% EO$_2$). The usual methods of sulfur capture in the combustion process involve the reactions of a sorbent such as calcined limestone, CaO, with SO$_2$ to

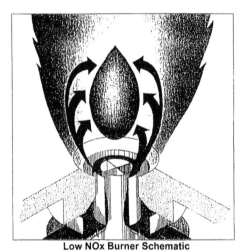

Low NOx Burner Schematic

Fig. 8. Low NO$_x$ burner schematic.

produce CaSO$_4$, a stable, disposable solid waste. However, in the high temperature fuel-lean environment of pulverized coal flames ($T_{peak} \approx 2000$ K), CaSO$_4$ is unstable; it decomposes, leaving flue gas desulfurization as the viable option of sulfur capture from pulverized coal combustion.

The development of fluidized bed combustion provided the opportunity to effectively retain sulfur in the combustion process because CaSO$_4$ is stable at the FBC operating temperature of 1050–1170 K [19]. One of the difficulties of FBC technology is that it does not lend itself well to scale-up to the 700–1000 MW range, mainly because of the large number of feed points it requires to ensure the uniform distribution of the coal in the bed. In circulating fluidized bed combustion, the gas velocities are higher than in the conventional bubbling fluidized bed as the bed cross-sectional area for the same heat release rate is smaller. This helps to reduce the number of coal feed points, which is an operational convenience. Also, smaller size limestone particles can be used in the feed which improves the sulfur capture and reduces the Ca/S mole ratio necessary for reaching a target value of sulfur capture.

Nitrogen oxides as pollutants deserve special attention because of their wide ranging effects on the environment, including contribution to acid rain, reduction of atmospheric visibility, production of tropospheric ozone, and, in the case of N$_2$O, depletion of stratospheric ozone. It is also noteworthy that NO$_x$ emissions are well amenable to reduction by combustion process modifications (Beér [20]).

Nitric oxide, NO, is formed in fuel-lean flames by the attack of O atom on molecular nitrogen ("thermal NO"). In fuel-rich flames it forms via capture of N$_2$ by hydrocarbon radicals, ("prompt NO"), and by the pyrolysis and oxidation of heterocyclic nitrogen compounds in coals and petroleum fuels ("fuel NO"). Fig. 7 illustrates chemical pathways of nitrogen compound interconversions in fuel lean and fuel rich flames. Examination of the chemical reaction paths of nitrogen oxides formation and destruction in flames led to the formulation of guidelines for primary measures of NO$_x$ emissions reduction in boilers:

- reducing the peak flame temperature by heat extraction, and/or by flue gas recirculation;
- diluting the reactant concentrations by flue gas or steam mixed with gaseous fuels and recirculated burned gas mixed with the combustion air;
- staging the combustion air to produce fuel-rich/fuel-lean sequencing favorable for the conversion of fuel bound nitrogen to N$_2$; and
- staging the fuel so that the NO formed earlier in the flame is getting reduced by its reactions with hydrocarbon radicals ("NO reburning") (Wendt et al. [21]).

The reduction of NO$_x$ emission by combustion process modification, a science based technology, has been successfully applied in industry. More than 188 GW of electric power generating capacity currently in operation

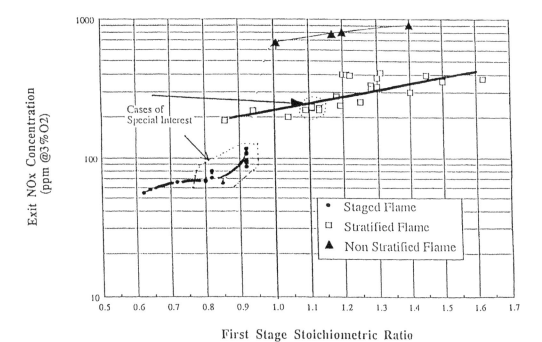

Fig. 9. RSFC burner NO$_x$ emissions with HV bituminous coal [27].

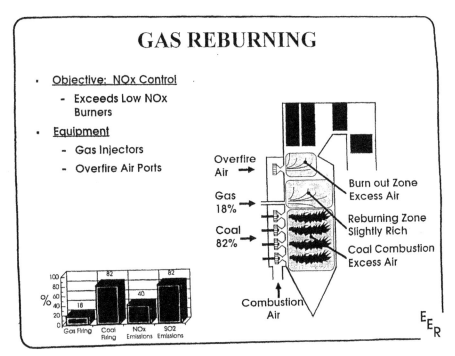

Fig. 10. NO$_x$ gas reburning [29].

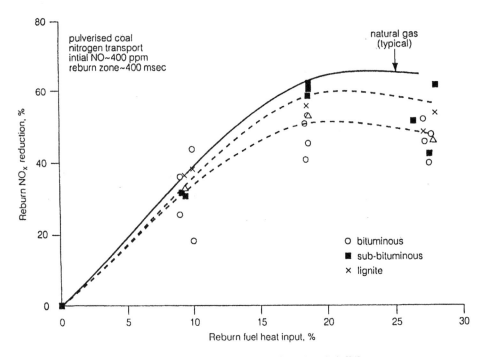

Fig. 11. Comparison of coal types and NG as reburn fuels [34].

internationally has been fitted by these "primary combustion measures" (Table 1 IEA [22]).

Low NO$_x$ burners (LNB) (Fig. 8) represent the most cost effective method of achieving reduced NO$_x$ emissions from new plant boilers, and also from existing boilers by retrofit. Air staging in these burners is achieved by the aerodynamically tailored mixing of the fuel jet with air streams supplied through the burner, rather than by the use of over-fire air.

One of the problems of LNBs is the requirement of maintaining a fuel-rich environment close to the burner for the pyrolysis reactions to run their course, followed by the admixing of the residual combustion air to complete combustion. An example of an engineering solution of this problem based on first principles is the radially stratified flame core burner (RSFC). The process of turbulence damping through radial density stratification in rotating flows demonstrated by Emmons and Ying [23], and Chigier et al. [24] has been employed in the design of this LNB. Premature air-fuel mixing is prevented by the damping of turbulence in the near burner region followed by the vigorous admixing of the residual burner air issuing from the outer annulus of a triple annular burner. Toqan et al. [25], Shiadeh et al. [26], and Barta et al. [27] reported NO$_x$ emissions achieved with the RSFC burner burning natural gas, heavy fuel oil, and coal, respectively, in the 1.5 MWth MIT combustion Research Facility. In Fig. 9, measured NO$_x$ concentrations are plotted for pulverized coal [27]. The

straight line in the middle of the graph represents the case without external air staging. The top line refers to uncontrolled conditions, and the lowest line to the use of the RSFC burner with over-fire air.

The RSFC burner has been scaled up and commercialized by ABB under license from MIT. LaFlesh et al. [28] have recently given an update on ABB-CE's RSFC Low NO$_x$ wall burner technology.

In "*NO reburning*," the secondary fuel is usually natural gas (Fig. 10) [29],but fuel oil or even coal can also be used. In the latter case, the coal volatiles are the main reactants, but carbonaceous solids may also react to reduce NO to N$_2$ (Chan et al. [30]). The reburning technology has been successfully applied in the USA for slagging cyclones (Borio et al. [31]), for wall and tangentially fired pulverized coal combustion with dry ash removal (Folsom et al. [29]), and in both oil and coal fired boilers in Italy following computational studies, and laboratory and pilot plant experiments at ENEL's R&D Laboratories in Pisa (DeMichele et al.[32]; LaFlesh et al. [33]). NO$_x$ reductions reported with coal and natural gas as reburn fuel in a coal fired boiler by Payne et al. [34] are shown in Fig. 11.

The modeling of the reburn process represents a special challenge because it requires detailed descriptions of both the nitrogen chemistry and the controlled mixing of relatively small mass flows of reburn fuel and tertiary air with the bulk flow of the combustion products in the furnace.

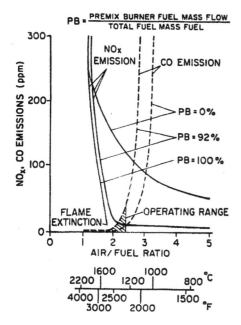

Fig. 12. NO$_x$ and CO emissions; diffusion flames, partially,-and fully premixed combustion in gas turbines [39].

Ehrhardt et al. [35] have developed a model in which following a CFD calculation of the spatial distributions of flow, major species-concentrations, and temperature, the combustion space is subdivided into a relatively small number (say one hundred) of volume zones which then permit the application of more detailed chemistry than would have been possible in CFD models. Comparison with experiment showed good agreement with computations for an axisymmetric oil fired reburn system [35].

Because of the low combustion temperature in fluidized combustion, the NO is formed mainly by the conversion of coal-nitrogen, a process which lends itself for minimizing emission by the application of staged air introduction (Gibbs et al. [36]). However, a difficulty due to the low combustion temperature is that nitrous oxide, N$_2$O, an intermediate product of NO formation, survives and is emitted from FBC at concentrations ranging from 40 to 100 ppm [22]. Nitrous oxide is a specially unpleasant pollutant; it is a greenhouse gas which also depletes stratospheric ozone. The raising of the gas temperature before the convective section of the boiler to above 1200 K could eliminate N$_2$O emissions, but the temperature rise adversely affects sulfur capture in the fluidized bed (Tullin et al. [37]). It is noteworthy that pressurized fluidized beds also emit N$_2$O, except the second generation pressurized fluidized bed in which the gas temperature is raised by a topping combustor before entry to the gas turbine, thereby eliminating the N$_2$O in the combustion products [38].

In *gas turbine* (GT) applications, large amount of excess air is used to cool the combustion products before entry to the gas turbine to a temperature limited by the structural integrity of gas turbine blades. In the conventional method of combustion, the fuel and air are separately injected into the combustor and mix in the course of combustion (diffusion flame). This process is prone to the formation of thermal NO because of the near stochiometric conditions which prevail on the boundaries of fuel-rich and fuel-lean eddies in such flames. To overcome this problem, the fuel gas and air are premixed prior to their entry to the combustor creating a strongly fuel-lean mixture corresponding to the combustor exit gas temperature (presently about 1573 K).

This so called ultralean premixed combustion gives very low NO$_x$ emissions, typically less than 15 ppm at 15% O$_2$, with natural gas as fuel, but is left with the problem of flame stability. The latter is generally solved by the injection of a small percentage of the fuel, say 10%, to produce a fuel jet pilot flame as a stable source of ignition. Fig. 12 illustrates the opportunities and some of the problems of ultralean premix combustion (Maghon et al. [39]). The figure shows

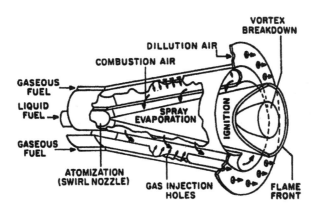

Fig. 13. ABB's double cone GT combustor [41].

214

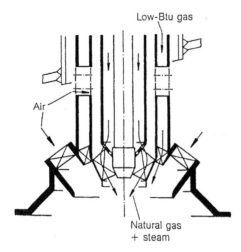

Fig. 14. Siemens' low NO$_x$ GT combustor [42].

the measured variations of NO$_x$ and CO emission data as functions of the air/fuel ratio and of the adiabatic flame temperature of a natural-gas/air diffusion flame, and for 92% and 100% premixing. In the case of 92% premixing, the rest of the fuel (8%) burns in the form of a pilot diffusion flame. NO$_x$ and CO emissions are very low in the range of state of the art turbine entry temperatures (1573 K), but the flame stability is not ensured without the pilot flame. As the air/fuel ratio increases, the CO emission rises steeply. Most of the major gas turbine manufacturers are offering lean burn premix combustors capable of 15 ppm NO$_x$ (15% O$_2$) dry or with minimal water injection (Beér) [40]. As examples, ABB's Double Cone burner is illustrated in Fig. 13 (Sattlemayer et al. [41]), and Siemens' KWU in Fig. 14 (Becker et al. [42]). As new materials permit raising the turbine inlet gas temperature beyond 1800 K, the opportunity to reduce NO$_x$ emissions by lean premixed combustion will become limited. This will present a new challenge to combustion R&D in gas turbine applications.

3. Coal mineral matter transformations

The transformation of coal mineral matter during combustion has been the subject of a rich literature, several texts, and conference proceedings, e.g. CEGB Marchwood Conf. [43], Reid [13], Wall [44], Couch [45], and Engnr. Found. Conferences [46–48].

A combination of computer controlled scanning electron microscopy and mathematical modeling made it possible to follow the changes in the properties of the coal ash (particle size, chemical composition, viscosity) during combustion of the coal particle, and to determine the propensity of the ash for deposition on heat exchange surfaces in the boiler plant (Barta et al. [49]).

At the high peak temperatures of pulverized coal combustion (2000 K) and especially under the conditions of slagging combustion (liquid ash removal), the coal mineral matter melts and partially vaporizes. The vaporization of the mineral matter in the flame is of great interest because its emission represents a health hazard due to the submicrometer size inorganic particulate formed on condensation of the vaporized ash (Gumz [50]). Another part of this inorganic vapor may condense on superheater tubes causing operational problems. Experimentally determined fractions of coal minerals which remained unvaporized at various temperatures are shown in Fig. 15 (Sarofim et al. [51]).

Environmental interest in the utilization of fly ash as an additive to clinker in the cement making process drew attention to the level of residual carbon in the ash, normally restricted to 5%. It has been the experience with some of the low NO$_x$ (staged air) combustion systems that the residual carbon in the fly ash has risen well above this concentration. Examinations of the structure of residual carbon samples by means of X-ray diffraction, optical reflectance, and high resolution transmission electron microscopy (HRTM) (Hurt and Hardesty [52], and Beeley et al. [53]) have shown that the coal undergoes significant structural changes on the atomic and mesoscale leading to changes in surface area or available surface sites as it passes through high temperature combustion regions. Reactivity measurements show a propensity for deactivation, which correlates with the extent of crystalline order development as observed by HRTM fringe imaging. They also find that deactivation affects more the more reactive vitrinite-rich sample and argue that this is the reason for the small difference between the reactivity of residual carbon samples originating from vitrinite-rich and inertinite-rich coal.

4. Power cycles of high thermodynamic efficiency

Pollutant emission from electric power generating plant can be further reduced by the improvement of the thermodynamic cycle of power generation. The power cycles chosen for discussion from a great variety of such schemes include:

- pulverized coal combustion (PCC) in supercritical steam boiler in a single Rankine cycle (e.g. 250 atm 3 × 853 K);
- natural gas fired gas turbine–steam combined (Brayton–Rankine) cycle (GCC);
- indirectly coal and natural gas fired (air heater) combined GT-steam cycles;
- pressurized fluidized bed (PFBC) without and with topping combustor; and
- integrated gasification combined cycle (IGCC) with coal, refinery waste, or biomass as fuel.

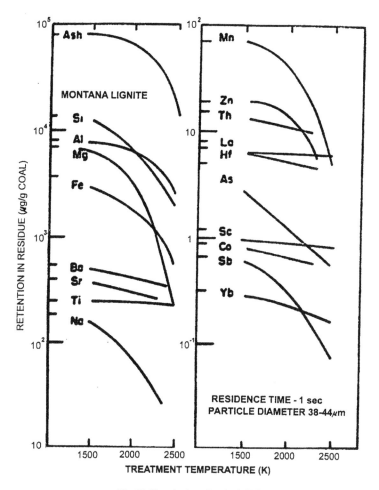

Fig. 15. Vaporization of coal ash [51].

Pulverized coal fired supercritical steam boilers have been in use since the 1930's but improvements in materials and increasing demand for higher efficiency are making this system presently the choice of new coal fired utility plant world-wide. The increase in efficiency is due to the higher mean temperature of the heat addition as illustrated by the *T–S* diagram of a supercritical steam cycle in Fig. 16 (Büki [54]). Because of the increased moisture content of the high pressure steam at the last stages of the steam turbine, the steam has to be reheated in the course of its expansion. The reheat, single or double, is, however, chosen not only to solve the problem of the reduced dryness fraction of the steam but also to raise the thermodynamic efficiency of the power cycle. In the example, the thermodynamic mean temperature can be seen to increase from T1 for the basic supercritical cycle with single superheat to Tu1 and Tu2 for single and double reheat, respectively. Comparison of

design parameters of a 300 MW subcritical steam cycle plant with a supercritical plant of the same performance shows an efficiency gain of 1.7% with a fuel saving of 50.000 t/y and a CO_2 emission reduction of 137.000 t/y for the supercritical unit (Table 2) (Hauser [55]).

The use of the supercritical boiler, however, carries implications for the combustion process. The lower heat transfer rate on the steam side of supercritical pressure boiler tubes compared to those at subcritical pressure demands spatially more uniform heat release in the combustion chamber, and the higher surface temperature of the superheater and reheater tubes may cause increased tendency for coal ash deposition.

The Rankin cycle efficiency of a pulverized coal fired steam plant can be increased in small steps to beyond 45% using supercritical steam parameters as shown in Fig. 17 (Schilling [56]).

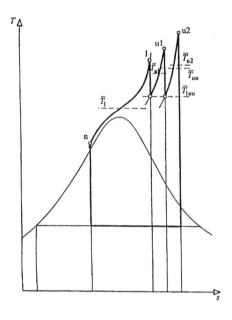

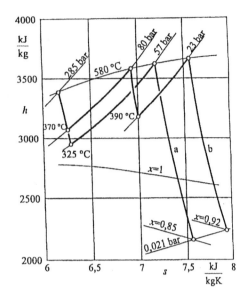

REVERSIBLE SUPERCRITICAL STEAM CYLE
WITH DOUBBLE REHEAT

SUPERCRITICAL STEAM TURBINE EXPANSION LINES
a.)Single, and b)Doubble reheat

Fig. 16. *T–S* diagram of a supercritical steam cycle [54]: left) reversible supercritical steam cycle with double reheat; right) supercritical steam turbine expansion lines for cases of single and double reheat.

4.1. Gas turbine–steam combined cycles

Because of the complementary temperature ranges of the Brayton (1600–900 K) and Rankine (850–288 K) cycles, their combinations can produce significantly improved thermodynamic cycle efficiency. A thermodynamic (temperature–entropy diagram) representation of the increased efficiency is shown in Fig. 18. In a more recent development of a reheat cycle (Meisl et al. [57]), in which additional fuel is injected downstream of the high pressure stage of the gas

Table 2
Design parameters for a 300 MWe coal-fired plant (Häuser [55])

	Subcritical	Supercritical
Steam generator	Controlled Circulation	Once Through
Steam turbine	Two casings	Three casings
Steam pressure (MPa)	16.5	25
Steam temperature (°C)	540	540
Reheat temperature 1 (°C)	540	566
Feedwater temperature (°C)	261	288
Cooling water temperature (°C)	0027	27
Heat rate (kJ/kWh)	7955	7522
Net efficiency (LHV) (%)	39.4	41.1
Coal consumption (t/y)	874,000	826,000

turbine (Sequential Combustion), combined cycle efficiency approaching 60% is achieved with ultralean premix low NO_x combustors.

Because of the low first cost and high efficiency, GCC is highly competitive with PCC despite the gas/coal price differential today and the uncertainty of future gas prices. The comparison of the levelized costs of electricity of a 500 MW coal and 225 MW gas fired combined cycle plant for 60% and 100% capacity factors in 2000 and 2010 are shown in Fig. 19a and b, respectively (Beamon and Wade [58]). As can be seen from this comparison, the present difference in cost of electricity between PCC and GCC is reduced at high capacity factors, and according to these estimates, the difference disappears in 2010 due to the rising gas prices and the expected reduction in the cost of PCC.

When coal or residual fuel oil is added as a supplementary fuel in the Heat Recovery Steam Generator of a GCC plant (Fig. 20), there is considerable gain in operational flexibility, albeit at the expense of a small reduction in the cycle efficiency. The gas turbine exhaust gas contains typically 12–16% O_2, depending on the thermal load. At strongly depleted O_2 concentrations, combustion stability and carbon burn-out of the supplementary fuel may present problems, while at higher O_2 levels and gas turbine exhaust gas temperatures, the NO_x emission needs special attention (Smart and deKamp [59]).

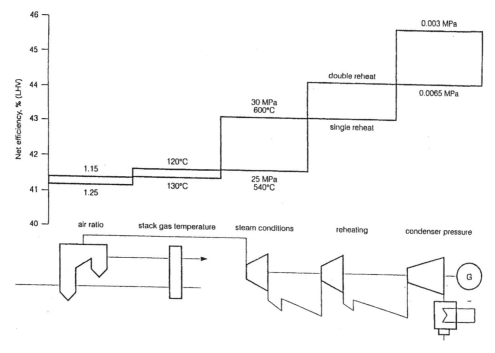

Fig. 17. Steam cycle improvements in incremental steps [56].

Gas turbines can be used for retrofitting existing steam plants (Hot Windbox Repowering) or applied to partial feedwater heating allowing thereby more of the steam flow in the steam turbine to expand to condenser pressure instead of being bled for regenerative feed water heating. This will then result in both higher efficiency and also increased electric power generating capacity.

The present gas/coal price ratio of 2.0 and the uncertainty of the gas price for the future motivate developments with partial coal/gas firing. In *indirectly fired gas turbine combined cycles* (Fig. 21), compressed air is preheated to high temperature in an atmospheric pressure coal burning furnace. A closed gas turbine cycle operating with high temperature air was suggested by Ackeret and Keller [60–63] in 1939. Mordell [64,65] further developed it into a semi-closed cycle in which the air from the gas turbine entered a pulverized coal fired furnace as combustion air. Bermann and Eustis [66] and Foster-Pegg [67] have used fluidized coal combustion in their analysis of a combined cycle, with steam also raised to generate power in a steam turbine and a natural gas fired topping combustor to increase the gas turbine entry temperature to 1600 K. The cycle efficiency was shown to exceed 50%. It is noteworthy that in contrast to the Combined Cycle with Supplementary Coal Combustion, coal makes a direct contribution to the heat input to the gas turbine in the Indirectly Fired Combined Cycle. An important advantage of this scheme is that it does not need flue gas cleanup as the gas turbine is entered by the clean medium of the combustion products of natural gas and air. The natural gas/coal energy input ratio depends on the level of air preheat, and in turn upon the air preheater material (superalloy steel or ceramic). The air preheat limit with presently available metallic heat exchanger materials is about 1100 K. Temperatures of 1400 K and higher are the aim of the High Performance Power Systems Program (HIPPS) of the USDOE (Ruth [11]; Seery [12]) (Fig. 22). The higher air preheat does not change the cycle efficiency but increases the fractional contribution of coal in the fuel mixture from 1/2 to 2/3. The problems of the higher temperature, however, affect the structural integrity of the heat exchanger which, on the furnace side, has to stand up to the corrosive attack of molten coal ash in the high temperature environment of slagging combustion of pulverized coal.

With the lower air preheat of 1100 K, the indirectly fired combined cycle could be demonstrated presently with CFBC. Fig. 23 illustrates such an air heater cycle designed for a 300 MW power plant for a subbituminous coal in Hungary (Beér and Homola [68]). The use of highly preheated air in the gas turbine entails problems concerning the cooling of metallic combustor walls, and NO formation and emission (see Topping Combustor).

Pressurized fluidized bed combustion (PFBC) [69] has grown out of the early development at BCURA in Leatherhead, UK in the 70's, the Grimethorp Experimental Facility in the UK in the 80's funded by Germany, the UK, the USA and the IEA., and R&D at the Stal Laval Company in

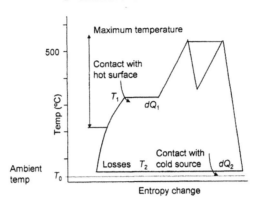

a Steam turbine

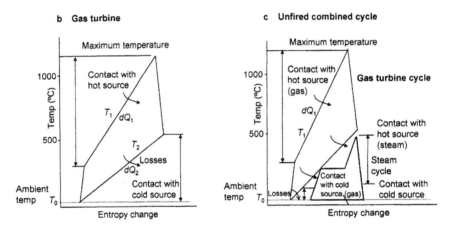

b Gas turbine

c Unfired combined cycle

Fig. 18. A thermodynamic representation of the improved efficiency of combined cycle plants [58]: (a) steam (Rankine cycle); (b) gas-turbine(Brayton cycle); and (c) unfired combined cycle.

Sweden in preparation of the 70 MWe demonstration plant to be built for the American Electric Power Co. Miniplants to study environmental performance of PFBC's were successfully operared at Exxon in Linden, New Jersey [70] and at the USDOE Argonne National Laboratory [71]. Massimilla's group at the University of Naples (Poletto [72]), and Jahkola's group at the University of Helsinki [73] have contributed valuably to the fundamental understanding of fluid flow, heat transfer, and combustion processes in pressurized fluid bed combustors.

A schematic of PFBC is shown in Fig. 24. Compared to FBC, the heat release rate per unit bed area in PFBC is several times higher and the bed height is 3–4 m instead of the typical bed height of 1 m in FBC. Under atmospheric pressure conditions, the bed height is limited by the acceptable pressure drop of about 100 mbar across the bed. In the

PFBC-GT Cycle, the 300 mbar pressure drop represents less than 3% of the total pressure ratio. A consequence of the increased bed height is a larger carbon inventory in the bed and lower NO emission due to the reduction of the formed NO by solid carbon in the bed (Chan et al. [30]). However, the high carbon load does not reduce the emission of the N_2O, which is still stable at the relatively low temperature of the PFBC. The low temperature is unfavorable also from the point of view of combined cycle thermodynamics; it is too low for an efficient gas turbine application.

The thermodynamic efficiency of the PFBC combined cycle units in operation is about 40%. Further increases in efficiency can be obtained by raising the turbine inlet temperature using a topping combustor.

The *second generation of PFBC* (Fig. 25) (Robertson et al. [74]) is a response to the needs of raising the gas

(a)

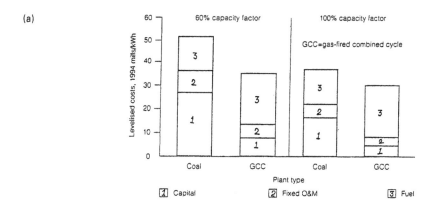

Forecast of levelised costs in 2000 at 60% and 100% capacity
factors in Florida, USA (Beamon and Wade, 1996)

(b)

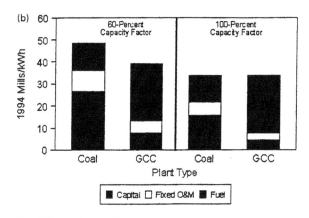

Note: GCC = natural-gas combined-cycle.
Source: Projection by Energy Information Administration, Office of Integrated Analysis and Forecasting.

Year 2010 Levelized Costs for Electric Power Plants in Florida at 60-Percent and 100-Percent

Fig. 19. Comparison of cost of electricity for PC and GCC plants in 2000 and 2010 at 60% and 100% capacity [58].

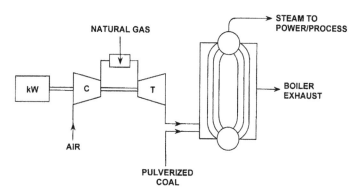

Fig. 20. Combined cycle with fired HRSG.

CIRCULATING FLUID BED SYSTEM WITH TOPPING COMBUSTOR

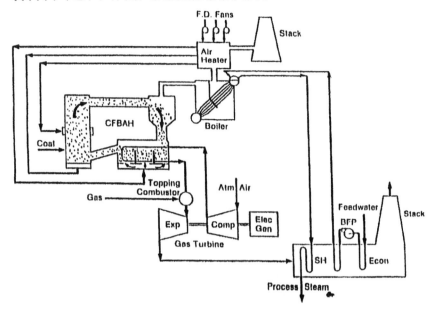

Fig. 21. Indirectly fired gas turbine combined cycle (CFBC air heater cycle with topping combustor) [66].

temperature at the inlet to the gas turbine. In this cycle, coal, usually in the form of coal-water slurry, is injected into a pressurized carbonizer where it undergoes mild gasification to produce a low calorific value gas and char. The char is burned in a pressurized circulating fluidized bed (PCFBC) with high excess air, and the flue gas is cleaned of particulates and alkali at high temperature. Sulfur is captured in the PCFBC by additive dolomite. The fuel gas from the

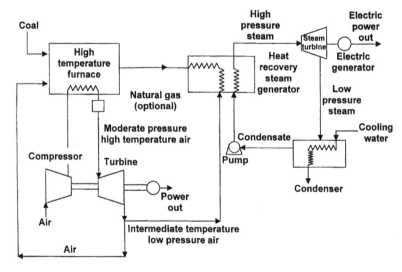

Fig. 22. HIPPS plant schematic [11,12].

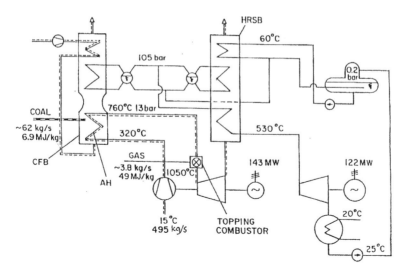

Fig. 23. Schematic of a 300 MW CFB air heater combined cycle for hungarian lignite with natural gas topping combustor [68].

carbonizer is cleaned of sulfur in the fluidized bed carbonizer, and of particulates and alkali by hot gas cleanup. It is then injected into the topping combustor where it raises the temperature at the inlet to the gas turbine to 1623 K. This temperature rise increases the cycle efficiency to about 47%. Further improvements in efficiency can be obtained by the application of advanced gas turbine technology, and on the steam side, by supercritical steam parameters with high temperature double reheat. An additional advantage of this cycle is that the N_2O emission is eliminated because the N_2O formed in the pressurized fluidized combustor decomposes

at the elevated temperature in the Topping Combustor (Beér and Garland [38]).

4.2. The topping combustor

Because the air entering the combustor is at 1144 K rather than the usual 644 K for gas turbines, the conventional type of combustor is not suitable. Both the emissions and cooling problems preclude the use of conventional designs. An all metallic multiannular combustor (MASB) patented by Beér [75] and developed by Westinghouse (Fig. 26) solves the

Basic PFBC Process

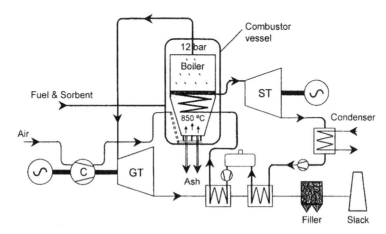

Fig. 24. Pressurized fluidized bed combined cycle.

222

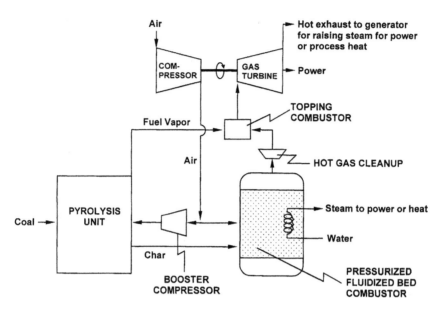

**Pressurized Fluidized
Bed Combined Cycle**

Fig. 25. Second generation PFBC with topping combustor [74].

cooling problem by creating thick layers of gas flow over the leading edges of overlapping concentric annular passages in the combustor. Due to the high convective heat transfer rates, the temperature difference between the high alloy metal combustor wall and the 1144 K PFBC exhaust gas is less than 200 K, even at 1625 K combustor outlet temperature.

The mixing of fuel and air follows the Rich-Fast Quench-

Lean sequence which gives very low NO$_x$ emissions, with less than 10% of the fuel nitrogen in the syngas converted to NO, and has a high combustion efficiency of 99.9%. The MASB pressure drop is less than 1% [38].

An additional demand of the topping combustor is that for cold start-up it has to be capable of operation with natural gas as a fuel. Experiments carried out by Westinghouse under a DOE Contract in 1986 using a 254 mm (10″)

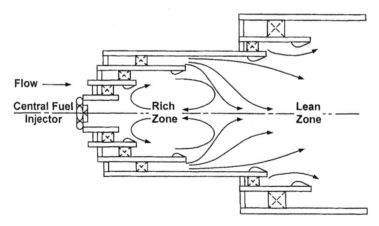

Fig. 26. Multi annular swirl burner (MASB) topping combustor [75].

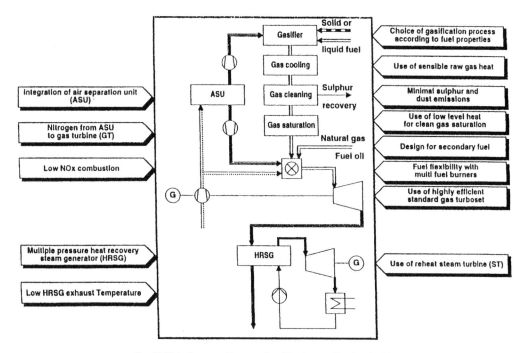

Fig. 27. Main features of integrated gasification combined cycle [76].

diameter MASB, with 1034 K (1400°F) combustion air and burning natural gas yielded wall temperatures less than 100K higher than the air temperature. The condition of the combustor was excellent even though the calculated rich zone adiabatic temperature was 2458 K (3964°F). It has to be verified experimentally, however, that these favorable results hold for the scaled-up, 457 mm (18″) combustor in which radiative heat transfer to the combustor walls is expected to increase on account of the higher gas emissivity.

4.3. Integrated gasification combined cycle

IGCC involves the total gasification of coal, mostly with oxygen and steam, to produce a high heating value fuel gas for combustion in a gas turbine. The gasifier also produces steam for a steam power cycle. The fuel gas has to be cleaned of particulates, alkali, and sulfur compounds; the fuel-nitrogen bearing gas is burned in a low NO_x combustor. The main features of IGCC are shown in Fig. 27 (Haupt and Karg [76]).

IGCC is the cleanest advanced coal technology. It is also demonstrated to be working with no major operational problems. The future of IGCC depends on whether it will be possible to reduce its first cost and to increase the cycle efficiency. The cost is presently high, mainly because of the oxygen plant necessary for the oxygen blown gasifier, and because of less than complete integration of

the various subsystems such as the gasifier air separation system, fuel gas cooler and cleanup, gas turbine, and steam plants.

Air blown gasifiers have also been developed for IGCC systems. Their drawback is that the reduced heating value of the fuel gas they produce makes the use of hot gas clean up, a technology still under development, indispensable. Also, because they have to process larger gas volumes, the costs of the fuel gas and flue gas treatment plants are higher. The existing IGCC demonstration plants in the USA have efficiencies below 40%, but two more recently commissioned European IGCC demonstration plants, the one in Buggenum in the Netherlands, and the other, the Puertollano plant in Spain (Huth et al. [77]), both of which began operation in 1993, have higher design efficiencies of 43% and 45%, respectively. The higher cycle efficiencies are mainly due to improved gas turbine and steam plant efficiencies and better sub-system integration.

An example of such an improved-sub-system integration occurs at the boundary of the oxygen blown gasifier and the gas turbine combustor. The air separation unit produces oxygen for the gasifier in the Portollano plant at 13 bar pressure. The nitrogen which is also available at high pressure is mixed with the clean gas to feed the gas turbine combustors; this helps to control NO_x emission and increases the mass flow rate through the turbine increasing thereby the turbine performance. The NO_x is further lowered through the saturation of the coal gas by water (condensate

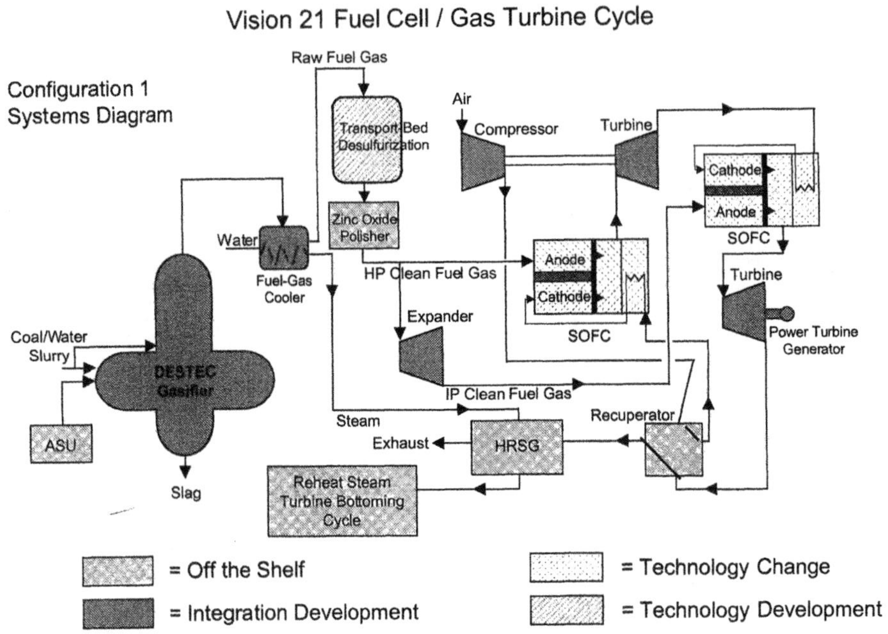

Vision 21 Fuel Cell / Gas Turbine Cycle

Fig. 28. US DOE's vision 21 fuel cell/GT cycle [80].

from the steam cycle) heated up by the sensible heat of the compressed air (673 K) at the compressor outlet.

Even with these improved cycle efficiencies, however, IGCC plants are presently not competitive with other advanced coal burning systems such as PCC fired supercritical steam plants because of their higher installation cost. Nevertheless, there are further considerations that may, in the future, tilt the balance in favor of IGCC applications:

- IGCC lends itself for the efficient removal of CO_2 from the high-pressure fuel gas (Pruschek et al. [78]).
- By broadening the fuel supply to the strongly increasing volume of Refinery Wastes (heavy residual oils, petroleum coke, Orimulsion), IGCC could become attractive for clean and efficient Central Power Generation by using fuels of very low or even "negative" cost (the waste fuel cost is negative if it stands against the cost of disposal (Empsperger and Karg [79])).
- The cycle efficiency could approach 60% in a scheme in which hydrogen is produced in the gasification process and the Brayton–Rankine Cycles are combined with fuel cells.

An example, the US DOE's Vision21 Gasification/Gas Turbine/Fuel Cell Cycle Schematic is shown in Fig. 28 and the Plant Performance Summary in Fig. 29 (Ruth [80]).

5. Comparison of clean coal technologies

Several studies attempted to compare the emerging Clean Coal Technologies with PCC + Flue Gas Treatment (FGT) from the points of view construction cost, operating cost, cost of electricity, fuel flexibility, pollutant emission, and cycle efficiency (CO_2 emission). Efficiency is becoming an important factor as it is the main determinant of CO_2 emission. Should there be a carbon tax in the OECD countries, this may become also an important cost item.

VISION 21 FUEL CELL/GT CYCLE

Plant Performance Summary

Gasifier	Destec
Coal Input to Gasifier, kg/h	116,288
Thermal Input, MW	875.8
HP Solid Oxide Fuel Cell Module MW, dc/ac	189.4/182.8
LP Solid Oxide Fuel Cell Module MW, dc/ac	121.4/117.2
Gas Turbine, MW	133.7
Steam Turbine, MW	118.0
Fuel Expander, MW	9.6
Gross Power, MW	561.3
Auxiliary Power, MW	40.4
Net Power, MW	520.9
Efficiency, % HHV	59.5

Fig. 29. Vision 21 projected performance summary [80].

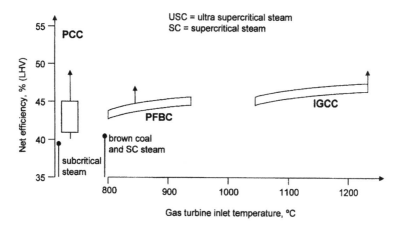

Fig. 30. Comparison of high efficiency cycles; effect of GT inlet temperature [81,82].

For a comparison of these technologies, it is also important to forecast the development of some enabling technologies such as new superalloys for gas turbine blading and for superheaters of supercritical steam boilers.

It is generally agreed that PCC + FGT has an edge over the coal fired combined cycle systems in the short and medium term mainly because it is a mature technology and also because competition and R&D in recent years has reduced its first cost to below US$ 1200 per kW installed capacity.

Potential efficiencies of PCC, PFBC and IGCC as a function of gas turbine inlet temperature are illustrated in Fig. 30 (Lovis et al. [81]; Couch [82]). As the GT inlet temperature rises, so does the combined cycle efficiency (the cycle efficiency increases about 2% for 100 K rise in gas turbine inlet temperature); the single cycle PCC remains, of course, unaffected. The diagram shows that PCC with supercritical steam promises comparable efficiency to those of the coal fired combined cycle systems. Table 3 contains data about the number and type of coal fired utility plants under construction world wide (based on Couch [82]). Clearly the great majority of these plants are PCC with supercritical steam boilers.

Table 3
Utility plants under construction 1997–2000 (based on IEA 1997 data)

Total number of plants worldwide	42	
Distribution by combustion system		
Pulverized coal	35 (29 supercrit. steam)	700–1000 MWe
CFBC	3 (all subcrit. steam)	200–235 MWe
PFBC3	3 (1 supercrit. steam)	80–360 MWe
IGCC	1 (subcrit. steam)	86 MWe

Comparisons of environmental performance data at full load estimated by Delot et al. [83] are shown in Table 4. The low temperature fluid bed systems, except of the second generation PFBC with the topping combustor, have a problem with N_2O emission, while PCC may face some future regulation on fine particulate (PM 2.5) and HAP (e.g. mercury) emissions. IGCC is the potentially cleanest of the advanced coal fired cycles. It has the special advantage that it is amenable to the efficient capture of CO_2 from its high pressure syngas stream. This is further underlined by the comparative illustration of environmental advantages of advanced power cycles in Fig. 31 [76].

Table 5 has the data on costs relative to that of PCC + FGT [75]. It is noted that the advantage of supercritical steam for the efficiency affects more favorably the single steam cycles, PCC and CFBC, rather than the combined cycles, because the latter benefit only partially from advanced steam parameters. Advances in gas turbines on the other hand, lead to higher thermodynamic efficiency for the combined cycle plants (see Fig. 30). The prospective net efficiency of IGCC at 44.5% LHV seems somewhat low. Improved subsystem integration is promising IGCC efficiency close to 50% [76].

5.1. CO_2 sequestration

In addition to current strategies of improving the efficiency of power generation, greenhouse gas emission can be mitigated by CO_2 sequestration (Herzog and Drake [84]. The discussion of the technology of CO_2 separation from the flue gas, and its disposal fall outside the scope of this review. There are, however, combustion measures, which by increasing the CO_2 concentration in the flue gas can reduce the costs associated with CO_2 separation. One such technique applicable to atmospheric pressure boiler-combustion, is oxygen enrichment of the combustion air and recirculation of high CO_2 bearing flue gas through the burners. Thambimutu

Table 4
Environmental performances at full load (mg/N m^3, 6% O$_2$)

Technology	PCC + FGT	CFBC	PFBC	PCFBC	IGCC	TC
Nitrogen oxides						
Intrinsic	800–1300	150–250	200–300	100–200	150–200	150–300
Low-NO$_x$ burners	400[a]					
In project		75–120	70–200	40–100		
With SCR	100–200					50–150
Nitrous oxide	0–5	20–100	20–100	20–100	0–5	0–5
		[b]	[b]	[b]		[b]
Sulfur dioxide (1.0% S)						
Intrinsic	2000	200	200	200	10–25	200
Ca/S ratio	1.05	2.5	2.2	1.5		2.0
With FGD	200					
Dust	50	50	50	10[c]	10[c]	10[c]
			10[c]			

[a] For new boilers 400 mg/N m^3; from 500 to 700 mg/N m^3 for existing ones.

[b] With ceramic filters.

[c] It is possible to reach emission levels below 200 mg/N m^3 by simply increasing the quantity of limestone injected into the furnace, and consequently the Ca/S ratio.

and Croiset [85] have shown that while such a scheme has to bear the cost of an air separation unit to produce the oxygen, there is some compensation in the reduced volume of the gas, which improves the boiler efficiency. Reduction of NO$_x$ emission is also demonstrated [85].

The adiabatic temperature is generally higher with O$_2$ enriched air, but enrichment can be increased further without increased temperature for the case of recirculated CO$_2$ due to its high heat capacity. Increased levels of O$_2$ concentration may promote the conversion of organically bound nitrogen compounds to NO but the NO recirculated with the flue gas is getting reduced in the flame. The laboratory

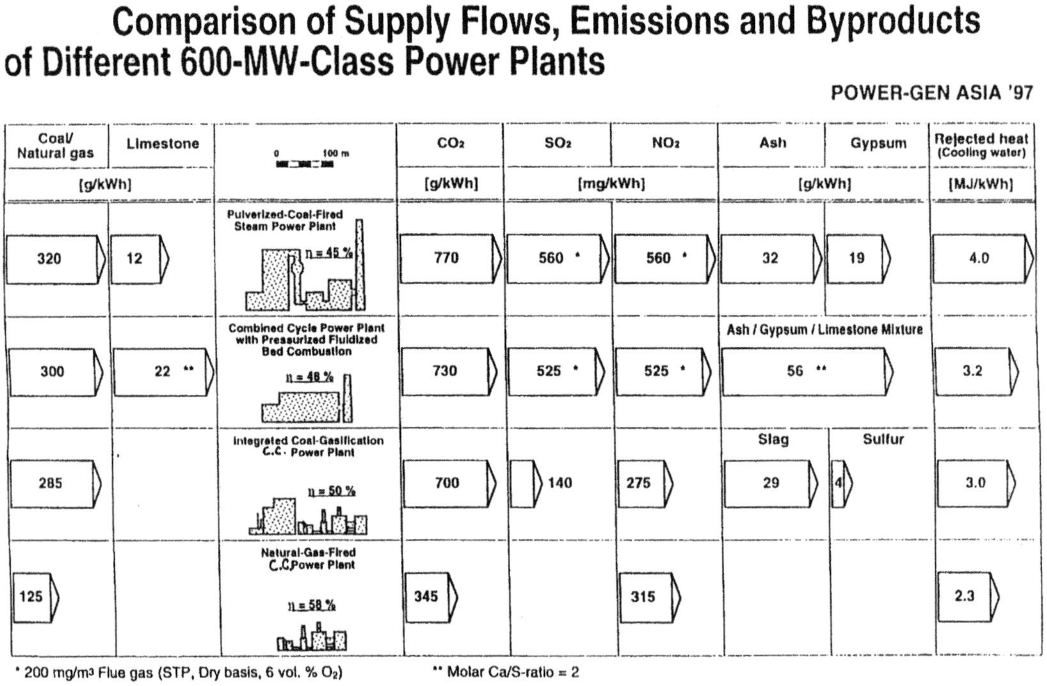

Comparison of Supply Flows, Emissions and Byproducts of Different 600-MW-Class Power Plants

POWER-GEN ASIA '97

* 200 mg/m³ Flue gas (STP, Dry basis, 6 vol. % O₂)   ** Molar Ca/S-ratio = 2

Fig. 31. Comparison of supply flows, emissions and byproducts of different 600 MW-class power plants [76].

Table 5
Costs in percent of PC + FGT technology costs (costs are estimated for a serial unit, without prototype effect, and assuming that each technology has become industrially mature)

Technology	PCC + FGT	CFBC	PFBC	PCFBC	IGCC	TC
Space requirement of the unit block (m^2)	9000	6600	5700	3600	28,000	3900–6900
Net efficiency (% LHV)	45	44.5	43	46	44.5	48
Construction duration (months)	45	43	42	42	48	46
Construction cost (%)	100	90	104	86	117	106
Capital cost (%)	100	90	103	85	118	106
Operating and maintenance cost (%)	100	90	130	123	155	145
Fuel cost (%)	100	102	104	97	101	94
Cost of electricity (%)	100	95	108	97	116	107

studies show lower NO emission from flames with coal as the fuel when O_2/CO_2 is the oxidant instead of air [85]. The resulting effluent is a CO_2 rich flue gas (up to 98% vol. dry) that can be removed for utilization or disposal with minimum flue gas treatment. Clearly, enriched oxygen coal fired combustion has potential for application in both new and existing power plants.

As an alternative to recycling CO_2, Anderson et al. [86], and Bilger [87] recommend the use of pure oxygen in a steam generator burning a hydrocarbon fuel with water injection and water-vapor recirculation while separating the CO_2 from water by condensation. The combustion products of 90% H_2O and 10% CO_2 enter the HP steam turbine; the gases flow through a second boiler where further fuel and oxygen is added before their expansion in the IP and LP turbines. This is a Rankine Cycle with a calculated efficiency for a 100 MWe plant of 63% compared with a corresponding combined cycle efficiency of 50%. The recommended cycle, however, depends on the development of future high temperature (2000 K) steam turbines.

6. Conclusions

- Coal is the prevailing fuel of power generation world wide, and it is likely to remain important, well into the next century.
- Natural gas in gas turbine combined cycle will, in the short and medium term, replace some coal fired capacity, especially in distributed power generation, and in heat and power cogeneration.
- Because of coal's pollution, and especially due to the expected future limitations on CO_2 emissions, clean coal utilization technology with high thermodynamic efficiency will have to be applied in the new generation of coal fired central power stations to be built in OECD countries.
- Pulverized coal combustion in supercritical steam boilers (240 atm $3 \times 565°C$) is the likely choice for new central power plants in the short and medium term because of the relatively high cycle efficiency (42–45% LHV) and the

long experience with pulverized coal combustion. Also, the cost of these plants is continually getting reduced (presently about US$ 1200 per kW). There are more than 30 SC plants being commissioned at present in Europe, Australia, Japan, S. Korea, and Taiwan. Further developments towards ultra-supercritical coal plants (300 atm $3 \times 650°C$) with 50% single cycle efficiency are dependent on progress in materials R&D; applications are expected past 2010.

- Indirectly fired air heater combined cycles with 50/50 coal/natural gas fueling and 52% efficiency are available now for demonstration or for repowering existing power plants. R&D in progress is aiming at less than 36% natural gas contribution using refractory coated high alloy steel or refractory air preheaters in a furnace fired with pulverized slagging coal combustion. The cost of this process and the technical risk/benefit ratio increases with preheat temperature. The air heater combined cycle with 1173 K air preheat is likely to be ready for demonstration by 2005.
- Pressurized fluidized combustion combined cycle: Several plants of 70 MWe have been operating satisfactorily since 1991 and a larger plant (360 MWe) is starting up in 1999 in Japan. Low alkali coal permits the use of higher pressure ratio of a GT23 with 1123 K gas turbine entry temperature. Efficiency is around 40% (LHV). Emissions are low, except N_2O (50–100 ppm).
- Topping cycle (2nd generation PFBC): Pregasification produced syngas raises turbine inlet temperature to 1573 K with efficiency increase to 47%. N_2O is eliminated at the elevated temperature in the topping combustor. Enabling technologies are: Hot gas cleanup and topping combustor. Demonstration is expected by 2010.
- IGCC has a long history of demonstrations. The demonstration plants in the US were designed to have relatively low cycle efficiency (~40%). Interest in IGCC technology has risen recently because it provides favorable conditions for CO_2 sequestration and possibilities for efficiencies exceeding 50% by the production of hydrogen and the combination of IGCC with fuel cell technology. Lack of sensitivity to fuel quality variation makes the IGCC capable

J.M. Beér / Progress in Energy and Combustion Science 26 (2000) 301–327

of using inexpensive oil refinery wastes as fuels which opens up a new cost effective mode of power generation.

- Combustion research and R&D have been instrumental to the development of advanced clean and efficient coal burning power plant technologies. These new technologies have to be demonstrated and further developed in order to make their contribution to minimizing emissions of pollutants, including of CO_2, from electric power generation worldwide. It is essential for the success of the development of these new high efficiency power generating systems that R&D and demonstration be accompanied by more fundamental modeling and laboratory scale experimental studies. This is the cost-effective way of interpreting, extending, and generalizing the data obtained in demonstration trials, data which are always scarce because of the high cost of full-scale experimentation.

Acknowledgements

Parts of the paper have been presented by the author in the International Workshop "Energy and the Environment" of the Accademia dei Lincei in Rome, Italy in December 1998; in a Seminar at ABB's Corporate Research Center in Baden-Daettwill, Switzerland in June 1999; and in a Plenary Lecture of the Mediterranean Combustion Symposium in Antalya, Turkey in July 1999. The author is grateful for the generous hospitality extended to him at these meetings. Thanks are due to Mr Don McGaffigan for his valuable assistance in the production of the manuscript and illustrations in this paper.

References

[1] Werkmeister H. Arch Warmewirt 1931;12:225–32.

[2] Thring MW. The science of flames and furnaces. Chapman and Hall, 1962.

[3] Beér JM. Sheffield University Fuel Soc J 1958:1–12.

[4] Field MA, Gill DW, Morgan BB, Hawksley PGW. Combustion of Pulverized Coal, BCURA, Leatherhead, England, 1967. 413pp.

[5] Dolezal R. In: Beér JM, editor. Large boiler furnaces. Fuel and energy science monographs, Amsterdam: Elsevier, 1967.

[6] Essenhigh RH. In: Lowry HH, editor. Chemistry of coal utilization, 2nd supplementary volume. New York: Wiley, 1981 (chap. 19).

[7] Beér JM, Chomiak J, Smoot LD. Prog Energy Combust Sci 1984;10:177–208.

[8] Smoot LD, Smith PJ. Coal combustion and gasification. New York: Plenum Press, 1985.

[9] Wall TF. In: Lawn CJ, editor. Principles of Combustion Engineering for Boilers, 1987 (chap. 3).

[10] Dolezal R. Mitt VGB 1957;49:223–33.

[11] Beittel R. In: Proceedings of advanced coal based power and environmental systems, 1998 conference DOE FETC, Morgantown, PA, 21–23 July 1998.

[12] Seery D, Sangiovanni J, Proceedings of advanced coal based power and environmental systems, 1998 conference DOE FETC, Morgantown, PA, 21–23 July 1998.

[13] Reid WT. In: Beér JM, editor. External corrosion and deposits,

boilers and gas turbines. Fuel and energy science monographs, New York: Elsevier, 1971.

[14] Borghi G, A F, Sarofim JM. Beér. Comb Flame 1985;61:1–16.

[15] Kerstein AR, Niksa S. The Twentieth Symposium (International) on Combustion, The Combustion Institute, Pittsburgh, PA, 1984. p. 941–9.

[16] Beér JM, Massimilla L, Sarofim AF. Fluidized combustion systems and applications, Inst. of Energy Symposium Series No.4., London, 1980.

[17] Chirone R, D'Amore M, Massimilla L. The Twentieth Symposium (International) on Combustion, The Combustion Institute, Pittsburgh, PA, 1984. p. 1505–11.

[18] Sundback CA, Beér JM, Sarofim AF. The Twentieth Symposium (International) on Combustion, The Combustion Institute, Pittsburgh, PA, 1984. p. 1495–503.

[19] Sarofim AF, Beér JM. The Seventeenth Symposium (International) on Combustion, The Combustion Institute, Pittsburgh, PA, 1978. p. 189–204.

[20] Beér JM, Hottel HC. Plenary lecture, Twenty-Second Symposium (International) on Combustion, The Combustion Institute, Pittsburgh, PA, 1988. p. 1–16.

[21] Wendt JO, Sternling CV, Mafovich MA. The Fourteenth Symposium (International) on Combustion, The Combustion Institute, Pittsburgh, PA, 1973. p. 897–904.

[22] Soud HN, Fukusawa K. Developments in NO_x abatement and control, IEACR 89, IEA Coal Research, London, UK, 1996.

[23] Emmons HW, Jing SJ. The Eleventh Symposium (International) on Combustion, The Combustion Institute, Pittsburgh, PA, 1967. p. 475–9.

[24] Chigier NA, Beér JM, Grecov D, Bassindale K. Combust Flame 1970;14:171–9.

[25] Toqan MA, Beér JM, Jansohn P, Sun N, Shihadeh A, Teare JD. The Twenty-Fourth Symposium (International) on Combustion, The Combustion Institute, 1992. p. 1391–7.

[26] Shihadeh AL, Toqan MA, Beér JM, Lewis PF, Teare JD, Jimenez JL, Barta L. ASME Fact 1994;18:195–200.

[27] Beér JM, Barta LE, Lewis PF, Wood V, Akinyemi O, Haynes J, Jimenez J, Manurung R, Rogers LW. In: Proceedings of EPRI/EPA 1995 Joint Symposium on Stationary Combustion NO_x Control, Book3, Session 7/A, Electric Power Research Institute, Palo Alto, CA, USA, 1995.

[28] LaFlesh R, Briggs O, Barlow D, Wessel G. ASME IJPG Conference, San Francisco, CA., USA 1999.

[29] Folsom B, Hong C, Sommer T, Pratapas JM. In: Proceedings of the 1993 Joint Symposium on Stationary Combustion NO_x Control, EPRI, Palo Alto, Ca., USA Rep. TR103265-V2, 1993. p. 7A, 15–34.

[30] Chan LK, Sarofim AF, Beér JM. Comb Flame 1983;52:37–45.

[31] Borio R, Lewis R, Keough MB, Booth RC, Hall RE, Lott RA, Kokkinos A, Gyorke DF, Durrant S, Johnson HJ, Kienkle JJ. In: Proceedings of the 1991 Joint Symposium on Stationary Combustion NO_x Control, EPRI, Palo Alto, CA, USA, Rep. TR103265-V2, 1991. p. 49–68.

[32] DeMichele G, Malloggi S, Merlini S, Passini S. SIAM Fourth Int Conf on Numerical Combustion, St. Petersburg, FL, 1991.

[33] LaFlesh R, Marion J, Towle D, Maney C, DeMichele G, Passini S, Bertecchi S, Piantanida A, Galli G, Manini G. ASME Int Joint Power Generation Conf, San Diego, CA, USA, 1991.

[34] Payne R, Moyeda DK, Maly P, Glavicic T, Weber B. In: Proceedings of EPRI/EPA 1995 Joint Symposium on

Stationary Combustion NO$_x$ Control, EPRI, Palo Alto, CA, USA, Book 54, Session 8, 1995.

[35] Ehrhardt KM, Toqan M, Jansohn P, Teare JD, Beér JM, Sybon G, Leuckel W. Combust Sci Tech 1998;131:131–46.

[36] Gibbs BM, Pereira FJ, Beér JM. The Sixteenth Symposium (International) on Combustion, The Combustion Institute, Pittsburgh, PA, 1976. p. 461–74.

[37] Tullin CJ, Sarofim AF, Beér JM. In: Proc Int Conf Fluidized Comb, Book No. 10344, 1993.

[38] Beér JM, Garland RV. Coal fueled combustion turbine cogeneration system with topping combustion. Trans ASME J Engng Gas Turbines Power 1997;119(1):84–92.

[39] Maghon H, Kreutzerand A, Termuehlen H. In: Proc American Power Conf 60, 1988.

[40] Beér JM. J Inst Energy 1995;LXVIII(474):2–10.

[41] Sattlemayer TH, Felchin MP, Haumann J, Hellat J, Styner D. ASME Paper 90-GT-162, 1990.

[42] Becker B, Berenbrink P, Brandtner H. ASME Paper 86-GT-157, 1986.

[43] Johnson HR, Littler DS. In: Johnson HR, Littler DS, editors. Proceedings of the Marchwood Conference, The Mechanism of Corrosion by Fuel Impurities, London: Butterworths, 1963.

[44] Wall TF. The Twenty-Fourth Symposium (International) on Combustion, The Combustion Institute, Pittsburgh, PA, 1992. p. 1119–26.

[45] Couch GR. Understanding slagging and fouling during PF combustion, IEACR/72, IEA Coal Research London, UK, 1994.

[46] Engineering Foundation Conferences, Bryers RW, Vorres KS, editors, published on behalf of the Eng Found ASME, New York, 1990.

[47] Engineering Foundation Conferences, Benson SA, editor, published on behalf of the Eng Found ASME, New York, 1991.

[48] Engineering Foundation Conferences, Benson SA, editor, published on behalf of the Eng Found ASME, New York, 1993.

[49] Barta LE, Beér JM, Sarofim AF, Teare JD, Toqan MA. Eng. Found. Conference. Washington, DC: Taylor and Francis, 1993. p. 177–88.

[50] Gumz W, Kirsch H, Mackowsky MT. Schlackenkunde. Berlin: Springer, 1958.

[51] Sarofim AF, Howard BJ, Padia AS. Combust Sci Tech 1977;16:187.

[52] Hurt RH, Hardesty DR. Sandia Technical Report, SAND 93-8230, 1993.

[53] Beeley T, Crelling J, Gibbins J, Hurt R, Lunden M, Man C, Williamson J, Yang N. The Twenty-Sixth Symposium (International) on Combustion, The Combustion Institute, Pittsburgh, PA., 1996. p. 3103–10.

[54] Büki G. Magyar Energetika 1998;6:33–42.

[55] Hauser U, OECD/IEA Paris France, 1993. p. 381–95.

[56] Schilling HD. VGB Kraftwerkstechnik 1993;73(8):564–76 (English Edition).

[57] Meisl J., Knapp K, Leuckel W, Wittig S. Forschungsberichte Verbrennungs Kraftmaschienen FVV, 1994.

[58] Beamon JA, Wade SH. Energy Equipment Choices, Energy Information Admin., US DOE, Washington, DC, USA, 1996.

[59] Smart JP. van de Kamp. J Inst Energy 1994;67:78–82.

[60] Ackeret J. Keller. Sweiz Bauzeitg 1939;113:229.

[61] Ackeret J, Keller. Z VDI 1941;85:491.

[62] Ackeret J. Engineering 1946;161:1.

[63] Keller, Trans ASME 1946:791.

[64] Mordell DL. Proc Inst Mech Engnrs 1955;169:163–80.

[65] Mordell DL. Engineer 1957;203:210–3.

[66] Berman PA, Eustis JN. Atmospheric Fluid Bed Air Heater for Coal Fired Cogeneration, Report to DOE by Westinghouse Electric Corporation, 1986.

[67] Foster-Pegg RW. Private communication, 1983.

[68] Beér JM, Homola V. Technological Initiatives to Upgrade Power Plants IEA–USDOE–USAID Conference, Budapest, Hungary, 1992.

[69] Roberts AG, Pillai KK, Stantan JJ. In: Howard JR, editor. Pressurized fluidized combustion in fluidized beds, London: Applied Science Publishers, 1983.

[70] Hoke RC, Bertrand RR, Nutkis MS, Kinzler DD, Ruth LA, Gregory MW. USEPA-600/7-76-011 Office of Research and Development, Washington, DC, 1976.

[71] Vogel GJ. A development program on pressurized fluidized bed combustion, Annual Report to ERDA Office of Fossil Energy, Argonne National Laboratory, 1975.

[72] Paletto M, Miccio M, DeMichele G. La Termotechnica, November, 1992.

[73] Horvath, A Hulkkonen S, Jahkola A. The Fourth International FBC Conference, London, UK, 1988.

[74] Robertson A, Garland R, Newby R, Rehmat A, Rebow L. Second Generation Pressurized Fluidized Bed Combustion Plant, Foster Wheeler Development Corp.Report, FWC/FWDC-TR 89/11 (3 volumes) to the DOE US, DE-AC21-86MC21023, 1989.

[75] Beér JM. Low NO$_x$ rich-lean combustion especially useful in gas turbines: British Patent 1968; US Patent 4,45940, 1989.

[76] Haupt G, Karg J. In: Proc of the 12th Conf on the Electric Supply Industry (CEPSI), Pattaya/Thailand, November 1998.

[77] Huth M, Vortmeyer N, Schetter B, Becker B, Karg J. In: Fourth International Conference on Technologies and Combustion for a Clean Environment, Lisbon, Portugal, 1997.

[78] Pruschek R, Goettlicher G, Oeljeklaus G, Haupt G, Zimmermann G. Power-Gen Europe 98 Milan, June 1998.

[79] Empsperger W, Karg J. ASME Power Engineering Journal, February 1997.

[80] Ruth LA, US DOE Vision 21 Workshop, FETC, Pittsburgh, PA, December 1998.

[81] Lovis M, Drdziok A, Witchow A. In: Proc. Power-Gen Europe 1994, Penn Well Utrecht, Netherlands, 1994. p. 327–49.

[82] Couch GR. OECD Coal Fired Power Generation, IEA Per/33, 1997.

[83] Delot P, DiMaggio I, Jaquet L, Roulet V. EDF Comparative Study of Clean Coal Technologies, Electricite de France Thermal Department, 1996.

[84] Herzog HJ, Drake EM. IEA/93/0E6, IEA Greenhouse Gas R&D Program, Cheltenham, UK, 1993.

[85] Thambimuthu KV, Croiset E. In: Proc. Advanced Coal Based Power and Environmental Systems 1998 Conference, USDOE, Morgantown, Paper 4.4, July 1998.

[86] Anderson R, Brandt H, Muggenburg H, Taylor H, Viteri F. In: Proc. Fourth International Conference on Carbon Dioxide Removal, IEA, Interlaken, Switzerland, 1998.

[87] Bilger RW. Fifth International Conference on Technologies and Combustion for a Clean Environment, Lisbon, Portugal, July 1999.

Emissions Reduction: NO$_x$/SO$_x$ Suppression
A. Tomita (Editor)

Alternative fuel reburning

Peter M. Maly, Vladimir M. Zamansky, Loc Ho, Roy Payne

Energy and Environmental Research Corporation, 18 Mason, Irvine, CA 92618, USA

Received 30 April 1998; accepted 7 August 1998

Abstract

Advanced reburning is a NO$_x$ control technology that couples basic reburning with the injection of nitrogen agents and promoter compounds. Pilot scale experiments were conducted in which efficiency of basic and advanced reburning processes were characterized with a wide range of reburn fuels. Test fuels included natural gas, pulverized coal, coal pond fines, biomass, refuse derived fuel, and Orimulsion. Process variables that were studied included reburn fuel type, reburn fuel heat input, reburn zone residence time, initial NO$_x$ concentration, nitrogen agent injection temperature, and promoter type and amount. Reburn fuel properties found to affect the performance most significantly include fuel nitrogen content, volatiles, and ash constituents. Basic reburning performance for the tested solid fuels was found to approach that of natural gas reburning, with over 70% NO$_x$ reduction achievable at reburn heat inputs above 20%. Advanced reburn tests were conducted in which reburning was coupled with injection of nitrogen agents and promoters. The most effective promoter compounds were found to be alkalis, most notably sodium compounds. At reburn heat input of 10%, NO$_x$ reductions in the range of 85%–95% were achieved with natural gas and biomass advanced reburning. © 1999 Elsevier Science Ltd. All rights reserved.

Keywords: Air pollution; Reburning; Nitrogen oxides; Acid rain

1. Introduction

Nitrogen oxide emissions from stationary source combustors are large contributors to a number of environmental hazards, including acid rain, high ground-level ozone concentrations, and elevated fine particulate levels. Therefore, there is increasing need for the development and application of cost effective technologies for controlling these emissions. Reburning is a mature NO$_x$ control technology that has been demonstrated on several coal-fired boilers in the United States [1,2] and world-wide [3]. Reburning is a three-step process, involving combustion of majority of the fuel under normal fuel lean conditions, followed by injection of a reburning fuel to establish a fuel-rich zone in which nitrogen oxides formed in the primary combustion zone are reduced to molecular nitrogen, and finally injection of overfire air to oxidize carbon monoxide, hydrogen, and any remaining hydrocarbons exiting the reburning zone. The basic reburning process typically provides 50%–60% NO$_x$ control.

To provide higher levels of NO$_x$ control, technologies are being developed based on hybrid schemes of reburning plus injection of a nitrogen agent, i.e., ammonia or urea. The N-agent can be added either to the reburning zone or to the overfire air. This approach is referred to as Advanced Reburning (AR). The AR tests were performed at a reburn

heat input of 10%, corresponding to a reburn zone stoichiometry of 0.99. At these near-stoichiometric conditions (instead of the more deeply fuel rich conditions normally used for reburning), CO and carbon-in-ash are controlled and the temperature window for NO$_x$ reduction is considerably broadened and deepened compared to selective noncatalytic reduction (SNCR) [4]. Furthermore, decreasing the stoichiometric ratio by adding additional reburn fuel has minimal AR performance benefits. Injection of inorganic salt "promoters" into the reburn zone along with ammonia or urea can significantly improve NO$_x$ control [5-7]. AR is also a flexible technology as additional elements, such as multiple N-agents, can be added over time to meet changing regulations.

To date, natural gas has been the primary fuel of choice for both basic and advanced reburning. Natural gas has advantages as a reburning fuel for retrofit applications with limited access and combustion space. However, using other fuels for reburning may have significant technical and cost benefits. Previous test work has shown that a variety of fossil fuels can be used for reburning, including fuel oil and pulverized coal [8-11]. The current work has extended these findings by applying basic and AR technology to a series of alternative fuels that offer potential cost and performance advantages. These alternative fuels include biomass, coal pond fines, pulverized coal, coal

Reprinted from *Fuel* **78 (3)**, 327-334 (1999)

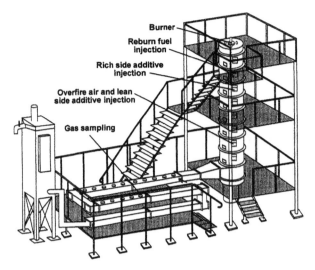

Fig. 1. Schematic representation of 300 kW Boiler Simulator Facility

water slurry (CWS), carbonized refuse derived fuel (CRDF), and Orimulsion.

This article describes the results of investigations into the impacts of fuel properties and process conditions on the effectiveness of the reburning process. The tests focused on evaluating the feasibility of using various test fuels for basic and advanced reburning in coal-fired utility boilers. The three primary objectives of the studies were: (1) to develop a database on the NO_x control performance of different types of reburning fuels; (b) to identify optimum reburning and AR process conditions for NO_x control and carbon burnout; and (c) to assess the potential impacts of reburn fuel characteristics on full scale boiler performance.

A series of different reburning fuels were evaluated in a pilot scale combustor designed to simulate a coal-fired boiler. The fuels were tested over a range of process conditions typical of utility boilers. Experimental conditions were defined following a brief survey of utility boiler design and operational characteristics, where key parameters such as uncontrolled NO_x emissions, furnace temperatures, thermal quench rates, and residence times were identified. The impacts of process parameters and fuel properties upon the performance of the reburning process were defined, along with implications for full scale boiler application.

2. Experimental

2.1. Test facility and methods

The reburning tests were conducted in a 300 kW boiler simulator facility, shown schematically in Fig. 1. The test facility is designed to provide an accurate subscale simulation of the flue gas temperatures and composition found in a full scale boiler. A variable swirl diffusion burner with an axial fuel injector is used to simulate the approximate temperature and gas composition of a commercial burner. Main facility components include a down-fired radiant furnace, convective pass, and baghouse. The cylindrical furnace section is constructed of eight modular refractory-lined sections with an inside diameter of 56 cm and a height of 7 m. The convective pass is also refractory-lined, and contains air cooled tube bundles to simulate the superheater, reheater, and economizer sections of a utility boiler. Numerous ports located along the axis of the facility allow supplementary equipment such as sampling probes, reburn injectors, cooling rods, and suction pyrometers to be placed in the furnace. The rate of heat extraction in the radiant furnace was set to provide a thermal quench rate of approximately 280 K/s.

Both pulverized coal and natural gas were tested as primary fuels. Primary burner excess air was 10%. Initial NO_x levels were controlled between 200 and 1300 ppm by mixing the combustion air with ammonia, which oxidized in the flame zone to generate NO_x. The reburn fuels were injected along the centerline of the radiant furnace. Reburn fuels tested included biomass, coal pond fines, pulverized coal, coal water slurry, carbonized refuse derived fuel, and Orimulsion. Natural gas was also tested for comparison. Twin-fluid atomizers were used to inject the liquid reburn fuels, with both compressed air and nitrogen tested as atomizing media. Solid reburn fuels were injected by means of a radial injector, with both air and nitrogen tested as transport media. Overfire air was injected at the end of the reburn zone to bring the final excess air level to 15%. Overfire air injection temperature was varied between 1200 and 1420 K. For the AR tests, the N-agent consisted of an aqueous urea solution. The solution was atomized either into the reburn zone or with the overfire air. To maintain

Table 1
Reburn fuel analytical properties

	Biomass	Coal pond fines	CRDF	Orimulsion	Bituminous coal	Low rank coal
Proximate (%, dry)						
Fixed carbon	19.92	53.34	22.80		53.89	51.78
Volatile matter	75.29	34.13	65.20		35.21	35.94
Ash	4.79	12.53	12.00		10.90	12.28
NCV (kJ/kg)	19 280	30 090	34 420	42 090	28 010	24 970
Ultimate (%, dry)						
Carbon	48.89	73.44	68.40	83.43	72.06	64.93
Hydrogen	5.75	4.45	7.70	10.24	4.59	4.00
Nitrogen	0.53	1.28	0.40	0.66	1.08	0.84
Sulfur	0.05	2.86	0.60	3.80	0.56	0.26
Ash	4.79	12.53	8.00	0.27	10.90	12.28
Oxygen	40.29	5.44	9.70	1.60	10.81	17.69

constant thermal conditions, the solution injection rate was maintained constant at 20 ml/min. The concentration of urea in the solution was varied to achieve the desired feed rate of N-agent. When promoters were employed, they were dissolved and premixed with the N-agent. Repeat tests were performed at several conditions. These tests indicated that the NO_x reduction results were reproducible within ± 3 percentage points.

2.2. Sampling and analysis

A continuous emissions monitoring system was used for an on-line analysis of the flue gas. Species analyzed, detection principles, and levels of precision were as follows:

- O_2: paramagnetism, 0.1%
- NO_x: chemiluminescence, 1 ppm
- CO: nondispersive infrared spectroscopy, 1 ppm
- CO_2: nondispersive infrared spectroscopy, 0.1%
- SO_2: nondispersive ultraviolet spectroscopy, 0.1%
- N_2O, nondispersive infrared spectroscopy, 1 ppm
- Hydrocarbons, flame ionization detection, 1 ppm

The analyzers were calibrated before and after each test run.

2.3. Fuel characteristics

Reburn fuels tested included natural gas, biomass, coal pond fines, carbonized refuse derived fuel (CRDF), Orimulsion, bituminous coal, and low-rank coal. Both the pond fines and the CRDF were processed as slurries. It is economically beneficial to avoid dewatering, and thus these fuels were tested in the form of slurries. Table 1 summarizes analytical properties of the test fuels.

The natural gas consisted of about 92% methane, 6% ethane, and 2% nitrogen. It had a heating value of 49 630 kJ/kg. The biomass tested was a fir lumber wood waste that was obtained in bulk form, with a branch top size of 20 cm. It was processed in a hammer mill to provide two size distributions: a fine grind with 88% passing through a 50 mesh sieve, and a coarse grind with 35% passing

through a 50 mesh sieve. The fine grind was used for most of the reburning test work.

The pond fines were obtained from a commercial coal processing facility. The parent coal for the pond fines was a medium sulfur, bituminous Kentucky coal. Slurries of fine size (less than 0.5 mm) coal particles are typically produced during coal processing operations. These slurries tend to have higher ash content than the parent coal and are difficult to handle due to their higher water content and small size. While recent work has focused on using these fines for power generation, they have historically been impounded in settling ponds and considered to be waste products [12]. Regarding the pond fines size distribution, 98% passed through a 50 mesh sieve, and 70% through a 200 mesh sieve. The pond fines were tested as a slurry consisting of 50 wt. % coal and 50 wt. % water. Multiple particle size distributions were not tested in the current work.

The CRDF consisted of municipal solid waste that was subjected to a carbonization procedure. As described in detail elsewhere [13,14], this procedure involves applying heat and pressure to the waste to remove oxygen functional groups from the structure of the RDF as carbon dioxide gas. The carbon dioxide that was removed comprises a significant weight percentage of the feed stream, but only a minimal percentage of the heating value. While the mass of total solids is reduced by 20%–70%, the carbonized product still contains 95%–98% of the energy content of the feed waste. For the reburning tests the CRDF was injected as an aqueous slurry consisting of 45 wt. % solids and 55 wt% water.

Orimulsion is an emulsion consisting of 70% bitumen and 30% water. It has similar viscosity and handling requirements as heavy fuel oil. Two coals, one bituminous and one low rank, were tested. Both were pulverized, such that 70% passed through a 200 mesh sieve.

3. Test results

The objective of the tests was to characterize NO_x reduction as a function of fuel analytical properties and combustor

234

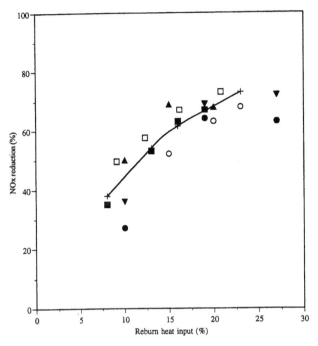

Fig. 2. Effect of reburn heat input on NO_x reduction for alternative reburning fuels. The main fuel is natural gas. Initial NO_x concentration is 400–600 ppm. Reburn zone residence time is 0.5 s. Reburning fuels: □, biomass; ■, Orimulsion; ▲, CRDF; ▼, low rank coal; ●, bituminous coal; ○, coal pond fines; +, natural gas

process variables. Basic reburning studies were performed with seven different test fuels. Four of these fuels were then selected for a more detailed AR and promoted AR tests.

Basic reburning tests were performed with each of the test fuels. Test variables that were found to have a direct on NO_x reduction include reburn heat input, reburn zone residence time, and initial NO_x concentration. Several grinds were tested with biomass. The finer grind improved both the NO_x reduction and carbon burnout. At constant initial NO_x level, the primary fuel type has been shown to have minimal impact on reburning performance. Therefore, most of the tests were performed with natural gas as the primary fuel to facilitate direct comparison of the different reburn fuels. CO emissions were monitored throughout the tests and were found to be below 50 ppm for all test conditions.

The percentage heat input of the reburning fuel controls the stoichiometry of the reburning zone. While slightly dependent on the composition of the reburn fuel, reburn zone stoichiometry is approximately 0.99 at 10% reburning and 0.88 at 20% reburning. Performance trends are generally similar whether presented as a function of reburn heat input or reburn zone stoichiometry. Fig. 2 shows the impact of reburn heat input on NO_x reduction for the different reburning fuels. Each of the test fuels proved to be effective for reburning, giving NO_x reduction comparable to that of

natural gas. At 18%–20% reburn heat input under the baseline furnace conditions, NO_x reduction for each of the test fuels was in the range of 64%–71%. Fuel parameters that appear to have a direct impact on NO_x reduction include fuel nitrogen, volatiles, fixed carbon, and ash constituents. Highest performance was obtained with biomass and CRDF, each of which has high volatiles, low nitrogen content, and high concentrations of sodium and potassium in ash.

The reburn fuels were injected into the furnace at 1700 K. The overfire air injection temperature was varied in order to vary the reburn zone residence time. Reburning NO_x reduction generally improves with increasing reburn zone residence time, i.e. with decreasing overfire air injection temperature. However, in full scale reburning applications, residence time is limited by boiler access and carbon burnout requirements. Typical full scale reburn zone residence times are in the range of 0.4–0.7 s. As shown in Fig. 3, NO_x reduction increased with increasing reburn zone residence time. Natural gas performs better than the other fuels at shorter residence times because it needs no time for devolatilization. At residence times above 0.5 s, similar performance was obtained for all reburning fuels.

Similar to other NO_x control technologies, percentage NO_x reduction increases with increasing initial NO_x concentration. As shown in Fig. 4, NO_x reduction drops rapidly at initial NO_x concentrations below 300 ppm, particularly for

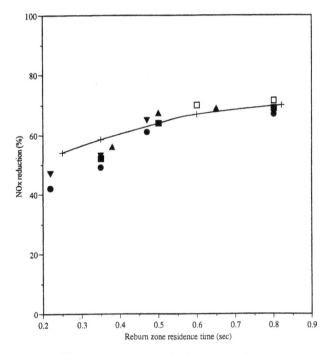

Fig. 3. Effect of reburn zone residence time on NO$_x$ reduction for alternative reburning fuels. The main fuel is natural gas. Initial NO$_x$ concentration is 400–600 ppm. Reburn heat input is 19%. Reburning fuels: □, biomass; ■, Orimulsion; ▲, CRDF; ▼, low rank coal; ●, bituminous coal; +, natural gas

pulverized coal reburning. Performance increases marginally at initial NO$_x$ concentrations above 400 ppm.

In AR, basic reburning is coupled with injection of an N-agent. The agent can be either added into the fuel rich reburning zone or with the overfire air. Based on technical and economic merits, fuels selected for the AR tests included biomass, CRDF, coal pond fines, and natural gas. Urea was used as the N-agent. The AR tests were performed at a reburn heat input of 10%, with N-agent injection temperatures ranging from 1200 K (i.e. co-injection with the overfire air) to 1560 K. At 10% reburn heat input, NO$_x$ reductions for reburning alone were 50% for biomass, 50% for CRDF, 44% for coal pond fines, and 47% for natural gas. The effect of AR injection temperature on NO$_x$ reduction for the different reburn fuels is shown in Fig. 5. Addition of urea significantly increased NO$_x$ reduction for all four reburn fuels. With natural gas reburning maximum NO$_x$ reduction was 81% at a urea injection temperature of 1200 K. With both CRDF and coal pond fines reburning maximum NO$_x$ reduction was 83% at a urea injection temperature of 1300 K. Best performance was obtained with biomass, for which maximum NO$_x$ reduction was 92% at a urea injection temperature of 1260 K. As with basic reburning, this performance is attributed to the fact that biomass has high volatiles, low nitrogen, and high concentrations of sodium and potassium in ash.

Alkali based promoter compounds that can enhance AR

performance via formation of radicals in chain reactions were identified [15,16]. The most reactive of these is sodium, which can be dissolved and injected along with the N-agent. Tests were conducted in which sodium carbonate promoter was injected with urea. The concentration of sodium in the flue gas was equivalent to 30 ppm. Promoted AR reburn fuels included biomass, CRDF, coal pond fines, and natural gas. With each fuel, initial NO$_x$ concentration was 600 ppm, reburn heat input was 10%, and reburn zone residence time was 0.6 s. Test conditions included reburning alone, AR (i.e. reburning plus urea injection), and AR with sodium promoter injection (Table 2).

Sodium promoted AR provided over 90% NO$_x$ reduction with natural gas, biomass, and CRDF. Best performance was obtained with biomass, followed closely by CRDF. With biomass, NO$_x$ was reduced from the baseline value of 600 ppm to below 30 ppm.

4. Discussion

The experimental data presented above shows the potential for achieving high levels of NO$_x$ control for each of the alternative reburning fuels evaluated. Previous success in translating the basic reburning process to full scale boiler systems lends confidence to expectations for applying AR to boilers. There are, however, a number of factors that must be

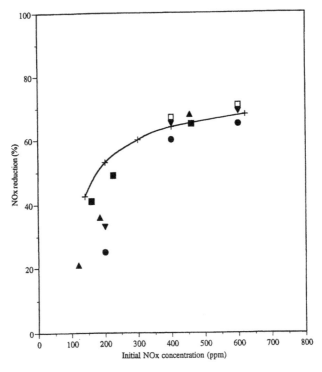

Fig. 4. Effect of initial NO$_x$ concentration on NO$_x$ reduction for alternative reburning fuels. The main fuel is natural gas. Reburn heat input is 19%. Reburn zone residence time is 0.5 s. Reburning fuels: □, biomass; ■, Orimulsion; ▲, CRDF; ▼, low rank coal; ●, bituminous coal; + , natural gas

considered in the translation of pilot scale results to full scale boiler systems. Experimental data obtained in test facilities, such as the 300 kW unit used in the current study, demonstrate the potential for NO$_x$ control under a given set of process conditions. They also indicate the relative performance expectations for different reburn fuels and process conditions (such as temperature, residence time, and initial NO$_x$ concentration). In translating such data to full scale systems, appropriate allowances must be made for these process conditions, as actual NO$_x$ reduction obtained at full scale will be dependent on site-specific considerations. Of particular importance are mixing effects and the physical access available for reburn fuel, nitrogen agent, and overfire air injectors. If access is not available at optimum boiler temperatures, then performance may be compromised. Similarly, excellent mixing between reagents and furnace gas is readily achievable at pilot scale, in which mixing distances are well below 1 m, whereas at full scale mixing distances may be 7 m or more. Increasing the momentum of the additive stream, and the use of a supplementary transport medium such as recirculated flue gas, can enhance mixing at full scale. However, due to scale effects and complex, three-dimensional flow fields it is difficult to obtain the degree of mixing possible at pilot scale. These scaling effects must be considered when making site-specific full scale boiler performance predictions.

Reburning can also have an impact on boiler thermal performance, by affecting slagging, fouling, and fly ash properties. In situations where the reburn fuel is the same as the primary fuel these impacts are generally minimal. However, if the reburn fuel is different from the main fuel, such as with biomass reburning, changes in particulate deposition can potentially occur. This may require changes in sootblowing cycles. Boiler impacts may be mitigated by the fact that the reburn fuel accounts for only a small fraction of the total heat input, typically 10% with AR. Carbon in ash can also be expected to increase, but generally by a minimal amount. For example, with coal used both as the primary and reburn fuel, carbon burnout was found to decrease from 99.7% for straight coal firing to 99.5% with 15% coal reburning. Fine grinding of the reburn fuel has also been found to improve carbon burnout.

As alternative fuels show similar basic reburning performance to that of natural gas, the process chemistry is also probably similar. A detailed kinetic model was developed recently and applied to predict NO$_x$ reduction for natural gas injection in basic and advanced reburning regimes [7,15,16]. The model predicts that injection of the reburning

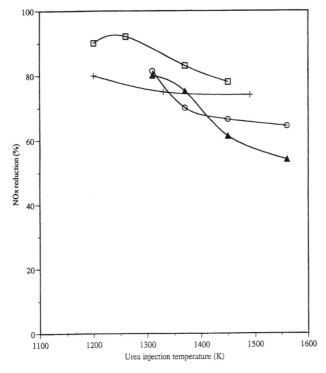

Fig. 5. Effect of urea injection temperature on NO_x reduction for biomass and natural gas advanced reburning. The main fuel is natural gas. Initial NO_x concentration is 600 ppm. Reburn zone residence time is 0.6 s. Reburn heat input is 10%. Reburning fuels: □, biomass; ▲, CRDF; ○ coal pond fines; +, natural gas

fuel (CH_4) forms C-containing radicals (HCCO, CH, CH_2 and CH_3) that rapidly react with oxygen from the main combustion zone. The C-radicals react with NO to form compounds with C–N bonds (HCN, CN, HNCO, etc.) which, in turn, form N-containing species (NH_2, NH, and N). The N-radicals are capable of reacting with NO to form molecular nitrogen, as shown below:

$$NH_2 + NO \rightarrow N_2 + H_2O \tag{1}$$

$$NH_2 + NO \rightarrow NNH + OH \tag{2}$$

Overfire air oxidizes combustibles from the reburning

Table 2
Results of promoted AR tests

Reburn fuel	NO_x Reduction		
	Reburning alone	AR	Promoted AR
Natural gas	47%	81%	91%
Biomass	50%	90%	96%
CRDF	50%	75%	94%
Coal fines	44%	70%	78%

zone to CO_2 and H_2O, as well as the N–H and C–N species to NO.

The model can also predict the chemical behavior in the AR mode. In this case, the amount of reburning fuel is smaller, and the C–N and N–H species are formed in the reburning zone in very low concentrations. When an N-agent (ammonia or urea) is added to the reburning zone, it can react with active radicals, e.g.

$$OH + NH_3 \rightarrow H_2O + NH_2 \tag{3}$$

$$H + NH_3 \rightarrow H_2 + NH_2 \tag{4}$$

The NH_2 species can then reduce NO under slightly fuel-rich conditions (reburn zone stoichiometry = 0.99) via reactions (1) and (2). This process (1–4) has a chain nature since NNH species eventually form N_2 and active radicals.

This mechanism qualitatively explains why some alternative fuels have higher AR efficiency than natural gas. Mineral compounds, such as alkalis, in the alternative fuels enhance the efficiency of NH_3–NO interaction via formation of additional radicals. In particular, it was found [15] that sodium carbonate is converted to NaOH at high temperature, which slowly decomposes to Na atoms and OH radicals. The Na atoms then react with water molecules to return NaOH:

$$NaOH + M \rightarrow Na + OH + M \qquad (5)$$

$$Na + H_2O \rightarrow NaOH + H \qquad (6)$$

The net effect of reactions (5) and (6) is the formation of active radicals ($H_2O \rightarrow H + OH$) which interact with N-agents (3–4) to form NH_2 and help to reduce additional portions of NO via reactions (1) and (2). If the alternative reburn fuels contain volatile nitrogen and alkali compounds, they increase the concentration of the N-agent or promote the interaction between the N-agent and NO in the reburning zone. Thus, certain mineral compounds contained in alternative fuels can enhance NO reduction via AR. Heterogeneous processes on the surface of solid particles can also contribute to enhanced NO reduction.

5. Conclusions

Reburning technology appears to be well suited to a variety of alternative fuels. Basic reburning performance for the alternative test fuels was found to be similar to or slightly better than that of natural gas, with over 70% NO_x reduction achievable at a reburn heat input of 20%. AR tests were conducted in which reburning was coupled with injection of N-agents and/or promoters. The most effective promoter compounds were found to be alkalis, most notably sodium compounds. At a reburn heat input of 10%, NO_x reductions in the range of 78%–96% were achieved with promoted AR. This places the technology in the same category as selective catalytic reduction in terms of NO_x control, at a cost that is lower by approximately a factor of three. While reburning with alternative fuels may have the potential to cause boiler impacts, such as increased slagging, these impacts are generally minimized because the reburn fuel comprises only a small fraction of total boiler heat input. The AR techniques are flexible since they can be readily applied to a variety of waste products and off-specification fuels having low cost. Therefore, alternative fuel AR has great economic benefits while offering the potential for high NO_x reduction, and has the potential to be the technology of choice for a wide range of new and retrofit power plant applications.

Acknowledgements

The subject, advanced reburning work was sponsored by the U.S. Department of Energy under contract no. DE-AC22-95PC95251, Project Manager Thomas J. Feeley, III.

References

[1] Folsom BA, Payne R, Lyon RK. Gas reburning and integrated NO_x and SO_x control: ready for commercial installations, ACS National Meeting. San Diego, March 1994.

[2] May TJ. Gas reburning in tangentially, wall-, and cyclone-fired boilers. An introduction to second-generation gas reburning:. Presented at the Third Annual Clean Coal Technology Conference, Chicago. 6–8 September 1994.

[3] Proceedings of the gas research institute/Swedish gas center/Danish gas technology center international gas reburn technology workshop, Malmö, Sweden. 7–8 February, 1995.

[4] Seeker WR, Chen SL, Kramlich JC. US Patent 5,139,755. 1992.

[5] Zamansky VM, Ho L, Maly P, Seeker WR. Second generation advanced reburning for high efficiency NO_x control, American Flame Research. Committee International Symposium, Baltimore. September 1996.

[6] Ho L, Chen SL, Seeker WR, PM Maly. US Patent 5,270,025. 1993.

[7] Zamansky VM, Ho L, Maly P, Seeker WR. Reburning promoted by nitrogen- and sodium-containing compounds. In: Proceedings of the 26th Symposium (Intern.) on Combustion. Pittsburg: The Combustion Institute, 1996.

[8] Chen SL. NO_x reduction by reburning with gas and coal- bench scale studies. In: Proceedings of the 1982 Joint Symposium on Stationary Combustion NO_x Control, vol. 1: Utility Boiler Applications, EPRI Report No. CS-3182, vol. 1. Palo Alto: Electric Power Research Institute, July 1983.

[9] Overmoe BJ. Pilot scale evaluation of NO_x control from pulverized coal combustion by reburning. Proceedings of the 1985 Joint Symposium on Stationary Combustion NO_x Control, vol. 1: Utility Boilers Applications, EPRI Report No. CS-4360, vol. 1. Palo Alto: Electric Power Research Institute, 1986.

[10] Chen WY, Lester TW. Effects of reburning fuel type on NO_x reduction. Presented at the Eighth Annual Coal Preparation, Utilization, and Environmental Control Contractors Conference, Pittsburgh, July 27–30 1992.

[11] Payne R, Moyeda DK, Maly PM. The use of pulverized coal and coal-water-slurry in reburning NO_x control. Presented at the EPRI/EPA 1995 Joint Symposium on Stationary Combustion NO_x Control, Kansas City, May 16–19 1995.

[12] Harrison CD, Akers DJ, Carson WR. Alternative fuels from slurry ponds. In: Proceedings of the 22nd International Conference on Coal Utilization & Fuel Systems, Clearwater. March 16–19 1997.

[13] Klosky MK. Clean liquid fuels from MSW. Proceedings of A and WMA's Conference and Exhibit on Solid Waste Management: Thermal Treatment & Waste-to-Energy Technologies, Washington. April 18–21 1995.

[14] Klosky MK. Pilot scale production and combustion of liquid fuels from refuse derived fuel (RDF) –Part II. In: Proceedings of 17th Biennial ASME Waste Processing Conference and Exhibit, Atlantic City, March 31–April 3 1996.

[15] Zamansky VM, Maly PM, Ho L, Lissianski V, Gardiner WC. Promotion of selective non-catalytic reduction of NO by sodium carbonate. 27th Symposium (International) on Combustion, The Combustion Institute, Pittsburgh, 1998.

[16] Zamansky VM, Sheldon MS, Maly PM. Enhanced NO_x reduction by interaction of nitrogen and sodium compounds in the reburning zone. 27th Symposium (International) on Combustion, The Combustion Institute, Pittsburgh, 1998.

Advanced reburning measurements of temperature and species in a pulverized coal flame

D.R. Tree*, A.W. Clark

435M, Crabtree Building, Department of Mechanical Engineering, Brigham Young University, Provo, UT, USA

Received 3 May 1999; received in revised form 22 November 1999

Abstract

An experimental program has been completed where detailed measurements of a pulverized coal flame with advanced reburning have been obtained. Maps of species (CO, CO_2, O_2, NO, HCN, and NH_3), temperature, and velocity have been obtained consisting of approximately 60 measurements across a cross sectional plane of the reactor. Two maps at a single operating condition were obtained and are compared. In addition to the mapping data, effluent measurements of gaseous products were obtained for various operating conditions, while investigating the affect of reburning zone stoichiometric ratio (SR), ammonia nitrogen to NO ratio (NSR), ammonia injection location, and burner swirl.

Advanced reburning was achieved by injecting natural gas downstream of the primary combustion zone to form a reburning zone followed by ammonia injection and then tertiary air. The data showed advanced reburning was more effective than either reburning or NH_3 injection alone. At one advanced reburning condition (SR = 1.05, Swirl = 1.5, NSR = 2.5) over 95% NO reduction was obtained. Ammonia injection was most beneficial when following a reburning zone which was slightly lean, SR = 1.05, but was not very effective when following a slightly rich reburning zone, SR of 0.95. In the cases where advanced reburning was most effective (reburning SR = 1.05), higher NSR values improved NO reduction, but the effect of NSR was secondary to NH_3 injector location. The optimal location for injection was found to coincide with changes in the temperature field.

The mapped temperature, species and velocity data for advanced reburning showed that the largest drops in NO occurred in a region where the O_2 concentration was between 0.7 and 3.0%, NH_3 was between 0 and 2961 ppm, and temperatures were between 1274 and 1343 K. These are similar to optimal conditions known for SNCR. Significant NO reductions were seen when NSR values were near one, suggesting NH_3 was very effective at NO reduction when surrounding temperature and species conditions were favorable. Because this was only one detailed set of data, it is difficult to conclude that these conditions are optimal or need to exist for optimal NO reduction. More detailed mapping data at other operating conditions would be useful in identifying optimal advanced reburning conditions. © 2000 Elsevier Science Ltd. All rights reserved.

Keywords: Advanced reburning; NO_x; Coal

1. Introduction

As a result of the need to reduce NO emissions in coal fired power plants, various NO reduction strategies have been investigated. Currently, air staging produced by low-NO_x burners is being implemented as a cost-effective technique but may face limited reduction capabilities due to carbon burnout and flame impingement problems [1]. When further NO reduction is required, *Selective Catalytic Reduction* (SCR) and *Selective Noncatalytic Reduction* (SNCR) have been implemented [2–4]. These techniques take advantage of an N-agent such as urea's or ammonia's ability to reduce NO to N_2 and O_2. Both of these techniques

require effluent gas processing which requires additional equipment cost and space as discussed by Pickens [5]. SCR requires an expensive catalyst while SNCR requires a narrow temperature window (1200–1350 K) and is only 50% effective in practice [6]. Reburning is a promising technology which has been demonstrated in laboratory reactors by Mereb and Wendt [7] and in full scale boilers such as reported by Folsom et al. [8]. In reburning, NO reduction occurs within the boiler region by the injection of fuel downstream of the primary combustion zone. The reburning fuel creates a rich zone where NO is reduced and is followed by downstream injection of air to burn out the excess fuel. Natural gas is the most common choice for reburning fuel.

Advanced reburning is a combination of reburning and N-agent injection downstream of the primary combustion zone where NO is formed. The reburning fuel produces a

* Corresponding author.
E-mail address: treed@et.byu.edu (D.R. Tree).

Reprinted from *Fuel* **79 (13)**, 1687-1695 (2000)

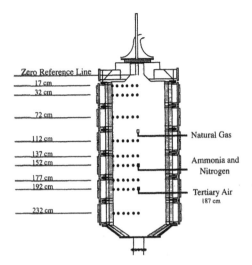

Fig. 1. Advanced reburning implementation in the CPR. Injector and measurement locations.

reduction in oxygen concentration from the typical lean conditions exiting the primary combustion zone to levels useful for SNCR or $DeNO_x$ chemistry (approximately 2%) [9]. In an attempt to obtain NO reduction within the boiler, Chen et al. [10] injected –NH and –CN compounds at various stoichiometric ratios within the reburning zone and within the burnout air. They found large NO reductions when N-agents were injected into slightly rich (SR = 0.99) reburning zones followed by burnout air. NO reduction was found to increase as the temperature at which the burnout air was injected was lowered to approximately 1200 K. They concluded that NO reduction was taking place as the fuel rich zone mixed with the air providing the appropriate O_2 and temperature combination. The temperature window was also found to be broadened by the presence of CO, which produced the necessary OH radicals needed to initiate NO reduction to N_2. Subsequent research [11–14] investigated the effect of CO on $DeNO_x$ chemistry and showed the temperature window could be lowered by 100°C and broadened if the appropriate level of CO (approximately 3800 ppm) were present. Higher CO levels tend to increase radicals such that NO formation is enhanced over NO reduction. Hemberger et al. [15] showed that the presence of hydrocarbon radicals produced from methane or ethane injection into the N-agent zone could also broaden and lower the temperature window.

Conclusions from much of the previous work concerning the role of CO, O_2 and temperature in advanced reburning rely on temperature and species measurements before and after the reaction zone and kinetic modeling within the reaction zone. Previous measurements leave an uncertainty if the temperature at the point of N-agent injection or the temperature at the point of burnout air injection controls

the reaction process. The kinetic models describing $DeNO_x$ reactions have only been compared to data at the beginning and end but not within the reaction zone. The objective of this work is to obtain a detailed map of species concentration and temperature throughout the reaction zone of an advanced reburning process. The detailed measurements will give insight into the NO reduction process and allow for a comparison of detailed combustion modeling with measurements.

2. Method

All measurements have been obtained in the BYU/ ACERC controlled profile reactor (CPR) which is a 0.2 MW, axisymmetric, down-fired, pulverized coal reactor. The reactor shown schematically in Fig. 1 is made up of six cylindrical sections 40 cm in length and 75 cm in diameter. Each section has four equally spaced access ports. The bottom-section houses a converging exit section, while the top sections consists of a diverging water cooled quarl and variable swirl burner. The burner consists of a centrally located, primary fuel/air tube 2.54 cm in diameter with a variable-block swirl generator. The reactor was operated at an overall stoichiometric ratio of 1.1 (fuel lean), and two different swirl number settings of 1.5 and 0.5 were used. These swirl numbers refer to the axial to tangential momentum flux calculated from the angle and flow rate of air in the swirl generator. A more precise measurement of swirl was obtained 5 mm downstream of the primary fuel inlet tube using axial and tangential velocity profiles using LDA. The swirl measured from LDA [17] was reported to be 0.87 and 0.36 at the nominal 1.5 and 0.5 swirl conditions respectively. The nominal values calculated from the swirl generator have been used in reporting the data presented here because it was thought that LDA data is not typically available to characterize swirl. Those wishing to compare this data to model calculations should use the measured LDA swirl to characterize the flow. A complete description of the reactor can be found in Nazeer et al. [16]

Temperature profiles were obtained using a suction pyrometer. Gas species of O_2, CO_2, CO, and NO were obtained using a water quenched suction probe and on-line gas analyzers. Concentrations of NH_3 and HCN were obtained by aqueous sampling and ion electrode analysis. Advanced reburning was achieved by injecting natural gas and ammonia into the CPR along the centerline using a water-cooled probe. The ammonia was initially stored in compressed tanks as 25% NH_3 and 75% N_2 (by volume). During injection, the NH_3/N_2 mixture (to be referred to as NH_3 injection only) was further mixed with N_2 as a carrier gas to increase the momentum and mixing of the injected flow. In an initial set of experiments, the locations of the natural gas and ammonia injection were varied and effluent gas concentrations were measured. This was used to investigate the effects of various injection locations, reburning

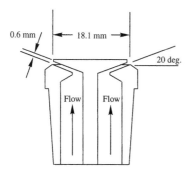

Fig. 2. Shematic diagram of injector nozzles.

zone equivalence ratios, and ammonia injection flow rates on the effluent NO concentration. Reburning zone stoichiometric ratio was defined as the total air–fuel ratio divided by the stoichiometric air–fuel ratio of all air and fuel in the primary and reburning combustion zones. Thus, a rebuning stochimetric ratio of 1.05 had 5% excess air as calculated using all fuel and air upstream of the reburning zone. After considering the effluent measurements, a location for each injector and flow rates for ammonia and natural gas were selected for a detailed map. Fig. 1 shows the location of natural gas, ammonia, and tertiary air injection selected for the advanced reburning map with dots representing specific measurement locations for the gaseous species. Temperature measurements were obtained at the same number of locations as the gas measurements shown but were located 5 cm below each gas species location due to differences in the geometry of the sampling probe tips.

All of the gas injectors (Air, Nitrogen and Ammonia) used water-cooled tubes with pintle type nozzles as shown in Fig. 2. The gap between the pintle and nozzle housing through which the gasses flowed was kept at 0.6 mm for each of the injectors. The diameter of the nozzle was 18.1 mm at the outlet and angled upwards at 20°. The slight upward angle was used because numerical calculations suggested the flow would penetrate toward the wall in a shorter axial distance when directed slightly upstream.

Operating conditions for reactor fuel and air feed for all conditions tested are shown in Table 1 along with reburning fuel, advanced reburning fuel and tertiary air flow rates used in the advanced reburning map to be described later. The

Table 1
Operating conditions for the detailed data map

Parameter	Flow rate (kg/h)	Temperature (K)
Primary air	16.2	310
Secondary air	176.5	600
Coal	25.6	300
Reburning fuel (natural gas)	0.53	–
Ammonia/Nitrogen	0.65	–
Tertiary air	18.9	–

secondary air was heated prior to entering the reactor. Primary air, coal and all injected gases were unheated. A Wyodak sub-bituminous coal was used in the study with properties shown in Table 2. Coal size was measured to have 60 μm mass mean size with a distribution similar to that shown in Pickett et al. [17]

3. Results and discussion

Before completing a detailed map of the reactor under an advanced reburning condition, effluent measurements of NO were obtained at numerous operating conditions which varied the reburning zone stoichiometric ratio, the swirl, and the nitrogen to ammonia stoichiometric ratio (NSR) in the advanced reburning zone. For these 1.5 swirl, effluent species tests, the natural gas was injected at 89 cm and the tertiary air at 187 cm. The results are shown in Fig. 3. The dashed lines show the NO reduction due to reburning alone. NO reduction due to reburning decreased as the reburning zone stoichiometric ratio increased. This was expected because there was less natural gas available for NO reduction at the leaner conditions. At a reburning zone SR of 0.95 the addition of ammonia did little to improve NO reduction, but as reburning zone SR increased, the total NO reduction increased for all NSR values. Because the mixtures which were slightly lean produced greater NO reduction, the data suggests that excess oxygen or reactive species created by small amounts of excess oxygen promote advanced reburning. At a reburning zone SR of 0.95 the affect of NSR on NO_x reduction could not be determined. At a reburning zone SR of 1.05, the data show that NH_3 injection at 130 cm produced higher NO reductions. The reason for this location producing higher NO reductions is suspected to be related to optimal temperatures as will be discussed later.

Effluent measurements taken at a swirl number of 0.5 are shown in Figs. 4 and 5. Fig. 4 includes data from various injection locations and at several NSR values where the NH_3 was injected with the assistance of additional nitrogen carrier gas. The trend of decreased NO reduction with increased SR for reburning alone was the same at this swirl as it was at 1.5. NO reduction due to NH_3 injection in comparison to NO reduction due to reburning alone was again seen to be negligible at a reburning zone SR of 0.95 but again increased with increasing reburning zone SR Reburning alone at a SR of 1.05 is shown to be better than the combination of reburning (SR = 0.95 and 1.0) and NH_3 injection except when a slightly lean (SR = 1.05) reburning zone was followed by injection of NH_3 at 169 cm. This implies again, as in the 1.5 swirl data, that the combination of natural gas followed by ammonia injection (advanced reburning) can provide greater NO reduction than reburning alone, but the location of the NH_3 injection is critical. At 0.5 swirl, the greatest NO reduction was achieved at an injection location of 169 cm while at 1.5 swirl the injection location of maximum NO reduction

Table 2
Ultimate and proximate analysis of the Wyodak sub-bituminous coal

Proximate					
Description	Moisture	Ash	Volatiles	Fixed Carbon	Heating Value (kJ/kg)
As Received	25.3	5.1	34.0	35.6	20,830
Ultimate					
Description	C	H	N	S	O By difference
As Received	53.32	3.47	0.74	0.36	11.74

was 130 cm. If residence time were a limiting factor in NO reduction, the injection location furthest upstream would tend to be the best. While this was the case at 1.5 swirl it was not true at 0.5 swirl. Concentration of O_2 and other gaseous species are expected to remain relatively constant downstream of the reburning zone suggesting that these species are not responsible for the dependence of NO reduction on injector location. The temperature is however changing with position as the gas flows down the reactor. In the

absence of additional heat release, the temperature drop in the reactor has been measured at approximately 1°C/cm.

Fig. 5 shows data repeating two of the conditions just previously presented (SR = 1.0 and 1.05) except in this case, the NH_3 injection was not assisted by additional carrier gas. The primary purpose of this data was to evaluate the NO reduction at a lower injection velocity of NH_3 into the product gases exiting the reburning zone. The results are similar to that of NH_3 injection with carrier gas except that at the 1.05 reburning zone SR, both the 149 and 169 cm injection locations produced high NO reduction, while without the carrier gas, only the 169 cm location was as effective.

One explanation of the data is to assume a zone within the reactor, starting just below 149 cm and extending below 169 cm which has an optimal environment for NO reduction when NH_3 is injected. When injected at low momentum, as was the case with no extra carrier gas, the NH_3 injected at 149 cm moved downstream into the optimal region. With high injection momentum at the 149 cm injection location, the injector, which directed the gases at approximately 20° upstream, forced the gases radially outward and slightly upstream, above the optimal zone. When injecting at 169 cm, the injector was centered in the optimal zone and thus either amount of injection momentum produced adequate mixing of the NH_3 with NO at the desired temperature. Throughout this region, it would be expected that oxygen and other species concentrations remained relatively constant while temperature was dropping due to heat transfer to the walls. This reasoning suggests that the sensitivity of NO reduction to location is actually sensitivity to temperature. Downstream of the ammonia injection, at the point of tertiary air injection, both temperature and O_2 concentration change. It is possible the tertiary air creates a desirable drop in temperature at that location which facilitates NO reduction, but if this were the case, NO reduction would not be sensitive to the NH_3 injection location.

The 1.5 swirl data showed the highest NO reduction when injecting the NH_3 at 130 cm, which was further upstream than the 0.5 swirl case. In order to determine what temperature and species conditions existed at the point of NH_3 injection in each case, a detailed map would be needed. The resources of this work allowed only one detailed map to be obtained. Detailed mapping of baseline conditions in this reactor without reburning or advanced reburning by Nazeer et al. [16], however, showed that the flame at 1.5

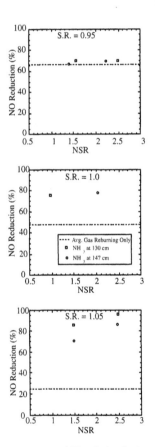

Fig. 3. Effluent measurements of NO reduction for three stoichiometric ratios as a function of NSR at 1.5 swirl.

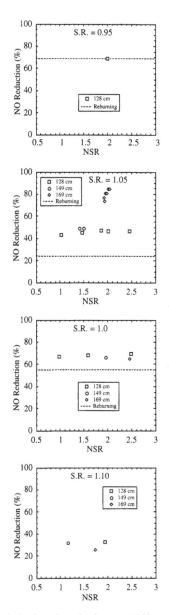

Fig. 4. NO reduction for various reburning zone stoichiometric ratios as a function of NSR at 0.5 swirl.

The conditions selected for the reburning map are shown below in Table 3. The primary consideration in selecting a condition for the map was to find an advanced reburning case which was useful for comparison with comprehensive combustion models including an advanced reburning submodel. Thus it was considered important to have a significant amount of NO reduction occurring at the simplest possible advanced reburning configuration. A 0.5 swirl case was selected because it was considered to be more realistic to industrial swirl levels and because 1.5 swirl created turbulence levels and particle penetration, which would be difficult to model. The 149 cm injection location was selected to allow the longest possible residence time prior to tertiary air injection while maintaining high NO reductions. The NH_3/N_2 injection without additional carrier gas was selected because it was felt that smaller NH_3 velocities would be easier to model and less dependent on the fluid modeling of the injector spray.

The mapped data were obtained during three separate test periods, two of which repeat the gaseous on-line measurements. In Test A, gaseous species and aqueous samples of the reactor were obtained. In Test B, only gaseous species were taken. Temperature data were obtained in Test C. The species and temperature are displayed below in the form of gray scale and line contour plots. The contours were generated by a commercial interpolation and plotting program and mirror imaged using a graphics package. This form of presentation offers the advantage of enabling the reader to visualize the combustion phenomena and rapidly identify important regions of change or high or low concentration. The contour plots, however, do not lend themselves well to an accurate determination of precise values because or the uncertainty of interpolation routines used in the plotting software. Complete tabular data for these maps are available on request with some of the most critical data in the advanced reburning zone being presented in Table 4.

Before presenting the advanced reburning profile data it is important to understand the nature of the flow in the reactor which has been measured using LDA and is reported in detail by Pickett et al. [17]. Above a swirl level of approximately 0.25, the swirl generator imparts tangential velocity on the flow sufficient that upon leaving the quarl, the flow expands outward radially toward the walls of the reactor. This produces a low pressure along the centerline of the reactor. The flow, therefore, exits the quarl moving down and out toward the walls and circulates up the centerline. As swirl is increased this recirculation becomes stronger moving the stagnation point of the axial flow at the centerline within one or two centimeters below the primary inlet tube. Solid sample carbon and ash measurements indicate that under baseline conditions at 1.5 swirl, carbon burnout is 95% complete at the 40 cm axial position and the velocity profile is flat at 80 cm and below.

Species data are shown comparing tests A and B in Figs. 6 and 7. Test B contains the same number of data points but extends a little further down the reactor which may cause

swirl was essentially higher in the reactor than at 0.5 swirl and that the temperature dropped more rapidly. The temperature at 130 cm and 1.5 swirl was 1420 K which was approximately the same temperature at 190 cm at 0.5 swirl in the baseline data. It is therefore reasonable to conclude that these two optimal injection locations had similar temperatures.

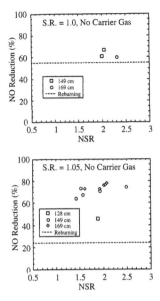

Fig. 5. NO reduction for various reburning zone stoichiometric ratios as a function of NSR at 0.5 swirl where the injected NH_3 has no carrier gas.

Table 4
A selection of species and temperature data in the advanced reburning zone

	Radial	0	8	15	23	30	38
Axial							
152[a]	NO (ppm)	420	430	440	555	490	500
	O_2 (%)	1.1	1.2	1.1	0.9	0.4	0.5
	NH_3 (ppm)	1268	1097	414	0	0	0
	Temp. (K)	1274	1326	1351	1343	1317	1248
177[a]	NO (ppm)	87	115	220	330	400	400
	O_2 (%)	0.7	0.8	0.8	1.0	1.0	1.0
	NH_3 (ppm)	2961	2271	1041	352	86	23
	Temp. (K)	1300	1300	1283	1265	1239	1106
192[a]	NO (ppm)	68	80	95	105	210	280
	O_2 (%)	6.2	4.6	3.2	3.0	2.3	1.6
	NH_3 (ppm)	1005	759	373	224	68	41
	Temp. (K)	1160	1195	1204	1213	1222	1160

[a] Temperature values were taken 5 cm downstream of the species measurements.

some difference in the way the data appears after it has been interpolated by the software. The two data sets show regions of high concentration and gradients which are generally the same in both data sets, and the magnitudes of the measurements are in fairly good agreement. Test A consistently showed for several species evidence of a stronger axial penetration of the secondary air jet at the top of the reactor and a weaker recirculation zone. The two data sets demonstrate the variability of the results for a given operating condition. The largest source of variation in the reactor was determined to be the variation of the coal feed rate. Coal was delivered using a gravity and auger feed system. Vibrators were used to avoid air pockets in the coal feed bin. Variations of 5% in the coal feed were caused by compaction of coal in the feeder bin over time. Coal was delivered to the feeder from barrels, which were emptied approximately every four hours. This created a 4 h cycle in the coal feed while the entire map required 8 h of sampling. Changes in coal feed were evaluated by turning off all but the baseline fuel and air feed and monitoring the exhaust O_2 concentration. When the O_2 concentration was not in good agreement with expected values ($\pm 2.5\%$), the reburning

Table 3
Conditions and injection locations used for mapping advanced reburning

Conditions	Swirl	Primary SR	Reburn SR	Overall SR	NSR
	0.5	1.10	1.05	1.15	1.7
Injector	Nat. Gas	$NH_3 + N_2$	Tert. Air		
	88 cm	149 cm	189 cm		

data from the previous measurements were not accepted and the data was retaken. This variation in the coal feed is one reason the two data sets are slightly different. Another reason is the possibility of moisture or other variations in coal content that can occur within a given load.

Concentrating on a set of data from a single test, Test A, the primary, reburning and advanced reburning zones can clearly be identified. Near the top of the reactor, O_2 is seen to penetrate to about 60 cm before reaching a level of approximately 3%. In this same region, CO_2 increases to 16%, and NO increases to 550 ppm (at the centerline) to 625 ppm (near the walls). CO in this primary combustion region has narrow regions where the concentration is very high, representing unburned fuel, and other areas where CO is low suggesting O_2 availability and more complete combustion.

The region from 140 cm down is of particular interest because it shows the conditions where the NO reduction was occurring. Before examining this region in more detail, the NH_3 and HCN species concentrations and temperature need to be considered. Contour plots for these species are seen in Fig. 8. Significant amounts of HCN were found directly below the quarl outlet in the primary combustion zone but not at any other location within the reactor. There is no evidence of either NH_3 or HCN in the reburning zone. This may have been because the fuel mixed readily with the surrounding gases, and the overall SR in the reburning zone was lean. NH_3 was found only in the vicinity of the ammonia injection. Ammonia was found primarily on the downstream side of injection suggesting that the jet, which was directed at approximately 20° upward, did not contain a significant amount of momentum relative to the downward flow. The temperature, which was highest at the top of the reactor, decreased more slowly in the top half of the reactor where the reburning fuel was added than the bottom half where ammonia and tertiary air were added. This was as expected because

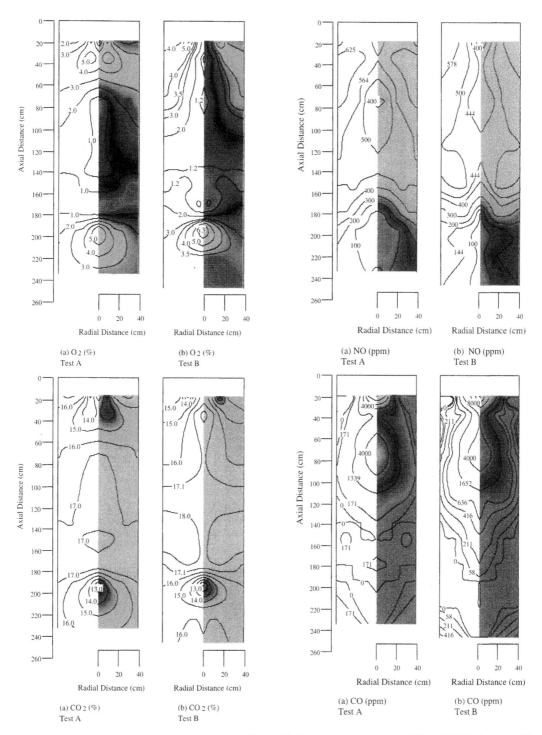

Fig. 6. Concentration contour maps of O_2 and CO_2 for both tests, A and B.

Fig. 7. Concentration contour maps of NO and CO for both tests, A and B.

246

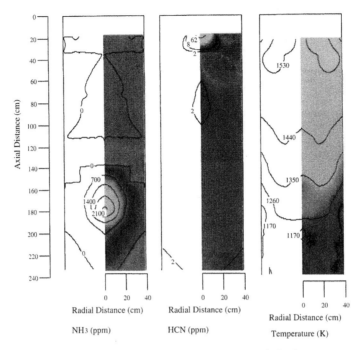

Fig. 8. Ammonia, HCN and temperature profiles for advanced reburning.

the natural gas provided an additional source of energy, while the ammonia and tertiary air, cooled the gasses by dilution.

It is interesting to look closely at the data in the region of rapid NO destruction where measurements were made at the 152, 177, and 192 cm axial locations and 6 radial locations. Tabular data for these points are given in Table 4. NH_3 injection occurred at 149 cm just above this zone, and air injection occurred at 189 cm, near the end of this zone. Assuming a plugged flow in the downward direction, the steepest reduction in NO occurred at the centerline between 152 and 177 cm where the concentration dropped from 420 to 87 ppm. In this same interval, NH_3 increased from 1268 to 2961 ppm, O_2 was present at 0.7%, and the temperature was approximately between 1274 and 1300 K (temperature measurements were collected 5 cm below the location of the species measurements). This means that the greatest NO reduction occurred in the presence of O_2 with a high ratio of NH_3 to NO.

At the 23 cm radial location, moving downstream from 152 to 177 to 192 cm axially, the NO decreased from 555 to 330 to 105 ppm, respectively. During the first segment between 152 and 177 cm, the O_2 concentration was still relatively low between 0.9 and 1.0%, NH_3 was between 0 and 352 ppm and the temperature was between 1343 and 1265 K. In the second segment between 177 and 192 cm, NH_3 was between 352 and 224 ppm, O_2 increased from 1.0 to 3.0% and tempera-

ture decreased from 1265 to 1213 K. The second segment shows evidence of mixing with tertiary air because of the O_2 increase. In both segments, the NH_3 levels are of the same order of magnitude as the NO. Thus, the NH_3 to NO ratio does not have to be significantly higher than one to achieve high NO reductions. The data also demonstrate that large NO reduction occurred under slightly lean (0.7–3% O_2) conditions at temperatures ranging from 1274 to 1343 K. These conditions are close to SNCR conditions but the temperatures and O_2 concentrations are on the higher end of the optimal range. It would be of interest to look at detailed maps where the ammonia was injected upstream at higher temperatures to see if NO increased initially and to determine under what conditions the NO was decreasing.

Because of the close proximity of the NH_3 injector and the tertiary air, it was uncertain from the mapped data whether or not the tertiary air helped create the temperature or species environment necessary for NO reduction. An additional effluent test was taken to determine the extent of NO reduction without tertiary air injections. The results showed a 61% reduction when no tertiary air was used in comparison to greater than 70% reduction with both NH_3 and tertiary air injection. This suggested that the bulk of the NO reduction was independent of the tertiary air, but in this configuration, the tertiary air was also a contributor to producing an environment conducive to NO destruction.

4. Summary and conclusions

Advanced reburning has been investigated in a 0.2 MW pulverized coal flame by injecting a gaseous mixture of NH_3 and N_2 down stream of natural gas injection used as a reburning fuel. Various amounts and combinations of reburning fuel and ammonia injection were investigated through gaseous effluent measurements to determine the effect of parameters such as reburning zone SR, advanced reburning NSR, injector location and carrier gas momentum. The results showed that while reburning and ammonia injection alone were both effective in reducing NO, the greatest reductions were realized when both were combined. Although the testing was not extensive enough to determine an optimal condition, the highest NO reductions of over 90% were observed when a reburning zone SR of 1.05 was followed by ammonia injection. NO reduction was very sensitive to the NH_3 injection location. The reason behind this sensitivity was thought to be variations in temperature at and below the injection location. NO reduction was found to be sensitive to NSR under conditions where NO reductions due to advanced reburning were high, but the effects of NSR were smaller than those of injection location. NO reductions of over 95% were seen at an NSR of 2.5, SR = 1.05, and swirl = 1.5.

A single operating condition was selected for mapping the species and temperatures in the reactor. The map recorded CO, CO_2, NO, O_2, HCN, NH_3 and temperature at 60 locations in a cross sectional plane of the reactor. The data showed the region of most rapid NO reduction occurred directly below the ammonia injection location and that high NO reductions were occurring in regions where the O_2 was between 0.7 and 3%, the temperatures were between 1274 and 1351 K and NH_3 was between 352 and 2961 ppm. Thus the reduction occurred at slightly lean conditions and temperatures slightly higher than typical optimums for SNCR of NO. Further detailed measurements need to be obtained to determine if the NO reduction at other global operating conditions and injection locations occurs at these same local conditions.

Acknowledgements

The authors would like to acknowledge funding received from the Department of Energy, Federal Energy Technology Center Grant Number DE-FG22-95PC95223 and from the Advanced Combustion Engineering Research Center (ACERC) at BYU which was used to support this work.

References

[1] Cooper CD, Alley FC. Air pollution control: a design approach. 2nd ed.. Wiley, 1994.

[2] Jodal M, Nielsen C, Hulgaard T, Dam-Johansen K. Pilot-scale experiments with ammonia and urea as reductants in selective non-catalytic reduction of nitric oxide. Twenty-third Symposium (International) on Combustion, The Combustion Institute, 1990. p. 237–43.

[3] Jodal M, Lauridsen TL, Dam-Jahansen K. NO_x removal on a coal-fired utility boiler by selective non-catalytic reduction. Environmental Progress 1992;11(4):296–301.

[4] Update on NO_x control technologies, PETC Review, Spring 1995.

[5] Pickens RD. Add-on control techniques for nitrogen oxide emission during municipal waste combustion. Journal of Hazardous Materials 1996;47:195–204.

[6] Teixeiria D, Lin C, Muzio L, Jones D, Okazaki S. Selective noncatalytic reduction (SNCR) demonstration in a natural gas-fired boiler. Proceedings, 1993 Joint Symposium on Stationary NO_x Control, Bal Harbor, FL, 1993. p. 6A-105–11.

[7] Mereb J, Wendt JOL. Reburning mechanisms in a pulverized coal combustor. Twenty-third Symposium (International) on Combustion, The Combustion Institute, 1990. p. 1273.

[8] Folsom BA, Sommer TM, Latham CE, Moyeda DK, Gaufillet GD, Janik GS, Whelen MP. Demonstration of advanced gas reburning for NO_x emission control, 1997 Joint Power Generation Conference, 3–6 November 1997. p. 1–7.

[9] Lyon RK. Thermal DeNO$_x$: how it works. Hydrocarbon Processing 1979;58(10):109–12.

[10] Chen SL, Cole JA, Heap MP, Kramlich JC, McCarthy JM, Pershing DW. Advanced NO_x reduction processes using –NH and –CN compounds in conjunction with staged air addition. Twenty-second (Symposium) International on Combustion, The Combustion Institute, 1988. p. 1135–45.

[11] Zamansky VM, Malay PM, Ho L. Family of advanced reburning technologies: pilot scale development. 1997 Joint Power Generation Conference, ASME, 1997, EC-vol. 5, vol. 1, p. 107–13.

[12] Suhlmann J, Rotzoll G. Experimental characterization of the influence of CO on the high-temperature reduction of NO by NH_3. Fuel 1993;72:175–9.

[13] Pont J, Evans A, England G, Lyon R, Seeker W. Evaluation of the CombiNO$_x$ process at pilot scale. Environmental Progress 1993;12(2):140–5.

[14] Brauwer J, Heap MP, Pershing DW, Smith PJ. A model prediction of selective noncatalytic reduction of nitrogen oxides by ammonia, urea, and cyanuric acid with mixing limitations in the presence of CO. Twenty-sixth Symposium (International) on Combustion, The Combustion Institute, 1996. p. 2117–24.

[15] Hemberger R, Muris S, Pleban KU, Wolfrum J. An experimental and modeling study of the selective noncatalytic reduction of NO by ammonia in the presence of hydrocarbons. Combustion and Flame 1994;99:660–8.

[16] Nazeer WA, Pickett LM, Tree DR. In-situ species, temperature, and velocity measurements in a pulverized coal flame. Combustion Science and Technology 1999;143:63–77.

[17] Pickett LM, Jackson RE, Tree DR. LDA Measurements in a pulverized coal flame at three swirl ratios. Combustion Science and Technology 1999;143:79–107.

Techno-economic analysis of NO$_x$ reduction technologies in p.f. boilers

S. McCahey*, J.T. McMullan, B.C. Williams

NICERT, University of Ulster, Cromore Road, Coleraine, Co. Londonderry BT52 1SA, Northern Ireland, UK

Received 26 October 1998; received in revised form 21 April 1999; accepted 21 May 1999

Abstract

The impact of NO$_x$ reduction technologies upon a supercritical coal fired p.f. power station has been investigated using the ECLIPSE process simulator. Technical, environmental and economic assessments were performed, based upon a model of the Amer 9 Power Station at Geertruidenberg, the Netherlands. Selective catalytic reduction (s.c.r.) achieves the largest reduction in NO$_x$ emissions to below 50 mg/N m^3, but at an additional electricity cost of 0.21 p/kWh, over the base case. The additional cost for coal-over-coal reburning is 0.03 p/kWh, reducing NO$_x$ emissions to below 200 mg/N m^3. Only high unburnt carbon losses or high priced reburn coal justify the use of a coal microniser. Natural gas-over-coal reburning requires an unrealisable natural gas price of £0.98/GJ to compete with coal-over-coal reburning. Gas prices between £1.76/GJ and £1.93/GJ are required for it to compete with s.c.r. © 1999 Elsevier Science Ltd. All rights reserved.

Keywords: Pulverised coal; NO$_x$ reduction; Techno-economic analysis

1. Introduction

Combustion of fossil fuels contributes significantly to the annual world-wide emissions of nitrogen oxides; generally termed NO$_x$, these include NO, NO$_2$ and N$_2$O. In pulverised coal-fired (p.f.) boilers the N$_2$O and NO$_2$ emissions are usually less significant than the emissions of NO. NO$_x$ emissions, in general, are considered a major pollutant of the atmosphere, N$_2$O is said to participate in global warming. NO and NO$_2$ are believed to play a major role in the formation of ground level ozone, photochemical smog and acid rain [1,2]. Increasingly stringent regulatory requirements provide impetus for the development and use of advanced technologies to reduce air pollutant emissions, including NO$_x$.

Conventional coal-based power generation technology relies heavily on PF boilers and over 1000 of these plants currently exist world wide [3]. It is unlikely that use of low-NO$_x$ burner technology on its own will be sufficient to meet projected European NO$_x$ emission standards. Therefore, additional NO$_x$ reduction technologies are required. NO$_x$ emissions can be controlled either during or after combustion. Post-combustion techniques, such as selective catalytic reduction (s.c.r.), address emissions after they form and are generally more expensive. Combustion methods, such as reburning, prevent NO$_x$ from forming in the first place.

---

* Corresponding author. Tel.: + 44-1265-324302.
E-mail address: s.mccahey@ulst.ac.uk (S. McCahey)

Reprinted from *Fuel* **78** (14), 1771-1778 (1999)

1.1. Selective catalytic reduction

In the s.c.r. process, NO$_x$ is reduced with NH$_3$ (or other nitrogen source like urea) in the presence of a catalyst to nitrogen and water. This technique can be used at temperatures between 150 and 450°C [2], depending on the catalyst employed. Zeolites and activated coals as well as titanium oxide, iron oxide and vanadium-based catalysts have been used [4]. Positioning the catalyst after the flue gas desulphurisation (f.g.d.) system is known as tail gas arrangement. After desulphurisation the flue gas is cleaner but at a reduced temperature. Therefore, reheating is necessary upstream of the s.c.r. system. Use of s.c.r. provides a high conversion of NO to N$_2$, of around 80–90 vol%.

1.2. Reburning technology

The term reburning was first introduced in 1973 to describe the process of secondary fuel injection [5] and is shown schematically in Fig. 1 [6]. In the main combustion zone, which generally operates under low excess air conditions, coal is burned releasing between 70 and 90% of the total heat input to the boiler. Downstream, in the reburning zone, additional hydrocarbon fuel is introduced without combustion air, creating a fuel-rich, oxygen-deficient environment wherein the secondary fuel breaks down to produce hydrocarbon radicals. These react with NO$_x$ which is produced by combustion of the main fuel, reducing it to atmospheric nitrogen. Overfire air completes the process

250

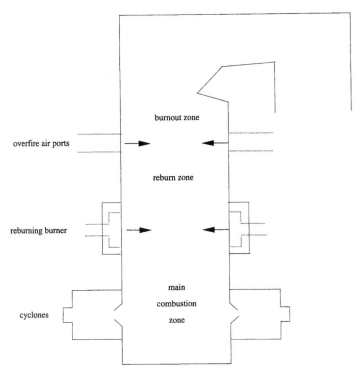

Fig. 1. Coal reburning boiler.

in the burnout zone, aiding combustion of remaining hydro-carbons and carbon monoxide.

This work is based on computer simulations of alternative NO_x reduction technologies when applied conceptually to the Amer 9 power station, which is a 600 MWe supercritical p.f. power station with low-NO_x burners [7]. The technical, environmental and economic evaluations have been performed using the ECLIPSE process simulator [8]. The first study looked at the effect of installing a s.c.r. NO_x reduction system after the f.g.d. system. The second and third studies examined the performance of the Amer 9 power station if either natural gas-over-coal reburning or coal-over-coal reburning were installed. The effect of using micronisers, instead of mills, in the coal-over-coal reburning system was also assessed and a number of sensitivity analyses were performed. These studies incorporate the results of reburning test work performed by Mitsui Babcock at their pilot plant facility in Renfrew, Scotland.

2. Process descriptions

The evaluated systems have all been based around the Amer 9 power station at Geertruidenberg in the Netherlands [7]; a description of the process follows. To provide a consistent basis for evaluation and comparison, the systems analysed were modelled using the ECLIPSE process simulation package [8–10]. ECLIPSE was developed for the European Commission and has been used by the Northern Ireland Centre for Energy Research and Technology at the University of Ulster since 1986 [11,12].

ECLIPSE is a personal-computer-based package containing all of the program modules necessary to complete rapid and reliable step-by-step technical, environmental and economic evaluations of chemical and allied processes. ECLIPSE uses generic chemical engineering equations and formulae and includes a high-accuracy steam-water thermodynamics package for steam cycle analysis. It has its own chemical industry capital costing program covering over 100 equipment types. The chemical compound properties database and the plant cost database can both be modified to allow new or conceptual processes to be evaluated.

A techno-economic assessment study is carried out in stages; initially a process flow diagram is prepared, technical design data can then be added and a mass and energy balance completed. Consequently, the system's environmental impact is assessed, capital and operating costs are estimated and an economic analysis performed. Whilst every effort is made to validate the capital cost estimation data, using published information and actual quotations from equipment vendors, the absolute accuracy of this type of capital cost estimation procedure has been estimated

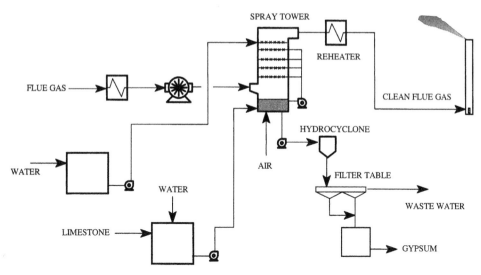

Fig. 2. Wet limestone f.g.d. Process.

at about 25–30%. However, as the comparative capital cost estimates are based on the accurate calculation by the mass and energy balance program of differences in basic design, families of similar technologies composed of similar types of equipment can be compared on a consistent basis.

2.1. P.f. coal fired power station

Coal for the power station is shipped by barges from the seaports. Normal coal storage facilities are provided from where the coal is pulverised in mills and then transferred pneumatically using preheated primary air to a two pass once through boiler, with a spirally wound single furnace and tangentially fired low NO_x burners. Most of the unburned coal and ash is removed at the base of the furnace, with the rest carried forward with the hot gases and removed in cold-side electrostatic precipitators. Before reaching the electrostatic precipitators the hot gases are cooled first by transferring heat to steam in the superheater tubes and the reheater tubes, then by transferring heat to condensate in the economiser section. Finally heat is transferred to combustion air in the air preheater section.

The steam cycle is a supercritical single reheat system. The steam leaving the superheater is sent to the turbine stop valve where it is expanded in the high-pressure turbine. The steam turbines have facilities for steam extraction and allow for steam to be tapped off to the regenerative feedwater heaters. Drains from the three high-pressure feedwater heaters are fed to the deaerator. The steam from the high-pressure turbine is then reheated before passing through one double-flow intermediate pressure and three double-flow low-pressure turbines. At the crossover from the intermediate- to the low-pressure turbines steam is extracted for the

deaerator. The steam from the low-pressure turbine is condensed and the condensate is pumped through the four low-pressure feedwater heaters to the deaerator. From the deaerator tank the boiler feed pump forces the condensate through the three high-pressure surface-type feedwater heaters and the economiser before entering the boiler and completing the steam cycle.

The cooled gases are exhausted via the induced draught fan to a wet limestone gypsum flue gas desulphurisation system where most of the SO_x is removed. This process is based on the Deutsche Babcock design [13], which is shown schematically in Fig. 2. Process design conditions are available [14]. The flue gas from the electrostatic precipitators is first cooled against the clean gas and then fed to the base of the spray tower. Limestone solution is circulated through the sprays in the tower and the SO_2 in the flue gas reacts to form calcium sulphite. In the base of the spray tower the calcium sulphite is oxidised to gypsum which then settles out. The gypsum solution is pumped through a hydrocyclone and then fed onto a filter table where most of the water and impurities are removed. Gypsum is then ready for sale for use in plasterboard manufacture and the wastewater is treated to separate out the impurities. The clean gas is then reheated before being vented up the stack to the atmosphere.

2.2. NO_x reduction modifications

Modifications to this basic model were investigated to evaluate alternative NO_x reduction technologies. Four variations involving s.c.r. and reburning technology were explored. The s.c.r. process that was evaluated is based on the Hitachi tail end deNO$_x$ design [15]. The flue gas from the

Table 1
Technical and environmental results

System	Base case	With s.c.r.	Coal/coal reburn
Thermal input (MW) HHV	1425	1444	1426
LHV	1367	1387	1369
Steam turbine power output (MWe)	648.7	648.7	648.7
Total auxiliary power consumption (MW)	47.2	51.0	52.3
Net power production (MWe)	601.5	597.7	596.4
Net electrical efficiency (%) (HHV)	42.2	41.3	41.8
(LHV)	44.0	43.1	43.6
Auxiliary power consumption (MW)			
Fans	10.96	10.96	16.36
Pumps	25.77	25.77	25.77
Coal-crusher	2.84	2.84	2.37
Cooling water	3.97	3.97	3.94
F.g.d. system	3.58	3.60	3.40
DNOx system fan	0.13	3.69	0
Other	0.13	0.13	0.11
Gaseous emissions (g/kWh)/(mg/N m^3 6% O_2)			
CO_2	759, 230,000	772, 230,000	767, 230,000
SO_x	0.93, 280	0.94, 280	0.94, 280
NO_x	0.96, 310	0.10, 30	0.49, 170
Solid wastes (t/h)	4.9	4.9	4.9
Flue gas (t/h) flow	2118	2118	2118
Temperature (°C)	80	115	80
Composition (% w/w)			
N2 + Ar	70.19	70.19	71.62
CO_2	21.58	21.58	20.16
O_2	3.09	3.09	3.09
H_2O	5.09	5.09	5.09
SO_x	0.03	0.03	0.02
NO_x	0.03	0.03	0.01

f.g.d. system is preheated with clean hot flue gas from the deNO$_x$ reactor and natural gas is burned to maintain a temperature of 300°C. Ammonia solution is evaporated and mixed with air before being injected into the flue gas upstream of the deNO$_x$ reactor. The deNO$_x$ reactor incorporates a TiO$_2$-based honeycomb which catalyses the reaction of ammonia with NO$_x$ to form nitrogen and water. The s.c.r. process reduces the NO$_x$ emissions by about 90%.

With the coal-over-coal and natural gas-over-coal reburn technologies a NO$_x$ emissions reduction of about 50% can be achieved by staging the combustion. In the Amer 9 power station the coal is burned in six levels of burners all with the same stoichiometry of about 1.2. In the reburn systems modelled, the bottom 5 sets of burners burn the main coal fuel but with a stoichiometry of 1.12. The top set of burners burn the reburn fuel at a stoichiometry of 0.9. Above this top set of burners overfire air is added, completing the combustion and maintaining overall stoichiometry at 1.2.

The final stage of combustion is the slowest step in the coal combustion process and involves the oxidation of residual carbon from ash. The complete combustion, or burnout, of carbon can be inhibited by the mineral matter in the ash and also, in this type of staged combustion, by the reduced levels of oxygen. High unburnt carbon losses can

result in a reduction in the boiler efficiency and an increase in slagging and corrosion. When coal is the reburn fuel, the difficulty in achieving a high carbon burnout can be assuaged by using a reburn coal of higher volatility or grinding the reburn coal more finely. The option of using a microniser to provide a fine reburn coal is assessed in the fourth study. A microniser is a jet or fluid energy mill, which uses high-velocity gas jets to entrain the coal particles and promote self-attrition. The gas jets use high-pressure steam taken from the boiler steam system and the steam travels along with the coal into the furnace. The design conditions for the microniser were supplied by James Howden Group Technology Limited [16].

3. Results

The technical and environmental results of the cases investigated are given in Tables 1 and 2. The results determined during the economic analysis are displayed in Table 3.

Use of s.c.r. involves burning natural gas to raise the flue gas temperature to that required by the deNO$_x$ catalyst, thus increasing the thermal input. The auxiliary power requirement

Table 2
Technical and environmental results

System	With microniser	Natural gas/coal reburn
Thermal input (MW) HHV	1420	1451
LHV	1363	1374
Steam turbine power output (MWe)	637.4	648.7
Total auxiliary power consumption (MW)	51.9	51.9
Net power production (MWe)	585.5	596.8
Net electrical efficiency (%) (HHV)	41.2	41.1
(LHV)	43.0	43.4
Auxiliary power consumption (MW)		
Fans	16.23	16.23
Pumps	25.77	25.77
Coal-crusher	2.37	2.27
Cooling water	3.77	3.94
FGD system	3.61	3.40
DNOx system fan	0	0
Other	0.11	0.08
Gaseous emissions (g/kWh)/(mg/N m^3 6% O2)		
CO_2	780, 230,000	704, 215,000
SO_x	0.95, 280	0.75, 226
NO_x	0.50, 170	0.48, 170
Solid wastes (t/h)	4.9	4.1
Flue gas (t/h) flow	2119	2081
Temperature (°C)	80	80
Composition (% w/w)		
N_2 + Ar	70.15	71.62
CO_2	21.58	20.16
O_2	3.08	3.09
H_2O	5.15	5.09
SO_x	0.03	0.02
NO_x	0.01	0.01

also rises due to the additional load on the induced draft fan. This reduces the net power production by 3.8–597.7 MWe and the efficiency from 42.2 to 41.3%, in terms of its higher heating value. The specific CO_2 emissions rise from 759 to 772 g/kWh due to this reduced efficiency, but the main environmental change is the reduction in specific NO_x emissions from 0.96 to 0.1 g/kWh. Case 1, with a coal price of £25/tonne and natural gas price of £2.0/GJ gives a break-even electricity selling price of 2.44 p/kWh, which compares with the 2.23 p/kWh for the base case. Case 3, the equivalent case with coal costing £40/tonne gives a break-even electricity selling price of 2.91 p/kWh compared with the base case of 2.70 p/kWh. The effect of natural gas price, Cases 2 and 4, is small due to the relatively small amount used. The additional capital cost required for the s.c.r. was calculated at £37M. There are several ways that s.c.r. can be applied, such as high dust, low dust or tail gas configuration. These affect the capital cost estimate and the catalyst life for the s.c.r. system. Cases 5 and 6 look at the sensitivity of the break-even electricity selling price to estimated capital costs for s.c.r., in the range £25M–£60M. For a coal price of £40/tonne and natural gas price of £3.0/GJ,

the break-even electricity selling price varies from 2.88 p/kWh to 3.02 p/kWh over this range of capital costs.

The main change in the technical and environmental results for coal-over-coal reburning is the increased power required by the fans. This results from the need to recycle flue gases to the reburn coal mill and to provide good mixing of the overfire air with the furnace gases to achieve good burnout. This reduces the net power production by 4.1–596.4 MWe and the higher heating valve efficiency from 42.2 to 41.8%. The specific CO_2 emissions rise from 759 to 767 g/kWh due to this reduced efficiency, but the main environmental change is the near 50% reduction in NO_x emissions from 0.96 to 0.49 g/kWh. As with s.c.r., use of coal reburn increases the cost of electricity but to a lesser degree. At a coal price of £25/tonne, Case 7, the break-even electricity selling price is calculated at 0.03 p/kWh above the base case. Similarly, in Case 8 where coal is priced at £40/tonne the break-even electricity selling price is raised by 0.04 p/kWh. With Case 9 a difference in cost between the main and reburn coal is introduced, in case a higher quality reburn coal is required to achieve good burnout. Here the main coal costs £25/tonne and the reburn coal costs £40/tonne, giving a break-even electricity selling price of 2.36 p/kWh, an increase of 0.13 p/kWh over the base case. This price difference is used to illustrate the sensitivity of the break-even electricity selling price to the relative price that could be paid for the main and reburn coal. Case 10 shows that a price of £54/tonne could be paid for the reburn fuel and still give a break-even electricity selling price similar to the s.c.r. Case 1. Cases 11 and 12 look at the sensitivity of the break-even electricity selling price to estimated capital costs. The capital cost for building a new plant to the coal-over-coal reburn technology specification was calculated as £365M compared to building a new conventional plant at £363M. However, the maximum additional capital cost required for retrofitting coal-over-coal reburn technology to an existing plant, Case 11, was estimated as £16M, depending on the extent of air heater modifications. This would produce a break-even electricity selling price of 2.32 p/kWh. Case 12 shows that an additional capital cost of up to £46M could be made before reaching the break-even electricity selling price determined for case 1, the equivalent s.c.r. system.

The third NO_x reduction technology studied was coal-over-coal reburning using a microniser to give a finer coal particle in the reburn zone. Compared with the previous coal-over-coal reburn technology, the thermal input is similar, but there is a reduction in power output from the low-pressure steam turbine of 11.3 MWe due to the steam required by the microniser. The total auxiliary power requirement is down 0.4 MWe due mainly to taking one of the coal mills out of service. This gives an overall higher heating value efficiency of 41.2% compared with 41.8% for the previous coal-over-coal reburn case. The specific CO_2 emissions rise from 767 to 780 g/kWh due to this reduced efficiency. The increased capital cost to £376M for new

Table 3
Economic results

Base case	(i)	(ii)				
Coal cost (£/tonne)	25	40				
Total cap cost (£M)	363	363				
Break-even electricity selling price (p/kWh)	2.23	2.7				
s.c.r.	Case 1	Case 2	Case 3	Case 4	Case 5	Case 6
Coal cost (£/tonne)	25	25	40	40	40	40
NG cost (£/GJ)	2.0	3.0	2.0	3.0	3.0	3.0
Total cap cost (£M)	400	400	400	400	388	423
Break-even electricity selling price (p/Kwh)	2.44	2.45	2.91	2.93	2.88	3.02
Coal-over-coal reburn	Case 7	Case 8	Case 9	Case 10	Case 11	Case 12
Main coal cost (£/tonne)	25	40	25	25	25	25
Reburn coal cost (£/tonne)	25	40	40	54	25	25
Total capital cost (£M)	365	365	365	365	379	409
Break-even electricity selling price (p/kWh)	2.26	2.74	2.36	2.44	2.32	2.44
With microniser	Case 13	Case 14	Case 15	Case 16		
Main coal cost (£/tonne)	25	40	25	25		
Reburn coal cost (£/tonne)	25	40	25	25		
Total capital cost (£M)	376	376	386	368		
Break-even electricity selling price (p/kWh)	2.34	2.82	2.39	2.31		
Natural gas-over-coal	Case 17	Case 18	Case 19	Case 20	Case 21	Case 22
Coal cost (£/tonne)	25	25	40	40	25	25
NG cost (£/GJ)	2.0	3.0	2.0	3.0	3.0	3.0
Total cap cost (£M)	361	361	361	361	368	379
Break-even electricity selling price (p/kWh)	2.45	2.64	2.84	3.02	2.67	2.71
Natural gas-over-coal	Case 23	Case 24	Case 25	Case 25	Case 27	Case 28
Coal cost (£/tonne)	25	40	25	25	25	25
NG cost (£/GJ)	0.98	1.53	1.93	1.76	2.0	2.0
Total cap cost (£M)	361	361	361	368	368	383
Break-even electricity selling price (p/kWh)	2.26	2.74	2.44	2.44	2.48	2.54

plant is associated with the purchase of two 20 tonnes/h micronisers to replace one coal mill, and the additional surface area required in the economiser section and the air heater. Case 13, with a main coal price of £25/tonne and a reburn coal price also of £25/tonne gives a break-even electricity selling price of 2.34 p/kWh, which compares with he 2.26 p/kWh for the coal-over-coal reburn case 7. Case 14, the equivalent case only with both the main and reburn coals costing £40/tonne gives a break-even electricity selling price of 2.82 p/kWh compared with the coal-over-coal reburn Case 8 of 2.74 p/kWh. Therefore, there is an additional charge on the break-even electricity selling price of 0.08 p/kWh to compensate for the loss of efficiency and the increased capital cost associated with the microniser. The capital cost calculated for the coal-over-coal reburn technology with microniser was £376M, which is the cost of building a new plant. However, if an existing plant was retrofitted, Case 15, then the total capital cost would be higher due to equipment modification and replacement costs. Alternatively, in Case 16, some of the existing equipment (heat exchangers) could be retained but with a loss in efficiency due to higher economiser and air heater exit temperatures. For Case 15 the maximum additional capital cost required, whilst maintaining efficiency, was estimated at £23M. This gives a break-even electricity selling price of 2.39 p/kWh. For Case 16, by not making changes to the economiser and air heater there is a potential reduction in efficiency of 0.25% points to 40.95% and a saving in capital cost of £18M. This gives a break-even electricity selling price of 2.31 p/kWh.

The final NO_x reduction technology studied was natural gas-over-coal reburning. Compared with coal-over-coal reburn technology the thermal input is higher due to an expected increase in unburnt carbon levels. Test results from the Mitsui Babcock pilot plant indicate that the unburnt carbon percentage increases by a factor of two when natural gas is used as the reburn fuel instead of coal. The total auxiliary power requirement is down 0.4 MWe due mainly to taking one of the coal mills out of service. This gives an overall higher heating value efficiency of 41.1% compared with 41.8% for the basic coal-over-coal reburn case. The CO_2 emissions fall from 767 to 704 g/kWh as 20% of the thermal input comes from natural gas. This also accounts for the reduction in SO_2 emissions from 0.94 to 0.75 g/kWh, with the NO_x emissions remaining the same. The capital cost of the system is slightly lower than for coal-over-coal reburning due to the reduced cost of coal storage and one less coal mill. With the Amer 9 power plant a natural gas supply is already available to the plant and so only a small natural gas connection cost is involved.

Case 17, with a coal price of £25/tonne and natural gas price of £2.0/GJ gives a break-even electricity selling price of 2.45 p/kWh, which compares with the 2.26 p/kWh for the basic coal-over-coal reburn case. Case 19, the equivalent case only with coal costing £40/tonne gives a break-even electricity selling price of 2.84 p/kWh compared with the basic coal-over-coal reburn case of 2.74 p/kWh. The effect of natural gas price, Cases 18 and 20, is now significant due to the larger quantity used. An increase in natural gas cost from £2.0/GJ to £3.0/GJ increases the break-even electricity selling price by ~0.19 p/kWh. Cases 21 and 22 look at a retrofitting situation and an estimated additional capital cost in the range £5M–£16M, depending on the extent of modifications to the air heater section. For a coal price of £25/tonne and a natural gas price of £3.0/GJ the additional capital investment increases the break-even electricity selling price by 0.03–0.07 p/kWh. Cases 23 and 24 examined the natural gas price that would be required in order to give the same break-even electricity selling price as the equivalent coal-over-coal reburn technology. For a coal price of £25/tonne the equivalent natural gas price is £0.98/GJ and for a coal price of £40/tonne the equivalent natural gas price is £1.53/GJ. The next two cases, 25 and 26, look at the maximum natural gas price that could be paid and still give the same break-even electricity selling price as the basic s.c.r. case, for both a new natural gas-over-coal reburn plant and a minimum cost retrofit plant. For Case 25, the new natural gas-over-coal reburn plant, the maximum allowable natural gas price is £1.93/GJ, and for Case 26, the minimum cost retrofit plant, the maximum allowable natural gas price is £1.76/GJ. The final two cases examine the situation where a natural gas connection does not already exist, which was not the case for Amer 9 power station. In Case 27 it is assumed that a 5 km pipeline connection is required, at a cost of £6.5M and in Case 28 it is assumed that a 20 km pipeline connection is required at a cost of £21.5M [17]. The effect is to increase the break-even electricity selling price from 2.45 p/kWh for Case 17 to 2.48 p/kWh for Case 27 and to 2.54 p/kWh for Case 28.

4. Conclusions

The assessment of alternative NO_x reduction technologies, involving modifications to the basic Amer 9 model, was successfully completed using the ECLIPSE process simulator. These studies provided a detailed technical, environmental and economic analysis from which the following conclusions can be drawn.

All of the technologies investigated provide considerable reductions in NO_x emissions. The coal-over-coal and natural gas-over-coal reburn systems both produced approximately 50% less NO_x than the base case, however the s.c.r. post-combustion method showed the most substantial change, producing a 90% reduction. Of the technologies considered, s.c.r. is the only method of achieving NOx emission levels below 50 mg/N m^3.

Use of these NO_x reduction technologies generally incurred a small increase in auxiliary power and consequently a slight increase in CO_2 emissions as net electrical efficiencies dropped. This increase in auxiliary power was mainly due to higher fan duties. Variation between the systems' auxiliary power requirements was slight and mainly resulted from the fan duty requirements of the reburn systems and the requirements of the optional sixth coal mill. Operating a microniser in a coal-over-coal reburning system reduced the power output as steam was taken from the turbine cycle and fed to the microniser. In consequence, whist this system had the lowest thermal input, it also had the highest specific emissions of CO_2, as well as SO_x and NO_x, due to its low electrical efficiency.

Use of natural gas reduced overall system efficiency slightly. In the natural gas-over-coal reburn system this was due to an expected difficulty in achieving complete burnout, and with s.c.r. this was due to the flue gas heating requirements. However, the natural gas-over-coal system, with 20% less thermal input from coal, showed a marked reduction in CO_2 and SO_x emissions.

At present the s.c.r. system could not compete economically with the coal-over-coal reburn system. However, it is possible that future legislation will prioritise environmental considerations over economic advantage, making this technology necessary as a feature of a very low NO_x system. The use of a microniser was regarded as unfavourable from an economic as well as an emissions point of view and should not to be followed unless the normal coal-over-coal reburn configuration was found to produce excessive unburnt carbon losses.

Assuming an onsite supply of natural gas, the natural gas-over-coal reburn system was estimated to have the lowest capital cost of the systems considered. However, to compete with the coal-over-coal reburn systems, an unrealisable natural gas price of £0.98/GJ was necessary. The natural gas-over-coal reburn plant was calculated to be competitive with s.c.r. at natural gas prices between £1.76/GJ and £1.93/GJ, but it was, of course unable to match the latter in reducing NO_x emission levels. Without a pre-existing gas connection, the economic viability of this natural gas-over-coal reburning system was considered to be further disadvantaged by distance from supply.

Acknowledgements

This work has been partly funded through the European Commission JOULE–THERMIE Programme—"Clean

Coal Technologies for Solid Fuels R&D (1996–1998)". I would like to acknowledge the assistance provided by Ing A.J.C Korthout, Plant Manager of the Amer 9 Power Station at Geertruidenberg, by Jim Cooper of James Howden Group Technology Limited, as well as the other partners in the Group, particularly KEMA, Mitsui Babcock Engineering Limited and ENEL SpA.

References

[1] Moore MJ. Proc Inst Mech Engng Part A–J Power Energy 1997;211(1):43.

[2] Muzio LJ, Quartucy GC. Prog Energy Combust Sci 1997;23(3):233.

[3] McMullan JT, Williams BC, Sloan EP. Proc Inst of Mech Engng Part A—J Power Energy 1997;211(1):95.

[4] van der Lans RP, Glarborg P, Dam-Johansen K. Prog Energy Combust Sci 1997;23(4):349.

[5] Wendt JOL, Sternling CV, Matovich MA. 14th Symp. (Int.) on Combustion, 1973. p. 881.

[6] Hjalmarsson AK, Soud HN. Systems for controlling NO_x from coal combustion. IEACR/30, International Energy Agency Coal Research, London. 1990.

[7] EPZ Reports. c1988. The 600 MW supercritical monotube steam generator Amer 9 and Amercentrale eenheid 9, technische gegevens.

[8] Williams BC. DPhil Thesis, Energy Research Centre, University of Ulster, Coleraine, N.I. 1994.

[9] Williams BC, McMullan JT. Int J Energy Res 1994;18(2):117.

[10] Williams BC, McMullan JT. Int J Energy Res 1996;20(2):125.

[11] Willams BC, McMullan JT. In: Imariso, Bemtgen, editors. Progress in synthetic fuels, London: Graham and Trotman, 1988. p. 183–9.

[12] ECLIPSE Process Simulator, Energy Research Centre, University of Ulster, Coleraine, N.I., 1992.

[13] Gramelt S. FGD system for 600 MWe coal fired power plant—process description, PFD, mass and energy balance, equipment specifications, 1994, Private Communication with Deutsche Babcock Anlagen GmbH. Postfach10 03 47-48, D-4200 Oberhausen 1, Germany.

[14] McCahey S, Campbell PE, McIlveeWright DR, Williams BC, McMullan JT. Joule–Thermie Programme, Clean Technologies for Solid Fuels, 1997, Contract No JOF3-CT95-0005, Bemtgen, J. M., European Commission.

[15] Breihofer, D., Mielenz, A., Rentz, O., Emission control of SO_2, NO_x and VOC at stationary sources in the Federal Republic of Germany, 1991, University of Karlsruhe, Germany.

[16] Cooper J. Coal Microniser—the Howden Micromill, James Howden Group Technology, Private Communication, 1997.

[17] Feines G. Private Communication from British Gas to Mitsui Babcock Energy Ltd., Renfrew Scotland, 1997.

NO$_x$ and N$_2$O emission characteristics from fluidised bed combustion of semi-dried municipal sewage sludge

M. Sänger[a], J. Werther[a,*], T. Ogada[b]

[a]*Technical University Hamburg-Harburg, Chemical Engineering I, D-21071 Hamburg, Germany*
[b]*Department of Production Engineering, Moi University, Eldoret, Kenya*

Accepted 19 May 2000

Abstract

Incineration is one of the major methods for the disposal of sewage sludge. Currently, several plants are incinerating mechanically dewatered (wet) sludge (20–40 wt.% d.m.) or semi-dried sewage sludge (30–55 wt.% d.m.), although some plants burn dry sludge (with more than 80 wt.% d.m.). Whereas significant information is available on NO$_x$ and N$_2$O emissions characteristics of wet and dry sludge, not much has been reported on semi-dried sludge. This paper presents some of the results obtained from the combustion of semi-dried sludge in a semi-pilot scale fluidised bed combustor (150 mm in diameter and 9 m high) together with some measurements from a large-scale FBC incineration plant (7 m^2 bed area, 9 m high and a capacity of 3 t/h dry sludge). The investigations have shown that semi-dried sludge exhibit emission characteristics which are similar to those of wet sludge. NO$_x$ decreases slightly whereas N$_2$O remains more or less the same with increase in oxygen concentrations. Just like wet sludge, staged combustion was not effective for the reduction of NO$_x$ and N$_2$O. However, increasing the freeboard temperature led to rapid reduction of N$_2$O and some NO$_x$ reduction was achieved using flue gas recycling technique. Comparison shows that the results from the test rig were more or less similar to those obtained from the large-scale plants. © 2000 Elsevier Science Ltd. All rights reserved.

Keywords: Incineration; Sewage sludge; Fluidised bed; NO$_x$ emission; N$_2$O emission

1. Introduction

Incineration has become one of the main disposal outlets for the ever-increasing quantity of sewage sludge in many developed countries and some newly industrialised nations. Already, incineration takes 24% of the sludge produced in Denmark, 20% in France, 15% in Belgium and 19% in Germany [1,2]. In the USA and Japan, 25 and 55% of the sludge produced, respectively, is incinerated [3]. It is expected that 38% of the sludge produced by the member states of the European Union will be incinerated by the year 2005 [1].

Depending on the water content, sludge for incineration may be generally classified into three groups, namely dry sludge (>80 wt.%, d.m.), semi-dried sludge (30–55 wt.% d.m.) and mechanically dewatered sludge (20–40 wt.%). Sewage sludge with solid contents of around 60–65 wt.% is normally avoided because of the sticky physical nature of sludge within this range.

Incineration of dry sludge granulates in mono-combustion plants is not very wide spread, although some cases have been reported, particularly for sludge smelting. In 1993, for example, 15 sewage sludge smelting plants were operating in Japan [4]. The Nanbu sewage sludge smelting plant in Tokyo South which is the largest, handles some 160 t/day of wet sludge (20 wt.% d.m.) which is dried to 80 wt.% d.m. and smelted in a cyclone furnace at 1400–1500°C [4]. Similarly, in Germany several smelting technologies are currently being introduced, in which dry sewage sludge, in granulates or pulverised form, is fired in cyclone furnaces [5,6]. Another area where dry sludge is often preferred is in co-combustion technology. Dry sludge has been successfully co-fired with coal in power plants, with municipal solid wastes in incineration plants and as supplementary fuel in cement kilns [7–9].

For mono-combustion, the preferred state of sludge has been mechanically dewatered (wet) sludge and of late, semi-dried sludge. The peculiar requirement of mono-combustion of wet sludge is the need for a supplementary fuel to support the combustion process. Normally, biogas, natural gas or furnace oil is used. The quantity of the support fuel required depends on many factors, including the moisture content

* Corresponding author. Tel.: +49-40-42878-3039; fax: +49-40-42878-2678.
E-mail address: werther@tu-harburg.de (J. Werther).

Reprinted from *Fuel* **80 (2)**, 167-177 (2001)

258

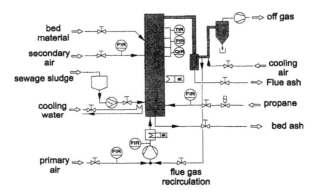

Fig. 1. Flow diagram of the bubbling fluidised bed test facility.

and the type of sewage sludge as well as the furnace operating parameters such as combustion temperature and excess air ratio [10,11]. Apart from the requirement for support fuel, the other shortcoming of burning wet sludge is the large quantity of vapour which is released during combustion. This increases the volume of flue gas and consequently leads to the need for larger sizes of flue gas cleaning equipment and fan power. Consequently, of late, there has been a trend towards eliminating the use of supplementary fuel. This is achieved by using more efficient dewatering equipment capable of producing sludge cakes with solid contents of 35 wt.% and above to enable auto-thermal combustion. The other variant of achieving auto-thermal combustion is to incorporate a drier after a dewatering equipment and produce partially dried sludge with moisture contents of 30–46 wt.%. Several incinerating plants are currently burning semi-dried sludge. In 1996, for example, five plants in Germany were reported to burn such sludge [11]. Another example is the 16,000 t/y d.m. (dry mass) incineration plant at Roundhill, supplied and commissioned by Dorr-Oliver to the Seven Trent Water, UK, where dewatered sludge with 23–25 wt.% d.m. is partially dried to around 34 wt.% and burnt in a fluidised bed [12]. There is also the 54,000 t/a d.m. capacity bubbling fluidised bed sludge incinerator at Dordrecht, The Netherlands, which was commissioned in 1993 by Lurgi Energie and Umwelt GmbH, [13]. At this plant, raw sludge is received at 20 wt.% dry matter and partially dried in a steam heated disk drier before being fed into the combustor.

An important aspect for consideration during sludge incineration is the control of the emissions of gaseous pollutants, especially N_2O and NO_x. The potential for N_2O and NO_x emissions is high during the combustion of sewage sludge due to its high contents of nitrogen. The nitrogen content of sewage sludge is generally in the range 6–8 wt.%, waf [14], but can be as high as 10 wt.% [15]. N_2O and NO_x emissions depend on nitrogen contents of the fuel [16,17].

Much information is currently available concerning the N_2O and NO_x emission characteristics of dry sludge [18–21] as well as for wet sludge [14,18,22,23]. It was found that dry sludge exhibits NO_x and N_2O emission characteristics similar to those of coals. High concentrations of NO_x (800–1200 mg/m^3) and N_2O (300–400 mg/m^3) were obtained from sewage sludge granulates due to the high nitrogen contents [19–21]. NO_x and N_2O increased with increase in excess air ratio whereas increasing the combustion temperature lead to an increase in NO_x and a decrease in N_2O. Furthermore, reduction of N_2O and NO_x could be achieved through application of staged combustion whereas non-catalytic $DeNO_x$ (i.e. NH_3 injection) was effective for NO_x control. Unlike dry sludge, wet sludge was found to exhibit totally different NO_x and N_2O emission characteristics [18,22,23]. Very low NO_x (<200 mg/m^3) was measured from wet sludge, although the nitrogen content was higher than that of the dry sludge. Furthermore, combustion parameters such as combustion temperature and excess air ratio as well as emission reduction strategies (e.g. staged combustion and non-catalytic $DeNO_x$) had little effect on the emissions. The low NO_x emissions from wet sewage sludge incineration were validated by the NO_x emission data collected from some large-scale incineration plants in Germany, burning wet sludge [11].

The above-mentioned differences in the NO_x and N_2O emission behaviour of dry and wet sewage sludges and the growing interest in large-scale combustion of semi-dried sludge have currently stimulated research on the combustion of semi-dried sewage sludge. The fundamental research question is whether semi-dried sludge would behave like wet sludge or dry sludge. In the current work, extensive investigations have been conducted with semi-dried sludge samples. The objectives were: first, to investigate the NO_x and N_2O emission characteristics and compare the results with those previously obtained for wet and dry sludge and secondly, to test the suitability of some known NO_x and N_2O emission reduction techniques such as staged combustion, freeboard temperature, furnace configuration

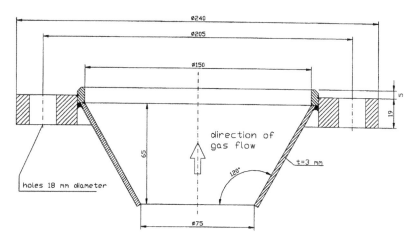

Fig. 2. Details of the mixing element installed in the freeboard of the bubbling fluidised bed test facility.

and flue gas recycling to the combustion of semi-dried sewage sludge.

2. Experimental

The incineration of semi-dried sewage sludge was conducted in a bubbling fluidised bed, 150 mm in diameter and 9 m high, the flow diagram of which is shown in Fig. 1. The required combustion temperature was maintained through electrical heating of the walls of the combustor. The test unit has the necessary facilities for automatic and continuous recording of combustion parameters such as temperatures, pressure drops, combustion air and fuel flow rate as well as facilities to feed wet sludge, solid, liquid and gaseous fuels. The height of the unit was chosen so as to ensure that the mean gas residence time could be adjusted to be similar to those applied in large-scale plants.

An important feature of some modern incinerators is a

reduction in the cross-sectional area in the upper section of the freeboard (e.g. sludge incineration plants at Ulm, Hoechst AG and Dordrecht, The Netherlands). The Ulm plant, for example, has a rectangular freeboard of a cross-sectional area of 15.6 m^2 with a cylindrical outlet chute of a cross-sectional area of 2.8 m^2. The purpose of such furnace configuration is to provide intensive mixing of the reactants in the freeboard by creating turbulence. To simulate this effect, a mixing device (Fig. 2) was recently installed in the test unit at 7000 mm above the distributor plate.

The combustor has several sampling ports along the riser. The details of the experimental set-up for gas sampling and analysis have been previously reported [18,22,23]. The gas sample is withdrawn from the centre line of the combustor and sucked through an electrically heated filter and sampling line by means of a gas pump. It is then directed into a cooler where it is dried before it is supplied to the various analysers at a rate of 60 l/h. Components of the flue gas such as O_2, CO_2, CO, NO, N_2O, and SO_2 can be analysed.

Table 1 shows the elementary analyses of the semi-dried sludge sample tested. For comparison, the analyses of dry and wet sludge from the samples investigated by Werther et al. [18,22] are also included. All the three sludge samples originated from different wastewater treatment plants in Germany. Generally, the compositions of the three sludge samples are more or less the same. They are all characterised by high contents of ash, volatile matter and nitrogen. However, the semi-dried sludge sample, being raw, had lower contents of ash and higher calorific value than the other samples, which were digested sludges. This is due to partial loss of the organic substance as CO_2 and CH_4 during sludge digestion.

Apart from the measurements in the authors' test rig, further measurements were undertaken in a large-scale incineration plant firing semi-dried sewage sludge. In a

Table 1
Elementary analyses of the semi-dried sludge (current work) compared with wet and dry sludges (previous work [18,22])

	Semi-dried sludge	Dry sludge	Wet sludge
Proximate			
Water, wt.% raw	68	6.9	76.0
Ash, wt.% wf	31	47.9	51.8
Volatiles, wt.% waf	90	89.2	92.4
NH_3–N, wt.% waf	0.9	0.97	3.2
LCV, kJ/kg raw	4408	8145	1098
wf	13,775	8758	4567
Ultimate, wt.% waf			
C	47.8	52.2	51.9
H	7.68	6.3	7.8
O	38.4	33.7	29.8
N	4.6	4.7	8.8
S	0.77	3.1	1.7

260

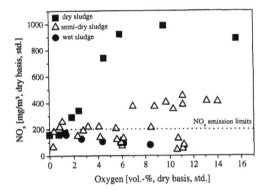

Fig. 3. NO$_x$ emissions as a function of oxygen concentrations—influence of the water contents (semi-dry sludge, this work; wet and dry sludges from Ref. [22].

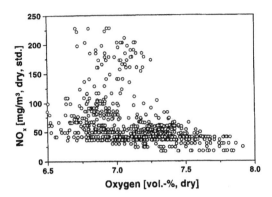

Fig. 4. NO$_x$ emissions as a function of oxygen concentrations measured at the large-scale sludge incineration plant VERA.

collaboration work between the Technical University of Hamburg-Harburg and the Sewage Water Treatment Authority of Hamburg, measurements were undertaken by the authors in the bubbling fluidised bed sewage sludge incinerating plant VERA operated by the authority at their Koehlbrandhoeft wastewater treatment plant. The plant was commissioned in 1997 by Deutsche Babcock Anlagen GmbH. The plant has three combustion lines, each incinerator has a grid area of 7 m^2 and a total height of 9 m. The plant burns semi-dried sludge with 39–46 wt.% dry solid contents. The sludge is fed at a mass flow rate of 3 t/h (dry sludge) with a belt charging machine on top of the bed. A total of 78,840 t of semi-dried sludge and a maximum 12,000 t of screens are burnt per year. The fluidised bed is operated in the bubbling mode at an average superficial velocity of 1.9 m/s (including the vapour from the sludge). Amongst the parameters investigated was the influence of oxygen concentrations on NO$_x$ emissions and the influence of freeboard temperature on N$_2$O emissions.

3. Results and discussions

3.1. Single stage combustion

3.1.1. NO$_x$ emissions as a function of excess air ratio

Fig. 3 shows the results of the emissions of NO$_x$ from semi-dried sludge samples during incineration in the test facility. The data were obtained from different samples of the same sludge collected and burnt on different days. For comparison, the results of NO$_x$ emissions from wet (water content around 75%) and dry (water content about 5%) sludge samples obtained from Werther et al. [22] are also included. The NO$_x$ emissions are given as a function of oxygen concentrations in the flue gas (i.e. excess air ratio). The excess air ratio was varied by adjusting the mass flow of the sludge while maintaining a constant combustion air supply. A static bed height of 560 mm was

used. The sludge was fed directly into the bed at 380 mm above the distributor plate and the gas sample was withdrawn at 9100 mm. The bed temperature was around 830°C whereas the freeboard was maintained at around 870°C. Unlike for wet sludge for which propane had been used as supplementary fuel [18,22], no support fuel was required for the experiments with semi-dried sludge.

First of all, there is a distinct difference between the emission behaviour of wet and dry sludge, respectively. With a wet sludge, the tendency of NO$_x$ is to decrease with increasing oxygen content in the flue gas whereas the reverse happens for a dry sludge. The reason is a difference in the reaction kinetics [22]: a dry sludge behaves much like a coal, or to be more precise like lignite. The lowering of the NO$_x$ emissions with decreasing oxygen content is the basis of the two-stage mode of combustion air supply. On the other hand, the wet sludge contains a lot of ammonia dissolved in water which helps in reducing NO as it is practised in the SNCR process. The result of the presence of ammonia in the sludge water is that the NO$_x$ emission is stabilised on a very low level. The decrease of NO with increasing oxygen concentration in the flue gas may at least partly be attributed to a dilution effect due to the increased air flow.

It can be seen that the NO$_x$ emission characteristics of semi-dried sludge samples are closer to those previously obtained for wet sludge samples. However, it is not possible to deduce a clear trend for the NO$_x$ emissions as a function of the oxygen concentrations. Since each measurement point is the result of an individual test run under otherwise constant operating conditions it can only be concluded that the scattering is due to fluctuations of sludge properties. The semi-dried sludge obviously varies in its NO$_x$ reducing capability. Its properties are sometimes closer to those of the dry sludge and sometimes resembling more to the wet sludge. Measurements conducted by Albrecht and Schelhaas [13] on the large-scale combustor of semi-dried sludge at Dordrecht, The Netherlands, showed also large variations

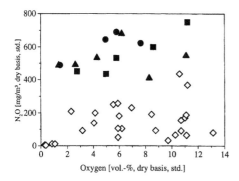

Fig. 5. Comparison of N_2O emissions from semi-dried sludge with those from wet sludge (open symbols—semi-dried sludge, this work; filled symbols—wet sludge, from Werther et al. [18] and Ogada [14].

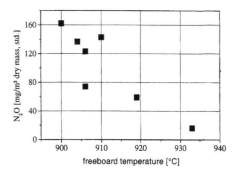

Fig. 6. Variation of N_2O emissions with freeboard temperature measured at the large-scale sludge incineration plant VERA.

of NO_x concentrations. For example, during a 24-h test period at the plant, NO_x emission could suddenly vary from 50 mg/m^2 to around 600–800 mg/m^3. However, during a more or less steady combustion condition, a slight decrease in NO_x emission was recorded as O_2 was increased from 8 to 11 vol.%. Measurements undertaken by the current authors at the VERA plant also showed a decreasing trend of NO_x with increasing flue gas oxygen concentrations when the latter varied between 6.5 and 8.0 vol.% (Fig. 4). In this respect, the semi-dried sludge exhibited NO_x emission characteristics similar to those obtained from wet sludge. Another important conclusion from these results is that, whereas it is possible to meet the NO_x limits of 200 mg/m^3 when burning wet sludge using single-stage combustion, this appears not to be guaranteed during the combustion of semi-dried sludge especially within the normal operating conditions (i.e. 6–11 vol.% O_2).

3.1.2. N_2O emissions as a function of excess air ratio

Fig. 5 shows the N_2O emissions from the combustion of semi-dried sludge as a function of oxygen concentrations in the flue gas. Here too, the data represent different samples of the same sludge collected and burnt on different days. Included in Fig. 5 for comparison, are the measurements from three different wet sludge samples previously reported by Werther et al. [18] and Ogada [14]. Compared with those obtained for wet sludge, the combustion of semi-dried sludge gave lower N_2O. N_2O concentrations less than 250 mg/m^3 were measured from semi-dried sludge compared with 500–700 mg/m^3 measured from wet sludge. Another observation is that, similar to wet sludge, there appears to be no dependence of N_2O emissions on oxygen concentrations in the flue gas. Apparently, the dilution effect of an increasing air supply is compensated for by an increased N_2O formation the reaction kinetics of which seem to be dependent on the partial pressure of oxygen. Which of the numerous known N_2O formation reactions [24] is altered remains highly speculative and cannot be answered by means of the present global

combustion experiments where all possible reactions occur simultaneously.

3.1.3. Effect of freeboard temperature on N_2O emissions

Measurements from the VERA incinerator have shown that the freeboard temperature is a very effective parameter for controlling N_2O (Fig. 6). N_2O concentrations decreased from 160 to less than 20 mg/m^3 as the freeboard temperature was raised from 900 to around 934°C. There was no significant effect of this temperature increase on the emissions of NO_x. In this respect, semi-dried sludge showed a similar trend to what was previously obtained from dry sludge [19] and wet sludge [18]. For example, Werther et al. [18] reported that the N_2O emission decreased from 560 to 110 mg/m^3 when the freeboard temperature was increased from 844 to 876°C during the combustion of wet sludge. The reason for this decrease of the N_2O emission is an increase in the rate of a homogeneous destruction reaction due to the increased temperature. This effect of the thermal decomposition of N_2O is well known [24].

During wet sludge combustion, high-freeboard temperature is generally achieved due to increased combustion of the volatiles in the freeboard. Ogada and Werther [25] have shown that volatile release during the combustion of wet sludge is delayed by the drying process. This leads to release of a larger part of the volatiles in the splash zone and in the freeboard, and consequently the combustion of the volatiles takes place there. In the case of the combustion of semi-dried sludge, the extent of volatile combustion in the freeboard may be less. Thus to maintain higher freeboard temperature, some large-scale plants (e.g. the VERA plant) incorporate natural gas or biogas burners in the freeboard.

3.1.4. Importance of effective mixing of reactants in the freeboard

In order to investigate the importance of enhanced gas mixing in the freeboard on combustion and emissions, gas concentrations were measured at locations 6500, 7500 and 9100 mm above the distributor plate (i.e. below and above

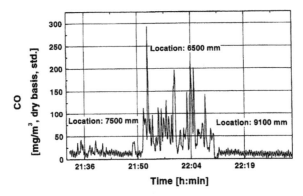

Fig. 7. Time resolved CO concentration signals before and after the mixing element which was located at 7 m above the distributor.

the location of the mixing element at the 7000 mm level). The analysis of the emissions signal indicates that indeed the mixing element has improved the combustion quality (Fig. 7). The measurements' signal drawn from the port located 0.5 m below the mixing element exhibited large fluctuations of CO concentrations with extreme peaks. Sampling from locations 0.5 and 2.1 m, respectively, above the mixing element, however, showed that the signals were much smoother. The combustion efficiency was also significantly improved. For example, at 6 vol.% outlet oxygen concentration, CO was rapidly reduced from 270 to less than 50 mg/m³ (Fig. 8). The CO-reduction is clearly attributable to the intense mixing generated by the mixing element.

With respect to the NO_x concentrations, the enhanced gas mixing appears to have not much effect. The NO_x level remained more or less the same (Fig. 8). However, measurements from a large-scale sludge incinerator have recently indicated that turbulence generated through injection of steam into the freeboard led to a reduction of NO_x emissions [26].

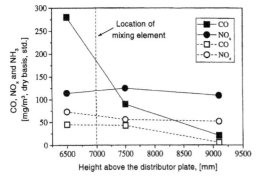

Fig. 8. Effect of gas mixing in the freeboard on the emissions of CO and NO (combustion temperature 850°C, filled symbols—6 vol.% O_2; open symbols—11 vol.% O_2).

3.1.5. Effect of flue gas recycling

Flue gas recycling (FGR) has been previously used as a technique for NO_x emissions reduction [27,28]. Michel and Belles [27], for example, applied FGR in 220 and 147 kW, respectively, gas-fired boilers and achieved a reduction of NO_x from 102 to 43–51 mg/m³ for the large boiler and a reduction of NO_x from 82 to 28–41 mg/m³ for the small boiler. Baltasar et al. [28] conducted their experiment in a 300 mm diameter laboratory-scale gas fired furnace equipped with facilities for FGR. They obtained a marked reduction in NO_x using FGR without affecting the flame stability, overall combustion efficiency and CO and C_{org} emissions. In both cases, the achieved reduction in NO_x emission using FGR was attributed to a decrease in partial pressure of O_2 and combustion temperature in the combustion zone. In the current work, some experiments were conducted with FGR during both single-stage and staged combustion. In both cases, the gas velocity in the combustor was maintained constant. The flue gas was withdrawn after the fabric filter at a temperature of around 200°C, mixed with the primary air and with the help of a compressor, supplied into the combustor. As the percentage of the primary air replaced by recycled flue gas increased, the primary air was reduced to maintain a constant gas flow through the distributor plate. Consequently, in order to maintain a given oxygen concentration in the flue gas, the mass flow rate of the sludge was also reduced.

Fig. 9 shows NO_x emissions as a function of the percentage of recycled flue gas in the fluidising gas. Here too, NO_x reductions were obtained. During single-stage combustion carried out with an excess air ratio of 1.4 (6 vol.% in flue gas) only a mild decrease in NO_x was obtained. The plot of the ratio of fuel nitrogen conversion to NO demonstrates that this is mainly a dilution effect. For staged combustion, on the other hand, a significant effect of the flue gas recycling is observed. For both 6 and 11 vol.% O_2 in the flue gas, a significant reduction in NO_x emissions is measured. The plot of the ratio of fuel nitrogen conversion to NO illustrates that this must be due to a reaction kinetic effect: increasing

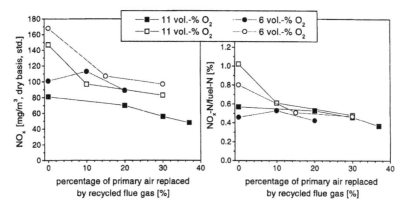

Fig. 9. Effect of flue gas recycling on the emissions of NO (bed and freeboard temperature 850°C, filled symbols—single-stage; open symbols—staged combustion).

the percentage of primary air which is replaced by the recycled flue gas from 0 to 10% decreases the conversion ratio by 20–30%. The chemical reason is that with the recycled flue gas, NO enters the reactor again and has the chance of being reduced. Flue gas recycling increases thus the residence time for NO_x in the reaction zone which leads to an overall reduction of the NO_x emission.

Fig. 10 compares the NO_x emission with the emissions of N_2O and NH_3 for the single-stage combustion case. The conversion ratio plot illustrates once again that under these conditions FGR is a mere dilution. However, the NH_3 emission drops sharply when the percentage of primary air replaced by recycled flue gas is increased from 0 to 10%. The reason for this effect is not clear. It may be a technically interesting effect in cases where the ammonia emission is too high.

The fact that FGR yielded a positive emission performance is a very important observation drawn from these experiments. Flue gas recycling can therefore be used as a strategy of achieving partial load in a sludge incineration

plant without affecting the fluidisation characteristics of the bed. However, a ratio of 10–15% recycled flue gas should not be exceeded since higher ratios are reported to lead to instability of combustion and higher CO emission [27,28].

3.2. Staged combustion

Experiments conducted with dry sludge have shown that staged combustion was very effective for the control of the emissions of CO, NO_x and N_2O. At O_2 concentration of around 6–11%, von Razceck [19] obtained a reduction of NO_x from over 1200 to less than 300 mg/m³ and N_2O could be reduced from 300 to less than 50 mg/m³. These positive results, however, could not be obtained during the combustion of wet sludge [18,22]. Therefore, the purpose of conducting staged combustion experiments with semi-dried sludge was aimed at establishing its behaviour in light of this diverging tendency between the wet and dry sludge samples.

The experiments were conducted in a similar manner as

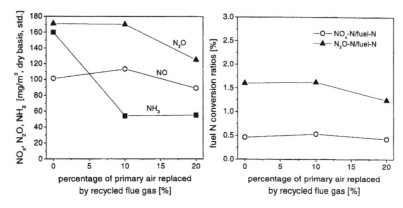

Fig. 10. Effect of flue gas recycling on the emissions of NO_x, N_2O and NH_3 (bed and freeboard temperature 850°C, single-stage combustion, 6 vol.% oxygen in flue gas).

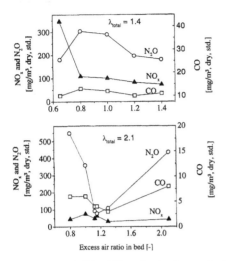

Fig. 11. Emissions of N_2O, NO and CO as a function of excess air ratio in the bed for staged combustion (combustion temperature 850°C, gas concentrations measured at the combustor exit).

whereas CO remained more or less the same. At an excess air ratio of the bed equal to 0.65, NO_x emissions jumped to over 300 mg/m³. Staged combustion experiments at $\lambda_{total} = 2.1$ gave emission reductions with an optimum air staging being in the range of $\lambda_{bed} = 1.0-1.3$. Although the NO_x emissions remained more or less the same, within the optimum range, it is seen that N_2O decreased from 438 to 77 mg/m³ and CO from 8 to less than 4 mg/m³. However, the performance gets worse as the excess air ratio in the bed was lowered further. In conclusion, like in the case of wet sludge combustion, staged combustion gave no positive results with respect to NO_x emission performance.

The failure of staged combustion to achieve the expected reduction in NO_x is another similarity of semi-dried sludge to wet sludge, and as before, is attributed to the unusual relationship between NO_x and O_2. Measurements of axial profiles by Ogada [14] during the combustion of wet sludge showed that higher NO_x emissions were formed within the bed at lower oxygen concentrations. The NO_x concentrations, although being rapidly reduced in the splash zone, remained high leading to a higher outlet emission for lower oxygen concentrations in the flue gas (Fig. 12). For example, the peak values of NO_x within the bed increased from 2350 to 4400 mg/m³ when the oxygen concentration in the flue gas was reduced from 10 to 1.5 vol.%. Near the point of secondary air injection, the NO_x values were 400 and 1000 mg/m³ at 10 and 1.5 vol.% O_2, respectively. This indicates that during staged combustion, the concentration of NO_x at the point of injection of secondary air is higher than during single-stage combustion (although the reverse would be expected from experience with coals). By injection of secondary air, an increase in NO_x will occur due to the combustion of the volatile nitrogen species carried into the freeboard. These two factors explain the increase in NO_x with decrease in the excess air ratio in the bed observed during the combustion of semi-dried sludge. If a large quantity of secondary air is injected or if the NO_x concentration at the location of secondary air injection is not high, the increase in NO_x may be offset by the effect of dilution.

before [18,22]. Both the primary and secondary air flows were adjusted simultaneously so as to establish the required oxygen concentration at the outlet. The experiments were done for outlet oxygen concentrations of 6 and 11 vol.%, respectively. The secondary air injection port was located at 4300 mm above the distributor plate and the gas samples were withdrawn at 9100 mm, after a retention time of more than 2 s, as required by the regulations for waste incineration in Germany (17.BImSchV). Fig. 11 shows the results obtained during staged combustion with total excess air ratios, λ_{total} of 1.4 and 2.1 corresponding to outlet oxygen concentrations of 6 and 11 vol.%, respectively.

For the staged combustion experiments at $\lambda_{total} = 1.4$, the emission performance deteriorates with a decrease in the excess air ratio in the bed. As the excess air ratio in the bed was decreased from 1.4 to 0.8, NO_x increased from 73 to 108 mg/m³, N_2O from 181 to 304 mg/m³

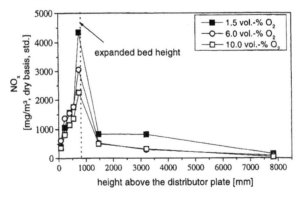

Fig. 12. Axial profile of NO_x concentrations as a function of oxygen concentration during single-stage combustion of wet sludge [14].

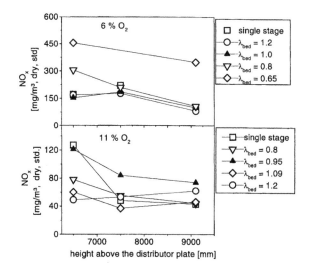

Fig. 13. Axial profile of NO$_x$ concentrations as a function of excess air ratio, λ_{bed} in the bed during staged combustion (outlet oxygen concentration 6 and 11 vol.%, i.e. λ_{total} = 1.4 and 2.1, respectively).

This interpretation can be understood better by analysing the axial profile concentrations of NO$_x$ from the current measurements (Fig. 13). It is seen that during staged combustion of semi-dried sludge, at a total excess air ratio, λ_{total} of 1.4, the NO$_x$ concentrations within the region above the feed point of secondary air increase from 150 to 450 mg/m^3 as excess air ratio in the bed was reduced from 1.4 to 0.65. At very low excess air ratio in the bed, apart from high NO$_x$ in the flue gas reaching the freeboard, the combustibles escaping into the freeboard contain sufficient nitrogen species such that the combustion reactions following injection of secondary air lead predominantly to the formation of NO$_x$. The reverse effect is seen during staged combustion at higher outlet oxygen concentration (11 vol.%) where above the secondary air injection point,

lower NO$_x$ concentrations were measured during staged combustion than for single-stage combustion. The decrease in NO$_x$ did not, however, correlate with the excess air ratio maintained in the bed, what is attributed to the competition between NO$_x$ formation and dilution. For example, at an excess air ratio of 1.2 in the bed, less combustion will be expected in the freeboard so that injection of 80% of the stoichiometric air requirement in the freeboard to achieve 11 vol.% at the outlet, would obviously lead to dilution.

3.3. Comparison with the performance of large-scale combustors

A comparison of the emission levels obtained from the test rig with measurements from large-scale combustors of

Table 2
The emission performance of some large-scale fluidised-bed incineration plants for semi-dried sludges in Germany (sources: plants nos. 1–4 from Ref. [11] and plant no. 5—measurements by the present authors, n.m. = not measured)

Plant and operation parameters						Emission performance			
Plant	Sludge type	Water content in feed sludge (wt.%)	Bed temp. (°C)	Freeboard temp. (°C)	Secondary air	CO (mg/m^3)	N$_2$O (mg/m^3)	NO$_x$ (mg/m^3)	O$_2$ (vol.%)
1	Raw	55–60	850	890	Yes	11–19	100–300	145	11
2	Digested	55	780	850	Yes	18	47–188	50	11
3	Raw	65	735	840	Yes	51	105	186	11
4	Digested	54	800	910	No	33	n.m.	72	11
5	Digested	54–61	800	920	Yes	0–5	20–160	20–100	6–8

266

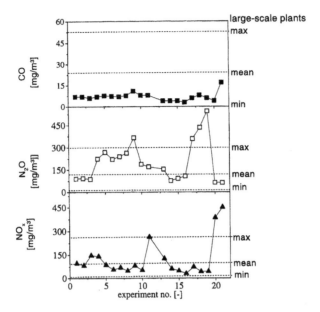

Fig. 14. Comparison of the CO, NO$_x$ and N$_2$O emission performance of the test rig with measurements from large-scale combustors of semi-dried sludge (11% oxygen in flue gas, maximum, minimum and mean values taken from Table 2).

semi-dried sludge has shown that the performance of the test rig was more or less the same as that of the large-scale plants. Table 2 shows the emission performance of five plants currently incinerating semi-dried sewage sludge in Germany [11]. Included in the table are our measurements from the VERA plant in Hamburg. All plants are operating according to the fluidised-bed principle. The semi-dried sewage sludge burnt had solid contents of 30–46 wt.%.

For comparison, the CO, N$_2$O and NO$_x$ concentrations from 21 measurements carried out in the bench-scale plant around 11 vol.% oxygen in the flue gas have been compiled in Fig. 14. The data are from various test runs, consisting of single-stage and staged combustion as well as flue gas recirculation experiments. In this way, the range of emission performance from the test rig at different operation conditions and including sludge properties fluctuations, is expressed. It can be seen that the measured CO, NO$_x$ and N$_2$O from the test facility are well within the range of the data from large-scale plants. There is good agreement between the test rig and the large plants as concerns NO$_x$ and N$_2$O although CO emissions from the test rig appear to be too low. Of the 21 tests, 17 were within the N$_2$O range of 0–300 mg/m^3 reported from the large-scale plants and 18 tests were in the NO$_x$ range of 40–230 mg/m^3.

4. Conclusion

In the current work, the emission characteristics of semi-dried sewage sludge have been investigated and the results

were compared with those previously obtained from mechanically dewatered and dry sludge, respectively. The results have shown that semi-dried sludge exhibits similar emission characteristics as wet sludge. NO$_x$ emissions were slightly higher than those from wet sludge but much lower than those from dry sludge. Like wet sludge, NO$_x$ exhibited a tendency to decrease with increase in O$_2$ whereas there was no strong dependence of N$_2$O emissions on O$_2$ concentrations in the flue gas. Staged combustion gave no positive effect on NO$_x$ and N$_2$O whereas FGR led to a decrease in NO$_x$. As known for wet sludge, increasing the freeboard temperature led to rapid reduction of N$_2$O. Measurements from the test rig compare very well with those from large-scale combustors. From the results, it can be concluded that for the combustion of dry sludge only, NO$_x$-reduction techniques need to be used to guarantee meeting the emission limits. A high freeboard temperature is recommended as a useful technique to achieve low emissions of N$_2$O, CO, NH$_3$ and C$_{org}$. Where high freeboard temperatures cannot be achieved through volatile combustion alone, auxiliary gas burners should be incorporated. Furthermore, the furnace configuration should be such as to promote gas mixing in the freeboard.

Acknowledgements

Fluidised bed combustion research at the Technical University of Hamburg-Harburg is financed by the Deutsche Forschungsgemeinschaft through Sonderforschungsbereich

238. This substantial support is gratefully acknowledged. One of the authors (Ogada, T.) further thanks the German Academic Exchange Programme (DAAD) and Moi University's Committee of Deans for their continued support. Further thanks are due to G. Hiller from the sewage water treatment plant Steinhaeule, Ulm/Germany, for initiating this study and supplying us with a lot of plant data.

References

[1] Hall JE, Dalimier F. Waste management-sewage sludge: survey of sludge production treatment, quality and disposal in the EC. EC reference no. B4-3040/014156/92, 1994.

[2] van Riesen S, Bringewski F. Press release, Abwassertechnische Vereinigung e.V. Germany, January, 1998.

[3] Loll U. Entsorgungspraxis-Spezial 1989;8:3.

[4] Imoto Y, Mori T, Takagi Y, Niwa C. In NGK Insulator, Ltd, Plant R&D Section, Environmental Equipment Div. 1, Maegeta-chyo, Handa-shi, Aichi, 475. Japan, p. 1.

[5] Company Information Booklet, Klein Energietechnik GmbH, Germany, 1997.

[6] Company Information Booklet, Seiler Trenn-Schmelzanlage Betriebs GmbH, Germany, 1997.

[7] Römer R. In: Klärschlamm Entsorgung 1, Daten—Dioxine, Entwässerung, Verwertung, Entsorgungsvorschläge, VDI-VERLAG GmbH, Düsseldorf. 1991, p. 250.

[8] Tejima H. In: Thome-Kozmiensky, Loll U, editors. Recycling von Klärschlamm. EF-VERLAG für Energie-und-Umwelttechnik GmbH, Berlin, 1987. p. 267.

[9] Lang T, Obrist A. In: Thome-Kozmiensky, Loll U, editors. Recycling von Klärschlamm, EF-VERLAG für Energie-und-Umwelttechnik GmbH, Berlin, 1987. p. 285.

[10] Mühlhaus L. In: Klärschlammentsorgung II, VDI-Seminar Düsseldorf, November 1991.

[11] Anonymous. ATV Arbeitsbericht, Korrespondenz Abwasser, no. 7, 1996. p. 1299.

[12] Goldsmith, P. The Chemical Engineer, January 1994. p. 13.

[13] Albrecht J, Schelhaas KP. In: Preto FDS, editor. Proceedings of the 14th international conference on fluidised bed combustion, Vancouver, Canada, 1977. p. 997.

[14] Ogada T. PhD thesis, Technical University Hamburg-Harburg, Germany, 1995.

[15] Wirsum M. PhD thesis, University of Siegen, Germany, 1997.

[16] Gulyurtlu IA. Fuel 1995;74:253–7.

[17] Hampartsoumian E, Gibbs BM. J Inst Energy 1984;403.

[18] Werther J, Ogada T, Philippek C. J Inst Energy 1995;68:93.

[19] von Raczeck. PhD thesis, Technical University Hamburg-Harburg, Germany, 1992.

[20] Vogel B, Lindau S, Busse U. Wissenschaft und Umwelt 1992;1:105.

[21] Hanssen H. Diploma thesis, Technical University Hamburg-Harburg, Germany, 1991.

[22] Werther J, Ogada T, Philippek C. In: Heinschel KJ, editor. Proceedings of the 13th international conference on fluidised bed combustion, Orlando, FloridaNew York: ASME, 1995. p. 951.

[23] Philippek C, Knöbig T, Werther J. In: Preto FDS, editor. Proceedings of the 14th international conference on fluidised bed combustion, Vancouver, Canada, New York: ASME, 1997. p. 983.

[24] Kilpinen P, Hupa M. Combustion and Flame 1991;85:94.

[25] Ogada T, Werther J. Fuel 1996;75(5):617.

[26] Ludwig P, Stamer F. In: Reuther RB, editor. Proceedings of the 15th international conference on fluidized bed combustion, Savannah, GA, New York: ASME, 1999 (paper no. FBC99-0053).

[27] Michel Y, Belles FE. Combus Sci Technol 1993;94(1-6):447.

[28] Baltasar J, Carvalho MG, Coelho P, Costa M. Fuel 1997;76(10):919.

Optimum NO_x abatement in diesel exhaust using inferential feedforward reductant control

H.C. Krijnsen, J.C.M. van Leeuwen, R. Bakker, C.M. van den Bleek, H.P.A. Calis*

DelftChemTech, Faculty of Applied Sciences, Chemical Reactor Engineering, Delft University of Technology, Julianalaan 136, 2628 BL Delft, The Netherlands

Accepted 8 November 2000

Abstract

To adequately control the reductant flow for the selective catalytic reduction of NO_x in diesel exhaust gas a tool is required that is capable of accurately and quickly predicting the engine's fluctuating NO_x emissions based on its time-dependent operating variables, and that is also capable of predicting the optimum reductant/NO_x ratio for NO_x abatement. Measurements were carried out on a semi-stationary diesel engine. Four algorithms for non-linear modelling are evaluated. The models resulting from the algorithms gave very accurate NO_x predictions with a short computation time. Together with the small errors this makes the models very promising tools for on-line automotive NO_x emission control. The optimum reductant/NO_x ratio (to get the lowest combined NO_x + reductant emission of the exhaust treating system) was best predicted by a neural network. © 2001 Published by Elsevier Science Ltd.

Keywords: Selective catalytic reduction; Reductant flow control; Soft-sensing

1. Introduction

One of the approaches for NO_x emission abatement is selective catalytic reduction (SCR), which involves adding a reductant to the exhaust gas to catalytically remove NO_x. The reductant flow rate that has to be injected into the exhaust gas depends on the NO_x concentration and the exhaust flow rate, the required NO_x reduction and the catalyst conditions. On-line NO_x analysis equipment is not only relatively very expensive but is also susceptible to soot plugging and has to be frequently calibrated and serviced. Therefore, an alternative for NO_x concentration measurement is needed. Furthermore, the optimum reductant/NO_x ratio has to be determined in order to keep the slip of both reductant and NO_x through the system as low as possible. This can be done by modelling the catalyst system in a classical way. However, almost all models describing the reaction between ammonia and NO over vanadia/titania monolithic catalysts [1–3] were only evaluated for a small temperature range (typically a range of 100 K) and for simulated exhaust gases, containing only nitrogen, water, NO and ammonia. The effect of the SO_2 concentration was only taken into account in several articles [4–7], while the effects of soot and hydrocarbons were not

taken into account so far, neither in the experiment nor in the models. Furthermore, the effect of ammonia oxidation was neglected in the models even though ammonia oxidation is significant at temperatures above 623 K. Only one author, [2] considered a range of NO concentrations within models and experiments. The result is that for all other models, the reaction order in NO is not verified within the model and therefore the model may show discrepancies when other NO concentrations are considered.

Earlier experiments have shown that the NO_x emission from the engine used in our study is a function of the intake air temperature, the intake air pressure, intake air humidity and the engine load, which is in agreement with literature [8–14].

The aim of this article is to show that it is possible to predict both the NO_x emission and the optimum reductant/NO_x ratio based on the diesel engine's and catalyst's time-dependent operating variables. For this so-called 'soft-sensing' or inferential measurement application, a non-linear black-box modelling algorithm is used that takes a large set of measurements to learn how to predict the NO_x emissions from the operating variables. Four candidates are evaluated in this paper:

1. An artificial neural network (ANN) [15–19].
2. The split & fit algorithm of Bakker — This split & fit algorithm (s&f) splits the input data into a number of

* Corresponding author. Tel.: +31-15-278-3516; fax: +31-15-278-4452.
E-mail address: h.p.a.calis@tnw.tudelft.nl (H.P.A. Calis).

Reprinted from *Fuel* **80 (7)**, 1001-1008 (2001)

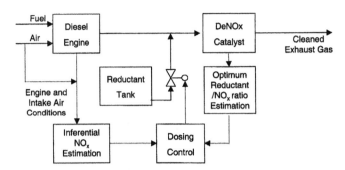

Fig. 1. Schematic representation of the inferential feedforward NO_x control system.

disjoint regions, to each of which is assigned a local linear model. To connect the various local linear models, sigmoid functions are used to give smooth transitions from one model to another. The bandwidth of the region to which the sigmoid function is applied, is optimised by the algorithm [20].

3. A polynomial NARX model that is linear in its parameters [21–23].

4. An engine map with humidity correction — This engine map with humidity correction is an extension of the traditionally used engine map (see below). Since the traditionally used engine map does not describe the effect of humidity on NO_x emissions, the NO_x emission estimated from the engine map is corrected for humidity. The correction factor is fitted to the data and is a function of the absolute humidity and the engine load. The control system is given in Fig. 1.

The models resulting from these black-box modelling algorithms are compared to an engine map and a linear fit through the measurement data. An engine map is an engine-specific database that lists (NO_x) emissions measured for a set of engine operating variables. Given the relevant operating conditions, e.g. engine load and rotation speed, the corresponding NO_x emission can be estimated through interpolation in the database, without actually measuring it.

A linear fit of the engine's NO_x emission is made by constructing only linear relationships between the input variables and the output variables.

2. Experimental set-up

The engine used for the experiments was a Lister–Petter LPW3, a three-cylinder water-cooled diesel engine fitted in a Wilson LD 12.5/W4 generator with a rated power of 8.0 kW and a constant engine speed of 1500 rpm. The fuel used during all the experiments was a summer quality diesel fuel containing 0.04 wt% sulphurs. The fuel was in accordance with the EN590:1996 specification.

Downstream of the diesel engine a Corning EX80 wall flow monolith was located in order to filter off the soot before deNO$_x$ experiments were performed in the deNO$_x$ reactor. The reactor contained either a 1.05L or 0.071L

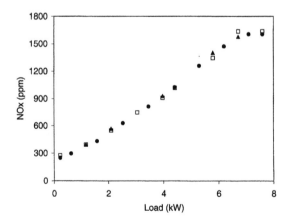

Fig. 2. Reproducibility of NO_x emission in times: (●) February 1998–November 1998; (□) April 2000.

Table 1
Lay-outs of the various models resulting from fitting the NO_x emission data by the algorithms

Modelling algorithm	Lay-out
ANN	1 hidden layer comprising 9 nodes, sigmoid transfer functions
S&f	19 clusters
NARX	3rd order polynome containing 12 terms
Engine map with humidity correction	4th order fit through the NO_x emission data as a function of the engine load, intake air temperature, relative humidity and pressure
Engine map	4th order fit through the NO_x emission data as a function of only the engine load. The data used for fitting were NO_x emissions measured at intake air relative humidities below 7%
Linear fit	Fitted linear relation between inputs and NO_x emission

catalytic honeycomb catalyst. The catalyst considered is fully extruded V_2O_5–WO_3–TiO_2 (Frauenthal).

In order to keep the inlet conditions of the combustion air constant during the experiments, dried air was used for $deNO_x$ experiments; to assess the influence of intake air temperature and intake air humidity, these variables were varied by injecting steam into the intake air and heating the intake air. Part of the exhaust gas of the engine was used for $deNO_x$ experiments, the rest of the exhaust gas was led to a vent which allowed us to vary the exhaust flow rate through the $deNO_x$ system independently from the engine load.

Sample gas streams to be analysed by a chemiluminescence NO_x analyser, were led through washing bottles containing 85 wt% phosphoric acid and 98 wt% sulphuric acid to remove water and ammonia. In addition to the NO_x analyser, a NO_x sensor (NGK) was placed directly downstream of the engine. A microwave NH_3 analyser (Siemens) measured the ammonia concentration in the exhaust gas.

The four input variables (intake air temperature, pressure and humidity and engine load) as well as the response variable (exhaust NO_x concentration, measured close to the

engine) were logged by a data acquisition system running on a PC. The intake air temperature was measured with a PT100 heat resistance. The intake air humidity was measured by a humidity sensor (Testo, Testo hygrotest 6337.9741). The pressure of the ambient (intake) air was measured by pressure sensor (SensorTechnics, 144SC0811BARO, range 800–1100 mbar). The engine load was measured with a load meter (FAGET, Active power transducer single phase EM168-K14CR, 0–20 kW). For reductant dosage control experiments the four input variables were made input to a 486 PC on which the dosing control program ran.

3. Results

First the reproducibility of the engine with respect to the NO_x emission was investigated. Comparison of data of February and November 1998 and April 2000 showed no significant change in NO_x emission characteristics over two years, see Fig. 2. The reproducibility of the engine measurements in time is of prime importance for any NO_x prediction application.

3.1. NO_x emission prediction

For training and testing the several fit algorithms, 13 h of data with a frequency of 1 Hz were used. Two thirds of the data set was used for training the algorithms and one third for validating the resulting models. The variables that affected the NO_x emission of the LPW3 engine were engine load, intake air temperature, intake air pressure and intake air humidity. These variables were taken into account in the fit algorithms. Based on cross-correlation data, the initial time delays of the variables were estimated. The time delays were further optimised by trial and error within the split and fit algorithm. The optimum set of input variables for the black-box models was found to be: (a) intake air temperature; (b) intake air pressure; and (c) intake air humidity; all three at the current time only, and (d) engine load not only at current time but also 2, 5, 8, 11 and 14 s in the past. The optimum lay-outs of the resulting models are given in Table 1. The averaged absolute errors of the various models are given in Table 2. The variance of the engine's NO_x emission is 3.0%. Compared to this natural variance of the NO_x

Table 2
Averaged absolute training and test errors of the various models resulting from fitting the NO_x emission data by the algorithms

Modelling algorithm	Training Winter 1998–1999	Testing Winter 1998–1999	Testing Winter 1999–2000
ANN	2.8	3.4	4.5
S&f	2.7	3.2	4.9
NARX	3.9	3.7	4.9
Engine map with humidity correction	5.5	5.2	4.2
Engine map	8.6	6.3	9.4
Linear fit	6.0	6.0	4.8

272

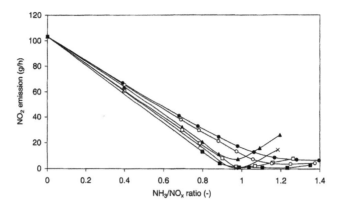

Fig. 3. Final NO_2 emission as a function of the NH_3/NO_x ratio for catalyst temperatures of 523–773 K, at a NO_x concentration of 1000 ppm at a GHSV of 4700 h^{-1}: (▲) 523 × 573 K; (□) 623 K; (■) 673 K; (○) 723 K; (●) 773 K.

emission, the prediction by the ANN and s&f model is so good that they cannot be further improved.

The reason for the linear fit model and the other sophisticated tools to have about the same error is that the relationship between engine load and NO_x emission is quite linear up to 7 kW (see also Fig. 2) and that the highest engine load used during the tests was 7 kW, in order to prevent problems due to soot plugging the equipment. Another reason is that, although the relative error at low engine loads is high, the error at low engine load hardly contributes to the overall NO_x prediction error due to the small absolute error. At low engine loads, the linear fit performance declined significantly.

3.2. Optimum NH_3/NO_x ratio estimation

To minimise the emission of pollutants (NO_x as well as reductant), the reductant/NO_x ratio resulting in the lowest impact on the environment had to be estimated and used for on-line reductant control. For that reason a control system was tested, in which the earlier described inferential NO_x estimation method determined the NO_x emission from the engine while another estimation method, based on the catalyst and exhaust conditions, estimated the optimum reductant/NO_x ratio. In this study, ammonia was used as a reductant.

The acidification of the environment is due to NO_x that is rapidly converted to NO_2 at the vent, and due to NH_3 that eventually is converted to NO_2. The sum of these NO_x and NH_3 emission was thus considered a measure of the acidification effect. The NH_3/NO_x ratio at the minimum acidification effect is, per definition, the optimum NH_3/NO_x ratio. It is a function of the NO_x concentration, the gas velocity and the catalyst temperature.

Measurements were performed at temperatures ranging from 523 to 773 K, at various engine loads (i.e. various NO_x concentrations), at NH_3/NO_x ratios varying from 0 to

1.6 and a GHSV of 4700 or 70,000 h^{-1}. This was achieved by keeping the exhaust gas flow through the catalyst at 5.0 m$_n^3$/h and using either a 1.05L or a 0.071L honeycomb catalyst. The effects of the catalyst temperature and NH_3/NO_x ratio on the final NO_2 emission are displayed in Figs. 3–5 for both GHSVs at an engine load of 4 kW (corresponding to approximately 1000 ppm NO_x).

The measurement data of final NO_2 emission versus the NO_x inlet concentration, the catalyst temperature and the NH_3/NO_x ratio, were used for training and testing a neural network, a s&f algorithm and a NARX model at a GHSV of 4700 or 70,000 h^{-1}. For future applications, in addition to the NO_x inlet concentration, the catalyst temperature and the NH_3/NO_x ratio, the effect of the GHSV should be further investigated and used for training and testing the algorithms. The optimal lay-outs of the resulting models are given in Table 3. The averaged absolute errors of the various models are given in Table 4.

With regard to the optimum lay-outs of the s&f model it should be noted that the s&f algorithm did not increase the number of clusters by one, but by a number that increased with the number of clusters.[1] Results on the estimation of the NH_3/NO_x ratio containing 10 or 11 clusters were not available. As a result, the s&f lay-outs, containing 12 or 9 clusters lie close to each other.

The prediction errors of the test set were very high due to the choice of the data in the test cycle and the small number of data used for training (99 data points for a GHSV of 4700 h^{-1} and 116 for a GHSV of 70,000 h^{-1}). Since the location of the optimum NH_3/NO_x ratio was of primary importance and not the absolute NO_2 emission the application of the NH_3/NO_x ratio in a control system, data points around the optimum NO_2 emission were taken into account

---

[1] The increase in the number of clusters is found by rounding up the value calculated, from the number of clusters divided by four, to the nearest integer. The reasons are that it saves calculation time.

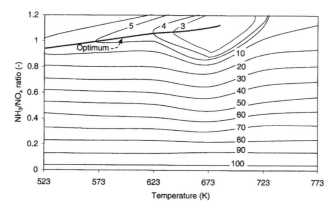

Fig. 4. Final NO₂ emission as a function of the NH₃/NOₓ ratio for catalyst temperatures of 523–773 K, at a NOₓ concentration of 1000 ppm at a GHSV of 4700 h⁻¹.

in the test set. Due to the small absolute values of the NO_2 emissions around the optimum NH_3/NO_x ratios, especially for the low GHSV of 4700 h⁻¹, the relative error of the prediction became quite large.

The three algorithms were also subjected to simulated input data (called 'evaluation data'), to check whether the trends in the final NO_2 emission prediction were correct. The evaluation set caused the s&f model to sometimes predict NO_2 maxima with increasing NH_3/NO_x ratios, instead of minima. The NARX model gave the good NO_2 emission trend, the location of the optimum NH_3/NO_x ratio was, however, incorrect. The same (undesired) trends were observed for the modelling results of the measurements at a GHSV of 70,000 h⁻¹.

3.3. Real-time NH₃ control

Resulting from the work described so far, six models were available for predicting the NO_x emission from the engine: ANN, s&f, NARX, engine map, engine map with humidity correction and linear fit. These were tested in combination with the ANN for prediction of the optimum NH_3/NO_x ratio.

Three different real-time NH_3 dosage control tests were performed. One test sequence was performed on all six models at a GHSV of 4700 h⁻¹ and a temperature of 573 K. The other two test sequences were performed at a GHSV of 70,000 h⁻¹, one measurement at 573 K, the other at 673 K. The engine load was varied during each test. The resulting overall emission reduction level was defined as

$$\text{Overall emission reduction} = \frac{NO_{x\,in} - NO_{2\,out}}{NO_{x\,in}} \times 100\%,$$

$$(1)$$

where $NO_{x\,in}$ is the engine's NO_x emission and $NO_{2\,out}$ the final NO_2 emission downstream of the catalyst, i.e.

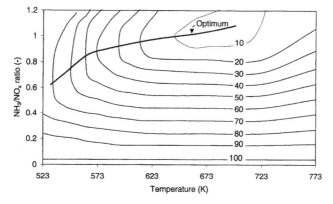

Fig. 5. Final NO₂ emission as a function of the NH₃/NOₓ ratio for catalyst temperatures of 523–773 K, at a NOₓ concentration of 1000 ppm at a GHSV of 70,000 h⁻¹.

Table 3
Lay-outs of the various models resulting from fitting the final NO$_2$ emission data by the algorithms

Modeling algorithm	Lay-out	
	GHSV of 4700 h^{-1}	GHSV of 70,000 h^{-1}
ANN	1 hidden layer comprising 10 nodes, sigmoid transfer functions	1 hidden layer comprising 10 nodes, sigmoid transfer functions
S&f	12 clusters	9 clusters
NARX	6th order polynome consisting of 13 terms	6th order polynome consisting of 17 terms

the sum of the NO, NO$_2$ and NH$_3$ emissions. The overall emission reduction levels are given in Table 5. During all three tests the intake air temperature was kept near 290 K and the intake air humidity was kept around 30%. During the tests the engine load was varied at four different loads. Results for the s&f model at a GHSV of 70,000 h^{-1} and 573 K are given in Fig. 6.

4. Discussion

NO$_x$ emission prediction by using s&f, ANN and NARX models showed very good agreement between prediction and measurement even with one year in between training and testing. The average absolute error of the NO$_x$ emission prediction showed values in the order of 4.2–4.9% for the linear fit, ANN, s&f and engine map with humidity correction. The traditionally used engine map performed much worse than the other models.

From the point of view of training time for NO$_x$ emission prediction, the linear fit, the s&f algorithm and the engine map performed best. The s&f algorithm needed only a few minutes, the engine map and the linear fit in the order of a quarter of an hour to be optimised on a Pentium III. The modelling of an engine map with humidity correction took about an hour and the polynomial NARX took a few hours for optimisation, while the neural network took up to about 24 h to be optimised. This large time needed for training the ANN is due to the relatively slow convergence of the network, followed by pruning and reduction of the size of the ANN, which is again followed by convergence, etc. until pruning does no longer improve the network's performance.

The time for calculating one NO$_x$ emission prediction, once a model has been optimised was in the order of 0.3 ms for the ANN, s&f, linear fit model and even somewhat less for both engine maps. For the polynomial NARX model a calculation time of 4.3 ms was estimated. All these calculation times are small enough for real-time NO$_x$ emission prediction.

The optimum NH$_3$/NO$_x$ ratio region flattened at increasing catalyst temperature (see Fig. 3) this was attributed to oxidation of ammonia: part of the ammonia fed to the catalyst was no longer used for converting the NO$_x$ but was itself converted to either NO$_x$ or N$_2$. This could be concluded from the slow increase in NH$_3$ emission at excess NH$_3$ dosage at high temperatures. As a result, the optimum NH$_3$/NO$_x$ ratio shifts to higher values.

Training the resulting NO$_2$ emission (i.e. NO + NO$_2$ + NH$_3$) as a function of gas throughput, catalyst temperature, the engine's NO$_x$ emission and NH$_3$/NO$_x$ ratio, showed that the ANN model could best predict the optimum NH$_3$/NO$_x$ ratio.

Real-time NH$_3$ dosage based on the integration of the NO$_x$ emission prediction and the estimation of the optimum NH$_3$/NO$_x$ ratio by the ANN were performed at a GHSV of 4700 and 70,000 h^{-1} and showed high conversions. Considering the GHSV of 70,000 h^{-1} results first, it can be seen that at 673 K the overall reduction levels were very close to one another. To give insight into the significance of the differences, the measurement errors were determined. The variance of the engine's NO$_x$ emission was 0.7% during the measurement day. Based on equipment specifications and measurement errors, the relative error in the overall emission reduction level was 3.2%. In combination with the overall emission reduction level of 80%, this gave an absolute error in emission reduction of 2.6%. The differences between the measurement results were smaller than the measurement error and therefore the results at a GHSV of 70,000 h^{-1} could not be distinguished.

At 573 K and a GHSV of 70,000 h^{-1}, the engine's NO$_x$ variance was 1.0%. The resulting maximum relative error in the overall emission reduction was 3.2%. With a maximum emission reduction of about 50%, this gave an absolute error of the emission reduction of 1.6%. As a result the s&f, engine map with humidity correction and linear fit showed highest overall emission reduction levels, followed by the ANN, engine map and NARX models.

Similar to the measurement at 673 K and 70,000 h^{-1}, the

Table 4
Averaged absolute training and test errors of the various models resulting from fitting the final NO$_2$ emission data by the algorithms

Modelling algorithm	GHSV of 4700 h^{-1}		GHSV of 70,000 h^{-1}	
	Training error	Testing error	Training error	Testing error
ANN	2.4	12	2.8	4.8
S&f	4.2	21	14	25
NARX	7.3	39	4.9	6.9

Table 5
Measured final NO_2 emissions of the integrated control structure combining a model for NO_x emission prediction with an ANN for estimation of the optimum NH_3/NO_x ratio

Model for NO_x emission prediction	Overall emission reduction level (%)		
	GHSV 4700 h^{-1} (573 K)[a]	GHSV 70,000 h^{-1} (573 K)[b]	GHSV 70,000 h^{-1} (673 K)[a]
ANN	92.2	48.2	81.8
S&f	93.4	51.1, 50.7[c]	81.5
NARX	93.7	47.0	81.0
Engine map with humidity correction	93.6	50.0	81.5
Engine map	94.0	47.7	80.7, 80.0[c]
Linear fit	92.6	50.2	80.8

[a] Measurements were performed within two months time.
[b] Measurements were all performed within 9 h of operating.
[c] Duplo measurement.

differences between the measurements at 4700 h^{-1} were smaller than the measurement errors. This disabled discrimination between measurement results. Since the measurements at a GHSV of 4700 h^{-1} were performed on several days, the engine's NO_x emission variance was 3.0%.

The main reason for the overall emission reduction levels to be this close is that the ANN for estimating the NH_3/NO_x ratio gives too high a value output when the NO_x emission prediction is too low and vice versa. As a result, the errors in NO_x emission prediction are levelled off giving similar actual NH_3 flows. It can be seen in Fig. 5 that especially for a GHSV of 70,000 h^{-1} the optimum range of the NH_3/NO_x ratio is quite large. As long as the NH_3/NO_x ratio is within 5% of the optimum value, hardly any effect on the overall emission reduction will be noticed. Since the engine load was varied during the tests, an average overall emission reduction was given, thus also levelling off the effects of modelling errors.

Not only the models used for NO_x emission prediction can be compared to one another, the integrated control

system can also be compared to a control system with a fixed NH_3/NO_x ratio. As can be seen from Figs. 4 and 5, the optimum NH_3/NO_x ratio increases with increasing catalyst temperature. Thus, a fixed NH_3/NO_x ratio may result in high overall emission reduction levels at one specific temperature, but will show lower emission reduction levels when the catalyst temperature is changed, making the real-time optimisation of the NH_3/NO_x ratio worthwhile.

In reality, one can choose between avoiding any NH_3 emission downstream of the catalyst and converting the NH_3 emission. As the objective function of this study was to maximise the emission reduction level, the occurrence of NH_3 emissions downstream of the deNO$_x$ catalyst was inevitable. Therefore, an oxidation catalyst has to applied for converting NH_3 to NO_x, but preferably to N_2, because the emission of NO_x is preferred above the emission of NH_3. Further research in this field is necessary. In case of implementation of an oxidation catalyst that (partially) converts NH_3 to N_2, the optimum NH_3/NO_x ratio will shift to higher values.

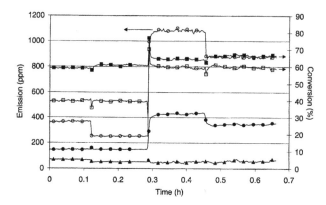

Fig. 6. Final conversions and emissions at 573 K and GHSV of 70,000 h^{-1} at various engine loads for the integrated control system, using the s&f model for NO_x emission prediction and the ANN for optimisation of the NH_3/NO_x ratio: (□) Overall emission reduction (right axis); (■) NO_x conversion (right axis); (○) NO_x emission engine; (●) NO_x emission downstream SCR catalyst; (▲) NH_3 emission downstream SCR catalyst.

5. Conclusions

The implementation of an optimisation strategy for the NH_3/NO_x ratio performs much better than when choosing a constant NH_3/NO_x ratio independent of gas flow, catalyst temperature and NO_x concentration in the exhaust gas entering the catalyst. This can be seen from Figs. 4 and 5, in which the optimum NH_3/NO_x ratio increases with catalyst temperature.

The s&f model, the engine map with humidity correction and the linear fit gave highest overall emission reduction levels, followed by the ANN, the engine map and the NARX model. From these results it can be concluded that not only the catalyst conditions were of importance for maximum emission reduction, but also the effect of humidity should be considered when controlling the reductant flow. Furthermore, the time consuming training of a neural network algorithm or a NARX algorithm did not give better results than the fast training s&f algorithm, linear fit and engine map with humidity correction. For all resulting models, the calculation time, needed for the prediction of the NO_x concentration (up to 4.3 ms) and optimum reductant/NO_x ratio (about 6 ms), was short enough to allow for real-time control.

It should be noted that the differences between measurements investigating the effect of the NO_x emission prediction model on the overall NO_2 emission reduction level were sometimes of the same order as the measurement errors. This can be explained by the fact that the emission reduction level is a weak function of the NH_3/NO_x ratio (see Fig. 5). Another reason of levelling off the differences between the various models is the fact that an over prediction is partially compensated by an under prediction of the optimum NH_3/NO_x ratio. This, once more, emphasises the need for optimisation of the NH_3/NO_x ratio. When a transient operating engine is considered for operating the above models [24], it is believed that the advanced models for NO_x emission prediction perform much better in combination with the optimisation of the NH_3/NO_x ratio than the linear fit and engine map.

Acknowledgements

The authors would like to thank Sander Baltussen for the implementation of the described fit algorithms into the diesel engine exhaust set-up and Frauenthal (Frauental, Austria) for supplying the vanadia deNO$_x$ catalyst.

References

[1] Buzanowski MA, Yang RT. Ind Engng Chem Res 1990;29:2074.
[2] Pinoy LJ, Hosten LH. Catal Today 1990;17:151.
[3] Svachula J, Ferlazzo N, Forzatti P, Tronconi E, Bregani F. Ind Engng Chem Res 1993;32:1053.
[4] Tronconi E. Catal Today 1997;34:421.
[5] Tronconi E, Forzatti P, Gomex Martin JP, Mallogi S. Chem Engng Sci 1992;47:2401.
[6] Tronconi E, Lietti L, Forzatti P, Mallogi S. Chem Engng Sci 1996;51:2965.
[7] Tronconi E, Cavanna A, Forzatti P. Ind Engng Chem Res 1998;37:2341–9.
[8] Lin C-Y, Jeng Y-L. J Ship Res 1996;40:172.
[9] Lin C-Y, Jeng Y-L, Wu C-S, Wu K-J. J Environ Sci Health 1996;31:765.
[10] Boot P. Int Congr Combust Engines, Proc 20th, London, May, CIMAC, Göteborg, 1993. vol. D67. p. 1.
[11] Kondoh H, Kawano T, Masuda K. Int Congr Combust Engines, Proc Conf CIMAC, Copenhagen, 1998. p. 803.
[12] Juva A, Rautiola A, Saikkonen P, Le Breton D. Fuels-gas oils 1996;16:1.
[13] Dodge LG, Leone DM, Naegeli DW, Dickey DW, Swenson KR. SAE technical paper 962060, 1996.
[14] Rakopoulos CD, Hountalas DT. SAE technical paper 981021, 1998.
[15] Bakker R, Schouten JC, Giles CL, Takens F, van den Bleek CM. Neural Comp 2000;12:2355.
[16] Haykin S. Neural networks, a comprehensive foundation. 2nd ed. New Jersey: Prentice Hall, 1999.
[17] Rumelhart DE, Hinton GE, Williams RJ. Nature 1986;323:533.
[18] Tsaptsinos D. Back-propagation and its variations. In: Bulsari AB, editor. Neural networks for chemical engineers. Amsterdam: Elsevier, 1995.
[19] Hornik K, Stinchcombe M, White H. Neural Networks 1989;2:359.
[20] Bakker R, Takens F, Schouten JC, Giles CL, Coppens M-O, Takens F, van den Bleek CM. In: Solla SA, Leen TK, Müller K-R, editors. Advances in neural information processing systems 12. Cambridge, MA: MIT Press, 2000 (in press).
[21] Chen S, Billings SA. Int J Control 1989;49:1013.
[22] Leontaritis IJ, Billings SA. Model selection and validation methods for non-linear systems. Int J Control 1987;45:311.
[23] Billings SA, Voon WSF. Int J Control 1986;44:803.
[24] Krijnsen HC, Bakker R, van Kooten WEJ, Calis HPA, Verbeek RP, van den Bleek CM. Evaluation of fit algorithms for NO$_x$ emission prediction for efficient deNO$_x$ control of transient diesel engine exhaust gas. Ind Engng Chem Res 2000;39:2992.

The reduction of gas phase air toxics from combustion and incineration sources using the MET–Mitsui–BF activated coke process

David G. Olson [a,*], K. Tsuji [b], I. Shiraishi [c]

[a] *Marsulex Environmental Technologies, 200 North Seventh Street, Lebanon, PA 17046 5006, USA*
[b] *Mitsui Mining, 2-1-1 Nihonbashi Muromachi, Chuo Ward, Tokyo 103, Japan*
[c] *Mitsui Mining, 1-3 Hibikimachi, Wakamatsu Ward, Kitakyushu City, Kukuoka 808, Japan*

Received 10 February 1999; accepted 30 June 1999

Abstract

The dry desulfurization, denitrification and air toxics removal process using activated coke (AC) was originally researched and developed during the 1960s by Bergbau Forschung (BF) [Knoblaugh, E. Richter, H. Juntgen, Application of active coke in processes of SO_x and NO_x removal from flue gases, Fuel 60, September 1981, p. 832.], now called Deutsche Montan Technologies. Mitsui Mining (MMC) signed a licensing agreement with BF in 1982 to investigate, test and adapt the system to facilities in Japan. Japanese regulations are stricter than in the United States toward SO_x/NO_x pollutants, as well as flyash emissions from the utility industry, oil refineries and other industries. This process is installed on four coal-fired boilers and Fluidized Catalytic Cracker (FCC) units. These plants were constructed by MMC in Japan and Uhde in Germany. Two additional plants on a utility boiler and cement kiln are scheduled to start operation in 1999 and 2000. MMC provided design, equipment supply and installation. Marsulex Environmental Technologies (MET) formerly General Electric Environmental Services (GEESI) signed a license agreement in 1992 with MMC and Mitsui of Tokyo. Under this agreement, MET will market, design, fabricate and install the Mitsui–BF process for flue gas cleaning applications in North America. MMC also developed a technology to produce AC used in the dry $DeSO_x/DeNO_x/$Air Toxics removal process based on their own metallurgical coke manufacturing technology. This paper provides information on the details of MMC's AC used in the dry $DeSO_x/DeNO_x/$Air Toxics removal process and of the $DeSO_x/DeNO_x/$Air Toxics removal process itself. © 2000 Published by Elsevier Science B.V. All rights reserved.

Keywords: Air toxics; Activated coke; Flue gas

---

* Corresponding author. Tel.: +1-717-274-7355; fax: 1-717-274-7145; e-mail: dolson@marsulex.com

Reprinted from *Fuel Processing Technology* **65-66**, 393-405 (2000)

1. Mitsui Mining's (MMC) activated coke (AC) for the dry DeSO$_x$/DeNO$_x$ process [3]

AC is a formed, carbonaceous material designed for a dry DeSO$_x$/DeNO$_x$/Air Toxics removal process used for flue gas cleaning (Table 1). AC has a high mechanical strength against abrasion and crushing during the circulation and handling process. AC is not metallurgical coke. It is the AC being specially designed for flue gas cleaning that is produced by steam activation at approximately 900°C. This AC has a surface area of 150–250 m^2/g that is less than that of conventional activated carbon. The catalytic activity for oxidative adsorption of SO$_2$ and NO is at 100–200°C. The catalytic active site is a surface oxide. Other catalytic carbons also have surface oxides on which some of the transition metal oxides or novel metals are doped. The Mitsui–BF AC has no doping materials.

1.1. Desulfurization with AC

With the fresh material, SO$_2$ adsorption capacity increases as the BET surface area of the adsorbent increases. Conversely, with the used material that has experienced several cycles of SO$_2$ adsorption and thermal desorption, the BET surface area of the materials tends to increase. SO$_2$ adsorption capacity of the used activated carbon decreases drastically, while those of AC, having a smaller BET surface area and being still less microporous than activated carbon, are less influenced.

The SO$_2$ removal efficiency in the DeSO$_x$ process is related to the SO$_2$ adsorption capacity. The AC maintains a somewhat stable DeSO$_x$ performance during the process. The system can operate efficiently with up to 50 ppm dry of hydrochloric acid (HCl) in the flue gas. Depending on flue gas conditions, close to 50% of the HCl is removed when it passes through the adsorber.

1.2. Denitrification with AC

The catalytic properties of carbon catalysts (AC and activated carbon) for denitrification do not always depend on their BET surface area. It is also true that their properties

Table 1
Chemistry of SO$_x$ removal in adsorption section
(ad.) = Adsorbed state; (g.) = gas phase.

1/2 O$_2$ (g.)	→ O (ad.)	Dissociative chemisorption
SO$_2$ (g.) + O (ad.)	→ SO$_3$ (ad.)	Oxidative chemisorption
SO$_3$ (ad.) + H$_2$O (ad.)	→ H$_2$SO$_4$ (ad.)	Dissolution of SO$_3$ (formation of sulfuric acid)
(Desulfurization into microporous structure of AC with NH$_3$; mainly proceeds in second stage adsorber.)		
SO$_3$ (ad.) + H$_2$O (ad.)	→ H$_2$SO$_4$ (ad.)	Formation of sulfuric acid
NH$_3$ (g.) + H$_2$SO$_4$ (ad.)	→ NH$_4$HSO$_4$ (ad.)	Formation of ammonium bisulfate
NH$_3$ (g.) + NH$_4$HSO$_4$ (ad.)	→ (NH$_4$)$_2$SO$_4$ (ad.)	Formation of ammonium sulfate
2NH$_3$ (g.) + H$_2$SO$_4$ (ad.)	→ (NH$_4$)$_2$SO$_4$ (ad.)	Formation of ammonium sulfate

are raised by the chemical modification of their microporous surface through SO_2 adsorption and desorption (called SO_x treatment). With both fresh and used material (after SO_x treatment), the catalytic properties of AC are superior to those of activated carbon. It should be noted that the chemical surface structure of carbon catalysts is a dominant factor for controlling the rate of denitrification over the physical micropore structure.

In the $DeSO_x$/$DeNO_x$ process, the catalytic properties of AC for denitrification are enhanced [1] by SO_x treatment joining with ammonia (NH_3) treatment forming oxide groups and nitrogen-species on its microporous surface. This has been confirmed with pilot and demonstration tests as mentioned later in this paper.

AC behaves as a catalyst for denitrification along with certain other non-metallic elements in the relatively low 80–180°C temperature range. In contrast, traditional transition element catalysts (such as Ti or V) require temperatures of 300–400°C to be effective. The denitrification reaction on AC is accelerated with NH_3 addition and O_2 presence. This is the same dependency with the V–Ti SCR catalyst. The adsorption sites for NO_x (oxidative adsorption on doubly bonded oxygen species) and NH_3 (acidic site) are similar with both the AC catalyst and the V–Ti catalyst. The SCR reaction occurs on these two catalysts with the same reaction for both denitrification reaction on AC and the denitrification reaction with V–Ti catalyst.

$$4NO + 4NH_3 + O_2 \rightarrow 4N_2 + 6H_2O$$

The catalytic active sites of these catalysts for SCR reaction are mainly surface oxides that are bonded to carbon in AC and doubly bonded to metal in vanadium catalyst. It is undesirable that the surface oxide on vanadium is also active for oxidation of SO_2 to sulfur trioxide (SO_3) at the working temperature of SCRs [about 350°C (662°F)] because of the formation of ammonium bisulfate deposition downstream of the flue gas flow. On the other hand, it is an advantage that the surface oxide on AC works in a relatively lower temperature range and it adsorbs SO_3 effectively.

1.3. Air toxics removal with AC

Recent performance tests have shown the Marsulex Environmental Technologies (MET)–Mitsui–BF process to be very effective in removing Hg, HCl, polychlorinated debenzo-*p*-dioxins (PCDD) and polychlorinated dibenzofurans (PCDF) as shown in the following:

Air toxic	Inlet conditions	Removal efficiency
Hg	0.16 mg/N m^3	90–99%
HCl	100 ppm dry	50%
PCDD	8.3 μg/N m^3	up to 99%
PCDF	23 μg/N m^3	up to 99%

Future development work will be devoted to optimizing system performance. Progress is anticipated in this area.

2. Mitsui–BF dry simultaneous DeSO$_x$/DeNO$_x$/Air Toxics removal process description [4]

A schematic of the Mitsui–BF DeSO$_x$/DeNO$_x$ process is provided in Fig. 1. This process consists of three sections: adsorption, AC regeneration and byproduct recovery.

2.1. Adsorption section

The adsorption section consists of two stages in simultaneous DeSO$_x$/DeNO$_x$/Air Toxics removal. AC moves continuously from top to bottom through the adsorber. First, AC enters in the top of the second stage, where NO$_x$ reduction with the addition of NH$_3$ occurs. The discharged AC from the bottom of the second stage enters the top of the first stage, where the majority of the SO$_x$ and air toxics adsorption occurs. The SO$_x$/Air Toxics-filled AC is discharged from the bottom of the first stage and sent to the regeneration section by conveyors and bucket lifts. The maximum flue gas pressure drop through both stages is approximately 8″ w.g. If only DeSO$_x$, DeNO$_x$, or air toxics removal (in case of no SO$_x$ in the flue gas) is required, a single-stage process can be designed.

2.2. SO$_x$ removal by adsorption in the first stage

Flue gas generated by burning coal with up to 4% sulfur by weight, ranging from to 100°C to 200°C (212°F to 392°F), passes through the first stage. During this stage, SO$_x$ (SO$_2$ and SO$_3$) and air toxics are removed by the AC mainly by adsorption. SO$_x$ is adsorbed and held as sulfuric acid (partially as ammonia salts) within the microporous structure of the AC. Chemistry of SO$_x$ removal is listed in Table 1.

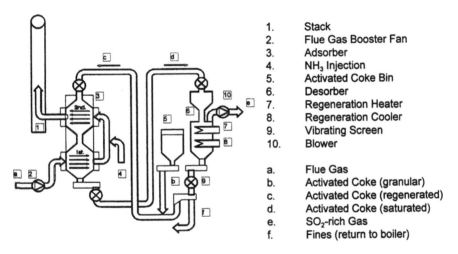

1.	Stack
2.	Flue Gas Booster Fan
3.	Adsorber
4.	NH$_3$ Injection
5.	Activated Coke Bin
6.	Desorber
7.	Regeneration Heater
8.	Regeneration Cooler
9.	Vibrating Screen
10.	Blower
a.	Flue Gas
b.	Activated Coke (granular)
c.	Activated Coke (regenerated)
d.	Activated Coke (saturated)
e.	SO$_2$-rich Gas
f.	Fines (return to boiler)

Fig. 1. MET–Mitsui–BF dry DeSO$_x$/DeNO$_x$/Air Toxics removal process.

2.3. NO_x removal by selective catalytic reduction in the second stage

When NH_3 is added while the SO_2 concentration is high, the NH_3 is consumed by SO_x forming ammonium salts and effective denitrification is not achieved. Hence, the optimum performance is achieved by injection of NH_3 at a location where less concentrated SO_2 exists between the first and second stages. This approach will provide a greater denitrification rate at a lesser NH_3 consumption. The flue gas, which has had the majority of the SO_x removed in the first stage, is then introduced into the second stage for denitrification.

During the second stage, mainly NO_x (NO and NO_2) is decomposed into N_2 and H_2O catalytically with the addition of NH_3 as a reductant. Surface oxides and nitrogen-species within the microporous structure of the AC act as the catalyst active sites for NO_x–NH_3–O_2 (SCR) reaction. Chemistry of the NO_x removal is listed in Table 2.

2.4. Flyash removal / polishing step in the adsorber

The third function of this dry process is the moving bed of AC that acts as a granular filter for removal of flyash in flue gas. MMC designed a special louver for effective removal and discharge of flyash. The system is designed for normal flyash loadings; however, the AC circulation rate can be increased by up to 150% to handle heavy dust overloading for short-term periods. The $DeNO_x$/$DeSO_x$/Dust Removal Test plant treated 1500 N m^3 N/h from high sulfur heavy oil combustion boiler flue gas. This plant, treating flue gas with dust concentration at 350 N m^3 N/h before ESP, indicated high performance of $DeNO_x$/$DeSO_x$/Dust removal. The flue gas condition and $DeNO_x$/$DeSO_x$/Dust removal efficiency results are shown in Table 3.

2.5. AC regeneration section

AC discharged from the bottom of the first stage is sent to a two-stage regeneration vessel, where sulfuric acid, its ammonium salts, and air toxics adsorbed into AC are

Table 2
Chemistry of NO_x removal in adsorption section
Nitrogen species are not identified.

(NO$_x$ selective catalytic reduction with NH$_3$ on AC: proceeds in second stage.)		
$1/4 \, O_2$ (g.)	$\rightarrow 1/2O$ (ad.)	Dissociative chemisorption
NO (g.) + O (ad.)	$\rightarrow NO_2$ (ad.)	Oxidative chemisorption of NO_x
NH_3 (g.) + OH (ad.)	$\rightarrow NH_4$ (ad.) + O (ad.)	Chemisorption of NH_3
NO_2 (ad.) + NH_4 (ad.)	$\rightarrow NH_4NO_2$ (ad.)	Surface reaction of adsorbed species
NH_4NO_2 (ad.) + $1/2O$ (ad.)	$\rightarrow N_2$ (g.) + $3/2H_2O$ (g.) + OH (ad.)	Desorption of decomposed products from AC microporous surface
NO + NH_3 + $1/4O_2$	$\rightarrow N_2 + 3/2H_2O$	Total SCR reaction
(NO$_x$ reduction with nitrogen species on microporous surface of AC)		
HN=C (surface) + NO_2 (ad.) $\rightarrow N_2$ (g.) + OH (ad.) + O–C (surface)		
(NO$_x$ direct reduction by AC)		
C + NO_2 (ad.) $\rightarrow 1/2N_2$ (g.) + CO_2 (g.)		

Table 3
DeNO$_x$/DeSO$_x$/Dust removal process for high sulfur heavy oil combustion boiler flue gas

Gas condition	
Gas volume	1500 N m^3 N/h
Temperature	140°C–150°C
NO$_x$	100–150 ppm
SO$_x$	1000–1500 ppm
Dust	350 N m^3 N/h
O$_2$	2–3%
H$_2$O	10–15%
Emission performance	
SO$_x$	100%
NO$_x$	$\geq$ 80%
Outlet dust concentration	$\leq$ 10 mg/m^3 N

thermally decomposed (at 400°C) and generate SO$_2$/Air Toxic concentrated gas. During the first stage, the AC is indirectly heated by an external furnace's combustion air. It is called SO$_2$-rich gas (SRG), which is sent to the byproduct section. Chemistry of the regeneration section is listed in Table 4.

After cooling during the second stage to 120°C, the regenerated AC is filtered to remove fine dusts (AC fine powder generated by abrasion plus flyash in the flue gas caught by AC) through a vibration screen, then recycled back to the adsorber. The AC fines can be recycled and disposed off by burning as fuel.

The AC lost by mechanical abrasion (approximately 1.0–1.4%) while circulating in the process (called mechanical loss — depends on AC moving velocity and distance) and by chemical consumption of SO$_2$ adsorption and desorption (approximately 0.9% for 1.2% sulfur coal — called chemical loss — depends on SO$_x$ load on AC) are made up by the addition of fresh AC after the vibrating screen.

While the SRG is being pretreated in the SRG scrubber for the downstream byproduct process, air toxics are simultaneously scrubbed and collected in the waste water. Aside from the conventional waste water treatment, the MET–Mitsui–BF system

Table 4
Chemistry of AC regeneration
Nitrogen species are not identified.

(Regeneration of AC: case without NH$_3$)		
H$_2$SO$_4$ (ad.)	$\rightarrow$ SO$_3$ (ad.)+H$_2$O (ad.)	Thermal degradation of sulfuric acid
SO$_3$ (ad.)+1/2C	$\rightarrow$ SO$_2$ (g.)+1/2CO$_2$	Reduction of SO$_3$ by carbon
SO$_3$ (ad.)+C	$\rightarrow$ SO$_2$ (g.)+O–C (surface)	Formation of surface oxides
(Regeneration of AC: case with NH$_3$)		
(NH$_4$)$_2$SO$_4$ (ad.)	$\rightarrow$ 2NH$_3$ (g.)+H$_2$SO$_4$ (ad.)	Thermal degradation of ammonium sulfate
H$_2$SO$_4$ (ad.)	$\rightarrow$ SO$_3$ (ad.)+H$_2$O (g.)	
3SO$_2$ (ad.)+2NH$_3$ (g.)	$\rightarrow$ 3SO$_2$ (g.)+N$_2$ (g.)+3H$_2$O (g.)	Reduction of SO$_3$ by NH$_3$
2NH$_3$ (g.)+6O–C	$\rightarrow$ N$_2$–6OH (ad.)	Decomposition of NH$_3$ by surface oxides
NH$_3$ (g.)+O–C (surface)	$\rightarrow$ H$_2$O+HN=C (surface)	Formation of surface nitrogen groups

can optionally be connected to an Hg recovery facility (for salable, pure Hg metal) or to a facility capable of solidifying the contaminant as HgSe for ultimate disposal.

2.6. Byproduct section

SRG, generated in AC regeneration section, contains approximately 20–25% SO_2. It can be converted into either salable elemental sulfur (purity $\geq$ 99.9%) either liquid or solid, sulfuric acid (purity $\geq$ 98%), or liquid SO_2.

3. Process applications

In 1987, a Mitsui–BF $DeSO_x$/$DeNO_x$ commercial plant started operation at Idemitsu Kosan, Aichi Oil Refinery (see Table 5). This plant treats flue gas (236,000 N m^3/h designed) from the catalyst regeneration section of a Residue Fluid Catalytic Cracking Unit (RFCC). Performance of this $DeSO_x$/$DeNO_x$ plant has been very successful. A removal efficiency of 100% for SO_x (SO_2 and SO_3) and over 80% for NO_x has been constantly achieved at an operating temperature of approximately 180°C (356°F). This plant is easy to operate, experiences almost no trouble and is almost maintenance-free.

Table 5
Summary of MMC's activities for AC production and dry $DeSO_x$/$DeNO_x$ process

Year	AC production and supply	Mitsui–Bergbau Forschung (BF) dry $DeSO_x$/$DeNO_x$ process
1982	· Pilot plant (0.5 ton/day)	· $DeSO_x$/$DeNO_x$ Pilot Plant (1000 N m^3/h) test operation at Tochigi
	· AC supply (10 ton) for Tochigi	
1984	· AC supply for Ohmuta	· $DeSO_x$/$DeNO_x$ plant (30,000 N m^3/h) starts to operate at Mitsui Coal Mining, Ohmuta
1985	· AC supply for Matsushima Power Station of EPDC	
1986	· Commercial AC production (3000 ton/year) starts to operate at Kitakyusyu	
1987	· AC supply for Idemitsu–Aichi Refinery	· $DeSO_x$/$DeNO_x$ plant (236,000 N m^3/h) starts to operate at Idemitsu–Aichi Refinery · $DeSO_x$/$DeNO_x$ plant (1,100,000 N m^3/h) starts to operate at Arzberg Power Station of EVO, GER
1988	· AC supply for Nippon Steel–Nagoya Works	
1989	· AC supply for Hoechst, Germany	· $DeSO_x$/$DeNO_x$ plant (323,000 N m^3/h) starts to operate at Hoechst, Frankfurt, Germany
1990	· AC supply for Wakamatsu Research Center of EPDC	· Low temperature $DeNO_x$ pilot plant (10,000 N m^3/h) at Wakamatsu of EPDC
1995	· AC supply for Takehara	· Low temperature $DeNO_x$ plant (1,163,000 N m^3/h) at Takehara Plant of EPDC

In 1990, MMC constructed an AC DeNO$_x$ pilot plant that was commissioned to EPDC by the Japanese government. This pilot plant was designed for treating 10,000 N m^3/h flue gas at an operating temperature of 140°C (284°F) from a fluidized bed combustion boiler.

In Germany, Uhde, also a licensee of BF, constructed two dry DeSO$_x$/DeNO$_x$ commercial plants. One started operating in 1987 for treating flue gas (totalling 1,110,000 N m^3/h designed) from lignite-fired power plant at EVO's Arzberg power station. Another began operation in 1989 for treating flue gas (323,000 N m^3/h designed) from hard coal-fired boiler of Hoechst at Frankfurt. MMC cooperated with Uhde with the engineering of these two plants.

The MET–Mitsui–BF process is applicable to a broad range of flue gas cleaning applications including boilers, sintering furnaces, incinerators, Fluidized Catalytic Cracker (FCC) catalyst regenerators and chemical plants.

3.1. SO$_x$/NO$_x$ removal process for coal-fired boiler exhaust gases

The AC SO$_x$/NO$_x$ removal process-installed Arzberg Power Station is one of three commercial plants that currently treat coal-fired boiler exhaust gases. This plant has been in constant service since operations began in 1987, attaining SO$_x$ and NO$_x$ removal rates of at least 99% and 80%, respectively (Table 6). This process completely recycles spent SRG wash water back to the previous stage of the EP unit and discharges no waste water.

3.2. Low-temperature NO$_x$ removal process for fluidized bed boiler exhaust gases

The fluidized bed combustion process has been attracting attention for its low pollution characteristics and applicability to a wide range of coal types. The exhaust gas treatment centers on NO$_x$ removal because SO$_x$ is removed in the combustion chamber. The Mitsui–BF process currently operating on a fluidized bed boiler unit at the Electric Power Development's Takehara Power Station Unit #2 treats 1,163,000 N m^3/h of exhaust gases for NO$_x$ reduction.

Some of the more important characteristics of the FBC boiler/Mitsui–BF combination are the ability to remove NO$_x$ at low temperatures (a desirable feature during boiler startup) and the overall dust collection capability. Before deciding to adopt the AC low-temperature NO$_x$ removal process for Takehara Unit No. 2, Electric Power Development conducted government-commissioned demonstration runs at the Wakamatsu Coal Utilization Test Station, on 10,000 N m^3/h of exhaust gases. The results indicated

Table 6
DeSO$_x$/DeNO$_x$/De toxics performance

Item	Inlet	Outlet	Efficiency (%)
NO$_x$ (ppm)	179	36	80
SO$_x$ (ppm)	634	3.5	99
Dust (mg/N m^3)	123.2	5.2	96

that the desired NO_x removal rate (80% or higher), SO_x removal rate (90% or higher) and exhaust dust concentration (30 mg/N m^3 or lower at 140°C) were all achieved.

The Takehara Unit 2 installation, in test operation since July 1994 with commercial operation scheduled for July 1995, shows $DeNO_x$ efficiency within design parameters for the number of SO_x-activation cycles on the coke. As expected, the SO_2 and SO_3 efficiencies are virtually 100%, while the outlet particulate concentration is less than the guaranteed value of 10 mg/N m^3. AC losses are lower than design, but expected to levelize at about 1.0%.

3.3. SO_x / NO_x removal process for RFCC exhaust gases

The Mitsui–BF process has also been in constant service at Idemitsu Kosan's Aichi Refinery since its startup in 1987. This process removes SO_x and NO_x from 200,000 N m^3/h of RFCC exhaust gas, satisfying all the required specifications. The easy operability and maintenance of this process greatly reduces maintenance-related expenses. The SRG evolved in the AC regeneration tower is sent to an existing Claus unit to recover elemental sulfur.

3.4. Low-temperature NO_x / Toxics removal process for incinerator exhaust gases

The exhaust gases from the initial treatment step applied to a waste incinerator typically have a temperature of 200°C or lower. Due to the expense of reheating these gases to employ conventional metallic catalyst for NO_x removal, alternative lower temperature treatment processes become more economically attractive. The Mitsui–BF process, characterized by low-temperature NO_x removal, can treat these gases without having to increase the gas temperature. Moreover, the process can also remove trace quantities of toxic compounds such as mercury and dioxins that are present in incinerator exhaust gases. One Japanese maker, after conducting demonstration tests to confirm the ability of the AC process to remove NO_x and trace quantities of toxic compounds from incinerator exhaust gases at low temperatures, has installed a 32,000 N m^3/h Mitsui–BF NO_x removal plant in this application.

3.5. NO_x removal efficiency of AC process treating incinerator exhaust gases

Because of higher moisture content of the incinerator's exhaust gases, the low-temperature AC process tends to remove NO_x less efficiently than that for exhaust gases from other combustion facilities such as boilers. Test results confirm that the low-temperature NO_x removal efficiency of the AC process for incinerator exhaust gases with a moisture content of approximately 20% and removal efficiency of 50–60% can be practically secured at 200°C or lower. Table 7 illustrates the actual performance data for the waste incineration gas at the Takehara Unit 2. Table 8 indicates the test results of Dioxin removal with a fixed bed of AC (surface area 150–300 m^2/g). After approximately 6000 h of treatment of the flue gas from a refuse incinerator, dioxin removal efficiency of approximately 98% has been achieved, indicating a very high adsorption capacity of the AC gaseous dioxin removal process. With the combination of the AC

Table 7
Mitsui–BF process performance — waste incineration application (polishing step)

Temperature		180°C			150°C		
Substances	Unit	Inlet	Outlet	Efficiency (%)	Inlet	Outlet	Efficiency (%)
PCDDs	ng/N m^3	38	8.9	76.6	40	4.7	88.3
PCDFs	ng/N m^3	110	6.6	94.0	130	2.6	98.0
Dioxine (TEQ)	ng/N m^3	2.50	0.24	90.4	3.30	0.06	98.2
W-soluble Hg	ng/N m^3	0.19	0.03	83.5	0.25	0.02	90.6
W-n-soluble HGg	ng/N m^3	0.13	0.03	79.6	0.22	0.02	90.9
Total Hg	ng/N m^3	0.31	0.06	81.9	0.47	0.04	90.8
NO$_x$	ppm	75.3	21.9	70.9	57.6	19.3	66.5
HCl	ppm	14.0	< 5	–	14.0	< 5	–
SO$_x$	ppm	< 5	< 5	–	< 5	< 5	–
Dust	ng/N m^3	0.0108	0.0008	92.6			–

bed and the dust and/or mist collector such as a bag filter, the total dioxin (gaseous plus particle phase) removal efficiency is expected to increase significantly. The AC maintained its removal performance for about 1 year in this life test.

3.6. Toxics removal efficiency of AC process treating incinerator exhaust gases

The AC process can remove mercury vapor by adsorbing it on pore surfaces (see note in Table 9). Mercury vapor adsorption is further enhanced when SO$_x$ is present, since sulfuric acid adsorbs incremental quantities of mercury. Table 9 shows the results of tests conducted to confirm the ability of the AC process to remove mercury from incinerator exhaust gases. The results show that the process removes mercury almost completely in a temperature range from 150°C to 180°C.

Tests were conducted in which incinerator exhaust gases were passed over the 720-mm-thick AC layer in the fixed bed-type NO$_x$ removal unit. AC samples were collected at 500 and 1000 h for analysis to determine the quantity of mercury adsorbed by the AC process. These samples confirmed that mercury is adsorbed mostly in the first half of the AC layer, even after 1000 h of exposure to exhaust gases.

Table 8
2Q'97 dioxin removal test results by a fixed bed of activated coke (AC)[a]

Adsorbents		Inlet dioxin (ng TEQ/N m^3)	Outlet dioxin (ng TEQ/N m^3)	Dioxin removal efficiency (%)
AC5 ⌀ — 150 m^2/g (test period: 1800 h)	Total	9.2	0.19	97.9
	Gas		0.043	
	Particle		0.15	
Acb ⌀ 300 m^2/g (test period: 1200 h)	Total	9.2	0.11	98.8

[a]Test conditions: flue gas volume: 22.3 N m^3/h; temperature: 150°C; SV: 2000 h^{-1}; AC volume: 11.3 L.

Table 9
Hg removal test result using AC

Temperature (°C)	Space velocity (h^{-1})	Inlet (mg/N m^3)	Outlet (mg/N m^3)	Removal efficiency (%)	Ref.
180	–	0.013	Tr	> 99	[1]
150	1000	0.064	Tr	> 99	[2]
		0.149	Tr	> 99	
		0.025	Tr	> 99	
180	1000	0.028	Tr	> 99	[2]
		0.019	Tr	> 99	
		0.049	Tr	> 99	

Elemental Hg(O) vapor is adsorbed effectively by "conditioned" AC as opposed to a fresh AC charge. It suggests that the oxygen species or the remaining trace sulfur species on the conditioned AC surface is the adsorption site for Hg(O) with bonding theorized as Hg–O (major part) and Hg–S (minor part). In the regenerator, the adsorbed Hg species begin to desorb over 400°C as well as CO and CO_2 gas which start to release over 600°C. This suggests that the bonding energy of C–O is higher than that of Hg–O; therefore, the adsorbed Hg species desorb as elemental Hg(O) over 400°C.

Tests results confirm the efficiency of the AC process in removing dioxins and chlorobenzene, a precursor of dioxin. It has been further confirmed that 70–98% of these compounds can be removed by adsorption under temperature conditions of 150°C to 180°C and an SV of 1000 h^{-1} or higher.

These results illustrate the ability of the process to treat incinerator exhaust gases for several air contaminants simultaneously including SO_2, SO_3, NO_x, Hg, dioxins, chlorobenzene and other volatile organic compounds.

3.7. Ultimate collection of elemental mercury

Once adsorbed on the coke, mercury must be collected in a form that is disposable. One example of a method to perform this operation is the selenium filter. The commercially proven selenium filter can treat typical SRG gas streams containing up to 4500 mg/N m^3 mercury. This enables the overall process to separate mercury from the flue gas, concentrate it several orders of magnitude, and treat it in a small, highly efficient filter. The mercury is absorbed by the filter and fixed in the form of HgSe, a chemically stable compound. Under these conditions, a collection efficiency of 98% is expected over a filter life of 4–5 years. Once spent, the selenium filter would be disposed in a licensed, hazardous waste facility. Other methods of mercury collection are also candidates for use depending on the stream treated; that is, SRG off-gas, sulfuric acid plant off-gas or SRG scrubber waste water.

4. Advantages of the MET–Mitsui–BF process

4.1. Simultaneous $DeSO_x$ / $DeNO_x$ / Air Toxics removal process

One of the primary advantages of the MET–Mitsui–BF process is that desulfuriza-tion, denitrification, flyash polishing and air toxics removal are performed in a single process. In addition, the process is highly effective at reducing SO_3 emissions that can be problematic under high sulfur oil conditions.

4.2. Dry process

Contact of flue gas with AC bed under dry conditions has simplified the process flow and reduced the installation space required. No large-scale waste water treatment, SO_3 mist separator or reheater for the treated flue gas is necessary in the MET–Mitsui–BF process resulting in less solid waste disposal. AC fines can be burned in the boiler as fuel. This process is ideally suited for locations where disposal of contaminated waste is a concern or is restricted.

4.3. Low-temperature process

High removal efficiencies of SO_x, NO_x and air toxics from flue gas are achieved at temperatures ranging from to 100°C to 200°C (212°F to 392°F) in the MET–Mitsui–BF process. This precludes reheating gases to achieve $DeNO_x$ as is required in metallic catalyst applications.

4.4. Optional byproduct

Elemental sulfur, sulfuric acid, liquid SO_2 or liquid mercury can be made from SRG generation in the AC regeneration section.

5. Conclusion

The MET–Mitsui–BF dry, low-temperature $DeSO_x$/$DeNO_x$/Air Toxics removal system was developed over many years, culminating with both pilot and demonstration plants. These tests have proven the system's capability of removing over 99% of the SO_x, including SO_3, up to 99% of selected air toxics and over 80% of the NO_x in coal-fired and fluidized bed boilers and RFCC units. In addition, the process is effective in reducing vaporous, elemental mercury by 99% + and dioxins/furans by 70–98% from combustion flue gases. Subsequent treatment of the highly concentrated contaminant stream from the regenerator makes ultimate collection and disposal less costly. The AC loss by mechanical abrasion can be recycled and disposed of by burning as fuel. The system can produce salable products such as elemental sulfur, sulfuric acid, liquid SO_2 or liquid mercury. This simple and flexible system can be adjusted for future changes in operating conditions and/or air toxics regulatory limits. The above is accomplished with a single step of operating equipment that does not require separate equipment for each toxic removed, thus resulting in a long train of process equipment. The system has a proven track record in four commercial applications. Finally, it can be retrofitted or installed in new applications, and does not require pre or post-treatment reheat.

In conclusion, the MET–Mitsui–BF dry process is a suitable candidate for simultaneously reducing emissions of SO_x, NO_x, air toxics and volatile organics from fossil fuel and waste combustion plants.

References

[1] I. Mochida, M. Ogaki, H. Fujitsu, Y. Komatsubara, S. Ida, Catalytic activity of coke activated with sulfuric acid for the reduction of nitric oxide, Fuel 62 (1983) 867.

[2] Dalton, Current Status of Dry NO_x–SO_x Emission Control Processes, Proceedings of the 1982 Joint Symposium on Stationary Combustion NO_x Control, 1982, pp. 32–41.

[3] K. Tsuji et al., Combined desulfurization, denitrification and reduction of air toxics using activated coke: 1. Activity of activated coke, Fuel 75 (1997) 549.

[4] K. Tsuji et al., Combined desulfurization, denitrification and reduction of air toxics using activated coke: 2. Process applications and performance of activated coke, Fuel 76 (1997) 555.

Emissions Reduction: NO_x/SO_x Suppression
A. Tomita (Editor)
© 2001 Elsevier science Ltd. All rights reserved

Design of scrubbers for condensing boilers

F. Haase*, H. Koehne

Lehr- und Forschungsgebiet für Energie- und Stofftransport, RWTH Aachen, Kopernikusstr. 16, 52056 Aachen, Germany

Received 16 April 1997; received in revised form 15 January 1999; accepted 15 January 1999

Abstract

Many fuels (oil, wood or process gases) contain components which can be found as acid-forming compounds in the flue gas after combustion and which are absorbed to a small degree in the condensate when the dew point is reached. This acid, and therefore highly corrosive condensate, increases the demands on the materials used for the areas affected by condensation in condensing boilers. The concepts described in the article "Design of Scrubbers for Condensing Boilers" are based upon use of the condensate for washing the flue gas. To achieve this, the flue gas is cooled below the dew point in contact with the already neutralized condensate. This process step allows the wet separation of noxious matter and avoids acid corrosion of the materials in the area of condensation, raising the choice of possible materials decisively. The review article also exemplifies the state-of-the-art for condensing boiler technology, as it gives a view of the fundamental principles of two-phase flows, of absorption of acid-forming gases and their neutralization. Using the examples of sulfuric oxides SO_x and nitrous oxides NO_x, the effects of different characteristic features of the gases on the reaction steps are described and possible process steps of the wet separation in the condensate are discussed. To achieve as high a separation degree as possible between the flue gas and the condensate, good heat and mass transfer conditions must be guaranteed between the two phases—flue gas and condensate. Dispersing one of the two phases leads to a strong increase of the interphase. Generally, fluid vaporizers (flue gas as coherent phase) and gas bubble washers (condensate as coherent phase) can be taken into consideration. Advantages and disadvantages of these absorbers are worked out for use as washers in combination with condensing boiler technology and the fluid-mechanical principles necessary for the design of a gas bubble washer. The sometimes contrary influences of constructive parameters on pressure changes in the flue gas, as pressure loss and mass transfer conditions between flue gas and condensate, are discussed. © 1999 Elsevier Science Ltd. All rights reserved.

Keywords: Condensing boiler; Scrubber; Desulfurization; NO_x absorption; Neutralization

Contents

* Corresponding author. Address for correspondence: Shell Research and Technology Centre, Hamburg, Deutsche Shell AG, PAE Labor, OGMPT/4, Hoh-Schaar-Str. 36, 21107 Hamburg, Germany. Tel.: + 49-40-7565-4739; fax: + 49-40-7565-4581.
E-mail address: frank.f.haase@ope.shell.com (F. Haase)

Reprinted from *Progress in Energy and Combustion Science* 25 (3), 305-337 (1999)

Nomenclature

Physical symbols

A	area (m^2)
a	volume specific interphase (m^{-1})
b	weir height (m)
c	concentration (mg/l, mol/m^3)
c	specific heat capacity (J/kg K)
d	diameter (m)
d_S	Sauter diameter (m)
E	specific emission (mg/MJ)
F	force (N)
f	power ratio of a radiator
Fr^*	modified Froude number
g	gravitation constant (m/s^2)
H	enthalpy (J)
h	height of water column above perforated bottom (m)
H	Henry constant (N m/mol)
h	specific enthalpy (J/kg)
h_{fg}	latent heat of vaporization (J/kg)
HHV	higher heating value (J/kg)
K	equilibrium constant $(mg/l)^{-1}$
k	velocity constant (s^{-1})
l	length (m)
LHV	lower heating value (J/kg)
m	mass (kg)

$\dot{m}$	mass flux (kg/s)
M	molar mass (kg/kmol)
$\dot{n}''$	transferred mole number per time and area ($mol/s\ m^2$)
N	mole number (mol)
P	power (W)
p	pressure (Pa)
pH	pH
Q	heat (J)
$\dot{q}''$	heat flow density ($J/m^2\ s$)
$\dot{Q}$	heat flow (J/s)
$\dot{Q}_{std}$	standard heat demand (J/s)
R	gas constant (J/mol K)
r	reaction velocity ($mol/m^3\ s$)
r	specific vaporization enthalpy (J/kg)
s	jet length (m)
T	temperature (K)
t	time (s)
t_{90}	time until 90% of the stationary value is reached (s)
$\dot{v}_d$	average velocity in tube (m/s)
$\dot{V}$	volume flow (m^3/s)

Symbols

w	average gas velocity (m/s)
We	Weber number

x	mole fraction		MgO	magnesium oxide
y	height of perforated bottom above boiler bottom (m)		MgO_2	magnesium peroxide
			$MgSO_3$	magnesium sulfite
z	number of holes		$MgSO_4$	magnesium sulfate
α	absorption degree		Mn	manganese
β	mass transfer coefficient (mol/N s, m/s)		MnO_2	manganese dioxide
Δ	difference		MOH	metal hydroxide
ε_i	Henry activity coefficient		$N(SO_3)_3^{3-}$	nitrilo trisulfonate
ε	volume fraction		N_2	nitrogen
η	efficiency degree		N_2O	dinitrogen monoxide
λ	air–fuel ratio		N_2O_4	dinitrogen tetroxide
ϑ	temperature (°C)		Na_2O_2	sodium peroxide
$\Delta\vartheta_{ln}$	logarithmic average temperature (K)		$Na_2C_2O_6$	sodium peroxocarbonate
ρ	density (kg/m^3)		Na_2SO_3	sodium sulfite
σ	interfacial tension (N/m)		Na_2SO_4	sodium sulfate
ξ	mass fraction		$NaOH$	sodium hydroxide
ψ	mole fraction		NH_3	ammonia
			$(NH_4)_2SO_4$	ammonia sulfate
Chemical symbols			NO	nitrogen monoxide
Ag	silver		NO_2	nitrogen dioxide
$AgCl$	silver chloride		NO_2^-	nitrite ion
Al	aluminum		NO_3^-	nitrate ion
Ca	calcium		NO_x	nitrogen oxide
$Ca(OH)_2$	calcium hydroxide		NTA	nitrilotriacetate
$CaCO_3$	calcium carbonate		O_2	oxygen (molecular)
CaO	calcium oxide		O_3	ozone
$CaSO_3$	calcium sulfite		OH^-	hydroxide ion
$CaSO_4$	calcium sulfate		PbO_2	lead dioxide
CH_4	methane		Pt	platinum
ClO_4^-	perchlorate ion		Rh	rhodium
CO_2	carbon dioxide		$S_2O_4^{2-}$	dithionite ion
CO_3^{2-}	carbonate ion		$S_2O_6^{2-}$	dithionate ion
e	electron		SiO_2	silicon dioxide
$EDTA$	ethylene diamine tetra-acetate		SO_2	sulfur dioxide
Fe	iron		SO_3	sulfur trioxide
H^+	hydrogen ion		SO_3^{2-}	sulfite ion
H_2CO_3	carbonic acid		SO_4^{2-}	sulfate ion
H_2O	water			
H_2O_2	hydrogen peroxide		*Index*	
H_2SO_3	sulfurous acid		*	state of saturation
H_2SO_4	sulfuric acid		0	standard conditions
$HN(SO_3)_2^{2-}$	imido disulfonate		aq	dissolved
HNO_2	nitrous acid		b	base
HNO_3	nitric acid		bu	gas bubble
HO_2^-	hydrogen dioxide ion		c	combustion
$HON(SO_3)_2^{2-}$	hydroxylamine disulfonate		c	continuously
H_3PO_4	phosphoric acid		crit	critical
HSO_3^-	hydrogen sulfite ion		d	dispersed
K	potassium		dp	dew point
$KClO_4$	potassium perchlorate		fg	flue gas
$KMnO_4$	potassium permanganate		fp	flow pipe
M^+	metal cation		g	gas side
M_2SO_3	metal sulfite		g	gaseous
Mg	magnesium		gran	granulated
$Mg(OH)_2$	magnesium hydroxide		h	heating medium
$MgNO_3$	magnesium nitrate		i	component

in	entering
kin	kinetic
l	lifting force
l	liquid side
lim	limit
max	maximum value
min	minimum value
out	leaving
p	pressure
r	return
σ	interfacial tension
std	standard state
wb	water bath

1. Introduction

Burning fossil energy sources, the acid-forming components sulfur dioxide (SO_2) and nitrogen oxides (NO_x) are formed, as well as others, and are emitted into the atmosphere. Following the mass flow of the nitrogenous and sulfurous compounds after their emission, it is necessary to differentiate between their effects on the air and on plants, earth and water.

As the average retention times of the trace gases SO_2 and NO_x are only very short (hours up to several days), their influence on air quality is regionally limited. SO_2 and NO_x are toxic to human beings at a certain concentration. The current discussion about anthropogene influences on the climate considers SO_2 and NO_x not in the first place. The major climate influencing gases in the atmosphere are steam (H_2O), carbon dioxide (CO_2), ozone (O_3), dinitrogen oxide (N_2O) and methane (CH_4). These gases cause the so-called greenhouse effect of the atmosphere. But SO_2, NO_x (and ammonia NH_3) lead to the formation of aerosols which are solid or liquid parts in the air outside of clouds with a diameter between 0.001 and 100 μm. They scatter and absorb solar irradiation and emit thermal radiation so they behave contrary to the anthropogene heating of the earth [1].

The largest part of the emission is chemically converted in the air before it is deposited via rain. At the chemical conversion of sulfur dioxide, sulfuric acid (H_2SO_4) is formed; its salts are called sulfates (SO_4^{2-}). Finally, gypsum or Na_2SO_4 is formed. Nitrogen monoxide is oxidized to nitrogen dioxide, which forms nitric acid; its salts are called nitrates (NO_3^-).

Before the acids formed by SO_2 and NO_x get into the ground, the protective wax coating on tree leaves and needles can be damaged. Nutritive substances are washed out. Also, the supply of nutritive substances of the ground decreases. With increasing acidity of the ground, first calcium magnesium and potassium, and in later stages, manganese and aluminum are dissolved out of the ground particles. They are washed out and lost to the ecosystem [2]. Together with the washed out alkali sulfates and nitrates

combine with drained rainfall into surface waters or via the sewage system into a sewage treatment plant.

Fig. 1 shows the effects of sulfurous and nitrogenous compounds in the air, on plants, earth and water for the emissions of SO_2 and NO_x.

This mass flow proves that the wet deposition of SO_2 and NO_x in condensing boilers under the formation of sulfate and nitrates and the necessary neutralization correspond to the processes in nature.

The wet deposition of SO_2 and NO_x, immediately after their formation in the combustion chamber of the washer for condensing boilers, is described here. The specialty of these washers is that the condensate formed by falling short of the steam dew point is used as washing water.

Thus, an almost complete absorption of the sulfur dioxides is possible in the condensate if the flue gas contacts immediately neutralized condensate under good heat and mass transfer conditions. Meeting these requirements results in washing the flue gas and preventing the boiler parts in the condensation area from corrosion.

Combining the condensing boiler technology (CBT) with flue gas washing increases the choice of usable materials for fuels with a very aggressive condensate and the operative expense is kept low at the same time. As a result of low energy prices, condensing boiler technology is difficult to realize economically for the fuels mentioned and a small power output. But the additional flue gas washing provides a motivation which should be encouraged by means of legislation.

2. State of science and technology

2.1. Condensing boiler technology

According to their energy content, fuels are described by the lower heating value (LHV) and the higher heating value (HHV). The LHV of a substance is the amount of heat set free at complete combustion when reactants and products show the same temperature and water formed at combustion exists as steam. The difference to the gross caloric value of the fuel is that it considers the water as liquid, so the gross caloric value is, by the condensation heat of the water, greater than the net caloric value (HHV/LHV = $h_{fg}\xi_h$/LHV with ξ_h = mass fraction of hydrogen in the fuel). Condensing boilers are those which make the latent heat contained in the flue gas usable by condensation.

2.1.1. Dew point and combustion efficiency

The condensation of water begins when the state of saturation for the steam in the flue gas is reached while cooling the flue gas. This point is called the "dew point", and its temperature is the dew point temperature ϑ_{dp}. At the dew point the steam pressure equals the saturation steam pressure of water, which is—as an approximation—for

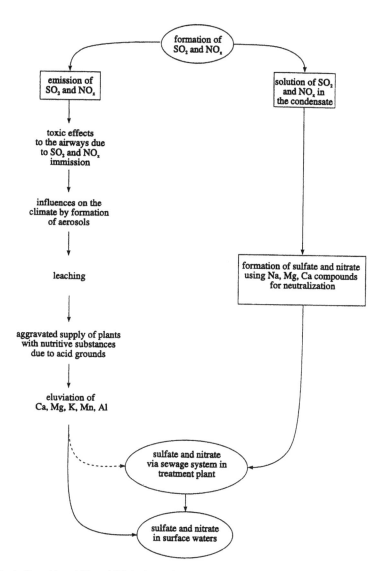

Fig. 1. Deposition of SO_2 and NO_x in the condensate of the flue gas as an alternative to their emission.

a total pressure below 10 bar, only dependent on the temperature.

For the combustion of fuels, the partial pressure of steam in the flue gas depends on fuel composition (hydrogen content), air–fuel ratio λ and the humidity of the combustion air. Disregarding the air humidity, the dependence between the dew point temperature ϑ_{dp} and air–fuel ratio λ is obtained, as shown in Fig. 2 for the standard heating oil in Germany called "EL" (extra light). With an increase in air–fuel ratio λ, the flue gas must be cooled to achieve steam condensation.

By burning sulfurous fuels, the hydrophile sulfur trioxide

(SO_3) in the flue gas acts as a condensation nucleus resulting in an increase of the dew point temperature. For an exact determination of the so-called "acid dew point", knowledge of the conversion of sulfur dioxide (SO_2) to sulfur trioxide (SO_3) is necessary. As an approximation, it is assumed that the volume concentration of SO_3 is about 1% of the concentration of SO_2 [3]. It can be seen in Fig. 3 that the dew point increases under these conditions for a sulfur content in the flue gas of $\xi_S = 0.2$ % by 72 K and lies for an air–fuel ratio $\lambda = 1.1$ at $\vartheta_{dp} = 120°C$.

As a measure against acid corrosion or sooting, conventional heating technology avoids falling short of a flue gas

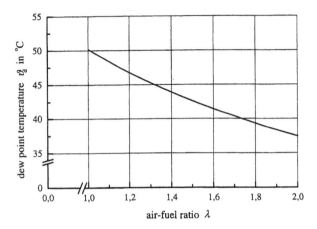

Fig. 2. Dew point temperature ϑ_{dp} of the flue gas from the combustion of heating oil EL versus air–fuel ratio λ.

temperature $\vartheta_{fg} = 150°C$ before leaving the flue gas system. By introducing condensing boiler technology into an existing heating system, a large fraction of the sensible heat connected to the high flue gas temperature of conventional heating systems can be used.

To evaluate a heat generator, the combustion efficiency

$$\eta_c = \frac{\text{supplied energy} - \text{energy at flue gas outlet}}{\text{supplied energy}} \quad (1)$$

can be used (the so called "boiler efficiency" takes into account in addition heat losses through the jacket).

Determining the delivery energy by the HHV, one obtains:

$$\eta_c = 1 - \frac{\sum_i \xi_i \left(h_i(\vartheta_{fg}) - h_i(\vartheta_0) \right) + \xi_{H_2O} h_{fg}}{H_g(\vartheta_0)} \quad (2)$$

where h_i is the specific enthalpy of the gaseous component i,

ξ_i the mass fraction of the gaseous component i in kg/kg$_{fuel}$, ϑ_0 the flue gas temperature and h_{fg} the latent heat of vaporization of water at standard temperature.

Here we assume that fuel and dry air are supplied at a standard temperature ϑ_0, which usually is the thermodynamic standard temperature $\vartheta_0 = 25°C$. The enthalpy of the condensed water is disregarded here. For heating oil EL, the result is dependence between combustion efficiency η_c, flue gas temperature ϑ_{fg} and air–fuel ratio λ shown in Fig. 4.

When falling short of the dew point temperature, a strong increase of the combustion efficiency η_c can be observed, which is attributed to the addition of condensation heat to the sensible heat.

The concept of flue gas washing, used with condensing boiler technology, dealt with here, aims at complete absorption of the sulfur oxides in the condensate of the

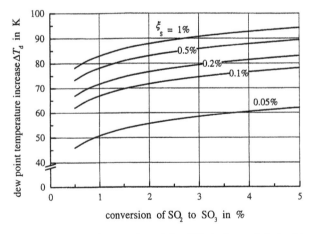

Fig. 3. Increase of the dew point caused by sulfuric acid in the flue gas (calculated data) [3].

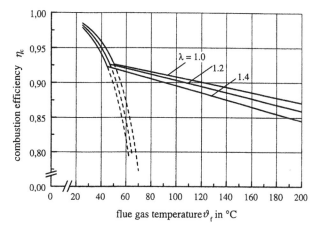

Fig. 4. Combustion efficiency η_c specific to HHV.

flue gas. Good mass transfer conditions between flue gas and necessary condensate result in complete saturation of the flue gas with steam, so that for only a slight increase of the temperature above the dew point, condensate will vaporize and combustion efficiency decreases compared to a dry flue gas system. This phenomenon is shown in dashed curves in Fig. 4 and provides the necessity of avoiding flue gas temperature above the dew point by appropriate control systems.

2.1.2. Integration of condensing boiler technology in the heating system

For the layout of a heating system, the so-called standard heat demand of a building, $\dot{Q}_{std}$ is used; it describes the heat power supplied to heat the rooms to comfortable temperatures at a low outdoor temperature [4]. Using the standard heat demand for the layout of the boiler (provided that the piping losses are negligible), the boiler has to supply the power

$$\dot{Q}_h = \dot{Q}_{std} = \dot{m}_h c_h \left(\vartheta_{fp} - \vartheta_r \right) \tag{3}$$

to the heating circuit. In the aforementioned equation, m_h represents mass flux of the heating medium, c_h the specific heat capacity of the heating medium, ϑ_{fp} the flow-pipe temperature of the heating medium and ϑ_r the return temperature of the heating medium.

The values of the process parameters m_h, ϑ_{fp}, ϑ_r and their combinations can only be chosen within certain limits. If the flue gas is cooled below its steam dew point by the heating medium, the maximum return temperature is then determined according to Fig. 2 depending on the air–fuel ratio λ chosen.

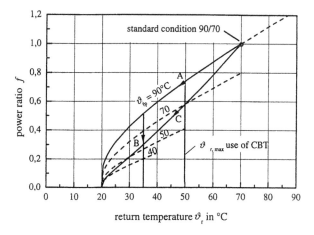

Fig. 5. Influence of flow-pipe and return temperature on the power ratio f of a radiator.

298

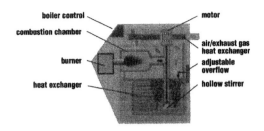

Fig. 6. Condensing boiler technology with flue gas desulfurization; concept I [13].

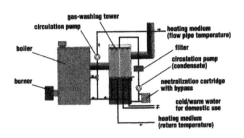

Fig. 7. Condensing boiler technology with flue gas desulfurization; concept II [14].

The value of the flow-pipe and the return temperature of the heating medium has a decisive influence on the heat supply of a radiator; its heat flux $\dot{Q}$ can be described by

$$\dot{Q} \sim A(\Delta \vartheta_{\mathrm{ln}})^n \qquad (4)$$

with the radiator surface A and the logarithmic average temperature

$$\Delta \vartheta_{\mathrm{ln}} = \frac{\vartheta_{\mathrm{fp}} - \vartheta_{\mathrm{r}}}{\ln\left(\vartheta_{\mathrm{fp}} - \vartheta_{\mathrm{room}}/\vartheta_{\mathrm{r}} - \vartheta_{\mathrm{room}}\right)} \qquad (5)$$

The value of the exponent n lies within the limits 1.1 to 1.5; for radiators, $n \approx 1.3$ is obtined [5], for floor radiators, $n \approx 1.1$ [6]. Making the heat flux from Eq. (4) specific to standard conditions ($\vartheta_{\mathrm{fp,std}} = 90°C$, $\vartheta_{\mathrm{r,std}} = 70°C$, $\vartheta_{\mathrm{room,std}} = 20°C$) [5], the so called power ratio of a radiator for equal-sized radiator surfaces

$$f = \frac{\dot{Q}}{\dot{Q}_{\mathrm{std}}} = \frac{\dot{q}}{\dot{q}_{\mathrm{std}}} = \left(\frac{\Delta \vartheta_{\mathrm{ln}}}{\Delta \vartheta_{\mathrm{ln,std}}}\right)^n = \left(\frac{\Delta \vartheta_{\mathrm{ln}}}{59.44\ \mathrm{K}}\right) \qquad (6)$$

This relation is valid for $\dot{m}_{\mathrm{H}}/\dot{m}_{\mathrm{H,N}} \geq 0.2$ and can be used to determine the heat a radiator supplies for various pairs of flow-pipe and return temperature. Fig. 5 shows the power ratio f dependent on the return temperature ϑ_{r} for different flow-pipe temperatures ϑ_{VL}. It can be seen from Fig. 5 that the supplied heat flux of a radiator decreases

- if the temperature difference ($\vartheta_{\mathrm{fp}} - \vartheta_{\mathrm{r}}$) is increased when the return temperature is lowered by reducing the heat medium mass flux and keeping the flow-pipe temperature constant (line A),
- if the temperature difference is decreased by lowering the flow-pipe temperature, keeping the return temperature constant (line B),
- if the temperature level is decreased, keeping the temperature difference constant (line C).

If the same heat flux is to be obtained as in the standard case, the radiator surface must be increased according to

$$A = \frac{1}{f} A_{\mathrm{std}}. \qquad (7)$$

The following rules can be summed up for the integration of condensing boilers into heating systems [7–10]:

- The temperature of the heating medium must be as low as possible, with a considerable difference between flow-pipe and return temperature. In new buildings, condensing boilers are often connected to floor radiators, or heating systems with low temperature radiators. Even in old buildings, there must be no limitations for appropriate integration of the condensing boiler technology. Heat demands of the building are less than during construction as a result of subsequent thermal insulation; and because the existing heating system—designed for temperatures $\vartheta_{\mathrm{fp}} = 90°C$ and $\vartheta_{\mathrm{r}} = 70°C$—is over-dimensioned because of formerly common security factors, lowering the heating medium temperatures is possible.
- In contrast to humidity-sensitive boilers (standard or low temperature boilers), measures leading to a raised return temperature (mixing valves) must be avoided generally when using condensing boilers.
- An over-dimensioned installed boiler power does not reduce the annual efficiency degree, in contrast to conventional boilers, because the boiler walls are cooled down intermittently so that lower flue gas temperatures are reached at the beginning of each phase of burner operation.
- To adapt to a lower heat demand (part load), not the heating medium flux, but the flow-pipe temperature should be reduced. A high heating medium flux is related to higher requirements on the thermostat valves [11].
- The use of liquid layer reservoirs with a separate heat exchanger has proved to be efficient in combination with the condensing boiler technology. The war water reservoir is connected via a priority-controlled three-way valve to the flow-pipe. The heat exchange takes place in counter-flow with the cold water for domestic use from the bottom of the reservoir, so condensing boiler technology is possible even for warm water temperatures of 60°C.

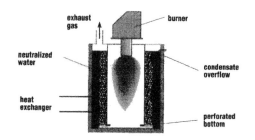

Fig. 8. Condensing boiler technology with flue gas desulfurization; concept III [15].

2.1.3. Concepts for condensing boiler technology for sulfurous fuels

Condensing boilers have been offered for sulfurous fuel for ten years. While the use of condensing boiler technology is state-of-the-art today for the nearly sulfur-free energy source gas, it has not found acceptance from planners, installers and users because of the higher standards on the boiler construction and the discharge of condensate into the public sewage system connected to the condensing boiler technology.

The boilers offered on the market offer the following constructive principles. The combustion chamber is cooled first by a heat exchanger, so the flue gas produces a temperature between 150 and 180°C. A lower flue gas temperature is avoided because condensation forming sulfuric and sulfurous acids would then require the use of non-corrosive materials. Nevertheless, a low temperature is targeted so as little energy as possible must be coupled out at a lower temperature level in a second heat exchanger. This second heat exchanger is constructed of non-corrosive material because the condensation of the steam contained in the flue gas takes place here, and sulfuric acid is formed. Gener-

ally, non-corrosive steels are used, but ceramics, glass or graphite can be used as well.

The heating medium either flows continuously through the heat exchangers, or as part of two separate heating circuits operating at different temperature levels. Pre-heating the combustion air with the second heat exchanger is a further variant.

The condensation of the steam which takes place, forming sulfuric and sulfurous acid, leads to an acid condensate with a pH value of about 2. Under these acid conditions, only a small part of the sulfur oxides contained in the flue gas remains in the condensate, so we cannot speak of a flue gas desulfurization [12]. Together with condensate from the flue gas system, the condensate is led to a neutralization unit and is finally discharged into the public sewage system. To achieve neutralization, either calcium or magnesium granulate or ion exchangers are used. The granulates loose their efficiency if the condensate contains metal ions which may arise from low temperature corrosion. In this case, metallic compounds settle down on the active surface of the granulate because of the low flow velocity and do not allow its further disassociation. To avoid this problem, ion exchangers are often used before the granulate.

Following are four concepts for the condensing boiler technology, which aim for desulfurization of the flue gas. The common points of these concepts is that the flue gas is brought in contact with the neutralized condensate when the temperature falls short of the dew point.

The condensate from the combustion chamber of the first boiler is drawn off via a drawing wheel and is dispersed in a neutralized water bath (Fig. 6). A good mixing of water and flue gas guarantees a complete heat transfer and the absorption of sulfur oxides. To achieve cooling of the flue gas, three heat exchangers are used. The return flow of the heating medium flows continuously through a heat exchanger in the cleaned flue gas, a coiled pipe which is arranged in the

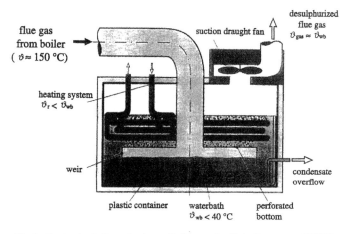

Fig. 9. Condensing boiler technology with flue gas desulfurization; concept IV [16].

water bath and at last through the heat exchanger in the combustion chamber. The electric motor used to power the drawing wheel has a power demand of 600 W [13]. For a nominal power of 17 kW, efficiency decreases by 3.5% without considering the low efficiency for the generation of electricity.

In the second concept for condensing boiler technology with flue gas desulfurization, the condensate is led through a neutralization cartridge via a circulation pump, is dispersed via nozzles in a gas-washing tower and is returned to the condensation heat exchanger (Fig. 7). In the gas-washing tower the condensate absorbs the sulfur oxides from the flue gas. In Fig. 7 are shown two heat exchangers, the first one for preheating the returned heating medium, the other one for heating the water for domestic use. The heat exchangers are arranged in a reservoir of neutralized condensate. Thus excellent heat exchange makes a compact apparatus possible.

The third concept for condensing boiler technology with flue gas desulfurization is a very compact burner/boiler setup. A vertically-arranged combustion chamber is cooled by surrounding water bath (Fig. 8). The flue gas leaves the combustion chamber at its open bottom end and rises in the water bath in the form of bubbles. Also in this concept, the contact of flue gas and water realizes a complete heat transfer and allows the absorption of sulfur oxides. The possibility of producing boilers of plastic materials is seen as a central advantage of this concept [15]. The complete combustion heat is transferred by one heat exchanger from the water bath to the heating medium. Preparing high flow-pipe temperatures and cooling the flue gas below its dew point at the same time is not possible, so this boiler is to be regarded only as a low-temperature heating system. As the combustion chamber is completely submerged in the water bath, a water column as high as the length of the combustion chamber must be overcome by a fan. The electric power demand is also connected to a decrease of efficiency.

The fourth concept for the condensing boiler technology, developed by the authors, is also based on a gas bubble washer. Fig. 9 shows a module which can be used in combination with a standard boiler or integrated in a condensing boiler. The heat changes by two temperatures, so the flow-pipe temperature above the dew point and condensation is possible simultaneously.

For raising efficiency of condensing boiler different measures are described. Variants on concept II [14] and concept IV [17] use the heat transfer and mass transfer (water steam) between flue gas and combustion air for reducing flue gas temperature.

2.2. Deposition of sulfur oxides from flue gas

Sulfur contained in the flue gas is burned together with oxygen to sulfur dioxide (SO_2). Sulfur dioxide reacts with oxygen to sulfur trioxide (SO_3) according to the following equilibrium:

$$2SO_2 + O_2 \leftrightharpoons 2SO_3 \qquad (8)$$

For gas temperatures higher than 1200°C, no relevant oxidation takes place because of the state of the equilibrium. Also, no relevant oxidation can be observed when cooling the flue gas in the boiler, because the velocity of reaction (8) is low [18]. SO_3 shows a very good solubility in water and reacts under the presence of steam to form sulfuric acid. Thus, the technology used for a deposition of SO_2 generally also achieves good results for SO_3, therefore in the following, the deposition of SO_2 will be seen as the central point of the investigations. In the USA and in Japan, flue gas desulfurization units have been used in power plants for over 30 years; in Germany, their use began about 20 years ago [19]. The long development time has led to a number of process variants which generally can be divided into washing, spray absorption and drying processes.

Washing processes
Washing processes usually work with suspensions of calcium oxide (CaO, caustic lime), calcium hydroxide ($Ca(OH)_2$, slaked lime) or calcium carbonate ($CaCO_3$, limestone). The primary product of the lime washing process is calcium sulfite ($CaSO_3$). In Germany, gypsum $CaSO_4 \cdot 2H_2O$ is usually gained via oxidation, which is either dumped or made usable by subsequent processes [20]. Alternatively, the Walther Company uses an ammonia solution for washing and acquires ammonium sulfate (($NH_4)_2SO_4$) as a product which can be used as synthetic fertilizer [21]. As a result of the good solubility of the ammonia salts, no fouling problems occur. Montan–Werke Brixlegg, in Austria, uses a washer which is operated with an $Mg(OH)_2$ suspension which is produced from MgO by adding water [22].

Spray absorption processes
During spray absorption processes (also called mixed methods), calcium hydroxide suspensions with a high solid fraction are sprayed into the hot flue gas stream [23]. The water vaporizes and the salts formed are drawn off, together with the excess lime and some fine dust, by textile or electric filters positioned further downstream. A disadvantage is the over-stoichiometric consumption of lime and the lower deposition degree of SO_2 as compared with wet processes.

Dry processes
During dry processes, we differentiate between lime absorption (dry additive process) and the absorption via activated coke (mining technology). During dry additive processes, small-grained lime is blown into the hot flue gas. As an alternative, lime may be added to the fuel. The calcium sulfite, calcium sulfate and the remaining lime which are formed are drawn off a dust filter. Low quality coal and brown coal naturally contain lime. In pulverized coal furnaces, which are common in boilers of large dimensions, lime reaches the combustion chamber as fine dust and absorbs part of SO_2 in the ash. A complete desulfurization seems possible with the fuel-own lime if a low combustion

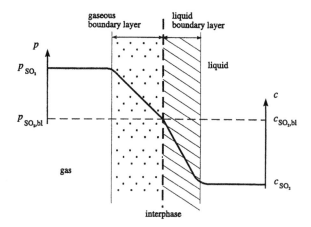

Fig. 10. Course of concentration and partial pressure of SO_2 according to the two-film theory [27].

temperature avoids burning the lime "to death" and realizes low flue gas temperatures [24].

The mining-technological process [25] binds SO_2 adsorptively to activated coke and converts it to sulfuric acid. Sulfuric acid is stored in the pores of the activated coke. The activated coke is then thermally regenerated so it can be used again. The gas mixture set free contains 20–30 vol.% SO_2, which is converted in further steps to sulfuric acid or elementary sulfur. This process can also be used for similar removal of sulfur and nitrogen oxides from flue gas, if ammonia is also used.

Both sorption processes used for flue gas desulfurization, adsorption and absorption, selectively take over one or more components from a gaseous mixture into a condensed phase, as happens during the simultaneous deposition of sulfur and nitrogen oxides. In the following, the solution of a gaseous phase in a liquid by absorption is examined because in the concept of flue gas washing with condensing boiler technology described here, the deposition of sulfur oxides and nitrogen oxides takes place in the liquid condensate of the flue gas.

2.2.1. Phase equilibria and oxidation at the absorption of SO_2

When dissolving a gas in a liquid, the physical and the chemical absorption are differentiated. In physical absorption, the absorbed gas molecules are surrounded by a hydrate cover, but no reaction takes place, so the gas molecules can leave the liquid very easily. In contrast, we speak of chemical absorption or chemisorption if the gas is engaged in a chemical reaction in the liquid.

To observe maximum load of the absorbent and the composition of the gaseous mixture after the cleaning, knowledge about the solubility of a gas in a liquid is necessary. Such knowledge is dependent upon temperature, pressure and composition of the gaseous and the liquid mixture. According to the classification of physical and chemical absorption, phase equilibria are described in different ways.

For physical absorbents, the phase equilibrium of a single component can be expressed by the Raoult's law. For ideal mixtures of gases and liquids, it reads

$$p_i = x_i p_i^* \tag{9}$$

where p_i is the partial pressure of the component i, x_i the mole fraction of the component i in the liquid and p_i^* the saturation partial pressure of the component i as function of temperature.

Raoult's law states that the partial pressure of the component i in the gaseous phase is proportional to the mole fraction of the component i in the liquid for a constant temperature. As the absorption of SO_2 takes place exothermically, the solubility decreases with increased temperature.

Solubilities determined with Raoult's law can be taken as approximate values but they do not consider the type of solvent. The law assumes that a certain gas has the same solubilities in all solvents for the same temperature and partial pressure [26].

To take into account the difference between the ideal behavior of the gas and the liquid mixture, Raoult's law can be extended by the Henry activity coefficient ε_i:

$$p_i = \varepsilon_i x_i p_i^* \text{ and } \varepsilon_i = \varepsilon_i(x_i, p, T) \tag{10}$$

where ε_i is the Henry activity coefficient, p the total pressure and T the temperature.

If the concentration of the absorbed component in the liquid is low, ε_i is constant and one obtains the Henry law found for dilute solutions:

$$p_i = \varepsilon_i x_i p_i^* \tag{11}$$

where $\varepsilon_i = \varepsilon_i(p, T)$. For a given temperature, the partial pressure of a component is connected via the Henry activity coefficient in a linear path to the mole fraction of the component in the liquid. If the determination must be even more exact than the Henry law, the dependence of the Henry activity coefficient on the mole fraction of the component in the liquid must be taken into account.

For the absorption of SO_2 in water, the hydration

$$SO_2(g) + H_2O \rightleftharpoons SO_2(aq) + H_2O \qquad (12)$$

is described by the physical absorption.

The following hydrolysis of the sulfur dioxide $SO_2(aq)$ corresponds to the chemical absorption. The chemical absorption of sulfur dioxide in water takes place as follows:

$$SO_2(aq) + H_2O \rightleftharpoons H_2SO_3 \qquad (13)$$

The law of mass action states that a chemical reaction stops at a given temperature (equilibrium), if the quotient from the product of the concentrations of the reaction products, and the product of the concentrations of the reactants, reaches a certain value K, which is characteristic for the reaction. The characteristic value K, is called "equilibrium constant" and can be determined for the hydrolysis of SO_2 according to Eq. (13) as follows:

$$K = \frac{c_{H_2SO_3}}{c_{SO_2(aq)}c_{H_2O}} \qquad (14)$$

Sulfurous acid (H_2SO_3) formed here dissociates in two steps. In the first step, hydrogen sulfite (HSO_3^-) and hydrogen ions (H^+) are formed, in the second steps, free sulfite ions (SO_3^{2-}) and H^+ are formed:

$$H_2SO_3 \rightleftharpoons HSO_3^- + H^+ \rightleftharpoons SO_3^{2-} + 2H^+ \qquad (15)$$

The hydrogen ions set free by the dissociation results in an acid reaction of the solution. Adding a hydroxide MOH to the absorbent which reacts with the hydrogen ions to achieve a neutralization

$$H^+ + MOH \rightleftharpoons H_2O + M^+ \qquad (16)$$

the concentration of the hydrogen ions decreases, and the reactions described in the Eqs. (12)–(14) react towards their right side in order to re-establish the equilibrium, so as a conclusion, $SO_2(g)$ can further be absorbed.

If the absorption of SO_2 and the neutralization are continued, the total concentration of the dissolved SO_2 ($SO_2(aq)$, H_2SO_3, HSO_3^-, SO_3^{2-}) increases until the state of saturation of the salt M_2SO_3 formed out of sulfite SO_3^{2-} and the cation M^+ in the solution is reached. Exceeding the state of saturation results in precipitation of the formerly dissociated salt as a solid.

Among the precipitation as a sulfite salt, it is possible that the dissolved sulfite reacts with dissolved oxygen to sulfate (SO_4^{2-}):

$$SO_3^{2-} + \frac{1}{2}O_2(aq) \rightleftharpoons SO_4^{2-} \qquad (17)$$

While the direction of the equilibrium reactions (11), (12) and (14) can be influenced easily by temperature changes, redox reactions show a strong dependence on the pH of the solution. The equilibrium of the oxidation reaction (17) lies in both—in acid and in alkaline solution—and for the temperature range in which condensing boiler technology is used on the side of the product.

The oxidation of a reductive (e.g. SO_3^{2-}) and the reduction of an oxidizer (here SO_4^{2-}) takes place according to the following equilibrium:

$$\text{reductive} \underset{\text{oxidation}}{\overset{\text{reduction}}{\rightleftharpoons}} \text{oxidizer} + \text{electron} \qquad (18)$$

A system which donates and accepts electrons according to this equation is called "redox system" or "redox pair".

As free electrons do not exist under normal chemical conditions, the donation of electrons only takes place in the presence of an appropriate reductant. The donated electrons are accepted by the reductant while it changes to the corresponding oxidant. Thus, a redox reaction always consists of two corresponding part reactions. The total reaction shown in Eq. (17) consists of the following part reactions:

$$SO_3^{2-} + 2OH^- \rightleftharpoons SO_4^{2-} + H_2O + 2e^- \qquad (19)$$

$$\frac{1}{2}O_2 + 2e^- + 2H^+ \rightleftharpoons H_2O \qquad (20)$$

$$SO_3^{2-} + \frac{1}{2}O_2 + 2OH^- + 2H^+ \rightleftharpoons SO_4^{2-} + 2H_2O \qquad (21)$$

2.2.2. Mass transfer at SO_2 absorption

At existing flue gas washers, gas mixture and washing liquid are in contact repeatedly with a limited contact area and for a limited contact time, so an almost complete mass transfer of SO_2 from the gas phase to the liquid phase takes place. The mass transfer model presented here is greatly simplified and will not be used for predicting the SO_2 concentration in the gas mixture but will permit a clear discussion of the influencing parameters.

The concentration differences between phase and interphase act as the moving force for the mass transfer. The two-film theory can be used to describe the mass transfer from fluid to fluid [27]. According to the two-film theory (Fig. 10), a laminar interphase is formed between the gas boundary layer and the liquid boundary. The boundary layers represent the main resistance for mass transfer. The interphase does not show a resistance if the boundary layers are in steady state.

Within the boundary layers, mass transfer takes place by diffusion. Turbulences which are based on the penetration model [28] result in a certain decrease of the boundary layer thickness, so an acceleration of the mass transfer is the result. The boundary layer that shows the highest resistance against diffusion determines the velocity of the mass transfer. If one of the resistances is dominant, the other can often be neglected. Gases which are easy to dissolve—such as SO_2—diffuse faster through the liquid boundary layer than through the gas boundary layer. Thus, the resistance of the gas boundary layer becomes dominant and can be regarded as the main resistance [28].

Generally, three kinds of phase distribution must be considered for the absorption:

(a) Both phases exist coherently (film absorber).
(b) The gaseous phase is dispersed in the coherent liquid phase (bubble column).
(c) The liquid phase is distributed dispersely in the coherent gaseous phase (spray washer).

A disperse distribution of a phase can achieve a larger contact surface per volume unit than a continuous distribution.

For the following model of a disperse distribution of the gas phase in the coherent liquid phase, several assumptions are made:

1. The SO_2 concentration in the coherent phase is homogeneous.
2. An equilibrium exists at the interphase.
3. Pressure variations are negligible.
4. The mass transfer is dominated by the resistance of the gaseous phase.

Regarding a single gas bubble, we can write for the amount of SO_2 transferred through the interphase:

$$\dot{n}''_{SO_2,g} = \beta_g(p_{SO_2} - p_{SO_2,int}) \tag{22}$$

$$\dot{n}''_{SO_2,l} = \beta_l(c_{SO_2,int} - c_{SO_2}) \tag{23}$$

$$\dot{n}''_{SO_2} = \dot{n}''_{SO_2,g} = \dot{n}''_{SO_2,l} \tag{24}$$

where $\dot{n}''_{SO_2}$ is the mole number per area and time of the transferred SO_2 β_g, β_l mass transfer coefficient for the gas side in mol/Ns, for the liquid side in m/s.

Further

$$\dot{n}''_{SO_2} = \frac{dn''_{SO_2}}{dt} = -\frac{dp_{SO_2}}{dt}\frac{1}{p_g}\frac{N_{bu}}{A_{bu}} \tag{25}$$

where N_{bu} is the mole number in the gas bubble, A_{bu} the surface of the gas bubble and $p_g$total pressure of the gas in the bubble.

Substituting Eq. (22) in Eq. (25) and integrating over time within the limits of entering the liquid and leaving the liquid, we obtain:

$$(p_{SO_2} - p_{SO_2,int})|_{t=t_{out}}$$
$$= (p_{SO_2} - p_{SO_2,int})|_{t=t_{in}}\exp\left[-\beta_g p_g \frac{A_{bu}}{N_{bu}}(t_{out} - t_{in})\right] \tag{26}$$

The partial pressure of SO_2 at the interphase $p_{SO_2,int}$ cannot be determined by experiments. Neglecting the resistance from the liquid side and assuming a homogeneous concentration of SO_2 in the liquid phase, as mentioned above, the following equations are valid:

$$c_{SO_2,gr} = c_{SO_2} = c_{SO_2}(t = t_{out}) = c_{SO_2}(t = t_{in}). \tag{27}$$

As an equilibrium exists at the interphase, according to the two-film theory (assumption 2), the partial pressure $p_{SO_2,int}$ is connected via the Henry law to the SO_2 concentration in the liquid:

$$p_{SO_2,gr} = H_{SO_2}c_{SO_2} \tag{28}$$

Substituting this in Eq. (26), one obtains:

$$p_{SO_2}(t = t_{out}) = H_{SO_2}c_{SO_2} + \left[p_{SO_2}(t = t_{in}) - H_{SO_2}c_{SO_2}\right]$$
$$\times \exp\left[-\beta_G p_G \frac{A_{bu}}{N_{bu}}(t_{out} - t_{in})\right] \tag{29}$$

The SO_2 partial pressure after leaving one stage of the washer, $p_{SO_2}(t = t_{out})$ is determined by

1. the concentration of SO_2 in the liquid c_{SO_2},
2. the SO_2 partial pressure before entering the liquid p_{SO_2} ($t = t_{in}$) as a function of the sulfur content of the used fuel and
3. the parameters β_G, $(t_{out} - t_{in})$, A_{bu}/N_{bu}, p_g which can be influenced by design and operation modus of the washer.

The concentration c_{SO_2} of the physically absorbed $SO_2(aq)$ is dependent on the state of the reaction equilibria according to Eqs. (13), (15) and (16). Assuming an ideal neutralization which guarantees a constant pH, a constant distribution of the various forms of sulfite ($SO_2(aq)$, H_2SO_3, HSO_3^-, SO_3^{2-}) is expected. In Section 2.2.3, the different possible neutralizers and their effects on the mentioned equilibrium chain are described.

Next to the equilibrium reactions, oxidation from sulfite to sulfate, according to Eq. (17) also takes place in the presence of dissolved oxygen. The reaction velocity is determined by the sulfite concentration, but not by the oxygen concentration, as long as enough dissolved oxygen is present [29]:

$$r = k(c_{SO_3^{2-}})^{1.5}c_{O_2}^0. \tag{30}$$

If gas is led stationary and under constant boundary conditions through the liquid, a quasi-stationary concentration of the total sulfite will ensue. Its experimental determination using the law of mass action (Section 3.2.2) allows conclusions to be drawn regarding the concentration $c_{SO_2}(aq)$ of the physically absorbed SO_2.

The parameters β_g, $(t_{out} - t_{in})$,...,p_g, which are influenced by design and operation modus of the washer, will be discussed in Section 4.3 for an ideally chosen dispersion unit. The choice of a long contact time ($t_{out} - t_{in}$) between gas mixture and liquid has, according to Eq. (29), the same influence as an increased contact area specific to the gas amount A_{bu}/N_{bu} or a large mass transfer coefficient β_g caused by turbulences in the gaseous phase. These variables determine the mass transfer from the gaseous phase to the liquid phase and have an exponential influence on the SO_2 partial pressure p_{SO_2} ($t = t_{out}$) after leaving the liquid.

Table 1
Solubility of the compounds at 20 and 50°C in 100 g water

Cation	Reactants		Products	
	Hydroxide (g)	Carbonate (g)	Sulfite (g)	Sulfate (g)
Na$^+$				
20°C	52.2 [30]	18.1 [30]	27.1 [30]	19.1 [30] as Na$_2$SO$_4$
50°C	59.2 [30]	32.2 [30]	34.7 [30]	46.4 [30] as Na$_2$SO$_4$ rhombic
Mg^{2+}				
20°C	0.000017^a [30]	0.000072^a [31]	0.6 [30] as MgSO$_4$(H$_2$O)$_3$	25.2 [30] as MgSO$_4$(H$_2$O)$_7$
50°C	0.000015^b [30]		0.844 [30] as MgSO$_3$(H$_2$O)$_3$	33.4 [30] as MgSO$_4$(H$_2$O)$_6$
Ca^{2+}				
20°C	0.126 [18]	0.0065 [30]	0.0043^a [30]	0.2 [32] as CaSO$_4$(H$_2$O)$_2$
50°C	0.0917 [30]	0.0038 [30]	0.0057 [30]	0.21 [32] as CaSO$_4$(H$_2$O)$_2$

^{a} At 18°C.
^{b} At 45°C.

Eq. (29) makes clear that even under ideal mass transfer conditions, the SO$_2$ partial pressure after the washer is not zero; so the conversion of the sulfite in the liquid is of a particular importance.

2.2.3. Neutralization

By absorption of sulfur dioxide or other acid gases, the pH decreases and the adsorbent will be quickly saturated with SO$_2$. Raising the pH to a neutral area (neutralization) will change the equilibria described in Section 2.2.2, so SO$_2$ can be further dissolved. The neutralizer must either be added continuously during the absorption of SO$_2$, or it must be in excess in the absorbent.

During neutralization, an acid reacts with a base under the formation of a salt and water. The main reaction is the synthesis of hydrogen ions and hydroxide ions:

$$H^+ + OH^- \leftrightharpoons H_2O \qquad (31)$$

Substances used as neutralizers must either contain hydroxide ions or form hydroxide ions in an aqueous solution. In choosing a neutralizer we must consider the following criteria:

1. affiliation of educates or products to a certain class of hazardous materials,
2. availability and price,
3. aperitif expense of the neutralizer dosage.

Considering these criteria, a large number of possible substances are reduced to oxides, hydroxides and carbonates of sodium, magnesium and calcium. While hydroxides and carbonates dissociate partly or (because of their high solubility), completely in an aqueous solution while tending toward a steady state, oxides are hydrated in an aqueous solution.

The characteristic features of magnesium oxide (MgO) and calcium oxide (CaO) depend on their preparation and their preparatory treatment. Magnesium and calcium oxides which were prepared at high temperatures are less reactive

and do not dissolve as well in water as those prepared at lower temperatures. MgO and CaO can be gained by dehydration of hydroxides or, as it is technically usual for large amounts, by heating (burning) the carbonates to temperatures between 800 and 1000°C [18].

If MgO is added to water, magnesium hydroxide is formed:

$$MgO + H_2O \rightarrow Mg(OH)_2 \qquad (32)$$

This hydration does not continue until the complete conversion of MgO because the active surface of the oxide decreases, pores are filled for compact material and magnesium hydroxide forms a protective coating.

Calcium oxide reacts with water in high temperatures, heat development and forms calcium hydroxide (slaked lime) [18]:

$$CaO + H_2O \rightarrow Ca(OH)_2 \qquad (33)$$

Magnesium hydroxide (Mg(OH)$_2$) is a compound which is difficult to dissolve. Ca(OH)$_2$ shows greater solubility compared to Mg(OH)$_2$ (Table 1), so dissolving this compound results in a higher pH.

Sodium hydroxide (NaOH) is achieved by the chlor-alkali electrolysis. Easily dissolved in water, its solution is caustic soda and is a strongly etching substance and therefore a hazardous material.

If carbon dioxide added to into an aqueous hydroxide solution, it forms dissolved CO$_2$, carbonic anhydride H$_2$CO$_3$, hydrogen carbonate HCO$_3^-$ and carbonate CO$_3^{2-}$ dependent on pH. With the absorption, the pH of the solution decreases until finally no carbon dioxide can dissolve any longer (equilibrium reactions). Adding a stronger acid (H$_2$SO$_3$, H$_2$SO$_4$), carbon dioxide exhales (replacement) while the pH remains constant; so the carbonates act as buffers:

$$M_2CO_3 + H_2SO_4 \rightarrow CO_2 + H_2O + M_2SO_4 \qquad (34)$$

The carbonates of magnesium and calcium are hardly

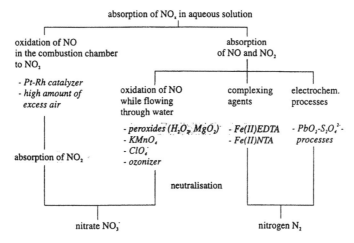

Fig. 11. Methods of absorbing nitrogen oxides in an aqueous solution.

soluble in water. Sodium carbonate dissolves in water with an alkaline reaction and shows greater solubility as the carbonates of magnesium and calcium. Handling sodium carbonate is much less dangerous than handling sodium hydroxide.

The different solubilities of the mentioned neutralizers shown in Table 1 require different dosage concepts:

Na compounds
Sodium hydroxide and sodium carbonate must be administered doses in corresponding to the absorbed amount of SO_2 to prevent the solution from becoming alkaline. If these compounds are added in excess, discharging the absorbent into the sewage system means a heavy loss of neutralizer. For such a dosage, a continuous pH measurement, a dosage valve or a dosage pump, a storage tank and a control system, is necessary.

Mg compounds
As a result of the low solubility of $Mg(OH)_2$, a dosage is not necessary when using MgO (which hydrates to $Mg(OH)_2$) or $Mg(OH)_2$ and the neutralizer can be added in excess to the neutralization solution. In the slightly alkaline solution, the equilibrium ($Mg(OH)_2 \rightarrow Mg^{2+} + OH^-$) is disturbed with the initial absorption of acid gases and a fast dissociation of $Mg(OH)_2$ begins, so the solution does not become acid. The reaction velocity decreases when approaching the equilibrium.

Ca compounds
Using calcium oxide and hydroxide in excess like the magnesium compounds, a higher pH in the solution will result because the slightly greater solubility of the calcium compounds. Calcium sulfite and sulfate show very low solubility in comparison to the sodium and magnesium salts. Therefore, the use of calcium compounds requires that the salts formed at the absorption of SO_2 be drawn off as solids to be used as products—as they are used in the power plant sector—or that they be dumped. To discharge the formed salts into the public sewage system, sodium and magnesium compounds are appropriate neutralizers.

The compounds described here are available in a very wide price range, depending on their method of preparation and their purity. Generally, calcium hydroxide is known as the least expensive industrially used neutralizer. As a result of its various applications in soap production, in dyes, in the preparation of cellulose and as a cleaner, caustic soda is very inexpensive.

2.3. Wet separation of nitrogen oxides from flue gases

In burning fossil energy sources, the noxious matter NO and NO_x are formed which together are spoken of as NO_x. As the Bodenstein reaction of NO

$$2NO + O_2 \leftrightarrows 2NO_2 \tag{35}$$

is exothermic, NO does not react with oxygen at temperatures above 650°C [18]. Nitrogen oxides emitted from furnaces primarily exist as NO.

2.3.1. Solution of NO and NO_2 in water
Gas is well absorbed in liquid, especially if the gas and the liquid show about the same polarity. The polarity of a molecule is determined by its structure and the difference between the electronegativity of the atoms in the molecule.

As a result of its angular structure, water shows a large dipole moment and therefore can easily dissolve molecules with a high polarity.

Nitrogen monoxide (NO) has a low solubility in water as a result of its low polarity. The linear structure of the molecule and the relatively small differences between the electronegativity of oxygen and nitrogen result in NO, which

306

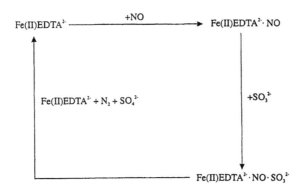

Fig. 12. Reaction scheme of absorption and reduction of NO using Fe(II)EDTA.

can only partially be absorbed under the formation of gas hydrates—similar to oxygen and nitrogen in water. The solubility of NO in water decreases greatly with increasing temperature until its minimum is reached at 81°C. Afterwards, it slowly increases. The solubility decrease is proportional to the concentration decreased of a certain kind of water molecule (probably $(H_2O)_6$) which reacts with NO under the formation of a gas hydrate [33].

Nitrogen dioxide (NO_2) shows a better solubility in water than NO because the additional oxygen atom leads to an angular structure similar to water and so increases the polarity of the molecule. Also, NO_2 is the anhydride of nitrous acid and nitric acid, whereas anhydrides generally show a good solubility in water.

Nitrogen dioxide is in equilibrium with dinitrogen tetroxide (N_2O_4):

$$2NO_2 \leftrightharpoons N_2O_4 \qquad (36)$$

As a result of the exothermic dimerization reaction, the equilibrium moves to its left side with increasing temperature. Thus 20% of NO_2 at 27°C, 40% at 50°C and nearly 100% at 140°C [18]. NO_2 (1/2 N_2O_4) reacts with water, transferring the nitrogen of a valence 4 to one part to a lower oxidation number of valence 3 (nitrous acid HNO_2) and to another part to a higher oxidation number with a valence 5 (nitric acid HNO_3):

$$N_2O_4 + H_2O \leftrightharpoons HNO_2 + HNO_3 \qquad (37)$$

Here the formed nitrous acid (HNO_2) reacts with a part of NO:

$$HNO_2 \leftrightharpoons \frac{1}{3}HNO_3 + \frac{2}{3}NO + \frac{1}{3}H_2O \qquad (38)$$

Under the presence of oxygen, NO_2 reacts with H_2O according to the following total reaction to nitric acid:

$$2N_2O_4 + 2H_2O + O_2 \leftrightharpoons 4HNO_3 \qquad (39)$$

Here, N_2O_4 is not oxidized with H_2O to HNO_3. First, N_2O_4

reacts with water according to Eqs. (37) and (38):

$$3N_2O_4 + 3H_2O \leftrightharpoons 3HNO_2 + 3HNO_3 \qquad (40)$$

$$3HNO_2 \leftrightharpoons HNO_3 + 2NO + H_2O \qquad (41)$$

Eqs. (40) and (41) result in following total reaction:

$$3N_2O_4 + 2H_2O \leftrightharpoons 4HNO_3 + 2NO \qquad (42)$$

Not before now, the reaction of oxygen begins by oxidizing the formed NO to NO_2

$$2NO + O_2 \leftrightharpoons 2NO_2 \leftrightharpoons N_2O_4 \qquad (43)$$

which now serves as reactant for the reaction in Eq. (40). The reaction in Eq. (43) is for gases with a small NO_2 fraction—the technical velocity determining reaction for the total reaction in Eq. (39).

Fig. 11 shows the selected methods for the absorption of nitrogen oxides which seem appropriate in combination with the concept of flue gas washing in the condensing boiler technique described here. The left branch shows the oxidation of NO to NO_2 in the gaseous phase and the subsequent absorption of the NO_2. In the middle, the method of NO absorption via NO oxidation in the water is shown. On the right side, we see methods of dissolving NO in water and finally reducing them. All methods neutralize the water necessary to convert the nitric and sulfuric acids produced to nitrate and sulfate. The methods, which are shadowed in gray represent the results gained experimentally and described in this paper, so the following explanations will deal with the principles of each process.

2.3.2. Oxidation of NO in aqueous solutions

The oxidation of gaseous NO flowing through water can be achieved by adding various substances to the water. NO_2 formed then will dissolve immediately according to Eq. (39). Using potassium permanganate ($KMnO_4$) as a strong oxidizer, brownstone (MnO_2) and dissolved manganate ions are formed. Thus, potassium permanganate only can be used

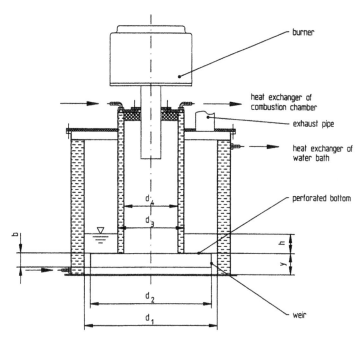

Fig. 13. Basic layout of the test boiler with water bath [37].

in closed systems, therefore using it in the condensate of the flue gas which serves as absorbent is not appropriate.

Perchlorates such as potassium perchlorate ($KClO_4$) are chemically unstable and tend to decompose explosively if kept in larger concentrations. As a result of the danger arising from incorrect handling, use in domestic situations is recommended only with special care.

The formation of ozone O_3 at a light bow or with help of an UV lamp requires electric energy. The costs of the apparatus and the electricity do not favor this technique in small plants. Additionally, ozone is toxic even in small concentrations so its use can only take place with corresponding safety systems. The use of peroxides is appropriate because neither they themselves nor the products formed are dangerous—using hydrogen peroxide produces water and oxygen. Magnesium peroxide (MgO_2) can also be used instead of H_2O_2.

The oxidation number of oxygen is -1 for peroxides, and lies in between that in O_2 (0) and H_2O (-2). Thus it is expected that aqueous solutions of peroxides show a spontaneous dismutation (the change of an elementary compound to a higher and a lower oxidation number). But H_2O_2 decomposes very slowly without catalyzers. Even though trace contaminants can lead to an explosive decomposition [34].

Magnesium peroxide reacts with water to hydrogen peroxide according to

$$MgO_2 + H_2O \leftrightharpoons MgO + H_2O_2 \tag{44}$$

For a systematic investigation of the use of peroxides in order to achieve an oxidation of NO, it is recommended that H_2O_2 be first considered. Using MgO_2 directly leads to a neutralization of the nitric acid, especially because MgO_2 is available only together with MgO.

2.3.3. Absorption and reduction of NO in aqueous solution

The lead dioxide oxide–hyposulfite method works on an electrochemical foundation [35]. The NO contained in the flue gas is chemically absorbed in a hyposulfiteous solution ($S_2O_4^{2-}$) and is reduced to nitrogen. This method requires expensive apparatus and a complex process control system.

Complexing agents, so-called chelates [e.g. EDTA (ethylene diamine tetra-acetate) or NTA (nitrilotriacetate)], are added to the water to bind NO to the flue gas reversibly to EDTA or NTA, which are used as Fe(II) complexes. Then it reacts with sulfite SO_3^{2-}, which dissolves in the water by absorption of SO_2 from the flue gas to molecular nitrogen N_2 and sulfate SO_4^{2-}. The Fe(II) complex is free to bind another NO after the reaction.

The main reactions

$$Fe(II)EDTA^{2-} + NO \rightarrow Fe(II)EDTA^{2-} \cdot NO \tag{45}$$

$$SO_2 + SO_3^{2-} + H_2O \rightarrow 2HSO_3^- \tag{46}$$

$$2Fe(II)EDTA \cdot NO + 2SO_3^{2-} \rightarrow 2Fe(II)EDTA^{2-} + N_2 + 2SO_4^{2-} \tag{47}$$

Table 2
Hole diameter d_0, number of holes z and minimum volume flow $\dot{V}_{bottom,std}$ of the perforated bottoms used

Perforated Bottom	d_0 in mm	z	$\dot{V}_{bottom,std}$ in m^3/h
1	2.5	80	12.0
2	2.5	120	18.0
3	1.5	192	11.0
4	3.5	40	11.0
5	2.5	128	19.2

of this absorption–reduction method are shown for the use of EDTA in Fig. 12.

As secondary reactions, we find:

1. Oxidation of sulfite ions by absorbed oxygen to sulfate and dimerization to dithionate $S_2O_6^{2-}$
2. Oxidation of Fe(II)EDTA to the far less active Fe(III)EDTA
3. Partial reduction of NO to various nitrogen–sulfur compounds such as $HON(SO_3)_2^{2-}$, $N(SO_3)_3^{3-}$, $HN(SO_3)_2^{2-}$ or N_2O.

The sulfate ions concentrating in the washing solution are deposited as calcium sulfate or ammonia sulfate in existing units. The concentration enrichment of the products mentioned in secondary reaction (3) is avoided by precipitation or hydrolysis. As the oxidation of sulfite by NO completes the reaction to the oxidation by oxygen, an SO_2–NO ratio of 3 or higher—dependent on the oxygen content of the flue gas—must be guaranteed to achieve an NO deposition degree of more than 70% [36].

The absorption–reduction reactions should be separated locally from the condensation of the steam in the flue gas. As complexing agents are not consumed, their use should take place in a closed system. Using condensate which is discharged afterwards into the sewage system, would mean an unnecessarily high consumption of complexing agent. Thus, the condensation of steam in the flue gas should take place in a heat exchanger arranged after the absorbing solution.

A result of the high stability of the complexing agents is that if EDTA is discharged it remobilizes barely soluble metal compounds in sewage sludge or river beds stronger than other complexing agents. It is a disadvantage that

EDTA (which has a higher efficiency at the NO absorption compared to NTA) is not decomposable biologically, so the metals remain dissolved in water.

3. Experimental apparatus and measuring techniques

3.1. Design of the test boiler

For the experiments on absorption and neutralization of SO_2 and NO_x, the boiler shown in Fig. 13 is used. Fluid-mechanical investigations are carried out on a model without a heat exchanger in the water bath. A transparent boiler wall allows the supervision of the bubble flow.

The burner and combustion chamber are arranged vertically. A perforated bottom is positioned horizontally at the end of the combustion chamber; its circumference is surrounded by a weir. The end of the combustion chamber and the perforated bottom are submerged in the water bath. During the burner operation, the water is displaced out of the combustion chamber already during the pre-ventilation time and a gas bubble is formed beneath the perforated bottom. The gas moving upward is dispersed into bubbles by the perforated bottom; it cools down on contact with water to water temperature while rising. Also, the water is used for absorption and neutralization of the acid-forming sulfurous and nitrous oxides. For this reason, the complete test boiler was made of stainless steel with the material number 1.4571 (commonly called V4A steel). Test gas portions can be taken at the end of the combustion chamber and immediately above the surface of the water bath. Test water portions are drawn at irregular intervals.

All experiments were carried out with an air pressure atomization burner, modified by a rotation-controlled fan. With rotation control, the fan pressure is adjusted to the counterpressure, which varies because of the use of different perforated bottoms and various water heights above the perforated bottom h. The water bath and the combustion chamber can be cooled separately. In the experiments, the heat exchangers are flown through one after the other.

The geometry of the perforated bottoms used is shown in Table 2. Perforated bottoms 1 to 4 are used for the fluid-mechanical investigations (Section 4). For this, the dimensions marked in Fig. 13 are $d_1 = 640$ mm, $d_2 = 440$ mm, $d_3 = 243$ mm, $d_4 = 231$ mm and $b = 60$ mm. Bottom 5 is used for experiments on absorption and

Table 3
Manufacturer's data of the gas analyzers used, NDIR [38], electrochemical measuring cell [39] and chemiluminescence [40]

Measuring value	ψ_{SO_2}		ψ_{NO}, ψ_{NO_x}	ψ_{CO_2}	ψ_{CO}
Measuring method	NDIR	Electro-chemical cell	Chemiluminescence	NDIR	NDIR
Measuring range	0–1000 ppm	0–20 ppm	0–100 ppm	0–20%	0–300 ppm
Linearity	±20 ppm	no data	±1 ppm	±0.4%	±6 ppm
Resolution	2 ppm	0.2 ppm	0.1 ppm	0.05%	2 ppm
Exactness	5 ppm	2% of measured value/month	1 ppm	0.1%	1.5 ppm
T_{90} time	1–60 s	60 s	1 s	1–60 s	1–60 s

neutralization of sulfurous and nitrous oxides (Section 5). The height of the weir remains unchanged; there, the dimensions are $d_1 = 570$ mm, $d_2 = 520$ mm, $d_3 = 285$ mm and $d_4 = 230$ mm.

3.2. Measuring techniques

3.2.1. Gas analysis

Table 3 shows the manufacturer's data on the gas analyzers used. For the measurement of the SO_2 concentration, two different measuring concepts are used. The non-dispersive infrared analysis (NDIR) is used to measure large concentrations and allows measuring fast concentration changes within an area between 1 and 60 s by the parametrizable t_{90} time. The electrochemical measuring cell is used to measure small, stationary SO_2 concentrations. The exactness of the electrochemical measuring cell refers to one month, as it is calibrated only by the manufacturer at the time of delivery and then is usable for several months, while the other analyzers are calibrated at the beginning of the daily measurements.

The concentrations of CO and CO_2 are also measured by NDIR. The measuring of the nitrogen oxide concentrations is carried out by a gas analyzer, based on the principle of chemiluminescence. The measuring apparatus used allows measurement of NO and NO_x concentrations. Both values are determined by the same detector, for the measurement of NO_x, the test gas portion is led through a converter first. The NO_2 concentration is determined via the difference between NO and NO_x concentration.

Using the electrochemical measuring cell, no preparation of the test gas portion is carried out. The detector is applied immediately to the flue gas channel. The measuring methods used in other analyzers show a high cross-sensitivity to steam. The test gas portion therefore is cooled down electrically before analysis to a temperature $\vartheta = 5°C$, so the largest part of water is condensed. The remaining humidity can be suppressed by moistening the test gas used for the calibration.

The danger caused by the high solubility of SO_2 in water arises when an indeterminate portion of SO_2 dissolves in the condensate at the analysis and an incorrect analysis results. This absorption is avoided by acidifying the condensation route from the point of taking the test gas portion until the end of the cooler. For this, phosphoric acid (H_3PO_4) is used, which is appropriate because of its very low steam pressure. Immediately at the point of taking the test gas portion, the 5% phosphoric acid is introduced into the test gas tube via a dosage valve. The test gas pipe has a very small inner diameter of $d = 2$ mm, so a condensate film will be established over the entire pipe wall. Thus, the condensate droplet will be reached immediately by the phosphoric acid and could not absorb SO_2.

3.2.2. Water analysis

The water analysis which is used covers the determination of the pH, the concentration of the dissolved oxygen and the concentration of the anions. The pH is measured by a pH combination electrode with Ag/AgCl reference system [41]; the oxygen concentration is measured with a Clark measuring cell.

An ion chromatograph with a conduction detector is used for the ion analysis. A phthalic acid eluent (3 mmol/l phthalic acid) is used with 5% acetonitrile for stabilization. The calibration is carried out by a multi-ion standard; for most of the measurements, it contains 5 mg/l of each sulfite, chloride and nitrite and 10 mg/l of each bromide, nitrate and sulfate. The multi-ion standard is prepared by weighing in the corresponding potassium or sodium neutral salts into distilled water. Sulfite is an exception. A separate sulfite standard is prepared in which the oxidation of sulfite with dissolved oxygen to sulfate is avoided by adding a caustic soda solution and formaldehyde. This sulfite standard is then mixed to the multi-ion standard.

Thus, the determination of sulfite, sulfate, nitrite, nitrate, chloride and bromide is possible with one chromatogram. If sulfite will be determined as well, the test water portion is brought to a pH = 9 to 10 and formaldehyde is added to the test water portion. The retention time of sulfite is shortened by the formation of an addition compound of formaldehyde and sulfite, and the parallel eluation with chloride is avoided.

Standard and test water portion are microfiltered. The test portions are also treated with a cartridge cation exchanger in H^+ form so that the front peak, caused by the cations (Mg^{2+}), does not cover the early eluating anions. A non-polar solid-phase extraction is also carried out by cartridges, so disturbances of the chromatograms and damage of the separation columns by hydrocarbons are avoided.

4. Fluid-mechanical layout of a gas bubble washer

To improve the heat and mass transfer between two phases, the interphase is increased by dispersing one of the two phases. For absorption—that is dissolving a gas in a liquid—either units with atomization of the liquid, or so-called gas bubble washers which disperse the gas, are used.

At liquid atomization, gas is led in counter-flow to falling droplets from the bottom to the top. This arrangement allows adaptation of the gaseous mass transfer coefficient β_g by changing the relative velocity between droplets and surrounding gas. In contrast, the liquid mass transfer coefficient β_l is small because of the lack of movement inside the small droplets. Thus, liquid atomization is appropriate for the absorption of easily soluble gases for which the liquid mass transfer resistance can be neglected as a result of their good solubility.

In bubble columns, the gaseous mixture and fluid are in contact at the surface of gas bubbles which rise through a standing or slow moving liquid. A high specific surface can

310

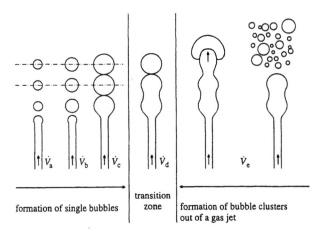

Fig. 14. Schematic presentation of different kinds of bubble formation [43].

be reached by producing small bubbles. This large surface, the simple design construction with small dimensions, and the good heat transfer conditions, are reasons for the universal usability of the bubble column as a reactor. Additionally, bubble columns show excellent transfer conditions in transversal direction, this means that concentration and temperature are constant in a cross-section [42]. The reason for this is so-called "exchange cells", which are formed by a counter-flow movement of liquid streams in the column. The rising gas bubbles cause a liquid stream flowing upward, which deviates when it reaches the liquid surface, resulting in a circular flow in the column. The circular flow of the liquid is subdivided into many small circular flows; the exchange cells guarantee a brisk longitudinal and transversal exchange.

The gas bubble washer was chosen as a absorbing apparatus because, among the advantages mentioned, it allows the possibility to achieve the dispersion of the flue gas with the burner fan in combination with the condensing boiler technology, and therefore no further aggregate is required to supply the dispersion energy.

4.1. Gas flow through the complete perforated bottom

A perforated bottom is used to achieve the dispersion of the flue gas in the condensate. First, the fluid-mechanical processes will be considered for a single hole before the description of the phenomena observed at the formation of bubbles is applied to the perforated bottom. Fig. 14 shows the bubble formation for a single hole for various gas volume flows.

The volume flow of the gas increases from $\dot{V}_a$ to $\dot{V}_e$. The separating bubbles get larger as the gas volume flow is increased. Ignoring the lifting force of the bubbles which increases with growing bubble diameter, the distance of the bubble clusters remains constant, so the bubbles touch each

other for the gas volume flow $\dot{V}_c$. They are separated only by a small liquid lamella.

Until this volume flow is reached, single bubbles are formed. If the volume flow is increased further, the thin liquid lamella is destroyed and a jet is formed which leads to the formation of bubble clusters with bubbles of different sizes.

The limit between the formation of single bubbles and bubble clusters can be determined by comparing the forces at the liquid lamella [44]. It is assumed that the liquid shows only a low viscosity. From the side of the hole, the kinetic force F_{kin} and the force of interfacial tension $F\sigma$ must be considered. From the side of the gas jet, we take the pressure force F_p, arising from the curved surface, and the lifting force F_l on the liquid lamella into account. These forces can be determined as follows:

$$F_{kin} = \frac{\pi}{4}\rho_d w_0^2 d_0^2 \tag{48}$$

$$F_\sigma = \pi\sigma d_0 \tag{49}$$

$$F_p = \frac{\pi}{2}\sigma d_0 \tag{50}$$

$$F_l = \frac{\pi}{4}(\rho_c - \rho_d)g\cdot s\cdot d_0^2 \tag{51}$$

where d_0 is the hole diameter, w_0 the average velocity in the hole, ρ_c density of the liquid (coherent phase), ρ_d the density of the gas (dispersed phase), g the gravitation constant, σ the interfacial tension and s the jet length.

The variable s describes the length of the gas jet at the time the bubble cluster tears off. Ruff [44] gives the following estimation:

$$s = \frac{1}{2}\left(\frac{\pi}{4}\right)^{2/5}\left(\frac{d_0^4\cdot w_0^2}{g}\right)^{1/5} \tag{52}$$

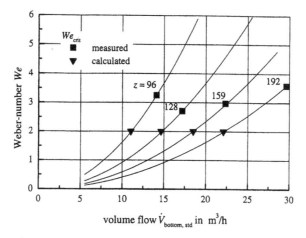

Fig. 15. Calculated Weber numbers We, We$_{crit}$ and measured critical Weber number We$_{crit}$ for a just completely passed perforated bottom, $d_0 =$ 2.5 mm [37].

The forces F_{kin} and F_p have a destructive effect on the liquid lamella, the forces $F\sigma$ and F_l are constructive. Setting up a balance for these forces, we obtain:

$$F_\sigma + F_l = F_{kin} + F_p \tag{53}$$

$$\Rightarrow \pi\sigma d_0 + \frac{\pi}{4}(\rho_c - \rho_d)g s d_0^2 = \frac{\pi}{4}\rho_d w_0^2 d_0^2 + \frac{\pi}{2}\sigma d_0 \tag{54}$$

$$\Leftrightarrow \frac{2\sigma}{\rho_d w_0^2 d_0} + \frac{\rho_c - \rho_d}{\rho_d}\left(\frac{d_0 g}{w_0^2}\right)^{4/5} 0.45395 = 1 \tag{55}$$

$$\Leftrightarrow \frac{2}{We} + \left(\frac{1}{Fr^*}\right)^{4/5} 0.45395 = 1 \tag{56}$$

where

$$We = \frac{\rho_d w_0^2 d_0}{\sigma} \tag{57}$$

and

$$Fr^* = \frac{w_0^2}{d_0 g}\left(\frac{\rho_d}{\rho_c - \rho_d}\right)^{5/4}. \tag{58}$$

The Weber number We is the ratio of kinetic force and interfacial tension force, the modified Froude number Fr* is the ratio of kinetic force and lifting force. The ratio of the forces is influenced by the size of the holes. In a small hole, the lifting force can be ignored so $1/Fr^* \rightarrow 0$. For a large

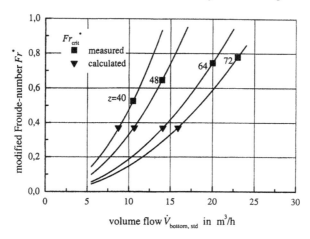

Fig. 16. Calculated modified Froude numbers Fr*, Fr$_{crit}^*$ and measured critical modified Froude number Fr$_{crit}^*$ for a completely passed perforated bottom, $d_0 = 3.5$ mm [37].

312

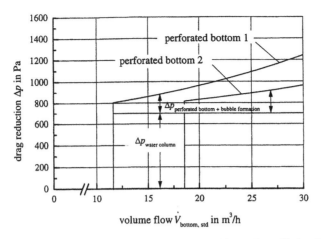

Fig. 17. Drag reduction of the flue gas flowing through the water bath [37] perforated bottom 1: $z = 80$, $d_0 = 2.5$ mm, $\dot{V}_{bottom,std} = 12$ m^3/h perforated; bottom 2: $z = 120$, $d_0 = 2.5$ mm, $\dot{V}_{bottom,std} = 18$ m^3/h.

hole, the influence of the interfacial tension force Fo becomes negligibly small and $1/We \rightarrow 0$.

For small holes, the limit between single bubble formation and bubble cluster formation is to be found for $We = We_{crit} = 2$. For $We > We_{crit}$, we observe bubble cluster formation, while for $We < We_{crit}$, single bubbles are formed. In the same way, for big holes the limit is to be found at $Fr^* = Fr^*_{crit} = 0.37$.

Equalizing the criteria $We_{crit} = 2$ and $Fr^*_{crit} = 0.37$, one obtains the determining equation for the hole diameter d_{gr} that separates small and large holes:

$$d_{gr} = \left(\frac{\sigma}{\rho_d g}\right)^{1/2} \left(\frac{\rho_d}{\rho_c - \rho_d}\right)^{5/8}. \tag{59}$$

For the dispersion of air in water, this diameter is $d_{gr} = 2.74$ mm for a temperature of 25°C; this changes very little within a range of flue gas and condensate temperatures between 25 and 50°C.

Comparing the phenomena of the gas flow through a single hole and a perforated bottom, it becomes evident that if the bottom is completely passed by the gas volume flow, each hole is in the state between single bubble formation and bubble cluster formation [44]. This means that every hole of the bottom shows bubble cluster formation if the perforated bottom is passed completely by gas. Falling short of this characteristic volume flow means that only a portion of the holes are passed by gas and inactive zones can develop in which no mass transfer takes place.

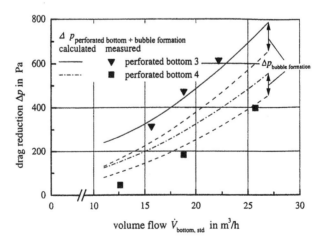

Fig. 18. Drag reduction of the flue gas by bubble formation and by the perforated bottom perforated bottom 3: $z = 192$, $d_0 = 1.5$ mm, $\dot{V}_{min} = 11$ m^3/h, perforated bottom 4: $z = 40$, $d_0 = 3.5$ mm, $\dot{V}_{min} = 11$ m^3/h.

At the layout of a perforated bottom for flue gas dispersion, amount and diameter of the holes must be chosen so that the bottom is passed completely by the gas. If too many holes (or too big in diameter) are chosen, the bottom is not passed completely by gas and inactive zones will result, while for too few holes, an increase of the gas volume flow means an additional drag reduction while passing the hole, which results in the formation of larger bubbles. Choosing burner power, air–fuel ratio and flue gas temperature before entering the water bath determines the volume flow of the gas.

Figs. 15 and 16 show the experimental results for completely passed perforated bottoms of different geometry. For these experiments, the test boiler shown in Fig. 13 was used. To allow good observation of the bubble formation, the heat exchanger was not built, and a plexiglass boiler wall was used.

In Fig. 15, the Weber number We for a completely passed bottom is the determining number, as small holes were chosen with a diameter $d_0 = 2.5$ mm. The curves show the calculated Weber number as a function of the number of holes and volume flow of the gas $\dot{V}_{bottom,std}$ which flows through the bottom under the conditions ($\vartheta_{std} = 25°C$, $p_{std} = 1$ bar). The measured points as shown represent the volume flow of the gas for which the bottom is passed completely by the gas for the chosen number of holes z.

In Fig. 16, the modified Froude number Fr* is used for the determination of a perforated bottom, as large holes were chosen with a diameter $d_0 = 3.5$ mm.

The values of the critical numbers We_{crit} and Fr^*_{crit}, determined in the experiments, are higher than the theoretically calculated values $We^*_{crit} = 2$ and $Fr^*_{crit} = 0.37$. The difference between theoretical and experimental values increases with a growing volume flow $\dot{V}_{bottom,std}$. This phenomenon can be explained because the perforated bottom does not cover the complete cross-section of the liquid. The liquid taken into motion by the rising gas bubbles is diverted when it reaches the liquid surface, and it flows back downward in the section which is not passed by the gas ($d_2 < d < d_1$). For such a distribution of the dispersed gas over a cross-section of the liquid, a stationary flow of the liquid will result, which we call "circulating flow". As a result of this circulating flow, a fifth force at the liquid lamella must be taken into account, which supports the lifting force F_1 and therefore is a constructive force for the lamella. Thus, bubble cluster formation begins at higher gas velocities, so We_{crit} and Fr^*_{crit} increase. With an increasing gas volume flow $\dot{V}_{bottom,std}$, the circulating flow becomes stronger and We_{crit} and Fr^*_{crit} increase.

The formation of the circulating flow is enforced by the geometrical arrangement of the perforated bottom to achieve a stable flow of the liquid. If the complete cross-section of the fluid is covered by the perforated bottom, rising and falling liquid flows come in contact with each other. Even the smallest disturbances can then lead to a complete breakdown of the stationary flow and can aggravate a locally constant gas flow through the perforated bottom. Fluctuations can be observed in both longitudinal and transversal direction [43]. This causes pressure variations in the gas beneath the perforated bottom which disturb the combustion. Our measurements have shown much lower pressure variations at a completely developed circulating flow compared to a gas distribution over the complete liquid cross-section. The size of the pressure variations does not show a dependence on the water height above the bottom h.

4.2. Drag reduction of the flue gas flowing through the water bath

The height of the water column above the perforated bottom h, and the drag reduction by the flue gas dispersion have to be overcome by the burner fan. Fig. 17 gives an example of drag reduction Δp for two perforated bottoms dependent on the flue gas volume flow $\dot{V}_{bottom,std}$.

To be added is the drag reduction, caused by bubble formation and by flow through the perforated bottom, and the pressure difference necessary to overcome the water column above the perforated bottom. With increased gas volume flow, the drag reduction by bubble formation and by flow through the bottom increases. The calculation are carried out according to Mersmann [45].

By choosing a larger number of holes (bottom 2), or a larger hole diameter, constant drag reduction can be achieved for increased volume flow $\dot{V}_{bottom,std}$, which may result either from a higher burner power or from external flue gas recirculation.

By using large holes, a decisive decrease in the drag reduction can be achieved. In Fig. 18, a comparison between the drag reduction ($\Delta p = \Delta p_{perforated\ bottom} + \Delta p_{bubble\ formation}$) of a bottom with small holes (bottom 3) and large holes (bottom 4) is shown. The number of holes z, is chosen so that both bottoms are passed completely by a gas volume flow of $\dot{V}_{bottom,std} = 11$ m³/h. Measured and calculated volume flow shows good agreement for bottom 3. For bottom 4, the measured values are lower than the calculated. This is attributed to the fact that the resistance coefficient for the calculation of the drag reduction by bubble formation was determined by Mersmann [45], based on small holes. The separate display of the calculated drag reduction by bubble formation and by perforated bottom, compared to our own measurements shows that for large holes the drag reduction by bubble formation can be ignored.

4.3. Mass transfer between dispersed gas and liquid

In Section 2.2.2, the mass transfer between dispersed gas and fluid was described by the two-film theory in the example of SO_2. Dispersing the flue gas by perforated bottom, the flow of the gas bubbles and of the liquid are strongly influenced by the geometric dimensions. Eq. (29)

314

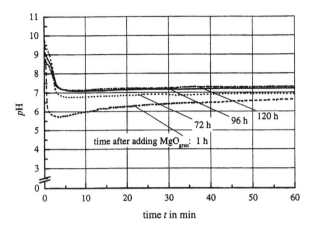

Fig. 19. Influence of MgO_{gran} hydration on the pH [37], $\dot{m}_{Oil} = 1.0$ kg/h, $\xi_S = 0.12\%$, $m_{MgO} = 1582$ g.

describes the SO_2 partial pressure in the flue gas after leaving the liquid as a function of following parameters:

1. gaseous mass transfer coefficient β_g,
2. contact time of flue gas and liquid Δt,
3. mole specific interphase A_{bu}/N_{bu} of the single bubble.

The bigger the bubbles, and the more turbulent the liquid flow around them, the bigger is their inner circulation and its mass transfer coefficient β_g. The mass transfer coefficient β_g is dependent on time, and decreases with increased retention of the bubble in the liquid. The reason for this is that each bubble—regarded as a closed system—does not show mass transition, therfore the concentration profile in the bubble interior changes with time. A very detailed description of experimental and theoretical results for mass transfer processes in bubbles is given by Hong and Brauer [46].

The contact time Δt can be influenced directly by the height of the water column above the perforated bottom h. The velocity of the bubbles rising in the liquid is primarily a function of the liquid density ρ_l and the gas density ρ_g and the interfacial tension σ.

The experimental determination of the interphase of dispersed gas and coherent liquid is carried out by determining the average bubble diameter d_s and the gas volume contained in the two-phase system. The gas volume is determined by measuring the difference of mixture volume and liquid volume. Thus the interphase a is specific to the gas volume and corresponds directly to the mole specific interphase A_{bu}/N_{bu}, via the ideal gas law.

We write for the volume specific interphase a [47]:

$$a = 6\frac{\varepsilon_d}{d_S} \qquad (60)$$

where d_s is the Sauter diameter and ε_d the volume fraction of the gas.

The Sauter diameter d_s is the average diameter of the bubbles at a greater distance from the perforated bottom

and is calculated in liquids with a low viscosity, as used here according to [47]:

$$d_S = \sqrt{\frac{6\sigma}{(\rho_c - \rho_d)g}}. \qquad (61)$$

If the dispersed gas only flows through a very low liquid column, the average bubble diameter cannot be expressed by only the Sauter diameter. In the immediate vicinity of the perforated bottom, the average diameter is also influenced by the size of the holes [43].

The mixture consists of the volume fraction of the gas ε_d and the volume fraction of the liquid ε_c ($\varepsilon_d + \varepsilon_c = 1$). The volume fraction of the gas ε_d determines the interphase; it is dependent on the composition and can additionally be influenced by the average velocity v_d. The average velocity v_d is determined immediately by the gas volume flow $\dot{V}_{bottom,std}$ for a given bubble column cross-section A:

$$v_d = \frac{\dot{V}_{bottom,std}}{A}. \qquad (62)$$

If the average velocity is low, the gas bubbles, which are strongly decelerated when entering the liquid, can distribute over the whole cross-section. Then a homogeneous structure exists and the volume fraction ε_d increases with growing average velocity. Further increase in the average velocity leads to a zone with a higher gas fraction as a result of the deceleration of the gaseous phase in which bigger gas jets and agglomerations of bubbles can be observed. These can rise faster than single bubbles because they show a higher lifting force as a result of their smaller volume specific surface. The formation of this heterogeneous surface is associated with a decrease in the volume fraction, ε_d.

Mersmann [48] gives an upper limit for the average velocity of a bubble column:

$$v_{d,max} = 0.2w_e \qquad (63)$$

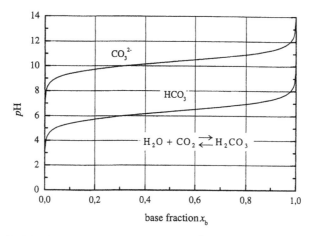

Fig. 20. Dissociation equilibria of CO_2, HCO_3^- and CO_3^{2-} versus pH and base fraction x_b.

w_e is the final rising velocity for bubbles that are end of being stable and is determined by the substance combination. For the combination of air and water, it is $w_e = 0.2$ m/s at 25°C. Increasing the average velocity above the limit mentioned results in a decisive increase of the mixture of the two phases, even so, the decrease of the interphase leads to a deterioration of the heat and mass transfer.

To enable a compact layout of the absorber, the aim is to receive a small cross-section A. To guarantee good mass transfer conditions, a minimum cross-section A must be achieved. By substituting Eq. (63) in Eq. (62), one obtains with $w_e = 0.2$ m/s:

$$A_{min} = \frac{\dot{V}_{bottom,std}}{0.04 \text{ m/s}} \tag{64}$$

The gas volume flow $\dot{V}_{bottom,std}$ is determined by the burner power.

For an ideal burner power of $P = 15$ kW and an air–fuel ratio $\lambda = 1.1$, we obtain for the combustion of heating oil EL, a flue gas volume flow $\dot{V}_{bottom,std} = 18.2$ m³/h and a minimum cross-section of the perforated bottom $A = 0.126$ m².

5. Experiments on the absorption and neutralization of SO_2 and NO_x in the water bath

During physical absorption of gases, the gas remains unchanged in the liquid, and no further gas is absorbed by the liquid when the saturation concentration is reached. For a continuous deposition of a gas component out of a gaseous mixture, the absorbed gas must react chemically in the solution to allow the absorption of further gas.

Thus, the chemical reaction is the key to a continuous absorption of a gas component in a limited portion of

water. Like SO_2, NO_2 reacts with water to form an acid. Its deprotonation is guaranteed by neutralization to allow a further absorption of gases and their reaction to acids.

5.1. Neutralization

Sulfur and nitrogen oxides react forming of acids when absorbed in water. The deprotonation of these acids leads to a strong increase of the concentration of H^+ ions, so the pH decreases. Thus, according to Eqs. (12), (13) and (15), a further absorption of SO_2 and NO_x is aggravated.

By adding a neutralizer, ideally, the same amount of OH^- ions are dissolved as H^+ ions are produced by deprotonation of the acids, so the solution stays neutral. The neutralization is a basic principle for a continuous absorption of acid-forming flue gas components in the condensate. For this reason, the experiments carried out on the neutralization shall be described before Sections 5.2 and 5.3, which will deal more closely with the absorption of sulfur and nitrogen oxides.

The water bath used for the absorption of acid-forming flue gas components is an open system. The condensation of the flue gas taking place during the flow of the flue gas through the water bath represents the water supply flow. An overflow regulating the water height above the perforated bottom corresponds to the water outlet. The continuous exchange of the absorbent allows a continuous, dosed solution of hydroxides (OH^-), otherwise the excess neutralizer would be discharged via the overflow.

In Section 2.2.3, the general suitability of oxides, hydroxides and carbonates of sodium, magnesium and calcium for the use as neutralizers, has been described. Here, we do not deal experimentally with calcium compounds as continuous discharge of the formed salts is not possible as a result of their low solubility.

The use of sodium hydroxide and sodium carbonate is generally favored because overflow resulting from

316

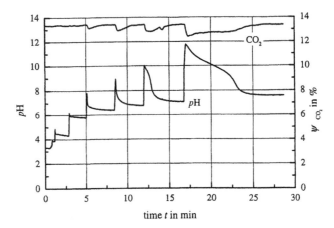

Fig. 21. Absorption of CO_2 from the flue gas in the water bath for a stepwise increasing pH by caustic soda.

discharging these highly soluble salts is not problematic. But the effective solubility of the neutralizer itself requires the use of a dosage system consisting of pH meter, dosage pump and storage tank.

By using magnesium compounds, expensive apparatus is not necessary. The very low solubility of magnesium hydroxide, in combination with the high solubility of the formed salts $MgSO_3$, $MgSO_4$ and $MgNO_3$, allows the discharge of these salts, together with the condensate, adding the neutralizer in excess at the same time, because the neutralizer is solid in water and is not drawn off via the overflow. In the following, experiments on neutralization with magnesium oxide (which forms magnesium hydroxide ($Mg(OH)_2$) when dissolved) are shown.

The experiments on neutralization and absorption were carried out in a test boiler whose design is shown in Section 3.1. Of the perforated bottoms described in this chapter, bottom 5 is used. The burner was operated at an oil mass flux $m_{oil} = 1.0$ kg/h and an air–fuel ratio $\lambda = 1.13$ for all the experiments. The fuel used was heating oil EL with a sulfur content $\xi_S = 0.12\%$.

5.1.1. Formation of $Mg(OH)_2$ via hydration of MgO

Before magnesium oxide shows a neutralizing effect in water, it has to react to hydroxide via hydration. Here it shall be investigated whether the hydration of the MgO takes place so slowly under unfavorable conditions that the neutralization of the water bath is not guaranteed because the consumption of $Mg(OH)_2$ for the absorption of acid-forming substances is greater than its production. The velocity of the hydration shown in Eq. (32) depends on the form of the MgO used. With a very compact granulate, a slower hydration occurs, compared to pulverized MgO.

Fig. 19 shows the course of the pH versus time for four different burner operation sequences, each lasting 1 h. At the start of the experiment, a mass $m_{MgO} = 1582$ g of granulated magnesium oxide (MgO_{gran})was added to the water bath.

One hour later, the first operation sequence begins. At the start of this operation sequence, a pH = 9.7 was measured. During the contact time of one hour, MgO and water have formed $Mg(OH)_2$, which has caused the increase of the pH tending to the dissociation equilibrium

$$Mg(OH)_2 \leftrightharpoons Mg^{2+} + 2OH^- \qquad (65)$$

With the absorption of the acid-forming flue gas components, a strong decrease of the pH, down to pH = 5.7, can be observed within the first three minutes of the first burner operation sequence. Having reached this minimum, the pH increases slowly until it reaches a value pH = 5.6, at the end of the operation sequence. Within the first 3 min of the burner operation sequence, the hydroxide formed before is consumed for the neutralization of the formed acid. The MgO which was covered with $Mg(OH)_2$ before the start of the burner is set free and the velocity of the hydroxide formation increases, so the pH increases as well.

After 72 h, the burner is started again. The pH decreases slower than at the first operation sequence and reaches its minimum at pH = 6.8 before increasing to its final value pH = 6.9. The reason for the slower decrease of the MgO and the less significant minimum is to be seen in the higher amount of $Mg(OH)_2$ that has been formed within the 72 h. Also, the following operation sequences, 96 and 102 h later, show a similar course of the pH. The stationary pH at the end of each operation sequence increases only slightly with increasing employment time of the MgO in the water bath, and remains below pH = 7.5 even for an operation time of several weeks.

The pH at the start of each burner operation sequence ($t = 0$) decreases with the total burner operation time. The reason for this is the increasing concentration of Mg^{2+} in the water bath. During the standstill times of the burner, magnesium hydroxide dissolves until the solubility product of Mg^{2+} and OH^- is reached. Thus, an increasing concentration of Mg^{2+} has a higher part on the solubility product and the OH^-

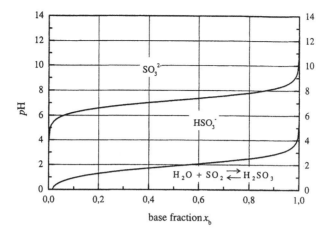

Fig. 22. Dissociation equilibria of SO_2, HSO_3^- and SO_3^{2-} versus pH and base fraction x_b.

concentration decreases. The difference between the pH at the burner standstill and burner operation therefore decreases with increasing employment time of the MgO.

5.1.2. Carbonate formation by CO_2 absorption

Flowing through the neutralized water bath, a part of the CO_2 contained in the flue gas also dissolves under the formation of carbonic acid. While the complete absorption of SO_2 is desired, this is different for the absorption of CO_2 ($m_{CO_2} = 3.2 m_{oil}$); for this reason, the system CO_2–H_2O shall be regarded here. The absorption of CO_2 is part of the following equilibrium chain:

$$CO_2(g) + H_2O \leftrightarrows CO_2(aq) + H_2O \leftrightarrows H_2CO_3 \leftrightarrows HCO_3^-$$

$$+H^+ \leftrightarrows CO_3 2^- + 2H^+ \tag{66}$$

CO_2 reacts with water to an aqueous complex $CO_2(aq)$. Only

a small part of the $CO_2(aq)$, about 0.2% [18], react to form carbonic acid (H_2CO_3), which dissociates into hydrocarbonate (HCO_3^-) and carbonate (CO_3^{2-}). Theoretically, carbonic acid is a weak acid. But as the largest part of the dissolved CO_2 exists as hydrated $CO_2(aq)$ and not as H_2CO_3, its solution acts as a weak acid. Fig. 20 shows the dissociation equilibrium in dependence on the pH and the base fraction x_b which is defined as

$$x_b = \frac{c_{base}}{c_{acid} + c_{base}} \tag{67}$$

Adding neutralizer to the water, the reaction equilibrium—according to Eq. (66)—is disturbed by the reaction of hydroxide and H^+ ions to H_2O. Tending towards the equilibrium, CO_2 dissolves again under the formation of hydrogen carbonate and carbonate.

If the pH in the water bath is 10 or higher, as can be found

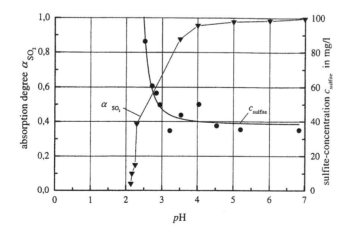

Fig. 23. Absorption of SO_2 versus pH and height of the water above the perforated bottom h [37], $\dot{m}_{Oil} = 1.0$ kg/h, $\xi_S = 0.12\%$.

318

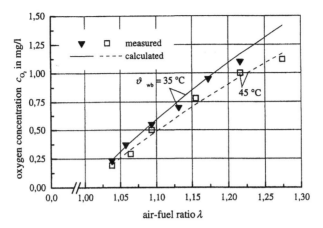

Fig. 24. O_2 concentration in the water bath versus temperature ϑ_{wb} and air–fuel ratio λ [16].

immediately after adding MgO to the water bath, a large amount of carbonate is formed. We speak of a carbonate buffer as this carbonate leaves the solution as CO_2 without changes in the pH when a strong acid like H_2SO_3 is dissolved. Thus the formation of carbonates by absorption of CO_2 from the flue gas does not have to be connected to an increased consumption of neutralizer, as long as no larger amounts are lost with the drawn-off water. If the pH lies between pH = 7 and 9, mostly HCO_3^- is formed.

Fig. 21 shows the course of the CO_2 concentration in the flue gas after the water bath and the pH of the water bath for the addition of caustic soda. The CO_2 concentration in the flue gas before entering the water bath is $\psi_{CO_2} = 13.5\%$. Starting with a low pH, a first decrease of the CO_2 concentration can be seen for a pH = 8. As a result of the absorption of CO_2, the pH decreases rapidly and the absorption of CO_2 is stopped.

Adding a large portion of caustic soda ($t = 17$ min), CO_2 is absorbed for several minutes. The pH which then ensues is pH = 7.7. A stationary pH in the water bath of pH > 8, due to the absorption of CO_2, is not expected for the given flue gas composition, and the used neutralizers. CO_2 does not seem to be absorbed for a pH < 7.5, but SO_2 is.

5.2. SO_2 absorption

Sulfur dioxide reacts to form acid when absorbed by water. The acid behavior is to be attributed to the formed sulfurous acid (H_2SO_3).

$$SO_2 + H_2O \leftrightharpoons H_2SO_3 \leftrightharpoons HSO_3 + H^+ \leftrightharpoons SO_3^{2-} + 2H^+ \tag{68}$$

Here, the equilibrium lies on the right side of the reaction Eq. (68), so almost all of the dissolved sulfur dioxide remains unchanged as $SO_2(aq)$ and only a small part is to be found as sulfurous acid, H_2SO_3. As a two-base acid, sulfurous acid dissociates according to Eq. (15) to hydrogen

sulfite (HSO_3^-), and sulfite (SO_3^{2-}), especially if the H^+ ions react to water H_2O with OH^-. Fig. 22 shows the dissociation equilibria dependent on the pH and the base fraction x_b.

5.2.1. Influence of pH on the SO_2 absorption

To investigate the influence of pH on the absorption of SO_2, the water bath was blown through with flue gas containing SO_2, without adding neutralizers. The concentrations of SO_2 before and after the water bath were measured. Fig. 23 shows the dependence of the absorption degree α_{SO_2} on the pH and the height of the water column above the perforated bottom h.

Beginning at neutral conditions in the water bath, an almost complete absorption of SO_2 can be observed up to a pH = 4 independent of the height h. At this pH, The absorbed SO_2 exists almost completely as hydrogen sulfite (HSO_3^-). With a further decrease of the pH, the absorption degree α_{SO_2} decreases rapidly. For a larger water height h, more SO_2 is absorbed compared to a small height. For a pH = 2.1, the absorption degree is smaller than $\alpha_{SO_2} = 0.05$ for the investigated water heights.

If falling short of pH = 4 is avoided by adding neutralizer to the water bath, sulfur dioxide is absorbed almost completely. In contrast, the pH quickly falls short of pH = 4 if the condensation of the steam is carried out without an immediate neutralization of the sulfurous acid, and no important absorption of SO_2 occurs any longer.

Fig. 23 shows the measured total sulfite concentration in the water bath. This total sulfite concentration consists of $SO_2(aq)$, H_2SO_3, HSO_3^- and SO_3^{2-}. The sulfite concentration is nearly constant between neutral conditions and a pH = 4 and its value lies between $c_{sulfite} = 35$ mg/l and $c_{sulfite} = 40$ mg/l. With a further decrease of the pH, the sulfite concentration increases strongly and thereby verifies the recognized decrease of the absorption degree α_{SO_2}.

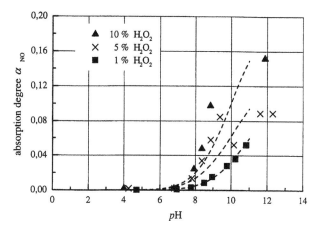

Fig. 25. NO absorption degree α_{NO} versus pH and H_2O_2 concentration at experiment start [37], $\vartheta_{wb} = 35°C$.

5.2.2. Sulfite oxidation

At the oxidation of sulfite to sulfate, oxygen dissolved in the water bath (O_2(aq)) is consumed. Fig. 24 shows the measured oxygen concentration in the water bath dependent on the air–fuel ratio λ and the water bath temperature ϑ_{wb}. Also, the measured oxygen concentration in the water bath is compared to the calculated saturation oxygen concentration. For a saturated solution, the O_2 partial pressure in the gas and the O_2 concentration in the water are in balance, which is described by the Henry law as a function of temperature. Here, the oxygen partial pressure in the flue gas is determined directly by the air–fuel ratio λ.

Independent of the air–fuel ratio λ, the measured O_2 concentration is in accordance with the saturation O_2 concentration. This means that the balance between dissolved oxygen O_2(aq) and oxygen in the flue gas is established quickly enough to compensate for the oxygen consumption by sulfite oxidation. In addition,

the SO_2 oxidation in an aqueous solution takes place with the order of zero according to the oxygen concentration [26,49]. It is not dependent on the oxygen concentration or on the oxygen transport. The accordance between measured and calculated O_2 concentration shown in Fig. 24 was observed within the limits pH = 3 and 7.5.

After the end of a burner operation sequence, the sulfite contained in the water is also oxidized, and the remaining dissolved oxygen is consumed. Nevertheless, enough oxygen is available for the sulfite oxidation at the beginning of the next burner operation sequence because enough oxygen is dissolved during a boiler pre-ventilation period during which air is led through the water bath.

Sulfite and oxygen show a negative standard potential difference with $\Delta\varepsilon = -1.31$ V (referred to pH = 14) in an alkaline solution and $\Delta\varepsilon = -1.12$ V in an acid solution

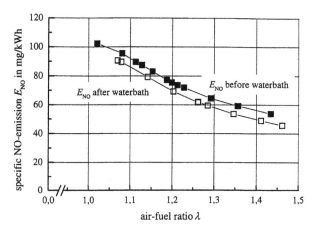

Fig. 26. Specific emissions E_{NO} before and after the water bath in dependence on the air–fuel ratio λ [37], $m_{MgO_2} = 270$ g, pH = 7.5, $h = 10$ cm.

320

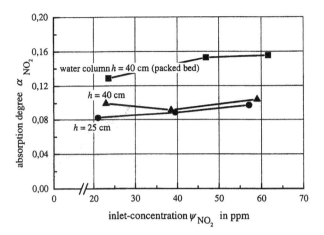

Fig. 27. Absorption of NO$_2$ versus height of the water column and the NO$_2$ start concentration when entering the water bath; pH = 7.

(referred to pH = 0), so the equilibrium of the reaction

$$SO_3^{2-} + \frac{1}{2}O_2 \leftrightharpoons SO_4^{2-} \qquad (69)$$

lies in both acid and alkaline solutions on the right side. In both cases, the sulfurization takes place completely.

Under neutral conditions the oxidation reaction from sulfite to sulfate takes place so that a reduction of the contact time between gas and water by decreasing the water height has unmeasurably small influence on the oxidation. Independent of the water height h and the sulfate concentration $c_{sulfate}$, a stationary sulfite concentration between $c_{sulfite} = 35$ mg/l and 40 mg/l can be observed.

In contrast, the acid condensate with a pH < 4 shows an increase of the sulfite concentration. This increase indicates that the oxidation of sulfite in an acid solution is kinetically aggravated. The aggravation of the sulfite oxidation and the lack of the deprotonation of the sulfurous acid have the result that only little SO$_2$ can be absorbed, compared to neutral or alkaline solutions.

5.3. Wet absorption of nitrogen oxides

Nitrogen monoxide (NO) does not react chemically in neutral water. The absorption of NO quickly leads to a saturated solution. As NO is the largest fraction of the nitrogen oxides (NO$_x$) in the flue gas, the NO$_x$ absorption can either be achieved by oxidizing NO to NO$_2$ or by using substances in the water bath which cause a chemical reaction of NO.

Section 5.3.1 describes the oxidation and the absorption of NO in the water bath by adding peroxides which are strongly oxidative. In Section 5.3.2, the absorption of NO$_2$, dependent on feed concentration and contact time between gas and liquid, are discussed.

5.3.1. Oxidation and absorption of NO in the water bath

In Section 2.3.3, peroxides of water and magnesium were described as appropriate oxidizers for the oxidation of NO in water, because no personal and environmental dangers arise from the substances or from their products. Hydrogen peroxide (H$_2$O$_2$) is used as an aqueous solution and is also a product of the reaction of magnesium peroxide (MgO$_2$) and water. Thus, first experiments were carried out with H$_2$O$_2$ before MgO$_2$ was used in the water bath, based on these experiments.

In the first experiments, gas is led via a thin glass tube into an acid solution and afterwards is led into a gas analyzer. A constant-temperature Wulff washing bottle was used as the water reservoir. The pH of the solution can be constantly measured by a pH meter. The test gas used consists of NO and N$_2$ with $\psi_{NO} = 90$ ppm and $\psi_{N_2} = 1 - \psi_{NO}$ and is led into the water with a volume flow $\dot{V} = 135$ l/h in a height $h = 50$ mm below the water surface.

Fig. 25 shows the dependence of absorption degree α_{NO} on pH and H$_2$O$_2$ concentration in water at the time of the start of the experiment. The absorption degree α_{NO} increases with pH and with the start concentration of H$_2$O$_2$.

We emphasize that Fig. 25 shows stationary absorption degrees α_{NO}. Far greater degrees of absorption have been reached for short time periods by beginning with a low pH and adding an alkaline solution; pH = 9 is reached. To explain this phenomenon, the following reaction scheme must be examined:

$$H_2O_2 + H_2O \leftrightharpoons H_3O^+ + HO_2^- \qquad (70)$$

$$HO_2^- + NO \leftrightharpoons NO_2 + OH^- \qquad (71)$$

$$2HO_2^- \leftrightharpoons O_2 + 2OH^- \qquad (72)$$

$$H_3O^+ + OH^- \leftrightharpoons 2H_2O \qquad (73)$$

By adding H_2O_2 to water, it decomposes according to Eq. (70). The formed H_2O can react according to Eqs. (71) and (72) forming hydroxide ions (OH^-), which react with hydronium ions (H_3O^+) to water, according to Eq. (73). If pH = 9 is reached by adding alkaline solutions, Eq. (70) is disturbed decisively towards the right side, and the increased HO_2^- concentrations causes a faster velocity of the reactions (71) and (72). Out of the weak base HO_2^-, the strong base OH^- is formed, and pH increases without adding further alkaline solution. As a result of the increasing reaction (71), the absorption degrees α_{NO} increases. Having passed the maximum of absorption, the equilibria of the reactions (70)–(73) are established again at a pH > 9.

To observe the durability of a washing solution with H_2O_2, the gas flow was cut off for 12 h. A higher pH and a lower absorption degree α_{NO} can be observed when the gas flow is continued. The reason for this phenomenon is the formation of O_2 according to Eq. (71), which is supported by catalytic materials such as steel or impurities during the 12 h in which no NO is led through the solution. This possible disproportion of H_2O_2 to H_2O and O_2 shows a consumption of neutralizer without NO absorption and does not favor the use of H_2O_2 in order to realize a reduction of the NO_x emission.

Using MgO_2, a decrease of the absorption degree α_{NO} cannot be observed, after a longer standstill time without leading NO through the washing solution. Magnesium peroxide is used as a 1:2 mixture with MgO.

With an increasing air–fuel ratio λ, the NO emission of the burner decreases in an almost linear manner (Fig. 26). The amount of absorbed NO is in contrast to this nearly independent on the air–fuel ratio λ with $\Delta E_{NO} = 6$ mg/kWh. Showing the same results by the absorption degree is not chosen here because the increase of α_{NO} with an increasing air–fuel ratio λ is only to be attributed to the decreasing NO emission of the burner with an increasing air–fuel ratio λ.

Further experiments have shown dependence of the absorbed amount of NO on the surface of MgO_2 (tested by varying amount of MgO_2) while the same amount of NO was absorbed between $h = 5$ cm and 30 cm. We draw the conclusion that the reaction of MgO_2 to H_2O_2 and MgO— which takes place more often with a larger amount of employed MgO_2—determines velocity but not the contact time between water and gas. The low solubility of MgO_2 suppresses the consumption of the peroxide during the times of burner standstill, as was observed for the use of the well-soluble H_2O_2.

5.3.2. Absorption of NO_2 in water

To observe NO_2 absorption in water, a gas mixture consisting of NO_2 and N_2 was added to the flue gas before entering the water bath. Thus, different concentrations of NO_2 up to $\psi_{NO_2} = 100$ ppm could be adjusted. A mix of this gas mixture with the combustion air is not possible because NO_2 decomposes at the high temperatures inside the combustion chamber.

Fig. 27 shows the dependence of absorption degree of NO_2 on the water height and the concentration of NO_2 when entering the neutral water bath (pH = 7). An increase of the inlet concentrations of NO_2, and of the contact time of the gas mixture and water over an increasing water column is connected to an increase of the NO_2 absorption degree. The absorption degree increases as a result of a higher contact time, caused by the NO_2 absorption. Specially when applying packed beds (Raschig rings) to the water bath for the experiments shown in Fig. 27, no increase of the NO concentration could be observed.

A further increase of the maximum absorption degree, for which a value of 15% has been reached in the experiments by a further increase of the contact time, is not supposed to be possible for a moderate apparative expense.

6. Summary

Many energy sources, like oil, wood or process gases, contain components which show acid-forming compounds in their flue gas after combustion, and which, to a small degree, are absorbed in the condensate when the dew point falls short. The acid so-formed, and highly corrosive condensate, raises the requirements on the materials used in the condensation area.

The concepts described in this paper are based on the idea of using the condensate for washing the flue gas; thus the flue gas is brought in contact with an already neutralized condensate and thereby is cooled below its dew point. This step process allows the wet separation of noxious matter and avoids acid corrosion of the materials used in the condensation area. Thereby, the choice of usable materials is increased.

With the example of a gas bubble washer, fluid-mechanical investigations of the flue gas dispersion in the water bath and experiments on the wet separation of SO_2 and NO_x are presented. The absorption of the gases and neutralization of the formed acids was regarded at the wet separation, whereas we primarily investigated the absorption of SO_2.

The fluid-mechanical investigations on the flue gas dispersion in the water bath were aimed at realizing good mass transfer conditions between flue gas and water, and a small expense of energy for the realization of the dispersion. The dispersion of the flue gas in the water is achieved by a perforated bottom. Its design, referring to the number of holes, is determined via the critical Weber number for small holes and via the critical modified Froude number for large holes. For the parameters chosen, a minimum gas volume flow is given.

To enforce a circulating flow of the water, only a part of the water bath cross-section is covered by the perforated bottom. Above the perforated bottom, water flows in an upward direction and is diverted when reaching the water surface before it flows downward in the area of the water bath, which is not blown through by gas. The stable hydrodynamic flow conditions decreases the pressure variations in the gas beneath the perforated bottom, which is directly connected to the combustion chamber. The critical Weber and modified Froude numbers are increased by the circulating flow, so it must be considered at the layout of the perforated bottom.

Flowing through the water bath, the pressure of the water column, as well as the pressure difference for flow through the perforated bottom and for bubble formation, act as counterpressure. A minimization of the total drag reduction can be carried out considering contact time and contact surface between gas and water, whereas for large hole diameters, the drag reduction by bubble formation can be neglected. Ideal mass transfer conditions are connected to a possibly high volume fraction of the gas in the two-component mixture gas–water, which can be influenced for a given gas volume flow by the cross-section of the perforated bottom.

Acids are formed at the wet separation of SO_2 and NO_x. The deprotonation of the acids and the neutralization of the H^+ ions formed by hydroxides (OH^-) necessary to achieve a continuous absorption of the acid-forming gases, is possible in a limited volume of water. In experiments, magnesium oxide (MgO) has proved to be especially adequate. MgO is hydrated to the hardly soluble magnesium hydroxide ($Mg(OH)_2$). Its low solubility is the reason why no dosage system is required. At the operation of the burner, a pH between 7.0 and 7.5 will ensue depending on the form of MgO used. During the standstill times of the burner, an increase of the pH takes place until the solubility product of $Mg(OH)_2$ is reached. The increasing concentration of Mg^{2+} with longer burner operation time cause a decrease of the dissolved amount of $Mg(OH)_2$, so the stationary pH of burner operation and burner standstill phases get closer to each other. CO_2 contained in the flue gas is dissolved only for a pH > 7.5 under the formation of a carbonate buffer. These carbonates are displaced when a stronger acid is dissolved and leave the solution as CO_2.

The aim of the investigations on the absorption of SO_2 in the water bath was a possibly complete desulfurization of the flue gas. Starting at neutral conditions in the water bath, an almost complete absorption of SO_2 can be observed down to pH = 4. At pH = 4, the absorbed SO_2 exists almost completely as hydrogen sulfite (HSO_3^-). With a further decrease of pH, the absorption degree of SO_2 decreases rapidly. For pH = 2.1, the absorption degree is smaller than 0.05.

For the investigations on the oxidation of sulfite to sulfate, the concentration of the oxygen dissolved in the water was determined. The measured oxygen concentration is the saturation concentration which depends directly on air–fuel ratio and water bath temperature. This means the equilibrium of dissolved oxygen and oxygen in the gas is established quickly enough to replace the oxygen consumed for the sulfite oxidation to sulfate. In the neutralized water bath, a stationary sulfite concentration of about 35 mg/l can be

observed, independent of the water height. The high sulfite concentration in the acid condensate that is not neutralized leads to the conclusion that the sulfatization is kinetically aggravated. A low SO_2 absorption is directly connected to a high sulfite concentration in the water bath.

Among the almost complete decomposition of SO_2, the washer offers the possibility to reduce the emission of nitrogen oxides. As NO—which is the dominating fraction of NO_x emitted from furnaces—is only physically dissolved in water, and is not engaged in a chemical reaction, a continuous absorption is not possible without further means. In contrast, NO_2 reacts in water under the formation of nitric and nitrous acid.

The oxidation and absorption of NO in water can be achieved by peroxides of hydrogen and magnesium. Hydrogen peroxide (H_2O_2) is not suitable for a long-term operation as a result of its low stability. Using the stable magnesium peroxide (MgO_2), a reduction of the NO_x emissions of $\Delta E = 6$ mg/kWh could be reached. While the water height was shown to have no effect on the absorbed amount of NO, a dependence on the employed amount of MgO_2 could be observed. We draw the conclusion that the reaction of MgO_2 with water to MgO and H_2O_2 is velocity-determining but not the contact time between gas and water. An increase of the absorbed amount of NO by a higher pH is not favorable, due to the absorption of CO_2 connected thereto.

The absorption of NO_2 in neutral water requires good mass transfer conditions between water and flue gas. By applying a packed bed in the water columns, an absorption degree of 15% could be achieved for a start concentration $\psi_{NO_2} = 62$ ppm, and a height of the water column of 40 cm.

Using gas bubble washers, a reduction of the SO_2 emissions of about 99% can be achieved within a wide range of sulfur content in the fuel and for a low energy expense for the dispersion of the flue gas. A corresponding reduction of the NO_x emission is not possible until now. The measures described showed a maximum reduction of the NO_x emission of about 25%.

References

[1] Kiehl JT, Briegleb BP. The relative roles of sulfate aerosols and greenhouse gases in climate forcing. Science, 260. Washington D.C: The American Association for the Advancement of Science, 1993. pp. 311.

[2] Enquete-Kommission Schutz der Erdatmosphäre des 12. Deutschen Bundestages: Schutz der grünen Erde— Klimaschutz durch umweltgerechte Landwirtschaft und Erhalt der Wälder. Bonn: Economica, 1994.

[3] Norm DIN 4705 Teil 1, Oktober 1993: Feuerungstechnische Berechnung von Schornsteinabmessungen.

[4] Norm DIN 4701, März 1983: Regeln für die Berechnung des Wärmebedarfs von Gebäuden.

[5] Norm DIN 4703 Teil 1, September 1988: Raumheizkörper.

[6] Recknagel H, Sprenger E, Schramek E-R. Taschenbuch für Heizung und Klimatechnik. 67. Aufl.. München: Oldenbourg, 1995.

[7] Handrock W. Installations- und Betriebstechnik von Brennwertkesseln. In: Asue e V, editor. Brennwertgeräte. Technik – Vorschriften—Erfahrungen, Essen: Vulkan, 1989 (Nr. 11 der ASUE-Schriftenreihe).

[8] Jannemann Th. Stand der Brennwertkessel-Technik in der Bundesrepublik Deutschland. In: Asue e V, editor. Brennwertgeräte. Technik - Vorschriften- Erfahrungen, Essen: Vulkan, 1989 (Nr. 11 der ASUE-Schriftenreihe).

[9] Marx E. Brennwerttechnik beim Einsatz von gasförmigen und flüssigen Brennstoffen. Stuttgart: Gustav Kopf, 1986.

[10] Göddeke H. (Verf.); Bundesministerium für Forschung und Technologie (Hrsg.): Rationale Nutzung und Bereitstellung von Energie. Bonn: 1984. Forschungsbericht im Auftrage des BMFT.

[11] Winkens HP. Bedeutung der Rücklauftemperatur für die Fernwärmeversorgung. Düsseldorf: VDI, 1980 (VDI-Berichte Nr. 388).

[12] Küchen C; Deutsche Shell AG, PAE-Labor: Untersuchungen zur Öl-Brennwerttechnik. Hamburg, 1993. Interner Bericht.

[13] Küchen C, Deutsche Shell AG, PAE-Labor: Strebel Neotherm Öl-Brennwertkessel. Hamburg, 1993. Interner Bericht.

[14] Farago Z. Brennwerttechnik mit Abgasentschwefelung für Hausheizungen. HLH Heizung Lüftung/Klima Haustechnik, 1. Düsseldorf: Springer VDI Verlag, 1999.

[15] Hinderer D. Heizkessel aus Kunststoff. Si-Informationen 1995;10:106–110.

[16] Haase F, Köhne H, Graf von Schweinitz H. Reduction of SO_2 and NO_x at an oil-fired condensing boiler. Proceedings of First European Conference on Small Burner Technology and Heating Equipment (ECSBT), pp. 425–432, vol. II. Zürich: ETH, 1996.

[17] Graf von Schweinitz H. Ein Beitrag zur Reduzierung der NO_x-Emission durch Inertisierung des Verbrennungsprozesses bei Einsatz von Brennwerttechnik. Aachen, RWTH, Diss., 1997.

[18] Holleman AF, Wiberg E, Wiberg N, editors. Lehrbuch der anorganischen Chemie. Berlin: de Gruyter, 1985. pp. 91–100.

[19] Igelbüscher H, Schröder R. Gegenwärtiger Stand und Zukunft des S-H-L-Verfahrens. Technische Mitteilungen, 1/2. Essen: Vulkan, 1985 pp. 45–49.

[20] Eggersdorfer R. Methodik zur Auswahl des optimalen Verfahrens zur Rauchgasentschwefelung unter Berücksichtigung der standortspezifischen Randbedingungen. Technische Mitteilungen, 1/2. Essen: Vulkan, 1985 pp. 67–71.

[21] Scholz F. Entwicklungsstand der Rauchgasreinigung. Brennstoff-Wärme-Kraft 1984;36(1/2):9–18.

[22] Wallner J, Stibich R, Veiter E. Erfahrungen mit dem Betrieb einer Entschwefelungsanlage hinter einem Kupfermetallkonverter. Zusammenfassung eines Vortrages auf einer Tagung des GDMB-Kupferausschusses, Hamburg.

[23] Pfeiffer K-D. Stand und Entwicklungsmöglichkeiten des Sprühabsorptionsverfahrens. Technische Mitteilungen, 1/2. Vulkan: Essen, 1985 pp. 16–22.

[24] Hofmann R. Entschwefelung wie von selbst. In: VDI-Nachrichten, 28.06.91.

[25] Richter E, Knoblauch K. Verfahren der Bergbau-Forschung zur Rauchgasentschwefelung und NO_x-Entfernung. Technische Mitteilungen, 1/2. Essen: Vulkan, 1985 pp. 13–15.

[26] Knapp H. Löslichkeiten von Gasen in flüssigen Lösungsmitteln. Staub—Reinhaltung der Luft, 36. Düsseldorf: VDI, 1976 Nr. 8, pp. 325–331.

[27] Vauck WRA, Müller HA. Grundoperationen chemischer Verfahrenstechnik. 10. Leipzig: Deutscher Verlag für Grundstoffindustrie, 1994.

[28] Mersmann A. Stoffübertragung. Berlin: Springer, 1986.

[29] Barron CH, O'Hern HA. Reaction kinetics of sodium sulfite oxidation by the rapid-mixing method. Chemical Engineering Science 1966;21:397–404.

[30] Linke WF. Solubilities. Inorganic and metal-organic compounds. 4th ed., vol. II. Princeton: D. van Nostrand, 1965.

[31] Pietsch G, editor. Gmelin Handbuch der anorganischen Chemie Weinheim: Verlag Chemie, 1939 System-Nr. 27 (Magnesium), Teil B, Lieferung 1.

[32] Pietsch G, editor. Gmelin Handbuch der anorganischen Chemie Weinheim: Verlag Chemie, 1961 System-Nr. 28 (Calcium), Teil B, Lieferung 3.

[33] Pietsch G. Gmelin Handbuch der anorganischen Chemie, Weinheim: Verlag Chemie, 1936 System-Nr. 4 (Stickstoff), Teil A1, Lieferung 1. 8.

[34] Greenwood NN, Earnshaw A. Chemie der Elemente, 1. Weinheim: VCH, 1990.

[35] Jüttner K, Kreysa G, Kleifges K-H, Rottmann R. Elektrochemisches Abgasreinigungsverfahren zur simultanen Entfernung von SO2 und NOx. Chem.-Ing.-Tech. 1994;66(1):82–85.

[36] Zeise W. Zur Reaktionskinetik der simultanen Absorption von SO_2 und NO_x in wäßrigen Eisen–EDTA-Lösungen in einer Füllkörperkolonne im Technikumsmaßstab. Aachen, RWTH, Diss., 1989.

[37] Haase F. Untersuchungen zur Naßabscheidung von SO_2 und NO_x bei Einsatz von brennwerttechnik. Aachen, RWTH, Diss., 1996.

[38] Hartmann&Braun A.G.: Uras 10 E - Uras 10 P. Frankfurt a. M., November 1994 (Druck). - Betriebsanleitung.

[39] Zellweger Analytics GmbH: Gas Scout. Dorset (GB), März 1995. - Betriebs- und Wartungsanleitung.

[40] Beckmann Industrial Prozeß-Geräte GmbH: NO/ NO_x -Analysator. München. - Firmenbulletin D4406B.

[41] Schott-Geräte GmbH: Gebrauchsanleitung für Elektroden zur Messung von pH-Werten und Redoxspannungen. Hofheim a. Ts., 1994. - Gebrauchsanleitung zur pH-Einstabmeßkette Typ Nr. N 6280.

[42] Grassmann P, Widmer F. Einführung in die thermische Verfahrenstechnik. 2. Berlin: de Gruyter, 1974.

[43] Brauer H. Grundlagen der Einphasen- und Mehrphasenströmungen. Aarau: Sauerländer, 1971.

[44] Ruff K, Pilhofer T, Mersmann A. Vollständige Durchströmung von Lochböden bei der Fluid-Dispergierung. Chem.-Ing.-Tech. 1976;48(9):759–764.

[45] Mersmann A. Druckverlust und Schaumhöhen von gasdurchströmten Flüssigkeitsschichten auf Siebböden. Beilage zu Forschung auf dem Gebiete des Ingenieurwesens, Ausgabe B, Band 28. Düsseldorf: VDI, 1962.—(VDI-Forschungsheft 491).

[46] Hong W-H, Brauer H. Stoffaustausch zwischen Gas und Flüssigkeit in Blasensäulen. Düsseldorf: VDI, 1984 (VDI-Forschungsheft Nr. 624).

[47] Mersmann A. Thermische Verfahrenstechnik. Berlin: Springer, 1980.

[48] Mersmann A. Auslegung und Maßstabsvergrößerung von Blasen- und Tropfensäulen. Chem.-Ing.-Tech. 1977;49(9):679–691.

[49] Van Dierendonck LL, Wilkinson PM, Doldersum B, Cramers PHMR. The kinetics of uncatalyzed sodium sulfite oxidation. Chemical Engineering Science 1993;48(5):933.

AUTHOR INDEX

Printed and bound by CPI Group (UK) Ltd, Croydon, CR0 4YY

03/10/2024

01040320-0012